A.P.H. POST GRADUATE LIBRARY
B09467

D1336739

Endocrine-Disrupting Chemicals

Contemporary Endocrinology

P. Michael Conn, *Series Editor*

Endocrine-Disrupting Chemicals: From Basic Research to Clinical Practice, edited by *Andrea C. Gore,* 2007
Androgens in Health and Disease, edited by *Carrie Bagatell and William J. Bremner,* 2003
Endocrine Replacement Therapy in Clinical Practice, edited by *A. Wayne Meikle,* 2003
Early Diagnosis and Treatment of Endocrine Disorders, edited by *Robert S. Bar,* 2003
Type I Diabetes: Etiology and Treatment, edited by *Mark A. Sperling,* 2003
Handbook of Diagnostic Endocrinology, edited by *Janet E. Hall and Lynnette K. Nieman,* 2003
Pediatric Endocrinology: A Practical Clinical Guide, edited by *Sally Radovick and Margaret H. MacGillivray,* 2003
Diseases of the Thyroid, 2nd ed., edited by *Lewis E. Braverman,* 2003
Developmental Endocrinology: From Research to Clinical Practice, edited by *Erica A. Eugster and Ora Hirsch Pescovitz,* 2002
Osteoporosis: Pathophysiology and Clinical Management, edited by *Eric S. Orwoll and Michael Bliziotes,* 2002
Challenging Cases in Endocrinology, edited by *Mark E. Molitch,* 2002
Selective Estrogen Receptor Modulators: Research and Clinical Applications, edited by *Andrea Manni and Michael F. Verderame,* 2002
Transgenics in Endocrinology, edited by *Martin Matzuk, Chester W. Brown, and T. Rajendra Kumar,* 2001
Assisted Fertilization and Nuclear Transfer in Mammals, edited by *Don P. Wolf and Mary Zelinski-Wooten,* 2001
Adrenal Disorders, edited by *Andrew N. Margioris and George P. Chrousos,* 2001
Endocrine Oncology, edited by *Stephen P. Ethier,* 2000
Endocrinology of the Lung: Development and Surfactant Synthesis, edited by *Carole R. Mendelson,* 2000
Sports Endocrinology, edited by *Michelle P. Warren and Naama W. Constantini,* 2000
Gene Engineering in Endocrinology, edited by *Margaret A. Shupnik,* 2000
Endocrinology of Aging, edited by *John E. Morley and Lucretia va den Berg,* 2000
Human Growth Hormone: Research and Clinical Practice, edited by *Roy G. Smith and Michael O. Thorner,* 2000
Hormones and the Heart in Health and Disease, edited by *Leonard Share,* 1999
Menopause: Endocrinology and Management, edited by *David B. Seifer and Elizabeth A. Kennard,* 1999
The IGF System: Molecular Biology, Physiology, and Clinical Applications, edited by *Ron G. Rosenfeld and Charles T. Roberts, Jr.,* 1999
Neurosteroids: A New Regulatory Function in the Nervous System, edited by *Etienne-Emile Baulieu, Paul Robel, and Michael Chumacher,* 1999
Autoimmune Endocrinopathies, edited by *Robert Volpé* 1999
Hormone Resistance Syndromes, edited by *J. Larry Jameson,* 1999
Hormone Replacement Therapy, edited by *A. Wayne Meikle,* 1999
Insulin Resistance: The Metabolic Syndrome X, edited by *Gerald M. Reaven and Ami Laws,* 1999
Endocrinology of Breast Cancer, edited by *Andrea Manni* 1999
Molecular and Cellular Pediatric Endocrinology, edited by *Stuart Handwerger,* 1999
Gastrointestinal Endocrinology, edited by *George H. Greeley, Jr.,* 1999
The Endocrinology of Pregnancy, edited by *Fuller W. Bazer,* 1998
Clinical Management of Diabetic Neuropathy, edited by *Aristidis Veves,* 1998
G Proteins, Receptors, and Disease, edited by *Allen M. Spiegel,* 1998
Natriuretic Peptides in Health and Disease, edited by *Willis K. Samson and Ellis R. Levin,* 1997
Endocrinology of Critical Disease, edited by *K. Patric Ober,* 1997
Diseases of the Pituitary: Diagnosis and Treatment, edited by *Margaret E. Wierman,* 1997
Endocrinology of the Vasculature, edited by *James R. Sowers,* 1996

Endocrine-Disrupting Chemicals

From Basic Research to Clinical Practice

Edited by

Andrea C. Gore, PhD

Division of Pharmacology, College of Pharmacy, The University of Texas, Austin, TX

Humana Press ✻ Totowa, New Jersey

999 Riverview Drive, Suite 208
Totowa, New Jersey 07512
www.humanapress.com

This publication is printed on acid-free paper. ∞

ANSI Z39.48-1984 (American National Standards Institute) Permanence of Paper for Printed Library Materials.

Cover design by Sarah M. Dickerson and Andrea C. Gore

For additional copies, pricing for bulk purchases, and/or information about other Humana titles, contact Humana at the above address or at any of the following numbers: Tel: 973-256-1699; Fax: 973-256-8341; E-mail: humana@humanapr.com or visit our website at http://humanapress.com

Printed in the United States of America. 10 9 8 7 6 5 4 3 2 1

Library of Congress Control Number: 2007928562

Dedicated to my husband, Dr. David Crews

Contents

Acknowledgments

My introduction to the field of endocrine disruption occurred while I was on the faculty of Mount Sinai School of Medicine. There, I met a group of colleagues in Community Medicine, particularly Drs. Mary Wolff, Phil Landrigan, Barbara Brenner, and Trudy Berkowitz, who invited me to submit a pilot project grant as part of an NIEHS Children's Center grant. This grant was funded, and I embarked upon a new research program to study effects of environmental toxicants found in New York City on reproductive neuroendocrine systems. When I moved to The University of Texas at Austin in 2003, I continued my NIH-supported research on effects of polychlorinated biphenyls (PCBs) on neuroendocrine physiology and reproductive behavior. This research program has continued to blossom, as UT-Austin has strong roots in understanding the mechanisms for how the environment contributes to disease and dysfunction. In my laboratory, three excellent graduate students, Rebecca M. Steinberg, Sarah M. Dickerson, and Deena M. Walker, are leading this research program at UT-Austin, and these experiments are some of the most exciting and rewarding that have ever been done in my laboratory. The other PhD students in my laboratory, Weiling Yin, Jackie Maffucci, and Di Wu, and a PharmD student, Theresa Wagner, also deserve my thanks. Although they did not work directly on the PCB project, they still put endless hours into this research project, which was a 24/7/365 endeavor and required considerable help from all Gore Lab members. Several undergraduate students, particularly Nygerma Dangleben, Tim Hsu and Sonya Hughes, also played important roles in assisting with these projects. Without the efforts of all the people in the Gore Lab, I could not have had the freedom or opportunity to work on this book and I am grateful to have such a personable, professional, and hard-working group. Finally, I thank Dr. P. Michael Conn for inviting and trusting me to produce this book as part of his "Contemporary Endocrinology" series and Richard Lansing of Humana Press for his support through the process.

Numerous agencies provided key funding for my endocrine disruption research and educational programs, and several individuals deserve special mention. I thank the NIH-NIEHS for its financial support of my laboratory work on endocrine disruption (NIH 1P50 ES09584, NIH ES07784, and NIH 1R21 ES12272). I am particularly grateful to Dr. Jerry Heindel who has strongly supported funding in this area in general. The Endocrine Society (TES) has been, and continues to be, a generous ideological and financial resource for endocrine disruption. TES was the home of a "Forum on Endocrine-disrupting Chemicals" that I organized and led and which took place in June 2005. This Forum proved to be a landmark activity that brought TES into the limelight as a leader for strong basic and translational research in endocrine disruption. TES continues its commitment to this area due in large part to the tireless efforts of Robert Bartel, MS. Financial support for the Forum and for the publication of a special supplement in *Endocrinology* (vol 147, June 2006) was generously provided by TES and from grants from the NIEHS (NIH 1R13 ES014258-01) and the EPA (X3-832341). Finally, two of my graduate students have received independent funding for their endocrine disruption dissertation research in my laboratory: Rebecca Steinberg received a 2-year fellowship from PhRMA and Sarah Dickerson received a 3-year NSF predoctoral award.

Finally, I acknowledge my ever-supportive and loving family. Although my parents, Norman and Caroline Gore, my sister, Audrey Gore, and my brother, Matthew Gore and his family (Ilene, Madison, and Sophie) live far from me, they make my world wonderful. My husband, David Crews, is a constant source of support and my number one fan. The past years in Austin with David and the dogs, Scylla and Cary, have been wonderful, and I continue my attempts to achieve that elusive and ephemeral balance between work and life.

Contributors

STUART R. ADLER, MD, PHD, *North County Endocrinology and BJC Health Systems, St. Louis, MO*

DAVID L. ARMSTRONG, PHD, *NIEHS, Research Triangle Park, NC*

JULIA S. BARTHOLD, MD, *Division of Urology, Alfred L. DuPont Hospital for Children/Nemours Children's Clinic, Wilmington, DE*

BARBARA BRENNER, DRPH, *Community and Preventive Medicine, Mount Sinai School of Medicine, New York, NY*

ANTONIA M. CALAFAT, PHD, *Division of Laboratory Sciences, National Center for Environmental Health, Centers for Disease Control and Prevention, Atlanta, GA*

PRUE A. COWIN, BBiomedsci, *Centre for Urological Research, Monash Institute of Medical Research, Monash University, Clayton, Victoria, Australia*

CHERYL A. DYER, PHD, *Department of Biological Sciences, Northern Arizona University, Flagstaff, AZ*

PAUL M. D. FOSTER, PHD, *National Institute of Environmental Health Sciences, Research Triangle Park, NC*

VICTOR Y. FUJIMOTO, MD, *Department of Obstetrics, Gynecology, and Reproductive Sciences, UCSF In Vitro Fertilization, University of California, San Francisco, CA*

MAIDA GALVEZ, MD, *Community and Preventive Medicine, Mount Sinai School of Medicine, New York, NY*

LINDA C. GIUDICE, MD, PHD, MSC, *Department of Obstetrics, Gynecology and Reproductive Sciences, University of California, San Francisco, CA*

ANDREA C. GORE, PHD, *Division of Pharmacology & Toxicology, College of Pharmacy, The University of Texas at Austin, Austin, TX*

CARLOS GUERRERO-BOSAGNA, PHD, *Division of Nutritional Sciences, Cornell University, Ithaca, NY*

RUSS HAUSER, MD, SCD, MPH, *Department of Environmental Health, Harvard School of Public Health, Boston, MA*

SARAH JANSSEN, MD, PHD, MPH, *Occupational and Environmental Medicine, Department of Internal Medicine, University of California, San Francisco, CA*

JOHN D. MEEKER, MS, SCD, *Department of Environmental Health Sciences, University of Michigan School of Public Health, Ann Arbor, MI*

LARRY L. NEEDHAM, PHD, *Division of Laboratory Sciences, National Center for Environmental Health, Centers for Disease Control and Prevention, Atlanta, GA*

GAIL P. RISBRIDGER, PHD, *Centre for Urological Research, Monash Institute of Medical Research, Monash University, Clayton, Victoria, Australia*

BEVERLY S. RUBIN, PHD, *Department of Anatomy and Cellular Biology, Tufts University School of Medicine, Boston, MA*

GINA SOLOMON, MD, MPH, *Senior Scientist, Natural Resources Defense Council, Assistant Clinical Professor of Medicine, University of California, San Francisco, CA*

CARLOS SONNENSCHEIN, MD, *Department of Anatomy and Cellular Biology, Tufts University School of Medicine, Boston, MA*

ANA M. SOTO, MD, *Department of Anatomy and Cellular Biology, Tufts University School of Medicine, Boston, MA*

JOSEPH THORNTON, PHD, *Center for Ecology and Evolutionary Biology, University of Oregon, Eugene, OR*

LUIS VALLADARES, PHD, *Universidad de Chile, Instituto de Nutrición y Tecnología de los Alimentos, Santiago, Chile*

DEENA M. WALKER, BS, *Institute for Neuroscience, The University of Texas at Austin, Austin, TX*

TRACEY J. WOODRUFF, PHD, MPH, *US Environmental Protection Agency, San Francisco, CA*

I

The Basic Biology of Endocrine Disruption

1 Introduction to Endocrine-Disrupting Chemicals

Andrea C. Gore, PhD

CONTENTS

1. WHAT IS AN ENDOCRINE-DISRUPTING CHEMICAL?

The United States Environmental Protection Agency (USEPA) defined endocrine-disrupting chemicals (EDCs) as "exogenous agents that interfere with the synthesis, secretion, transport, binding, action, or elimination of natural hormones in the body that are responsible for the maintenance of homeostasis, reproduction, development, and/or behavior." Although this definition may seem all-encompassing, it is becoming clear that it needs to be extended. The USEPA specifies "exogenous agents," implying that in order for a substance to be deemed an EDC, it must come from an external source. However, an organism's environment consists of not only external factors but also an organism's internal hormonal milieu. For example, if an organism's own endogenous hormonal systems are activated or inactivated at inappropriate times, it may disrupt endocrine processes. Another example is that of the mammalian fetus, in which the intrauterine environment may alter endocrine and homeostatic processes. Thus, the environment needs to be redefined to include not only exogenous environmental factors, as specified by the USEPA above, but also internal secretions such as inappropriate endogenous and maternal hormones.

As you will learn from the chapters in this book, EDCs comprise natural substances (phytoestrogens such as soy, alfalfa, and clover), pesticides [dichlorodiphenyl-trichloroethane (DDT)], fungicides (vinclozolin), substances used in production of plastics or as plasticizers (bisphenol A and phthalates), industrial chemicals [polychlorinated biphenyls (PCBs)], and metals (cadmium, lead, mercury, and uranium and arsenic, a metalloid; Chapter 5) *(1)*. The components of this list seem to have nothing

From: *Endocrine-Disrupting Chemicals: From Basic Research to Clinical Practice*
Edited by: A. C. Gore © Humana Press Inc., Totowa, NJ

in common, yet all are documented as endocrine disruptors. For a more comprehensive list of EDCs, and more details on their structures, functions, and mechanisms, I refer you to several chapters in this book. The chemical structures of specific EDCs are shown in Chapter 4, Fig. 2 *(2)*. Within Chapter 9, Table 1 presents a helpful list of EDC categorization and their effects on female reproductive health in humans *(3)*, and readers should note that one of those EDCs, bisphenol A, is discussed for its low-dose effects on female reproductive health in animal models (Chapter 2) *(4)*. The same EDCs in Table 1 of Chapter 9 are also discussed in the context of male reproductive health (Chapters 3 and 10) *(5,6)* and neuroendocrine systems (Chapter 4) *(2)*. Chapter 5 discusses how metals may act as EDCs *(1)*. Finally, the end of Chapter 13 *(7)* provides an important list of references and resources for additional information on EDCs that falls beyond the scope of this book.

2. EDC EFFECTS ON INDIVIDUALS AND POPULATIONS ACROSS THE LIFE CYCLE

Endocrine disruption needs to be considered in the context of both individuals and populations *(8)*. The importance of populations is made clear from epidemiologic studies demonstrating clear evidence for environmental endocrine disruption in humans. As an example, Chapter 10 of this book *(6)* provides a careful discussion of the epidemiologic evidence for a link between exposure to EDCs and male reproductive dysfunction. In addition, populations of wildlife are impacted by endocrine disruption *(9)*. However, not every individual within a population may be similarly affected *(10)*. Some may experience overt toxicity, others may experience more subtle dysfunctions, and still others will not have any evident phenotype. These differences in responsiveness among individuals are due to differences in genomes, in combination with an organism's entire life history of experiences. With the rare exception of monozygotic twins (humans) or identical littermates (animals), each organism has unique sequences of DNA, which may undergo mutations such as point mutations or deletions that compromise the gene product (protein) to result in disease and dysfunction. Moreover, even identical twins are uniquely and differentially affected by their environments, beginning in the uterus and throughout the rest of life *(11)*. The embryonic period is particularly important in this regard, as it is a developmental window when DNA becomes modified through methylation, demethylation, and/or remethylation, a process that is thought to play a key role in cellular differentiation. The final methylation patterns are not fixed until late in embryonic development and possibly into early neonatal life *(12)*. Differences among organisms in epigenetic modifications to the DNA, such as DNA methylation or acetylation, result in differential gene functions and potentially, differential vulnerability to EDCs. New evidence indicates that this is a biologically plausible mechanism of action for EDCs in the developing embryo *(8,13,14)*.

Although the developing fetus may be the most vulnerable to endocrine disruption (see Chapters 2–5 for specific examples) *(1,2,4,5)*, effects of EDCs may be exerted in other phases of the life cycle. Moreover, as EDCs themselves exert epigenetic effects that may be passed to subsequent generations, if expressed in the germ line, this may create a "vicious cycle" by which an organism's epigenome may predispose it to vulnerability to endocrine disruption, which in turn creates additional epigenetic modifications that are passed on to the offspring. This concept is discussed most extensively in Chapter 7 of this book *(11)*. Such a situation is further magnified if the descendants continue to live in contaminated environments. Thus, the genome, the epigenome, and the environment interact throughout the entire life cycle to influence the impact of EDCs. These life experiences, from the embryo through aging, are not only unique to each of us but need to be taken into account when considering cumulative effects of EDCs across the life cycle.

3. MECHANISMS OF EDCS

The potential mechanisms underlying the effects of EDCs are incredibly diverse, making studies on their biological effects daunting (for a review of traditional and non-traditional mechanisms of EDCs, see Chapter 6) *(15)*. Indeed, in the case of humans, the latency between early exposures to EDCs and adult dysfunction may be 60 years.

The first important challenge is the nature of hormonal systems. Endogenous hormones act through several mechanisms. The classical mechanism of action for hormones such as estrogens, androgens, thyroid, and progesterone involves the binding of the hormone to its receptor, the interaction of this hormone–receptor complex with other cofactors in a cell, and the activation or inactivation of transcription of a target gene. More recently, membrane steroid hormone receptors have been identified, and these appear to use different intracellular signaling pathways for activation of subcellular processes *(16)*. An important consideration is that the same ligand, for example, estradiol, activates a diversity of target receptors, signaling mechanisms, and may interact with completely different complements of cofactors depending upon the phenotype of the target cell. In addition, hormone signaling also involves the synthesis, degradation, or inactivation of hormones by specific enzymes, any or all of which may be targeted by EDCs. Another consideration is that endogenous hormones, particularly estrogens, androgens, and thyroid, bind to proteins in blood that reduce their bioavailability (i.e., their ability to act upon their receptors). EDCs may not bind to the same binding proteins, thereby increasing their bioavailability relative to endogenous hormones. This concept also applies to the enzymes that synthesize or degrade endogenous hormones. If EDCs are not as rapidly metabolized as endogenous steroid hormones, they may remain bioavailable far longer and get incorporated into the body burden, generally fat stores, as most EDCs are lipophilic. There are numerous other mechanisms for the regulation of steroid hormone actions that are beyond the scope of this introduction, but it is clear that the complexity is enormous. A specific example of the activation of diverse signaling pathways by EDCs is presented for the thyroid system in Chapter 8 of this book *(17)*. When put into the context of how an EDC may cause an effect on an endocrine system, all of these potential targets and pathways need to be considered.

The second and third challenges for endocrine disruption are, respectively, trying to reconcile and understand how extremely low doses of EDCs can exert potent effects on endocrine and homeostatic systems, and related to this, why EDCs exert non-traditional dose–response curves *(14,18)*. Although it is unclear how EDCs can act at such low levels, the implications are extremely clear, that low-dose EDC exposure, particularly at vulnerable developmental windows, can have long-term consequences on later health. Furthermore, the biological evidence for such low-dose effects is extremely strong from basic science studies (Chapter 2) *(4)*. In addition, the finding that EDCs, act in non-linear dose–response curves, often U or inverted U in shape, was initially puzzling to toxicologists. However, these findings were not a surprise to endocrinologists, as hormones often act in non-linear manners because of the diversity of target receptors, cofactors, and other signaling pathways that may not all be activated or inactivated at the same range of dosages. The overall shape of the dose–response curve thus reflects the cumulative action of EDCs upon a range of targets. As a whole, the non-monotonic dose–response curves emphasize the need for communication among basic scientists across disciplines, from toxicology to endocrinology *(14)*. Related to the low-dose phenomenon is the question of whether thresholds for different EDCs actually exist. Environmental toxicological protocols continue to use single doses of a single chemical at different concentrations, seeking the lowest dose at which no adverse effects are observed in the animal subject (the no-observed-adverse-effect level or NOAEL). The NOAEL for the chemical in question is then used as a threshold dose in risk assessments for human exposure. However, a power-analysis study revealed that thresholds may not exist for estrogenic EDCs, as any amount of the exogenous steroidal agent automatically exceeds the organism's threshold *(19)*.

A fourth challenge is that exposure to EDCs rarely occurs for a single substance, with the rare exception of toxic spills. For the most part, an environment that is contaminated by one industrial waste product will be contaminated by a complex mixture *(10)*. Again, this makes the design of experiments difficult, because there is no "typical" exposure. Studying the body burdens of humans, as discussed in Chapter 11 *(20)*, has proven to be very important in informing researchers about what we may be exposed to, but it is not possible to be absolutely comprehensive in analyzing these exposures. Thus, some animal studies are designed using single toxicants or phytoestrogens; others use more complex mixtures. There is rationale but also criticism for both of these approaches, making the perfect experiment elusive.

4. ENDOCRINE DISRUPTION: A TRANSLATIONAL APPROACH

Although the complexity of endocrine disruption makes understanding and mitigating exposures seem impossible, the chapters in this book help to clarify many of the key issues about endocrine disruption. Part I of this book focuses on the basic biology of EDCs, with greatest emphasis on animal models. The science of endocrine disruption depends on basic scientists to provide the fundamental and strong science that first identifies the EDCs; second, characterizes their effects; third, elucidates the mechanisms; and fourth, understands their functional implications. Chapters 2 through 7 provide such a basis by reviewing the scientific literature and demonstrating that there is conclusive evidence for biological and physiological effects of EDCs upon endocrine systems. In Part II, the biology of EDCs in humans is discussed in Chapters 8 through 11. These chapters discuss the evidence that humans are indeed exposed to EDCs, the mechanisms and implications for these effects, and how exposures are ascertained and measured. Finally, Part III of the book discusses the human health implications and provides information for actually dealing with the problem of EDCs. EDCs are already in our world, and we need to be able to talk about them with not only scientists and physicians but also industrialists, manufacturers, end users, and the community. Thus, this section of the book has chapters on public policy

(Chapter 12) *(21)*, on practical advice about what to do to avoid EDCs and provides resources on where to get more scientific information (Chapter 13) *(7)*. Importantly, Chapter 14 *(22)* talks about a community intervention to reduce exposure to some EDCs, and Chapter 15 *(23)* poses some potential solutions to the problem.

In the field of endocrine disruption, it is necessary to draw parallels between basic science in animal models and clinical implications in humans. Such a translational approach is the goal of this book, which synthesizes the field for readers such that basic scientists will learn about clinical relevance and clinicians may better understand the basic biology of EDCs for their practice or clinical research laboratories. As a whole, this book addresses the key themes in understanding the mechanisms of endocrine disruption, their relevance to humans, and how we may deal with a problem that is already widespread in our world.

REFERENCES

1. Dyer CA. (2007) Heavy metals as endocrine-disrupting chemicals. In: Gore AC, ed. *Endocrine-Disrupting Chemicals: From Basic Research to Clinical Practice.* Totowa, NJ: Humana Press.
2. Walker DM, Gore AC. (2007) Endocrine-disrupting chemicals and the brain. In: Gore AC, ed. *Endocrine-Disrupting Chemicals: From Basic Research to Clinical Practice.* Totowa, NJ: Humana Press.
3. Janssen S, Fujimoto VY, Giudice LC. (2007) Endocrine disruption and reproductive outcomes in women. In: Gore AC, ed. *Endocrine-Disrupting Chemicals: From Basic Research to Clinical Practice.* Totowa, NJ: Humana Press.
4. Soto AM, Rubin BS, Sonnenschein C. (2007) Endocrine disruption and the female. In: Gore AC, ed. *Endocrine-Disrupting Chemicals: From Basic Research to Clinical Practice.* Totowa, NJ: Humana Press.
5. Cowin PA, Foster P, Risbridger GP. (2007) Endocrine disruption in the male. In: Gore AC, ed. *Endocrine-Disrupting Chemicals: From Basic Research to Clinical Practice.* Totowa, NJ: Humana Press.
6. Hauser R, Barthold JS, Meeker JD. (2007) Epidemiologic evidence on the relationship between environmental endocrine disruptors and male reproductive and developmental health. In: Gore AC, ed. *Endocrine-Disrupting Chemicals: From Basic Research to Clinical Practice.* Totowa, NJ: Humana Press.
7. Solomon G, Janssen S. (2007) Talking with patients and the public about endocrine-disrupting chemicals. In: Gore AC, ed. *Endocrine-Disrupting Chemicals: From Basic Research to Clinical Practice.* Totowa, NJ: Humana Press.
8. Crews D, McLachlan JA. (2006) Epigenetics, evolution, endocrine disruption, health, and disease. *Endocrinology* 147:S4–10.
9. Zala SM, Penn DJ. (2004) Abnormal behaviours induced by chemical pollution: a review of the evidence and new challenges. *Anim Behav* 68:649–64.
10. Crews D, Willingham E, Skipper JK. (2000) Endocrine disruptors: present issues, future directions. *Q Rev Biol* 75:243–60.
11. Guerrero-Bosagna C, Valladares L. (2007) Endocrine disruptors, epigenetically induced changes, transgenerational transmission of characters, epigenetic states. In: Gore AC, ed. *Endocrine-Disrupting Chemicals: From Basic Research to Clinical Practice.* Totowa, NJ: Humana Press.
12. Baylin SB, Herman JG, Graff JR, Vertino PM, Issa JP. (1998) Alterations in DNA methylation: a fundamental aspect of neoplasia. *Adv Cancer Res* 72:141–96.
13. Anway MD, Skinner MK. (2006) Epigenetic transgenerational actions of endocrine disruptors. *Endocrinology* 147:S43–9.
14. Gore AC, Heindel JJ, Zoeller RT. (2006) Endocrine disruption for endocrinologists (and others). *Endocrinology* 147:S1–3.
15. Adler SR. (2007) Cellular mechanisms of endocrine disruption: traditional and novel actions. In: Gore AC, ed. *Endocrine-Disrupting Chemicals: From Basic Research to Clinical Practice.* Totowa, NJ: Humana Press.

16. Thomas P, Pang Y, Filardo EJ, Dong J. (2005) Identity of an estrogen membrane receptor coupled to a G protein in human breast cancer cells. *Endocrinology* 146:624–32.
17. Armstrong DL. (2007) Implications of thyroid hormone signaling through the phosphoinositide-3 kinase for xenobiotic disruption of human health. In: Gore AC, ed. *Endocrine-Disrupting Chemicals: From Basic Research to Clinical Practice.* Totowa, NJ: Humana Press.
18. Welshons WV, Nagel SC, vom Saal FS. (2006) Large effects from small exposures. III. Endocrine mechanisms mediating effects of bisphenol A at levels of human exposure. *Endocrinology* 147:S56–69.
19. Sheehan DM, Willingham EJ, Bergeron JM, Osborn CT, Crews D. (1999) No threshold dose for estradiol-induced sex reversal of turtle embryos: how little is too much. *Environ Health Perspect* 107:155–9.
20. Calafat AM, Needham LL. (2007) Human exposures and body burdens of endocrine-disrupting chemicals. In: Gore AC, ed. *Endocrine-Disrupting Chemicals: From Basic Research to Clinical Practice.* Totowa, NJ: Humana Press.
21. Woodruff TJ. (2007) Policy implications of endocrine-disrupting chemicals in humans. In: Gore AC, ed. *Endocrine-Disrupting Chemicals: From Basic Research to Clinical Practice.* Totowa, NJ: Humana Press.
22. Brenner B, Galvez M. (2007) Community interventions to reduce exposure to chemicals with endocrine disrupting properties. In: Gore AC, ed. *Endocrine-Disrupting Chemicals: From Basic Research to Clinical Practice.* Totowa, NJ: Humana Press.
23. Thornton JW. (2007) Solutions to the problem of endocrine disruption. In: Gore AC, ed. *Endocrine-Disrupting Chemicals: From Basic Research to Clinical Practice.* Totowa, NJ: Humana Press.

2 Endocrine Disruption and the Female

Ana M. Soto, MD, Beverly S. Rubin, PhD, and Carlos Sonnenschein, MD

CONTENTS

1. THE WINGSPREAD CONFERENCE AND THE CONCEPT OF "ENDOCRINE DISRUPTOR"

The production and release of synthetic chemicals into the environment have been a hallmark of the "Second Industrial Revolution" and the "Green Revolution." Soon after the inception of these chemicals, evidence emerged linking environmental exposure with a variety of developmental and reproductive abnormalities in wildlife species. Laboratory studies revealed that some of these compounds had estrogenic activity. Although the agricultural and industrial use of dichlorodiphenyltrichloroethane (DDT) and polychlorinated biphenyls (PCBs) was banned in the USA in 1973 and 1977, respectively, new and different cases of wildlife disturbances were reported over the next few decades due to biopersistence of the banned chemicals and the uninterrupted introduction of new chemicals. The mounting evidence suggested that the hormonal activity of these synthetic chemicals was detrimental to wildlife, and potentially humans, and led to the endocrine disruptor hypothesis at the Wingspread Conference held in Racine, Wisconsin, in 1991. The term "endocrine disruptor" was coined there as the participants proposed that the developmental alterations observed in diverse wildlife species were due to exposure to multiple chemicals that, through different modes of action, disrupted the endocrine systems of developing metazoan organisms *(1)*.

From: *Endocrine-Disrupting Chemicals: From Basic Research to Clinical Practice*
Edited by: A. C. Gore © Humana Press Inc., Totowa, NJ

Although exposures during adulthood were considered potentially deleterious, the main concern of the conferees was the developing organism because some effects documented in the genital tract of wildlife were comparable to those seen in the daughters and sons of women who had been exposed during pregnancy to the synthetic estrogen diethylstilbestrol (DES) *(2)*. The conference participants recognized that the human DES syndrome was an extreme expression of the plasticity of the fetus in response to environmental cues, and furthermore, it provided a template for the potential effects that other hormonally active chemicals could have on human health. Ever since, experimental and epidemiological data have shown that exposure of the developing fetus or neonate to environmentally relevant concentrations of certain synthetic chemicals causes morphological, biochemical, physiological, and behavioral anomalies in both vertebrate and invertebrate species.

1.1. The Complexity of Endocrine Disruptors: From One Chemical Displaying Multiple Actions to Non-Monotonic Dose–Response Curves

Environmental "endocrine disruptors" are defined by the US Environmental Protection Agency as "exogenous agents that interfere with the synthesis, secretion, transport, binding, action, or elimination of natural hormones in the body that are responsible for the maintenance of homeostasis, reproduction, development, and/or behavior." This definition evokes a diversity of targets and pathways that may be affected by exogenous agents and implies that endocrine disruptors are heterogeneous agents encompassing diverse chemical structures. This initial complexity increased as some of these compounds were found to have diverse hormonal activities. For example, the pesticide DDT is an estrogen agonist, whereas one of its metabolites is antiandrogenic *(3)*; bisphenol A (BPA) is a full estrogen agonist, which also binds to the thyroid hormone receptor *(4)*. Furthermore, a mixture of diverse endocrine disruptors is present in both human and wildlife tissues. This complexity suggests that wildlife and humans, who are exposed to mixtures at all developmental ages, will display more florid and diverse syndromes than laboratory animals exposed to a single chemical at a well-defined developmental stage. To begin to understand the health problems posed by endocrine disruptors, however, it is useful and desirable to examine single agents during specific developmental periods.

Historically, the first environmental endocrine disruptors identified were those with estrogenic activity. Still today, the vast majority of these chemicals are estrogen mimics; hence, this chapter will focus on the effects of xenoestrogens on the developing female genital tract and mammary gland. Endogenous estrogens are involved in the development and maintenance of the female reproductive tract and secondary sexual characteristics and the regulation of the menstrual/estrous cycle, pregnancy, and lactation. At the cellular level, these endogenous hormones mediate cell proliferation and the synthesis and secretion of cell type-specific proteins in reproductive tissues such as the ovary, oviduct, uterus, vagina, hypothalamus, pituitary, and mammary gland. These effects are mediated, for the most part, by estrogen receptors (ER-α and ER-β) in the female, although the expression of the same receptors in the male reproductive tract and in the non-reproductive organs such as the thyroid, cardiovascular system, and bone indicates the potentially vast reach of these synthetic estrogenic chemicals *(5)*.

Although xenoestrogens are usually less potent than estradiol regarding their binding affinity toward classical nuclear ERs, it is now clear that they act *additively* with endogenous estrogens. This may explain how low, seemingly insignificant levels of xenoestrogens have an impact when added to the already significant levels of endogenous steroidal hormones *(6)*. In addition, xenoestrogens bind to plasma carrier proteins with significantly lower affinities than those of natural estrogens and thus are more readily available to target cells than their endogenous counterparts *(7)*. The same xenoestrogens that seem to be weak agonists for nuclear ER are strong agonists when acting through membrane ERs; this may explain their ability to produce biological effects at low doses *(8)*.

Until recently, the issue of whether hormonally active agents at environmentally relevant doses could alter development was highly controversial. At the request of the US Environmental Protection Agency, the National Toxicology Program (NTP) met to consider this issue, and a final NTP Endocrine Disruptors Low-Dose Peer Review was published in 2001 *(9)*. This report stated that there was "credible evidence for low-dose effects." Another controversy identified by the NTP was the shape of the dose–response curve, which was reported to be non-monotonic for some of the effects of prenatal exposure to xenoestrogens. For example, prenatal exposure to methoxychlor altered the response of the adult uterus to estradiol; low doses increased the response and higher doses reduced it *(10)*. This type of response has also been observed in other end points with other estrogenic chemicals *(11–13)*.

Toxicologists typically assume a monotonic curve in response to an environmental exposure; however, hormones display diverse types of response curves, including those that have an inverted U-shape. For example, at low physiological levels, androgens increase the proliferation rate of prostate cell lines, whereas at high physiological levels, they induce proliferative quiescence *(14,15)*. In the mammary gland of ovariectomized prepubescent mice, estrogens also invoke a non-monotonic, inverted U-shaped response in a variety of morphometric parameters *(16)*. The frequent occurrence of non-monotonic dose–response curves in biological phenomena underlies the importance of understanding how these complex biological phenomena are regulated *(17)*. These patterns ultimately highlight the infeasibility of using the response to high doses of a natural or environmental hormone or other toxicant to extrapolate or predict the effects of exposure to low doses of the compound *(10,17–19)*.

1.2. Endocrine Disruptors Within a Developmental Context

The deleterious effects of endocrine-disrupting chemicals vary according to the age at which an organism is exposed. The developing organism is critically sensitive to both endogenous and exogenous hormones, a phenomenon that led Dr. Howard Bern to coin the phrase the "fragile fetus" *(20)*. The "critical window" of exposure differs depending upon the time at which specific developmental events occur in particular tissues or organs. For example, although it was later found to be ineffective, high DES doses were administered to pregnant women to prevent miscarriages. Clear cell adenocarcinomas of the vagina were observed in daughters of pregnant women exposed in utero to DES, but only if exposure occurred before the 13th week of gestation *(21,22)*. In addition to vaginal cancers, women exposed in utero to DES experienced diverse malformations of the genital tract and functional deficits including infertility. Parenthetically, DES produced a similar syndrome in mice *(23,24)*, thus establishing the mouse as a useful model for assessing the developmental toxicity of xenoestrogens and the mechanisms underlying these effects.

Studies in rodents revealed that the fetus is exquisitely sensitive to estrogen at doses much lower than those of DES to which human fetuses were exposed *(25)* or doses that produce estrogenic effects in adults *(26)*. For example, in mice, fetal exposure to doses of BPA that are four million fold lower than those required to induce an uterotropic effect in the prepubertal animal resulted in morphological alterations affecting the hypothalamus, the reproductive tract, and the mammary gland and altered estrous cyclicity and behaviors *(27–29)*.

Within this developmental context, in addition to the timing of exposure, other factors also impinge on the observed outcome. As mentioned previously, some xenoestrogens have additional hormonal activities. For example, the active metabolites of the pesticide methoxychlor are both antiandrogenic and estrogenic *(30)*, BPA is both an estrogen agonist and a thyroid hormone antagonist *(4)*, and dioxin has both agonistic and antagonistic estrogenic effects, although these actions do not involve dioxin binding directly to the ER *(31)*. Also discussed above, the shape of the dose–response curve influences the outcome, so that the same molecule may produce opposite results depending on the dose administered. Together, these factors limit a comprehensive synthesis of the available data. To circumvent these problems, we have chosen to focus this review on the effects of environmentally relevant levels of the xenoestrogen BPA during development of the female reproductive tract and mammary gland. BPA provides one pertinent model for study, because the chemical is ubiquitous in our environment, has high potential for fetal exposure, and the literature contains multiple examples of developmental effects.

2. BPA: A UBIQUITOUS XENOESTROGEN

BPA is widely used in the manufacture of polycarbonate plastics and epoxy resins. BPA is present in a multitude of products including the interior coating of food cans, wine storage vats, water carboys, milk containers, food storage vessels, baby formula bottles, water pipes, and dental materials, automotive lenses, optical lenses, protective coatings, adhesives, protective window glazing, compact disks, thermal paper, paper coatings, and as a developer in dyes. Halogenated derivatives of BPA, tetrabromobisphenol A (TBBPA), and tetrachlorobisphenol A (TCBPA) are widely used as flame retardants for building materials, paints, synthetic textiles, and plastic products including epoxy resin electronic circuit boards and other electronic equipment *(32)*. In 2003, the worldwide production of BPA exceeded 6 million pounds *(33)*. About 100 tons of BPA are released into the atmosphere each year during production *(32)*. BPA (4,4′-isopropylidenediphenol) is a diphenyl compound that contains two hydroxyl groups in the “para” position making it remarkably similar to the synthetic estrogen DES.

Although since 1936 BPA has been known to be estrogenic *(34)*, it was in the 1990s that this chemical was serendipitously discovered to leach from polycarbonate plastics in concentrations that were sufficient to up-regulate the expression of progesterone receptor (PR) and induce cell proliferation in estrogen-target, serum-sensitive MCF-7 breast cancer cell lines through binding to the ER *(35)*. In fact, studies have shown that incomplete polymerization of BPA during manufacture and/or depolymerization because of increased temperatures (induced either intentionally for sterilization/heating purposes or unintentionally during storage in warehouses) causes BPA and its derivatives to leach into foods *(36)*, beverages *(37)*, infant formula *(37)*, and saliva after application of dental sealants *(38)*. BPA has been found in aerosols and dust particles *(39,40)* including such far-reaching locales as in air and dust samples from certain residential and commercial environments in Massachusetts *(41)*, in leachates from waste water treatment plants and river water in Japan *(42)*, and in surface and drinking water in Spain *(43)*.

2.1. Human Exposure to BPA

The ubiquitous use of BPA provides great potential for exposure of both the developing fetus indirectly through maternal exposure and the neonate directly through ingestion of tinned food, infant formula, or maternal milk *(44)*. Indeed, BPA has been measured in maternal and fetal plasma and placental tissue at birth in humans *(45,46)*. The BPA concentrations in fetal plasma ranged from 0.2 to 9.2 ng/ml. A recently published study, the first using a reference human population (394 samples), reported that BPA was found in 95% of urine samples in the USA *(47)*. From these data, the mean exposure was estimated to be 30–40 ng/kg/day, and the 95th percentile was 180–230 ng/kg/day. In a smaller study, Arakawa et al. *(48)* reported a median daily urinary excretion of BPA of 1.2 μg/day and a maximum daily intake of BPA per body weight (bw) to be 0.23 μg/kg bw/day. BPA has also been measured in the milk of lactating mothers *(49)*. These data indicate that the developing human fetus and neonate are readily exposed to this chemical.

In rodents, BPA has been shown to readily cross the placenta *(50,51)* and bind α-fetoprotein (a binding protein that reduces the bioavailability of estrogens) with negligible affinity relative to estradiol; this results in enhanced bioavailability during neonatal development *(7)*. BPA is present in the mouse fetus and amniotic fluid during maternal exposure in higher concentrations than that of maternal blood. These results suggest that the pharmacokinetics and pharmacodynamics of BPA may exacerbate the impact of this chemical on the developing fetus and neonate *(51)*.

3. ER EXPRESSION DURING DEVELOPMENT

It is reasonable to assume that the presence of ERs during embryogenesis is crucial for the manifestation of toxic effects mediated by xenoestrogens. Hence, the expression of ER-α and ER-β during development in the female genital tract and mammary gland will be reviewed.

3.1. Development of the Female Genital Tract and ER Expression

Fetal developmental patterns are similar in the mouse and rat, the two animal models used most extensively for the study of endocrine disruption. ER-α mRNA was found in oocytes and fertilized eggs, although message concentration declined with successive cell divisions and became undetectable at the morula stage before reappearing at the blastocyst stage. There is little information available about ER expression prior to the period of organogenesis. In the mouse and rat, primordial germ cells enter the genital ridges at E11, and sexual differentiation of gonads occurs at E12.5 *(52)*. In the indifferent/undifferentiated gonad, expression of ER-α mRNA starts at E12.5, whereas ER-β becomes detectable at E16.5 in both sexes *(53)*.

In the developing fetus, the urogenital system originates from the intermediate mesoderm. The precursors of both the male and female genital tract are situated within the genital ridge. The Wolffian duct, from which the male reproductive tract derives, is first observed at E9.5. The Müllerian duct, from which the female reproductive tract is formed, is first observed at E11.5. Both ducts grow caudally and fuse with the cloaca wall. In genetic males, regression of the Müllerian duct is mediated by Müllerian inhibiting substance, produced by the fetal testes. In genetic females, regression of the Wolffian duct is due to the absence of androgens *(54)* (Fig. 1).

Expression of ER-β mRNA in the mesenchyme surrounding the Wolffian ducts in both sexes is first observed at E10.5. ER-α becomes detectable at E12.5. ER-α is still present at E16.5 in males, but it becomes undetectable at E14.5 in females *(53)*, coincident with regression of the Wolffian ducts. Beginning at E12.5, ER-α is expressed in the mesenchyme of the Müllerian ducts. In females, expression continues to be detected at E16.5 but ceases to be detected at E14.5 *(53)* in males, coincident with the regression of the Müllerian ducts.

In the oviduct of rats, expression of ER-α begins at E15 in the epithelial tissue compartment and at E19 in the tissue stroma. During neonatal development, the oviduct coils and its epithelium differentiates into two main cell types, that is, the ciliated and the secretory cells. ER-α is expressed mostly in the secretory cells of the epithelium but is also expressed in the lamina propria and the muscularis. ER-β is not detected in any compartment of the oviduct. At birth, ER-α is expressed in the epithelial compartment in the oviduct, cervix, and vagina, but not in the uterus, whereas it is expressed in the stroma of all these organs. In the uterus, epithelial expression of ER-α is observed at postnatal day four (PND4) in mice and PND5 in rats *(55)*.

3.2. Development of the Mammary Gland and Expression of the ER

In the mouse, the presumptive mammary lines, running between forelimbs and hindlimbs on each flank of the embryo, appear at E10.5. At E11.5, five placodes appear along each presumptive mammary line as lens-shaped ectodermal structures. One day later, the mammary buds appear as protruding structures that are apparent until E13.5, when they gradually invaginate into the dermis. During this period of invagination, the mammary epithelium displays an almost complete arrest of proliferation. At the

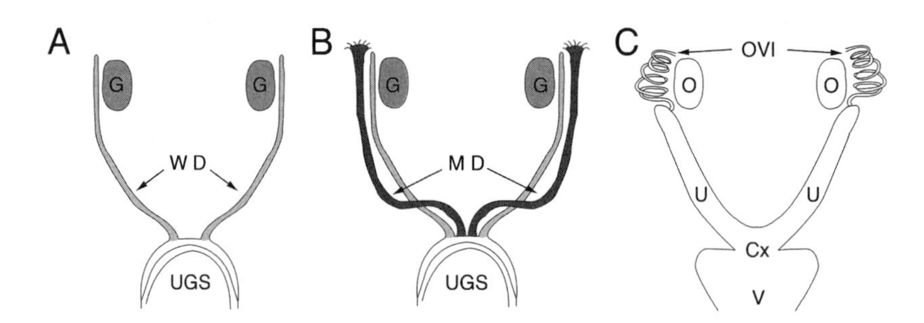

Fig. 1. Schematic representation of the development of the female genital tract in the mouse. **(A)** At E9.5, the Wolffian ducts (WD) are first visualized and grow toward the urogenital sinus (UGS). At this stage, the gonad is sexually undifferentiated. **(B)** The Müllerian ducts (MD) are first detectable at E11.5 and grow toward the UGS. **(C)** At E13.5, the WD starts to degenerate as the gonad differentiates into the ovary (O). The Müllerian ducts differentiate forming the oviduct (OVI), the uterus (U), the cervix (Cx), and the upper part of the vagina (V), whereas the lower part of the vagina originates from the UGS.

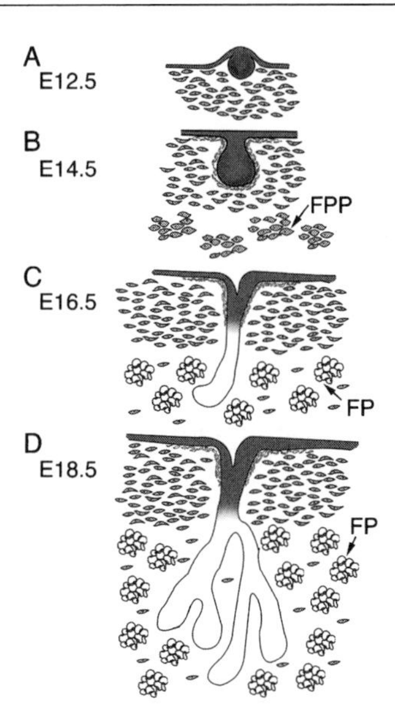

Fig. 2. Prenatal development of the mouse mammary gland. (**A**) The epithelial bud is surrounded by dermal mesenchyme (E12.5). (**B**) The bud invaginates into the surrounding stroma the fat pad precursor (FPP) starts to differentiate (E14.5). (**C**) The epithelial compartment elongates, forming the sprout (E16.5). (**D**) The fat pad (FP) has now differentiated, a prerequisite for ductal growth and branching. The epithelial compartment has now branched forming solid epithelial cords. Lumen formation starts at this point (E18.5).

same time, the mesenchyme abutting the mammary epithelium becomes denser than the surrounding mesenchyme with several concentric layers of fibroblasts aligning themselves around the epithelial compartment *(56)*. At E15.5, the bud elongates to become a cord, the mammary sprout, which invades the underlying fat pad precursor. Branching of the epithelial cord commences at E16 *(57)*. By E18, branching is apparent and the ductal lumen starts to form *(58)*. Growth continues at the same rate as the rest of the body until puberty *(57)* (Fig. 2).

ER-α and ER-β are first expressed at E12.5 in the mesenchyme surrounding the bud *(53)*. Autoradiographic experiments also revealed specific ^{3}H-DES binding at E16, suggesting the presence of functional receptors only in the mesenchyme surrounding the epithelial anlagen at that time *(59)*. However, ER-α expression is mainly localized to the epithelium at postnatal time points *(60)*.

4. BPA: A MODEL OF CHEMICALLY INDUCED DEVELOPMENTAL DISRUPTION

Numerous studies have described the effects of BPA exposure in rodents. We will concentrate on those studies that reported low, presumably environmentally relevant exposures during fetal or neonatal development; however, we will also include a few examples of high exposures to BPA and other xenoestrogens during fetal and neonatal life and in adulthood to illustrate specific points.

4.1. The Ovaries and the Oocytes

Perinatal exposure to 25 and 250 ng BPA/kg bw/day for 2 weeks, through osmotic pumps implanted into pregnant dams at gestational day 9, induced changes in the gross anatomy of the mouse ovaries *(61)*. A significant increase of either unilateral or bilateral blood-filled ovarian bursae, a sign of reproductive aging, was observed in the perinatally exposed mice at 6 months of age. There was also a significant increase in the percentage of ovarian tissue occupied by antral follicles and a trend toward a decrease in the percentage of tissue occupied by corpora lutea in the 3-month-old BPA-treated groups compared with the controls. These data are suggestive of a reduced number of ovulated oocytes in BPA-exposed females. Prenatal exposure of CD-1 mice to significantly higher BPA doses (0.5 to 10 mg BPA/kg bw/day) resulted in a frank reduction of the percentage of animals with corpora lutea at 30 days of age *(62,63)*, a phenomenon also observed after exposure to 6.7 and 67 μg DES/kg bw/day *(62)*. Polyovular follicles, a frequent finding in DES-exposed animals, were also found in animals exposed neonatally (PND1–5) to 100 mg BPA/kg bw/day *(62)*.

After an accidental exposure to BPA, Hunt et al. *(64)* reported a dramatic increase in meiotic disturbances including aneuploidy in oocytes collected from mice in their colony. They traced the source of the BPA exposure to cages that housed the animals as well as plastic water bottles. Both had been damaged after being washed with a detergent. The dose of BPA these juvenile female mice were exposed to was estimated to be in the range of 14–72 ng/gram of body weight. When the experiment was repeated using BPA exposures of 20, 40, or 100 ng BPA/g for 6–8 days, the congression failure (gross disturbances in chromosome alignment on the meiotic spindle) increased, and a dose-related increase in the level of abnormalities was observed among the treated animals *(64)*. These findings are of significant concern, because average levels of 2.4 ng BPA/ml have been reported in human follicular fluid *(46)*.

Aneuploidy is considered as one of the leading causes of miscarriages in humans *(65)*. Recently, a link has been suggested between BPA exposure and recurrent miscarriages *(66)*. BPA levels in the blood of women who miscarried were higher than in women who carried their pregnancies to term. Moreover, examination of tissues from women who had miscarriages revealed abnormal karyotypes that appeared to coincide with exposure to the highest BPA levels in these embryos.

4.2. The Uterus and the Vagina

Exposure of pregnant dams to 25 and 250 ng BPA/kg bw/day for 2 weeks, beginning on gestational day 9, induced alterations in the genital tract of female offspring that were revealed during adulthood (3 months of age). At the tissue level, the absolute volume of the lamina propria of the endometrium was significantly decreased in the animals exposed in utero to 250 ng BPA/kg bw/day, whereas the remaining compartments of the uterus showed a decreasing trend that was not statistically significant. At the organ level, the wet weight of the vagina was significantly decreased. Other studies have reported morphological changes in reproductive tract tissues after exposure to significantly higher doses of estrogenic substances. For example, in rats, prenatal treatment with 100 μg BPA/kg bw/day resulted in decreased thickness and reduced cornification of the vaginal epithelium *(67)*. In mice, neonatal exposure to 2 μg DES/day (PND1–5) has been shown to induce hypertrophy of luminal epithelial cells, a decrease in endometrial glands, and a disorganization of the stroma and muscularis, associated with an overall decrease in the size of the uterus *(68)*.

At the cellular level, perinatal exposure to 25 and 250 ng BPA/kg bw/day resulted in a significant increase in DNA synthesis within the glandular epithelium of the uterus that manifested at 3 months of age. The lack of concomitant apoptotic changes suggests an accrual of new cells. Additionally, increased expression of ER-α and PR was observed in the luminal epithelium of the endometrium and subepithelial stroma (Fig. 3). BPA-induced changes in the expression of sex steroid receptors may exacerbate the response of the uterus to subsequent hormonal administration, such as that imposed by hormonal replacement therapy or the contraceptive pill. For example, neonatal hamsters that were exposed to DES and then subsequently exposed to estradiol later in adulthood exhibit hyperplasia and increased apoptosis of the uterine luminal epithelium, which subsequently developed into neoplasms *(69)*. Neonatal exposure of mice (PND1–5) to 2 μg DES/pup/day induced ovary-dependent uterine adenocarcinoma in 18-month-old mice, a pathology that did not occur in unexposed controls *(70)*. Remarkably, transgenic mouse models in which ER-α expression is up-regulated in the uterus by approximately 25 % exhibited an earlier onset and a higher incidence of uterine adenocarcinoma following exposure to DES at PND1–5 than wild-type mice *(71)*. The finding that BPA induced a subepithelial clustering of ER-α-intense stromal cells within the uterus is also of interest because expression of ER-α in the stroma is a prerequisite for estradiol-mediated cell proliferation of the epithelium *(72)*.

4.3. Puberty and Cyclicity

A sharp decrease in the age of puberty has occurred over the last 150 years primarily due to changes in nutrition. Although the timing of puberty seemed to have reached a plateau in the 1960s, an advance in the age of both thelarche and menarche has been reported in recent years *(73)*. Current publications suggest that exposure to endocrine disruptors may be one of the factors underlying this trend *(74)*.

In female mice, puberty encompasses vaginal opening caused by the rise of estrogen levels and first ovulation caused by ovarian cyclic activity due to a gonadotropin surge. It is useful to relate vaginal opening to human thelarche, because they are both signs of estrogen stimulation, and first estrus with menarche, because both reflect central activity of the hypothalamic–pituitary–gonadal (HPG) axis. In rodents, programming of the central mechanisms that control ovulation occur between the last week of prenatal

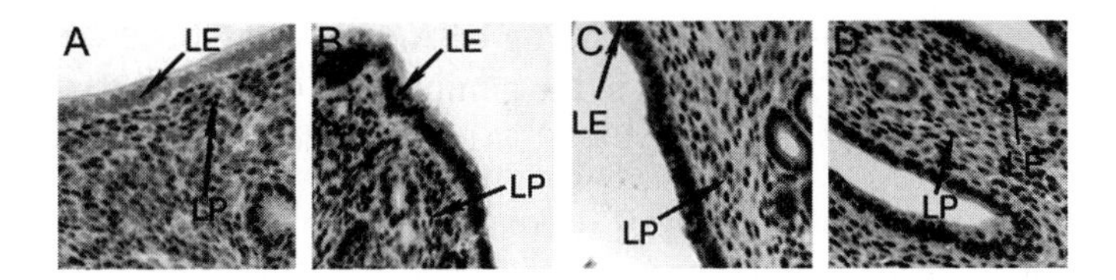

Fig. 3. Prenatal bisphenol A (BPA) exposure alters, the expression of sex steroid receptors in the endometrium of 3-month-old mice. Vehicle-treated (**A** and **C**) and BPA-treated (**B** and **D**) animals. Estrogen receptor alpha expression is depicted in panel **A** and **B**. Note the increased percentage of ER-positive cells in the luminal epithelium (LE) and the lamina propria (LP) right beneath the luminal epithelium in BPA-treated animals. Progesterone receptor expression is depicted in panels **C** and **D**. Notice the increased percentage of PR-positive cells both in the LE and the LP of BPA-treated animals.

life and the first week of postnatal life *(75)*. This maturation process can be influenced by exposure to exogenous sex steroids.

Exposure of pregnant mice to 20 μg BPA/kg bw/day from gestational day 11 through 17 has been shown to induce both vaginal opening and first vaginal estrus at a significantly earlier age in female offspring *(76)*. Exposure to 2.4 μg BPA/kg/day reduced the number of days between vaginal opening and first vaginal estrus in female offspring placed in utero between two females (0M), but not in those situated between two males (2M) *(77)*. This developmental precociousness of the pups was associated, not surprisingly, with an increased body weight at the time of weaning (day 22). The weight of those BPA-exposed females situated between two females (0M) increased by 22 % relative to controls, the body weight of 1M females increased by 9%, and that of 2M females remained unaffected *(77)*.

Mice exposed in utero to 25 and 250 ng BPA/kg bw/day (through the pregnant mother) exhibited an increased number of consecutive days in estrus or metestrus at 3 months of age *(61)*, and exposure of pregnant mouse dams to 2 and 20 μg BPA/kg bw/day on gestational days 11 through 17 increased the duration of the estrous cycle of the offspring. The lengthened cycles were due to an increase in the number of days showing predominantly cornified cells in the vaginal smears *(76)*.

Exposure of pregnant rats from gestational day 6 through lactation to 100 μg BPA/kg bw/day in drinking water significantly increased the body weight in female offspring from birth through 110 days of age and in male offspring from birth to 54 days of age. In the same study, exposure to 1.2 mg BPA/kg bw/day also disrupted the regularity of estrous cyclicity in females aged 4 and 6 months *(11)*. In addition, these animals showed reduced hypersecretion of plasma luteinizing hormone (LH) in response to long-term ovariectomy relative to vehicle-treated controls.

4.4. The Mammary Gland

During prenatal and neonatal development, the mouse mammary gland grows isometrically with respect to body growth until plasma estrogen levels rise during the 3rd week of postnatal life *(78)*. Estrogens then drive massive peripubertal ductal growth. The terminal end buds (TEBs) become bulbous and show both high-proliferative and high-apoptotic activity. Death of the body cells in the TEBs is essential for the formation of the lumen on the proximal side of the TEBs and to the growth of the subtending duct *(79)*. Thus, the ductal tree migrates into the stroma, led by a front of large TEBs. The ductal tree then reaches the edge of the fat pad and eventually establishes a network of ducts, terminal ducts, and a few alveolar buds *(80)*. Once again, this morphology remains relatively quiescent; minor fluctuations occur with each estrous cycle, adding and removing alveolar buds, until pregnancy. During pregnancy, the entire epithelial compartment undergoes dramatic proliferation resulting in a plethora of alveolar buds and lobuloalveolar units in preparation for lactation. Once the period of lactation is over, the mammary gland undergoes rapid involution, a process associated with widespread apoptosis, to return to its prepregnancy state *(81,82)*.

Exposure of the pregnant mouse dams to 25 and 250 ng BPA/kg bw/day for 14 days beginning on day 8 of gestation has been shown to impact certain aspects of development in their female offspring. When examined on embryonic day 18, fetuses of mothers exposed to the higher dose of BPA exhibited altered growth parameters of the mammary gland anlagen. Changes in the appearance of the mammary epithelium were observed such as decreased cell size and delayed lumen formation, as well as increased ductal area. In the stroma, BPA exposure promoted advanced maturation of the fat pad and altered localization of fibrous collagen *(58)*. Because maturation of the fat pad is the driving event for ductal growth and branching, it is likely that the increased ductal area in BPA-exposed animals is due to the accelerated formation of their fat pads.

By PND10, in the offspring born to mothers exposed to either dose of BPA, the percentage of epithelial cells undergoing incorporation of bromodeoxyuridine into DNA was significantly decreased relative to those exposed to vehicle. At 30 days of age, the area and number of TEBs relative to the gland ductal area increased, whereas apoptotic activity in these structures decreased in BPA-exposed offspring compared with controls. There was a positive correlation between ductal length and the age of first proestrus in control females. This correlation was reduced as the BPA dose increased, suggesting that BPA exposure slows down ductal invasion of the stroma. It is likely that the reduced apoptotic index in the TEBs of BPA-exposed females may be the cause of this ductal growth delay, as apoptosis is essential for both the hollowing and the outward growth of the subtending duct. Collectively, these effects observed at puberty may be attributed to an increased sensitivity to estradiol that has been observed in the BPA-exposed animals *(28)*.

In animals exposed perinatally to BPA, there was also a significant increase of ductal epithelial cells that were positive for PR at puberty. These positive cells were localized in clusters, suggesting future branching points. Indeed, lateral branching was significantly enhanced at 4 months of age in offspring born to mothers exposed to 25 ng BPA/kg bw/day *(28)*. These results are compatible with the notion that increased sensitivity to estrogens drives the induction of PRs in epithelial cells, leading to an increase in lateral branching. By 6 months of age, perinatally exposed virgin mice exhibit mammary glands that resemble those of a pregnant mouse. This is reflected by a significant increase in the percentage of ducts, terminal ends, terminal ducts, and alveolar buds *(27)*. In conclusion, these results indicate that perinatal exposure to environmentally relevant doses of BPA results in persistent alterations in mammary gland morphogenesis (Fig. 4). Moreover, the altered growth parameters noted in the developing mammary gland on embryonic day 18 suggest that the fetal gland is a direct target of BPA and that these alterations cause the mammary gland phenotypes observed in perinatally exposed mice at puberty and adulthood.

4.4.1. Perinatal Xenoestrogen Exposure: Relevance to Breast Cancer

Estrogen exposure throughout a woman's life is a major risk factor for the development of breast cancer, as has been demonstrated by the increased risk associated with early age of menarche and late age of menopause. The positive correlation between increased intrauterine levels of estrogens (a phenomenon observed in twin births) and breast cancer in daughters born from such pregnancies also supports this link *(83–85)*. The aforementioned studies describing how fetal exposure to low doses of BPA induces modulations in cell proliferation, apoptosis, and the timing of developmental events fits the notion that this chemical can predispose the mammary gland to carcinogenesis. In the mouse model, increased sensitivity to estrogens may represent a functional equivalent to this known risk factor. Moreover, increased ductal density in the mice

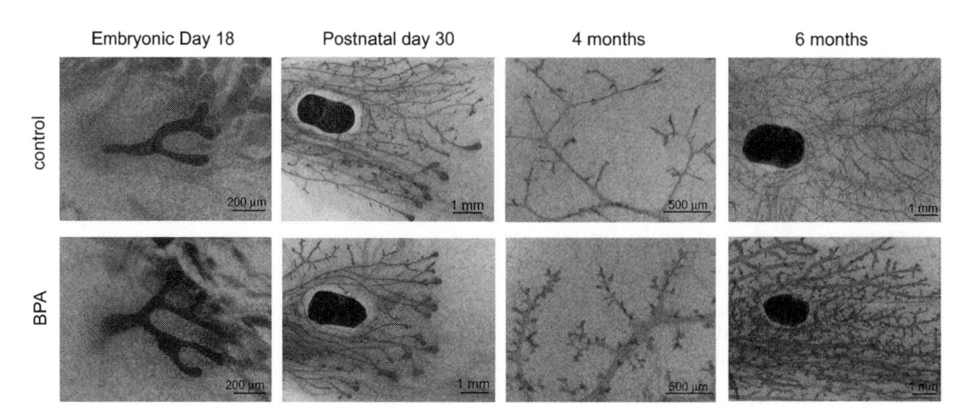

Fig. 4. Perinatal exposure to environmentally relevant doses of bisphenol A (BPA) alters the mammary gland histoarchitecture. Whole mounts of the 4th (inguinal) mammary gland reveal an increased ductal area at E18, delayed ductal growth at puberty [postnatal day (PND) 30], increased lateral branching at 4 months of age, and increased ductal density and lobuloalveolar development at 6 months of age in BPA-exposed animals.

exposed perinatally to BPA may be considered equivalent to another acknowledged risk factor in humans, namely, increased mammographic density *(86)*.

Various epidemiological case–control and cohort North American and European studies have revealed a positive correlation between blood serum levels of other endocrine disruptors in women, such as dieldrin, DDT, and PCBs, and breast cancer incidence *(87–89)*. Controversy abounds on the interpretation of this data, mainly because none of these chemicals can be construed to be a marker of a total xenoestrogen exposure *(90)*. Additional assessments of total xenoestrogen bioactivity in the adipose tissue correlated positively with breast cancer incidence *(91)*. It is worth noting that women who were exposed therapeutically to DES while pregnant between the years 1948 and 1971 now show a higher incidence of breast cancer *(92)*. Their daughters, the cohort of women exposed to DES in utero, are now reaching the age at which breast cancer is diagnosed. A recent publication reported that the rate ratio for the incidence of breast cancer in DES-exposed versus unexposed women now age 40 and older was 2.5, indicating a statistically significant increase in the DES-exposed women. This result "raises a concern calling for continued investigation" *(93)*. The multitude of environmental chemicals to which we are all exposed involuntarily, in addition to medically used hormones (hormonal contraceptives or hormone replacement therapy), may be a cumulative cause for the increase in breast cancer incidence that has been observed during the last 50 years. At present, epidemiological studies have shown that, at all stages of development, the human breast and the rodent mammary gland are vulnerable to carcinogenic stimuli. However, in humans, the impact of exposure to radiation is maximal during puberty when compared with adult age *(94)*, and in rats, sensitivity to chemical carcinogens also peaks at puberty *(95)*. To date, the vulnerability of the prenatal stage in humans to carcinogenic stimuli remains to be studied. Given the current evidence of exposure of the human fetus to BPA and the alterations in mammary gland development documented in rodents following in utero and neonatal exposure to BPA, the need for further study of prenatal vulnerability and breast cancer risk is apparent.

5. WHAT MECHANISMS UNDERLYING THE FETAL/NEONATAL ORIGIN OF BPA-INDUCED ALTERATIONS DURING ADULT?

All of the morphological and functional changes described above suggest both direct action of BPA on the target reproductive organ and BPA-induced alterations in the HPG axis that would in turn affect the peripheral organs.

5.1. Direct Effects of BPA on the Genital Tract and Mammary Gland

Direct effects on target reproductive organs and the mammary gland are believed to be mediated by regulation of the expression of estrogen-target genes involved in tissue patterning, histodifferentiation, and cytodifferentiation (Fig. 5). For example, neonatal exposure to DES or PCBs exerts an estrogenic effect through repression of the Wnt7a-signaling pathway in the female reproductive tract *(96,97)*. Prenatal exposure to DES as well as neonatal exposure to methoxychlor altered the expression of several Hox genes in the mouse Müllerian duct and uterus *(98,99)*. The uterine phenotype observed in mice carrying null mutations of these genes is similar to that resulting from fetal exposure to DES that alters the expression of these genes *(97)*. In the words of Block et al. *(99)*, "Estrogens are novel morphogens that directly regulate the expression pattern of posterior Hox genes in a manner analogous to retinoic acid regulation of anterior Hox

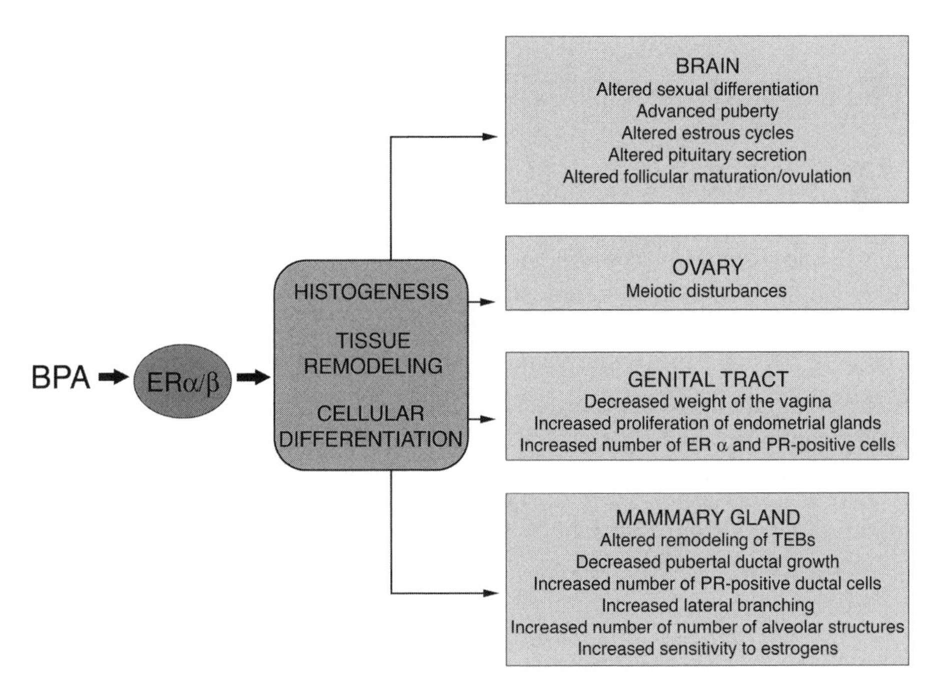

Fig. 5. Schematic representation of the effects of perinatal bisphenol A (BPA) exposure. BPA would act through the estrogen receptor alpha (ER-α) and ER-β present in the different target tissues. This activation of gene expression would lead to altered cell–cell and cell–extracellular matrix interactions and thus altered histogenesis and tissue sculpting and remodeling. The list of effects is compartmentalized according to the organ where BPA is thought to act primarily and directly to generate these effects.

genes." Additionally, some effects of xenoestrogen exposure may involve epigenetic modifications such as DNA methylation and chromatin reorganization *(100,101)*.

5.2. HPG Axis Function

As mentioned previously, there is evidence that perinatal BPA exposure may alter HPG axis function later in life. Initial evidence of this possibility came from the observation that estrous cyclicity was disrupted in females exposed perinatally to BPA relative to offspring exposed to vehicle alone. Studies in rats revealed that dams exposed orally to 100 μg/kg bw/day exhibited fewer regular estrous cycles than control offspring when they reached 6 months of age. Significant disruption of estrous cycles was noted even earlier in the offspring of dams exposed to a 10-fold higher dose of BPA in their drinking water *(11)*. Estrous cycles of mice born to dams exposed to 25 or 250 ng BPA/kg bw/day were also altered relative to controls; at 3 months of age, daily vaginal smear records revealed an increase in the number of consecutive days of estrus or metestrus in BPA-exposed offspring *(61)*.

Further evidence of the ability of BPA to exert lasting alterations at the hypothalamic–pituitary level was suggested by observations that female rats born to BPA-exposed dams exhibited decreased plasma LH levels following long-term ovariectomy relative to control offspring *(11)*. The exposure of male rats to far higher doses of BPA, that is, 500 μg/rat, from PND2 to 12 resulted in increased levels of plasma FSH by PND25 *(102)* further suggesting that BPA may affect regulation of gonadotropin levels. Additional evidence of the ability of BPA to act at the hypothalamus and/or pituitary is provided by the finding that both male and female rats injected with 100 and 500 μg BPA/rat from PND1 to 5 showed a progressive increase in serum prolactin levels that reached 3-fold that of the controls by PND30 *(103)*. This latter finding may be relevant to the previously discussed alterations in the BPA-exposed mammary gland, as the striking increase in alveolar development in BPA-exposed females is suggestive of a hyperprolactinemic state.

Prenatal and neonatal exposure to BPA can alter the expression of sex steroid receptors in the hypothalamus and pituitary. The study described above, in which increased prolactin levels were observed following neonatal exposure to 100 and 500 μg/rat during PND1–5, also documented a concomitant increase in the expression of ER-α and ER-β mRNA in the anterior pituitary of male, but not female, rats at PND30. In addition, female rats in this study exhibited an increase in ER-α mRNA in the medial basal hypothalamus at this same time point *(103)*. Changes in the expression of sex steroid receptors in the pituitary as well as in brain regions important for the regulation of LH release reveal how developmental exposure to BPA might impact steroidogenesis and estrous cyclicity.

5.3. Direct Effects of BPA in the Developing Brain

The developing brain appears be a primary target for BPA action. In addition to altering ER-α mRNA levels in the female hypothalamus as discussed above, BPA has other effects on the brain that may influence the function of the female reproductive axis.

The rostral periventricular preoptic area, and in particular the antero-ventral periventricular nucleus (AVPV), is essential for estrous cyclicity and estrogen positive feedback for the induction of the preovulatory LH surge. The volume of this nucleus is sexually dimorphic, with increased volume in females relative to males. Neurons in this region also exhibit sexually dimorphic patterns of steroid receptor distribution *(104)* and peptide expression in rats *(105–107)*. One robust sex difference that has been observed in the AVPV of both rats and mice is the sexually dimorphic population of tyrosine hydroxylase (TH) neurons *(108–110)*. TH is the rate-limiting enzyme for dopamine synthesis. The number of TH-positive neurons in the AVPV is significantly higher in female rats and mice relative to males, and the sexual dimorphism of this population of neurons appears to be dependent on perinatal levels of gonadal steroids. Studies of estrogen receptor knockout (ERKO) mice indicate that the significant decline in TH neuron number in the male AVPV is dependent upon the presence of ER-α *(110)*. Studies of mice born to mothers exposed to 25 or 250 ng BPA/kg bw/day indicate that these low levels of BPA can obliterate sex differences in AVPV volume and TH neuron number in this nucleus. Consistent with its potential actions as an estrogen, the number of TH neurons in the AVPV of BPA-exposed females is decreased relative to control females *(29)*. Although the role of the TH neurons in the AVPV is not yet known, there is evidence that TH neurons in this region may project to gonadotropin-releasing hormone (GnRH) neurons in the rostral preoptic area and therefore could play a role in the regulation of GnRH/LH release *(111,112)*.

In other studies, offspring born to rats treated with BPA in doses ranging from 30 μg to 1.5 mg/kg bw/day throughout pregnancy and lactation exhibited measurable alterations in the volume and number of cells in the locus coeruleus, a major noradrenergic nucleus of the brainstem *(113,114)* that is important for LH surge induction *(115,116)*. The volume of this nucleus and the number of cells in this nucleus are sexually dimorphic. Therefore, it is conceivable that the ability of perinatal BPA exposure to permanently alter the sexual differentiation of this nucleus may have an impact on cyclic gonadotropin release in adulthood.

It is clear that developmental exposure to BPA can exert direct effects on the reproductive tract and other estrogen-target tissues in the periphery, it is also clear that BPA can exert direct effects on the developing brain and, in particular, in regions of the brain known to be involved in the regulation of cyclical gonadotropin release. Therefore, both peripheral and central actions of BPA during development may contribute to the alterations documented in the uterus and the mammary glands of females long after the period of BPA exposure has ended.

Additional information on endocrine-disrupting chemical effects on the brain, including the neuroendocrine HPG axis, is provided in Chapter 4 of this book by Walker and Gore.

5.4. Mechanisms Underlying the Developmental Origin of Neoplasia

A majority of researchers support the idea that cancer is due to the accumulation of mutations in a cell (somatic mutation theory) *(117)*. In contrast, supporters of the novel theory of fetal origins of adult disease are proposing that changes in the epigenome play a central role in carcinogenesis. A recent study revealed that neonatal exposure of rats to estradiol benzoate or BPA increased the propensity for development of prostate neoplastic lesions following subsequent exposure to a carcinogen *(118)*. The authors postulate that the permanent alterations in the DNA methylation patterns of multiple cell-signaling genes identified in the estrogen and BPA-exposed prostates may be the underlying cause of later neoplastic development.

Both the genetic and epigenetic theories of carcinogenesis imply that cancer originates in a cell that has undergone genetic and/or epigenetic changes, which ultimately result in dysregulated cell proliferation *(119)*. Alternatively, the tissue organization field theory postulates that carcinogenesis represents a problem of tissue organization, comparable to organogenesis gone awry, and that proliferation is the default state of all cells *(120,121)*. According to this theory, carcinogens, as well as teratogens, would disrupt the normal dynamic interaction of neighboring cells and tissues during early development and throughout adulthood *(122,123)*.

During postnatal life, the mammary gland and uterine endometrium undergo massive architectural changes, comparable with those usually associated with organogenesis. These changes occur in response to alterations in endogenous hormone levels such as those associated with puberty and pregnancy and can be induced experimentally by endocrine manipulation. Many studies of endocrine disruptors have illustrated that developmental exposure to these exogenous hormone mimics can alter normal patterns of tissue organization and hence disrupt stromal–epithelial interactions *(28,124)*. These changes may disturb important regulatory mechanisms and enhance the potential for neoplastic lesions.

6. CONCLUSIONS

The organizational and functional changes reported to date provide important pieces of evidence for the understanding of how xenoestrogen exposures, and BPA exposure in particular, affect fetal development and adult function of the female reproductive system in mammals. Although low-level exposure to BPA or other xenoestrogens during adulthood may not have dramatic effects in females, it is clear that when exposure occurs in utero or during the perinatal period, they can exert significant and lasting effects on the development of the female reproductive tract, estrogen-sensitive targets, and reproductive axis function.

The findings reviewed above have both practical and theoretical implications. From a practical perspective, it is now evident that wildlife and humans are affected by environmental exposure to hormonally active chemicals at levels previously considered to be irrelevant. The data that has been collected in the field of environmental toxicology are sufficient to raise concerns about the potential deleterious impact of endocrine-disrupting chemicals on human development. Extrapolating data from animal studies to humans should be done cautiously, as differences among strains and species have been reported regarding a variety of parameters, yet the mouse has been shown to be an excellent model for the understanding of the DES syndrome. Thus, it would be derelict to ignore the increasing evidence coming from controlled experiments in the laboratory and from chemically exposed wildlife, alongside the increasing incidence of comparable issues in human populations exposed to the same chemicals during different developmental stages. All of this evidence should encourage us to apply the precautionary principle and thus ban or substitute those chemicals that are likely to be harmful to the normal development of humans and wildlife.

One impediment to achieving the aforementioned goal is the general belief that environmental endocrine disruptors, particularly those that are steroid hormone agonists and antagonists, are less potent than their natural counterparts, that is, endogenous sex steroids, and therefore do not have the capacity to cause health-threatening effects. However, there is convincing evidence that endocrine-disrupting chemicals, acting through the same receptor pathways as endogenous sex steroids, act additively. In addition, recent research has shown that xenoestrogens, previously considered to be weak agonists of the classical nuclear ERs, are strong agonists when acting through ERs localized in the plasma membrane. Furthermore, environmental estrogens that bind poorly to plasma steroid binding proteins may gain access to tissues normally protected from estrogen exposure during critical periods of development. Moreover, both endogenous hormones and their environmental mimics show non-monotonic dose–response curves, which result in unpredictable and different effects at low and high doses. Finally, it is now irrefutable that the susceptibility and sensitivity of the organism varies depending on the developmental stage at the time of exposure. Although environmentally relevant doses of xenoestrogens may be insufficient to evoke an uterotropic effect in prepubescent animals, they are certainly capable of inducing dramatic morphological, biochemical, physiological, and behavioral changes in laboratory animals exposed in utero.

From a theoretical perspective, these results suggest that the prevalent view of development as the mere unfolding of a genetically determined program is incorrect. The contamination of our environment with endocrine-disrupting chemicals is providing evidence that mammalian development is far more malleable than previously thought, as estrogen exposure during development results in morphological and functional effects that persist into adulthood. The emerging field of environmental endocrine disruption is poised to contribute to the understanding of the mechanisms that underlie the development of hormone-target organs. This quest will require the use of both bottom-up approaches (from genes to organisms) and top-down approaches (from organisms to genes), as well as a new conceptual framework that would take into account the existence of emergent properties *(125)*—that is, properties that cannot be explained from the properties of their components. The properties at one level of complexity (for instance, tissues) cannot be ascribed directly to their component parts (cells, extracellular matrix) but arise only because of the interactions among the parts. Developmental biology, guided by this integrative thinking, now has the tools to successfully revisit the old tradition of ecological regulation of development (phenotype plasticity) *(61,125,126)*.

ACKNOWLEDGMENTS

The authors acknowledge the assistance of Laura N. Vandenberg and Cheryl M. Schaeberle in the preparation of this manuscript. The work outlined herein was made possible by NIEHS grants ES012301 and ES08314.

REFERENCES

1. Colborn T, Clement C, eds. *Chemically Induced Alterations in Sexual and Functional Development: The Wildlife/Human Connection.* Princeton, NJ: Princeton Scientific Publishing, 1992.
2. Colborn T, vom Saal FS, Soto AM. Developmental effects of endocrine-disrupting chemicals in wildlife and humans. *Environ Health Perspect* 1993; 101:378–384.
3. Kelce WR, Stone CR, Laws SC, Gray LE, Kemppainen JA, Wilson EM. Persistent DDT metabolite p,p′-DDE is a potent androgen receptor antagonist. *Nature* 1995; 375:581–585.

4. Moriyama K, Tagami T, Akamizu T, Usui T, Saijo M, Kanamoto N, et al. Thyroid hormone action is disrupted by bisphenol A as an antagonist. *J Clin Endocrinol Metab* 2002; 87:5185–5190.
5. Couse JF, Korach KS. Estrogen receptor null mice: what have we learned and where will they lead us. *Endocr Rev* 1999; 20:358–417.
6. Silva E, Rajapakse N, Kortenkamp A. Something from "nothing" – eight weak estrogenic chemicals combined at concentrations below NOECs produce significant mixture effects. *Environ Sci Technol* 2002; 36:1751–1756.
7. Milligan SR, Khan O, Nash M. Competitive binding of xenobiotic oestrogens of rat alpha-fetoprotein and to sex steroid binding proteins in human and rainbow trout (oncorhynchus mykiss) plasma. *Gen Comp Endocrinol* 1998; 112:89–95.
8. Wozniak AL, Bulayeva NN, Watson CS. Xenoestrogens at picomolar to nanomolar concentrations trigger membrane estrogen receptor-α\-mediated Ca++ fluxes and prolactin release in GH3/B6 pituitary tumor cells. *Environ Health Perspect* 2005; 113:431–439.
9. NTP. *National Toxicology Program's Report of the Endocrine Disruptors Low Dose Peer Review*. Research Triangle Park, NC: National Toxicology Program, 2001. Available at http://ntp.niehs.nih.gov/ntp/htdocs/liason/LowDosePeerFinalRpt.pdf, accessed on March 5, 2006.
10. Alworth LC, Howdeshell KL, Ruhlen RL, Day JK, Lubahn DB, Huang TH-M, et al. Uterine responsiveness to estradiol and DNA methylation are altered by fetal exposure to diethylstilbestrol and methoxychlor in CD-1 mice: effects of low versus high doses. *Toxicol Appl Pharmacol* 2002; 183:10–22.
11. Rubin BS, Murray MK, Damassa DA, King JC, Soto AM. Perinatal exposure to low doses of bisphenol-A affects body weight, patterns of estrous cyclicity and plasma LH levels. *Environ Health Perspect* 2001; 109:675–680.
12. vom Saal FS, Timms BG, Montano MM, Palanza P, Thayer KA, Nagel SC, et al. Prostate enlargement in mice due to fetal exposure to low doses of estradiol or diethylstilbestrol and opposite effects at high doses. *Proc Natl Acad Sci USA* 1997; 94:2056–2061.
13. vom Saal FS, Timms BG. The role of natural and man-made estrogens in prostate development. In: Naz RK, editor. *Endocrine Disruptors: Effects on Male and Female Reproductive Systems*. Boca Raton, FL: CRC Press, 1999: 307–328.
14. Geck P, Maffini MV, Szelei J, Sonnenschein C, Soto AM. Androgen-induced proliferative quiescence in prostate cancer: the role of AS3 as its mediator. *Proc Natl Acad Sci USA* 2000; 97:10185–10190.
15. Sonnenschein C, Olea N, Pasanen ME, Soto AM. Negative controls of cell proliferation: human prostate cancer cells and androgens. *Cancer Res* 1989; 49:3474–3481.
16. Vandenberg LN, Wadia PR, Schaeberle CM, Rubin BS, Sonnenschein C, Soto AM. The mammary gland response to estradiol: monotonic at the cellular level, non-monotonic at the tissue-level of organization. *J Steroid Biochem Mol Biol* 2006; 101.
17. Conolly RB, Lutz WK. Nonmonotonic dose-response relationships: mechanistic basis, kinetic modeling, and implications for risk assessment. *Toxicol Sci* 2004; 77:151–157.
18. vom Saal FS, Hughes C. An extensive new literature concerning low-dose effects of bisphenol A shows the need for a new risk assessment. *Environ Health Perspect* 2005; 113:926–933.
19. Welshons WV, Thayer KA, Judy BM, Taylor JA, Curran EM, vom Saal FS. Large effects from small exposures. I. Mechanisms for endocrine-disrupting chemicals with estrogenic activity. *Environ Health Perspect* 2003; 111:994–1006.
20. Bern HA. The fragile fetus. In: Colburn T, Clement C, editors. *Chemically-Induced Alterations in Sexual and Functional Development: the Wildlife/Human Connection*. Princeton, NJ: Princeton Scientific Publishing Co., Inc, 1992: 9–15.
21. Mittendorf R. Teratogen update: carcinogenesis and teratogenesis associated with exposure to diethylstilbestrol (DES) in utero. *Teratology* 1995; 51:435–445.
22. Herbst AL. Behavior of estrogen-associated female genital tract cancer and its relation to neoplasia following intrauterine exposure to diethylstilbestrol (DES). *Gynecol Oncol* 2000; 76:147–156.
23. McLachlan JA, Newbold RR, Bullock BC. Long-term effects on the female mouse genital tract associated with prenatal exposure to diethylstilbestrol. *Cancer Res* 1980; 40:3988–3999.
24. Newbold RR. Diethylstilbestrol (DES) and environmental estrogens influence the developing female reproductive system. In: Naz RK, editor. *Endocrine Disruptors: Effects on the Male and Female Reproductive Sytems*. Boca Raton, FL: CRC Press, 1999: 39–56.

25. Newbold RR, Jefferson WN, Banks EP. *Developmental Exposure to Low Doses of Diethylstilbestrol (DES) Results in Permanent Alterations in the Reproductive Tract*. The Endocrine Society, Abstract, Annual Meeting, 1999.
26. Markey CM, Michaelson CL, Veson EC, Sonnenschein C, Soto AM. The rodent uterotrophic assay: response to Ashby and Newbold et al. *Environ Health Perspect* 2001; 109:A569–A570.
27. Markey CM, Luque EH, Munoz de Toro MM, Sonnenschein C, Soto AM. *In utero* exposure to bisphenol A alters the development and tissue organization of the mouse mammary gland. *Biol Reprod* 2001; 65:1215–1223.
28. Munoz de Toro MM, Markey CM, Wadia PR, Luque EH, Rubin BS, Sonnenschein C, et al. Perinatal exposure to bisphenol A alters peripubertal mammary gland development in mice. *Endocrinology* 2005; 146:4138–4147.
29. Rubin BS, Lenkowski JR, Schaeberle CM, Vandenberg LN, Ronsheim PM, Soto AM. Evidence of altered brain sexual differentiation in mice exposed perinatally to low environmentally relevant levels of bisphenol A. *Endocrinology*. 2006; 147:3681–3691.
30. Gaido KW, Maness SC, McDonnell DP, Dehal SS, Kupfer D, Safe S. Interaction of methoxychlor and related compounds with estrogen receptor alpha and beta, and androgen receptor: structure-activity studies. *Mol Pharmacol* 2000; 58:852–858.
31. Ohtake F, Takeyama K-I, Matsumoto T, Kitagawa H, Yamamoto Y, Nohara K, et al. Modulation of oestrogen receptor signalling by association with activated dioxin receptor. *Nature* 2003; 423:545–550.
32. Markey CM, Michaelson CL, Sonnenschein C, Soto AM. Alkylphenols and bisphenol A as environmental estrogens. In: Metzler M, editor. *The Handbook of Environmental Chemistry. Vol 3. Part L, Endocrine Disruptors - Part I*. Berlin and Heidelberg: Springer Verlag, 2001: 129–153.
33. McLeese DW, Zitko V, Sergeant DB, Burridge L, Metcalf CD. Lethality and accumulation of alkylphenol in aquatic fauna. *Chemosphere* 1981; 10:723–730.
34. Dodds EC, Lawson W. Molecular structure in relation to oestrogenic activity. Compounds without a phenathrene nucleus. *Proc Royal Soc Lon B* 1938; 125:222–232.
35. Krishnan AV, Starhis P, Permuth SF, Tokes L, Feldman D. bisphenol-A: an estrogenic substance is released from polycarbonate flasks during autoclaving. *Endocrinology* 1993; 132:2279–2286.
36. Brotons JA, Olea-Serrano MF, Villalobos M, Olea N. Xenoestrogens released from lacquer coating in food cans. *Environ Health Perspect* 1994; 103:608–612.
37. Biles JE, McNeal TP, Begley TH, Hollifield HC. Determination of bisphenol-A in reusable polycarbonate food-contact plastics and migration to food simulating liquids. *J Agric Food Chem* 1997; 45:3541–3544.
38. Olea N, Pulgar R, Perez P, Olea-Serrano F, Rivas A, Novillo-Fertrell A, et al. Estrogenicity of resin-based composites and sealants used in dentistry. *Environ Health Perspect* 1996; 104(3):298–305.
39. Matsumoto H, Adachi S, Suzuki Y. Bisphenol A in ambient air particulates responsible for the proliferation of MCF-7 human breast caner cells, its concentration changes over 6 months. *Arch Environ Contam Toxicol* 2005; 48(4):459–466.
40. Berkner S, Streck G, Herrmann R. Development and validation of a method for determination of trace levels of alkylphenols and bisphenol A in atmospheric samples. *Chemosphere* 2004; 54(4):575–584.
41. Rudel RA, Brody JG, Spengler JD, Vallarino J, Geno PW, Sun G, et al. Identification of selected hormonally active agents and animal mammary carcinogenesis in commercial and residential air and dust samples. *J Air Waste Manage Assoc* 2001; 51:499–513.
42. Behnisch PA, Fujii K, Shiozaki K, Kawakami I, Sakai S. Estrogenic and dioxin-like potency in each step of a controlled landfill leachate treatment plant in Japan. *Chemosphere* 2001; 43:977–984.
43. Rodrigues-Mozaz S, Lopez de Alda M, Barcelo D. Analysis of bisphenol A in natural waters by means of an optical immunosensor. *Water Res* 2005; 39:5071–5079.
44. Yoo SD, Shin BS, Lee BM, Lee KC, Han SY, Kim HS, et al. Bioavailability and mammary excretion of bisphenol A in Sprague Dawley rats. *J Toxicol Environ Health A* 2001; 64:417–426.
45. Schonfelder G, Wittfoht W, Hopp H, Talsness CE, Paul M, Chahoud I. Parent bisphenol A accumulation in the human maternal-fetal-placental unit. *Environ Health Perspect* 2002; 110:A703–A707.
46. Ikezuki Y, Tsutsumi O, Takai Y, Kamei Y, Taketani Y. Determination of bisphenol A concentrations in human biological fluids reveals significant early prenatal exposure. *Hum Reprod* 2002; 17:2839–2841.

47. Calafat AM, Kuklenyik Z, Reidy JA, Caudill SP, Ekong J, Needham JL. Urinary concentrations of bisphenol A and 4-Nonylphenol in a human reference population. *Environ Health Perspect* 2005; 113:391–395.
48. Arakawa C, Fujimaki K, Yoshinaga J, Imai H, Serizawa S, Shiraishi H. Daily urinary excretion of bisphenol A. *Environ Health Prev Med* 2004; 9:22–26.
49. Sun Y, Irie M, Kishikawa N, Wada M, Kuroda N, Nakashima K. Determination of bisphenol A in human breast milk by HPLC with column-switching and fluorescence detection. *Biomed Chromatogr* 2004; 18:501–507.
50. Takahashi O, Oishi S. Disposition of orally administered 2,2-bis(4-hydroxyphenyl) propane (bisphenol A) in pregnant rats and placental transfer to fetuses. *Environ Health Perspect* 2000; 108:931–935.
51. Zalko D, Soto AM, Dolo L, Dorio C, Ratahao E, Debrauwer L, et al. Biotransformations of bisphenol A in a mammalian model: answers and new questions raised by low-dose metabolic fate studies in pregnant CD1 mice. *Environ Health Perspect* 2003; 111:309–319.
52. Hogan B, Beddington R, Costantini F, Lacy E. Summary of mouse development. In: *Manipulating the Mouse Embryo: a Laboratory Manual*. Plainview, NY: Cold Spring Harbor Laboratory Press, 1994: 21–113.
53. Lemmen JG, Broekhof JLM, Kuiper GGJM, Gustafsson JA, Van Der Saag PT, van der Burg B. Expression of estrogen receptor alpha and beta during mouse embryogensis. *Mech Dev* 1999; 81:163–167.
54. Yin Y, Ma L. Development of the mammalian female reproductive tract. *J Biochem (Tokyo)* 2005; 137:677–683.
55. Okada A, Sato T, Ohta Y, Iguchi T. Sex steroid hormone receptors in the developing female reproductive tract of laboratory rodents. *J Toxicol Sci* 2005; 30:75–89.
56. Robinson GW, Karpf ABC, Kratochwil K. Regulation of mammary gland development by tissue interaction. *J Mammary Gland Biol Neoplasia* 1999; 4:9–19.
57. Veltmaat JM, Mailleux AA, Thiery JP, Bellusci S. Mouse embryonic mammogenesis as a model for the molecular regulation of pattern formation. *Differentiation* 2003; 71:1–17.
58. Vandenberg LN, Maffini MV, Wadia PR, Sonnenschein C, Rubin BS, Soto AM. Exposure to the xenoestrogen bisphenol-A alters development of the fetal mammary gland. *Endocrinology* 2007; 148:116–127.
59. Narbaitz R, Stumpf WE, Sar M. Estrogen receptors in the mammary gland primordia of fetal mouse. *Anat Embryol (Berl)* 1980; 158:161–166.
60. Saji S, Jensen EV, Nilsson S, Rylander T, Warner M, Gustafsson J-A. Estrogen receptors α and β in the rodent mammary gland. *Proc Natl Acad Sci USA* 2000; 97:337–342.
61. Markey CM, Coombs MA, Sonnenschein C, Soto AM. Mammalian development in a changing environment: exposure to endocrine disruptors reveals the developmental plasticity of steroid-hormone target organs. *Evol Dev* 2003; 5:1–9.
62. Suzuki A, Sugihara A, Uchida K, Sato T, Ohta Y, Katsu Y, et al. Developmental effects of perinatal exposure to bisphenol-A and diethylstilbestrol on reproductive organs in female mice. *Reprod Toxicol* 2002; 16:107–116.
63. Nikaido Y, Yoshizawa K, Danbara N, Tsujita-Kyutoku M, Yuri T, Uehara N, et al. Effects of maternal xenoestrogen exposure on development of the reproductive tract and mammary gland in female CD-1 mouse offspring. *Reprod Toxicol* 2004; 18:803–811.
64. Hunt PA, Koehler KE, Susiarjo M, Hodges CA, Ilagan A, Voigt RC, et al. Bisphenol A exposure causes meiotic aneuploidy in the female mouse. *Curr Biol* 2003; 13:546–553.
65. Hassold T, Hunt P. To err (meiotically) is human: the genesis of human aneuploidy. *Nat Rev Genet* 2001; 2:280–291.
66. Sugiura-Ogasawara M, Ozaki Y, Sonta S-I, Makino T, Suzumori K. Exposure to bisphenol A is associated with recurrent miscarriage. *Hum Reprod* 2005; 20:2325–2329.
67. Schonfelder G, Flick B, Mayr E, Talsness C, Paul M, Chahoud I. In utero exposure to low doses of bisphenol A lead to long-term deleterious effects in the vagina. *Neoplasia* 2002; 4:98–102.
68. Yoshida A, Newbold RR, Dixon D. Effects of neonatal diethylstilbestrol (DES) exposure on morphology and growth patterns of endometrial epithelial cells in CD-1 mice. *Toxicol Pathol* 1999; 27:325–333.
69. Hendry WJ, Zheng X, Leavitt WW, Branham WS, Sheehan DM. Endometrial hyperplasia and apoptosis following neonatal diethylstilbestrol exposure and subsequent estrogen stimulation in both host and transplanted hamster uteri. *Cancer Res* 1997; 57:1903–1908.

70. Newbold RR, Bullock BC, McLachlan JA. Uterine adenocarcinoma in mice following developmental treatment with estrogens: a model for hormonal carcinogenesis. *Cancer Res* 1990; 50:7677–7681.
71. Couse JF, Davis VL, Hanson RB, Jefferson WN, McLachlan JA, Bullock BC, et al. Accelerated onset of uterine tumors in transgenic mice with aberrant expression of the estrogen receptor after nenatal exposure to diethylstilbestrol. *Mol Carcinog* 1997; 19:236–242.
72. Cooke PS, Buchanan DL, Young P, Setiawan T, Broody J, Korach KS, et al. Stromal estrogen receptors mediate mitogenic effects of estradiol on uterine epithelium. *Proc Natl Acad Sci USA* 1997; 94:6535–6540.
73. Parent AS, Teilmann G, Juul A, Skakkebaek NE, Toppari J, Bourguignon J-P. The timing of normal puberty and the age limits of sexual precocity: variations around the world, secular trends, and changes after migration. *Endocr Rev* 2003; 24:668–693.
74. Rasier G, Toppari J, Parent AS, Bourguignon JP. Female sexual maturation and reproduction after prepubertal exposure to estrogens and endocrine disrupting chemicals: a review of rodent and human data. *Mol Cell Endocrinol* 2006; 254–255:187–201.
75. Naftolin F. Brain aromatization of androgens. *J Reprod Med* 1994; 39:257–261.
76. Honma S, Suzuki A, Buchanan DL, Katsu Y, Watanabe H, Iguchi T. Low dose effects of in utero exposure to bisphenol A and diethylstilbestrol on female mouse reproduction. *Reprod Toxicol* 2002; 16:117–122.
77. Howdeshell KL, Hotchkiss AK, Thayer KA, Vandenbergh JG, vom Saal FS. Exposure to bisphenol A advances puberty. *Nature* 1999; 401:763–764.
78. Nandi S, Guzman R, Yang J. Hormones and mammary carcinogenesis in mice, rats, and humans: a unifying hypothesis. *Proc Natl Acad Sci USA* 1995; 92:3650–3657.
79. Humphreys RC, Krajewska M, Krnacik S, Jæger R, Weiher H, Krajewski S, et al. Apoptosis in the terminal end bud of the murine mammary gland: a mechanism of ductal morphogenesis. *Development* 1996; 122:4013–4022.
80. Richert MM, Schwertfeger KL, Ryder JW, Anderson SM. An atlas of mouse mammary gland development. *J Mammary Gland Biol Neoplasia* 2000; 5:227–241.
81. Hennighausen L, Robinson GW. Think globally, act locally: the making of a mouse mammary gland. *Genes Dev* 1998; 12:449–455.
82. Daniel CW, Smith GH. The mammary gland: a model for development. *J Mammary Gland Biol Neoplasia* 1999; 4:3–8.
83. Ekbom A, Trichopoulos D, Adami HO, Hsieh CC, Lan SJ. Evidence of prenatal influences on breast cancer risk. *Lancet* 1992; 340:1015–1018.
84. Weiss HA, Potischman NA, Brinton LA, Brogan D, Coates RJ, Gammon MD, et al. Prenatal and perinatal risk factors for breast cancer in young women. *Epidemiology* 1997; 8:181–187.
85. Braun MM, Ahlbom A, Floderus B, Brinton LA, Hoover RN. Effect of twinship on incidence of cancer of the testis, breast, and other sites (Sweden). *Cancer Causes Control* 1995; 6:519–524.
86. McCormack VA, Dos Santos Silva I. Breast density and parenchymal patterns as markers of breast cancer risk: a meta-analysis. *Cancer Epidemiol Biomarkers Prev* 2006; 15:1159–1169.
87. Cohn B, Wolff M, Cirillo P, Sholtz R, Christianson R, van den Berg B, et al. Timing of DDT exposure and breast cancer before age 50. Proceedings of the International Society for Environmental Epidemiology. *Epidemiology* 2002; 13:S197.
88. Hoyer AP, Jorgensen T, Brock JW, Grandjean P. Organochloride exposure and breast cancer survival. *J Clin Epidemiol* 2000; 53:323–330.
89. Hoyer AP, Grandjean P, Jorgensen T, Brock JW, Hartvig HB. Organochloride exposure and risk of breast cancer. *Lancet* 1998; 352:1816–1820.
90. Soto AM, Fernandez MF, Luizzi MF, Oles Karasko AS, Sonnenschein C. Developing a marker of exposure to xenoestrogen mixtures in human serum. *Environ Health Perspect* 1997; 105:647–654.
91. Ibarluzea JM, Fernández MF, Santa-Marina L, Olea-Serrano MF, Rivas AM, Aurrekoetxea JJ, et al. Breast cancer risk in the combined effect of environmental estrogens. *Cancer Causes Control* 2004; 15:591–600.
92. Calle EE, Mervis CA, Thun MJ, Rodriguez C, Wingo PA, Heath CWJ. Diethylstilbestrol and risk of fatal breast cancer in a prospective cohort of US women. *Am J Epidemiol* 1996; 144:645–652.
93. Palmer JR, Hatch EE, Rosenberg CL, Hartge P, Kaufman RH, Titus-Ernstoff L, et al. Risk of breast cancer in women exposed to diethylstilbestrol in utero: preliminary results (United States). *Cancer Causes Control* 2002; 13:753–758.

94. Land CE, Tokunaga M, Koyama K, Soda M, Preston DL, Nishimori I, et al. Incidence of female breast cancer among atomic bomb survivors, Hiroshima and Nagasaki, 1950–1990. *Radiat Res* 2003; 160:707–117.
95. Gullino PM, Pettigrew HM, Grantham FH. N-nitrosomethylurea as mammary gland carcinogen in rats. *J Natl Cancer Inst* 1975; 54:401–414.
96. Ma R, Sassoon DA. PCBs exert an estrogenic effect through repression of the Wnt7a signaling pathway in the female reproductive tract. *Environ Health Perspect* 2006; 114:898–904.
97. Kitajewski J, Sassoon DA. The emergence of molecular gynecology: homeobox and Wnt genes in the female reproductive tract. *BioEssays* 2000; 22:902–910.
98. Ma L, Benson GV, Lim H, Dey SK, Maas RL. Abdominal B (AbdB) Hoxa genes: regulation in adult uterus by estrogen and progesterone and repression in mullerian duct by the synthetic estrogen diethylstilbestrol (DES). *Dev Biol* 1998; 197:141–154.
99. Block K, Kardana A, Igarashi P, Taylor HS. *In utero* diethylstilbestrol (DES) exposure alters Hox gene expression in the developing Müllerian system. *FASEB J* 2000; 14:1101–1108.
100. Newbold RR, Padilla-Banks E, Jefferson WN. Adverse effects of the model environmental estrogen diethylstilbestrol are transmitted to subsequent generations. *Endocrinology* 2006; 147(Suppl 6):S11–S17.
101. Anway MD, Cupp AS, Uzumcu M, Skinner MK. Epigenetic transgenerational actions of endocrine disruptors and male fertility. *Science* 2005; 308:1466–1469.
102. Atanassova N, McKinnell C, Turner KJ, Walker M, Fisher JS, Morley M, et al. Comparative effects of neonatal exposure of male rats to potent and weak (environmental) estrogens on spermatogenesis at puberty and the relationship to adult testis size and fertility: evidence for stimulatory effects of low estrogen levels. *Endocrinology* 2000; 141:3898–3907.
103. Khurana S, Ranmal S, Ben-Jonathan N. Exposure of newborn male and female rats to environmental estrogens: delayed and sustained hyperprolatinemia and alterations in estrogen receptor expression. *Endocrinology* 2000; 141:4512–4517.
104. Orikasa C, Kondo Y, Hayashi S, McEwen BS, Sakuma Y. Sexually dimorphic expression of ER beta in the anteroventral periventricular nucleus of the rat preoptic area: implication in luteinizing hormone surge. *Proc Natl Acad Sci USA* 2002; 99:3306–3311.
105. Simerly RB. Prodynorphin and proenkephalin gene expression in the anteroventral periventricular nucleus of the rat: sexual differentiation and hormonal regulation. *Mol Cell Neurosci* 1991; 2:473–484.
106. Herbison AE. Identification of a sexually dimorphic neural population immunoreactive for calcitonin gene-related peptide (CGRP) in the rat medial preoptic area. *Brain Res* 1992; 591:289–295.
107. Okamura H, Yokosuka M, Hayashi S. Induction of substance P-immunoreactivity by estrogen in neurons containing estrogen receptors in the anteroventral periventricular nucleus of female but not male rats. *J Neuroendocr* 1994; 6:609–615.
108. Simerly RB. Hormonal control of the development and regulation of tyrosine hydroxylase expression within a sexually dimorphic population of dopaminergic cells in the hypothalamus. *Mol Brain Res* 1989; 6:297–310.
109. Simerly RB, Swanson LW, Gorski RA. The distribution of monoaminergic cells and fibers in a periventricular preoptic nucleus involved in the control of gonadotropin release: immunohistochemical evidence for a dopaminergic sexual dimorphism. *Brain Res* 1985; 330:55–64.
110. Simerly RB, Zee MC, Pendleton JW, Lubahn DB, Korach KS. Estrogen receptor-dependant sexual differentiation of dopaminergic neurons in the preoptic region of the mouse. *Proc Natl Acad Sci USA* 1997; 94:14077–14082.
111. Simonian SX, Spratt DP, Herbison AE. Identification and characterization of estrogen receptor α-containing neurons projecting to the vicinity of the gonadotropin-releasing hormone perikarya in the rostral preoptic area of the rat. *J Comp Neurol* 1999; 411:346–358.
112. Le WW, Berghorn KA, Rassnick S, Hoffman GE. Periventricular preoptic area neurons coactivated with lutenizing hormone (LH)-releasing hormone (LHRH) neurons at the time of the LH surge are LHRH afferents. *Endocrinology* 1999; 140:510–519.
113. Kubo K, Arai O, Ogata R, Omura M, Hori T, Aou S. Exposure to bisphenol A during the fetal and suckling periods disrupts sexual differentiation of the locus coeruleus and of behavior in the rat. *Neurosci Lett* 2001; 304:73–76.
114. Kubo K, Arai O, Omura M, Watanabe R, Ogata R, Aou S. Low dose effects of bisphenol A on sexual differentiation of the brain and behavior in rats. *Neurosci Res* 2003; 45:345–356.

115. Martins-Afferri MP, Ferreira-Silva IA, Franci CR, Anselmo-Franci JA. LHRH release depends on Locus Coerulius noradrenergic inputs to the medial preoptic area and median eminence. *Brain Res Bull* 2003; 61:521–527.
116. Anselmo-Franci JA, Franci CR, Krulich L, Antunes-Rodrigues J, McCann SM. Locus Coeruleus lesions decrease norepinephrine input into the medial pre-optic and medial basal hypothalamus and block the PH, FSH and prolactin preovulatory surge. *Brain Res* 1997; 767:289–296.
117. Hahn WC, Weinberg RA. Modelling the molecular circuitry of cancer. *Nat Rev Cancer* 2002; 2:331–342.
118. Ho S-M, Tang WY, Belmonte de Frausto J, Prins GS. Developmental exposure to estradiol and bisphenol a increases susceptibility to prostate carcinogenesis and epigenetically regulates phosphodiesterase type 4 variant 4. *Cancer Res* 2006; 66:5624–5632.
119. Weinberg RA. *The Biology of Cancer.* New York: Taylor & Francis, 2006.
120. Soto AM, Sonnenschein C. The somatic mutation theory of cancer: growing problems with the paradigm. *BioEssays* 2004; 26:1097–1107.
121. Sonnenschein C, Soto AM. *The Society of Cells: Cancer and Control of Cell Proliferation.* New York: Springer Verlag, 1999.
122. Maffini MV, Soto AM, Calabro JM, Ucci AA, Sonnenschein C. The stroma as a crucial target in rat mammary gland carcinogenesis. *J Cell Sci* 2004; 117:1495–1502.
123. Maffini MV, Calabro JM, Soto AM, Sonnenschein C. Stromal regulation of neoplastic development: age-dependent normalization of neoplastic mammary cells by mammary stroma. *Am J Pathol* 2005; 67:1405–1410.
124. Markey CM, Rubin BS, Soto AM, Sonnenschein C. Endocrine disruptors from Wingspread to environmental developmental biology. *J Steroid Biochem Mol Biol* 2003; 83:235–244.
125. Soto AM, Sonnenschein C. Emergentism as a default: cancer as a problem of tissue organization. *J Biosci* 2005; 30:103–118.
126. Gilbert SF, Sarkar S. Embracing complexity: organicism for the 21st century. *Dev Dyn* 2000; 219:1–9.

3 Endocrine Disruption in the Male

Prue A. Cowin, BBmedSc (Hons),
Paul M. D. Foster, PhD,
and Gail P. Risbridger, PhD

Contents

1. INTRODUCTION

Men and women are different, and the saying "Men are from Mars and Women are from Venus" is often used in reference to the mysterious disparities between the sexes. Nevertheless, it is well known that sex steroid hormones, particularly androgens and estrogens, play a significant role in defining some of these differences. Androgens are commonly considered "male hormones" and estrogens "female hormones," but both men and women produce androgens and estrogens, and it is the mix of the "hormonal cocktail" that is critical in defining the differences between the sexes. This differing cocktail means that whilst men and women both produce androgens and estrogens, they do so in differing amounts and ratios. Furthermore, during an individual's lifetime, the ratio of androgens to estrogens is not static, changing from development to death. It is this changing balance of hormones that is critical during early development and throughout life and, if disrupted, has the potential to lead to aberrant health outcomes.

This review examines environmental endocrine-disrupting chemicals (EDCs) and their effects on male health, specifically reproductive tract abnormalities, male fertility, and prostate health. EDCs are defined by the World Health Organization as "exogenous substances or mixtures that alter function(s) of the endocrine system and consequently cause adverse health effects in an intact organism, or its progeny, or (sub) populations." During fetal or neonatal life, some environmental pollutants or industrial chemicals disrupt and have the potential to alter the action of gonadal steroid hormones by virtue of their anti-androgenic or estrogenic properties and in doing so, alter the

From: *Endocrine-Disrupting Chemicals: From Basic Research to Clinical Practice*
Edited by: A. C. Gore © Humana Press Inc., Totowa, NJ

hormonal balance. Their actions are particularly profound, because development of the gonads and urogenital tract in fetal and neonatal life is hormonally regulated. The reproductive tract during development is in an undifferentiated state and lacks compensatory homeostatic mechanisms to prevent adverse effects of EDCs. During early development, and before establishment of the hypothalamo–pituitary–gonadal (HPG) axis, the effects of EDCs on the developing tract are direct and can be permanent and irreversible. This chapter will focus on the basic biology of the effects of EDCs on male reproduction. We refer readers to Chapter 10 by Hauser, Barthold, and Meeker for an epidemiologic consideration of testicular dysgenesis syndrome in men.

2. SEX DETERMINATION, DEVELOPMENT, AND DIFFERENTIATION

2.1. Male Reproductive Tract Development and the Essential Role of Sex Steroids

Mammalian sex is genetically determined at the time of fertilization. During early human development, there is a short period immediately prior to sexual differentiation when the gonad is sexually indifferent, and it is not until the 7th week of gestation that male and female morphological characteristics begin to develop *(1)*. In rodents, the embryo remains sexually indifferent and possesses both male and female reproductive tract primordia until embryonic day 13.5 regardless of its genetic sex.

Gonadogenesis begins with proliferation of the mesodermal (coelomic) epithelium that invades the underlying mesenchyme, resulting in a longitudinal thickening on the medial side of the mesonephros, known as the gonadal ridge *(2,3)*. The invading epithelium begins to form primitive sex cords in the gonadal ridge, which are surrounded by undifferentiated mesenchyme *(4)*. Primordial germ cells, or primitive sex cells, are first visible in the 4th week in the caudal region of the yolk sac near the origin of the allantois and migrate along the hindgut, up the dorsal mesentery and into the gonadal ridges *(5,6)*. The primordial germ cells divide mitotically during migration and continue to proliferate as they migrate under the underlying mesenchyme and are incorporated into the primary sex cords *(1)*. As the primitive cords begin to form, the mesenchyme is invaded by capillaries. The indifferent gonad now consists of an outer cortex and an inner medulla. In the rodent, formation of the gonadal cords is a rapid process that occurs at gestational day 13 through transitory epithelial cell aggregates along the length of the gonadal ridge *(7)*.

Gonadal differentiation is dependent on signals from the Y chromosome, which contain the genes necessary to induce testicular morphogenesis. One of these signals is the SRY gene (the sex-determining region on the short arm of the Y chromosome *(8)*) and acts as a "switch" to initiate transcription of other genes that contribute to testicular organogenesis. In the absence of the SRY protein, the gonad remains indifferent for a short period of time before differentiating into an ovary.

The first morphological sign of testis formation is the aggregation of primordial germ cells and somatic cells (primitive Sertoli cells) *(9)*. These aggregates develop from the gonadal blastema into plate-like structures that then develop into simple arches of elongated testicular cords *(7,10)*. Throughout differentiation, the testicular cords remain connected to the basal portion of the mesonephric cell mass. The cords gradually transform and extend into the medulla of the gonad, where they branch and anastomose to form a network of cords, known as the rete testis *(1)*. A characteristic and diagnostic feature of testicular development is the development of a thick fibrous capsule, the tunica albuginea. As this capsule develops, the connection between the prominent testicular (seminiferous) cords and the surface epithelium is disrupted. Gradually, the testis separates from the regressing mesonephros, becoming suspended by its own mesentery. Concurrent with testicular cord formation, fetal Leydig cells differentiate from loosely packed, undifferentiated mesenchymal cells in the interstitium *(4)*. These interstitial Leydig cells produce the male sex hormone testosterone, which induces masculine differentiation of the Wolffian duct and external genitalia. Intratesticular vasculature differentiates in the gonadal mesenchyme along with the growth of epithelial components. A testis-specific distribution of blood vessels is obvious from an early phase of testicular development *(4)*.

The fetal testis is composed of testicular cords containing supporting immature Sertoli cells and centrally placed spermatogonia, derived from the surface epithelium and primordial germ cells, respectively. These cords are surrounded by a highly vascularized interstitium containing fetal Leydig cells and mesenchyme *(11)*. The testicular cords remain without a lumen throughout the fetal period. The seminiferous cords turn into tubules when the Sertoli cells undergo terminal differentiation. This occurs after birth when they finish dividing (roughly at the onset of puberty). They develop tight junctions between adjacent cells, and apical secretion of fluid begins as these cells become highly polarized. Thus, the lumen forms as Sertoli cells develop their mature phenotype.

In the rodent and human species, fetal testicular androgen production is not only necessary for proper testicular development and normal male sexual differentiation but also necessary for differentiation of the Wolffian ducts into the epididymides, vasa deferentia, and seminal vesicles *(12–19)*. Androgens derived from the Leydig interstitial cells stimulate the mesonephric (or Wolffian) ducts to form the male genital ducts, whereas Sertoli cells produce Müllerian inhibiting substance [MIS or anti-Müllerian hormone (AMH)] that suppresses the development of paramesonephric (Müllerian) ducts or female genital ducts.

There is differential maturation of the mesonephric ducts depending on location. Near the testis, some tubules persist and are transformed into efferent ductules, which open into the mesonephric duct, forming the ductus epididymis. Distal to the epididymis, the mesonephric ducts acquire a thickening of smooth muscle to become the ductus deferens or vas deferens *(1)*.

Development of the external genitalia is similar in the two sexes. In the human, the external genitalia are indistinguishable until the 9th week of gestation and not fully differentiated until the 12th week of development. Development of the external genitalia coincides with gonadal differentiation. Early in the 4th week of gestation, the sexually undifferentiated fetus develops a genital tubercle at the cranial end of the cloacal membrane. Labioscrotal (genital) swellings and urogenital (urethral) folds then develop on each side of the cloacal membrane. The genital tubercle then elongates forming a phallus. In response to testicular androgens, the phallus enlarges and elongates forming the penis, whereas the labioscrotal swellings ultimately form the scrotum. At the end of the 6th week of gestation, the urorectal septum fuses with the cloacal membrane dividing the membrane into a dorsal anal and a ventral urogenital membrane. Approximately a week following, these membranes rupture forming the anus and urogenital orifice, respectively *(1)*.

Fetal testicular androgens are responsible for the induction of masculinization of the indifferent external genitalia. The testis remains caudally positioned during the 10–15th week until entry into the inguinal canal and transabdominal descent. Testicular descent through the inguinal canal begins in the 28th week, and the testes enter the scrotum by the 32nd week. At birth, the testes reach the bottom of the scrotum *(1,20)*. There are two critical phases of testis descent, transabdominal and inguinoscrotal, essential to move the testes into the scrotum. Although the precise mechanisms of testicular descent, and causes of cryptorchidism, remain unclear, the insulin-like peptide hormone INSL3 and fetal testicular androgens are known to play a critical role *(21)* (Fig. 1). Cryptorchidism or undescended testes occurs in about 3 % of full term and 30 % of preterm males making it the most common human birth defect *(20)*. However, a study comparing the prevalence of cryptorchidism in cohorts of children in Denmark and Finland observed a higher prevalence of cryptorchidism in Denmark, with a 9 % incidence rate reported at birth *(22)*. These data add further evidence to the concept that there is a significant geographical difference in male reproductive health in two neighboring countries and therefore potential exposure to similar environmental effects. As the major difference was found in the milder forms of cryptorchidism, an environmental rather than a genetic basis for effect is favored. If this is indeed the case, then there is a need to determine the nature of the environmental agents responsible, because similar agents may well be implicated in the trends noted in other geographically diverse countries where an increasing frequency of cryptorchidism and testicular cancer has been found *(23,24)*.

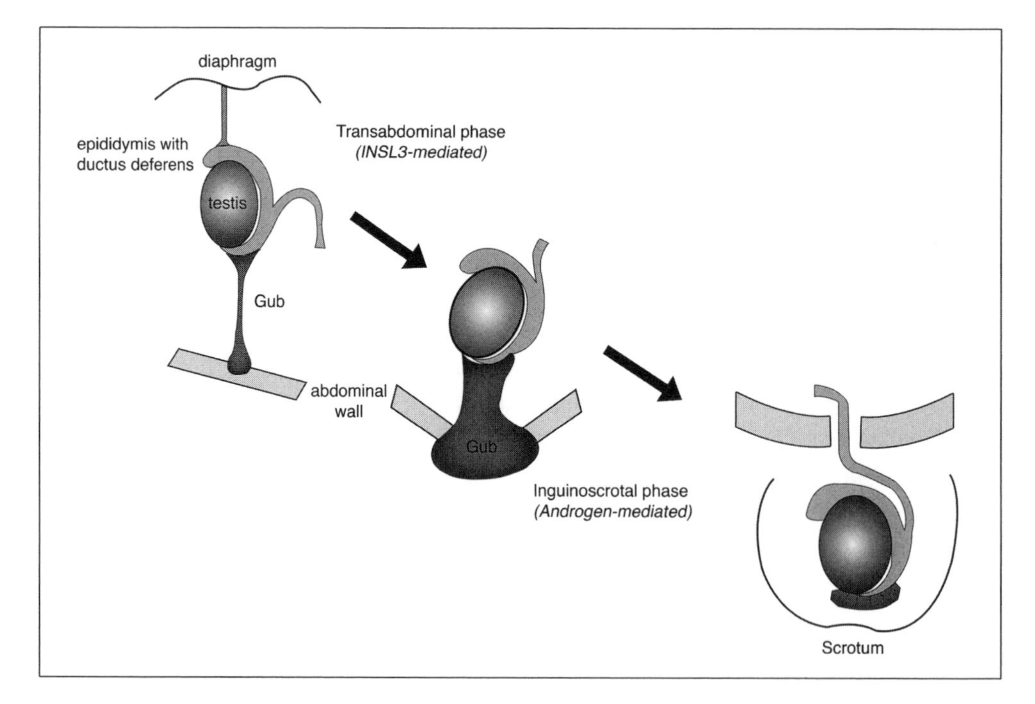

Fig. 1. Testicular descent is an essential developmental step in male reproduction. Two critical phases of testis descent transabdominal (left) and inguinoscrotal (right) travel are essential to move the testes into the scrotum. The intermediate stage is shown in the middle. Adapted from *(21)*.

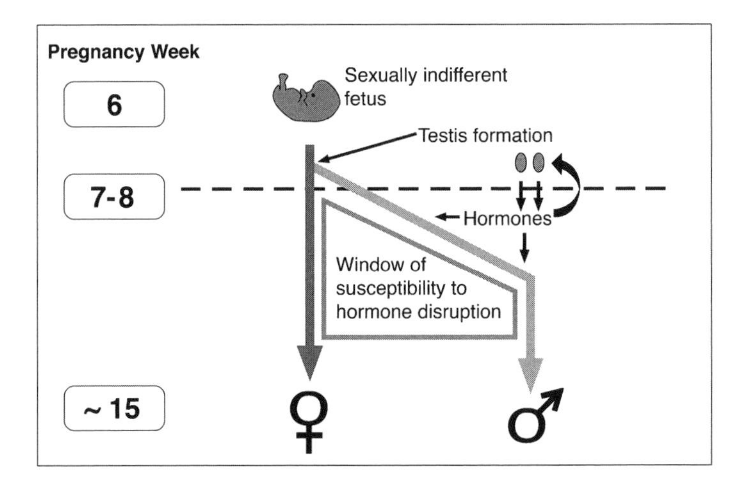

Fig. 2. Sexual differentiation and determination in the human. Female development is largely hormone independent and considered the "default" pathway. Formation of the testes and the resultant hormone production are responsible for diversion from this pathway and masculinization of the fetus. Male development is thus totally hormone dependent. This hormone-dependent masulinization continues into postnatal life, and as such, the male fetus is inherently more susceptible to endocrine disruption. Modified from *(25)*.

In summary, male, but not female, reproductive tract development is hormonally dependent and thus inherently more susceptible to endocrine disruption (Fig. 2).

2.2. Evidence that Environmental Endocrine Disruptors Alter Male Reproductive Tract Developmental Processes

A majority of the focus on EDCs in the past has been directed to study chemicals with estrogen-like activity. However, based on our summary of the basic biology of male reproductive tract development in mammals, it is evident that chemicals with anti-androgenic activity will also constitute a mechanism of equal, if not greater, importance, because androgens are fundamental and critical to establishing the male phenotype.

The first environmental anti-androgen pp-DDE, the metabolite of op-DDT, was reported in 1995 *(26)*. Since then, there have been numerous other androgen receptor (AR) antagonists reported, such as linuron *(27–29)*, vinclozolin, procymidone *(30,31)*, and prochloraz *(32,33)*. The Endocrine Disruptor Screening and Testing Advisory Committee (EDSTAC) was established as an independent committee to advice the US Environment Protection Agency (EPA) on how it should fulfill its mandate to screen and test for EDCs. It realized that it was possible to modify androgen signaling by modes of action other than blocking the AR, including blocking fetal androgen production. From the environmental standpoint, there are anti-androgens that are insecticides (DDE from DDT and fenitrothion), fungicides (vinclozolin, procymidone, and prochloraz), and herbicides (linuron) plus potent drugs that may get into the environment (such as finasteride, a 5α reductase inhibitor). There are also a multitude of chemicals that can interfere with androgen action, such as phthalates.

Phthalates are a ubiquitous group of environmental EDCs. Certain phthalate esters have been shown to produce reproductive toxicity in male rodents with age-dependent

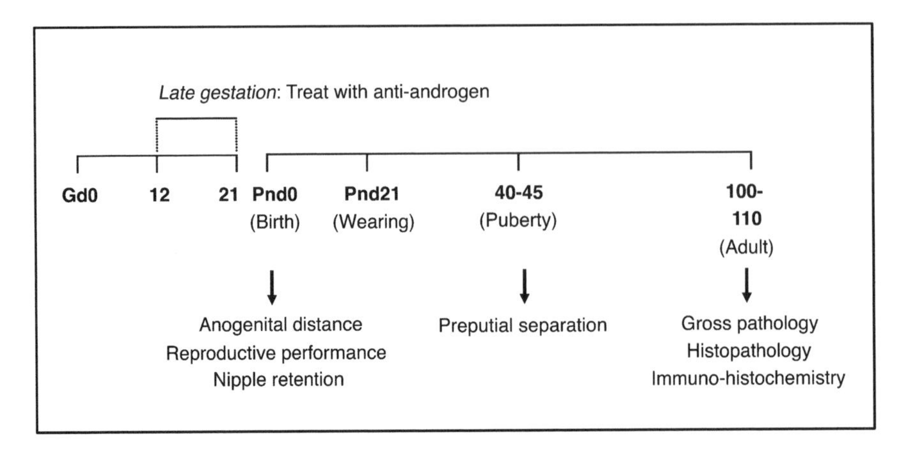

Fig. 3. Schematic diagrammatic representation of the experimental design commonly employed in the investigation of anti-androgen action on male reproductive tract development. Modified from *(34)*.

sensitivity in effects. Fetal animals are more sensitive than neonates who are in turn more sensitive than pubertal and adult animals. Although the testicular effects of phthalates in rodents have been known for more than 30 years, recent attention has focused on the ability of these chemicals to produce effects on reproductive tract development in male offspring following in utero exposure. The common experimental model employed in the study of anti-androgen action on male reproductive tract development in rats is indicated in Fig. 3.

Briefly, pregnant rats are exposed to chemicals of interest by oral gavage during the critical period of male reproductive tract development (gestational days 12–21). On the day of delivery, pups are examined for signs of clinical toxicity. Anogenital distance (AGD) is recorded, and definitive sex of offspring is not determined until weaning. At puberty, male pups are examined for malformations of the external genitalia, testicular descent, and preputial separation. Gross pathology of sexually mature animals is examined, as well as histopathology and immunohistochemistry analysis *(34)*.

2.2.1. Phenotypic Effects Following in Utero Exposure

In the rat, exposure to anti-androgenic compounds during late gestation alters androgen-dependent reproductive tract development *(29,35–40)*. Importantly, agents that interfere with androgen signaling all give the same spectrum of reproductive tract malformations such as cryptorchidism, hypospadias, and effects on the Wolffian ducts, seminal vesicles, and prostate. However, a critical difference is that they show remarkable tissue selectivity in response depending upon whether it is T-mediated or dihydrotestosterone (DHT)-mediated reproductive developmental parameter. For example, the fungicide vinclozolin drastically affects the development of the prostate and induces hypospadias but does little to the normal development of the epididymis. However, almost the opposite effect is observed following phthalate exposure, with the development of the epididymis being most sensitive to adverse effects, with effects on the prostate and induction of hypospadias only seen at low incidence in the highest dose levels examined that do not induce fetal loss. Thus, considerable interest has focused on the developmental and reproductive toxicities of a number of important phthalate esters with anti-androgenic properties: di-(2-ethylhexyl) phthalate (DEHP), di-*n*-butyl phthalate (DBP), and butylbenzyl phthalate (BBP). The general potency of response on male reproductive tract development is DEHP > DBP > BBP *(39)*.

The disturbance of androgen-mediated development is characterized by malformations of the epididymis, vas deferens, seminal vesicles, prostate, external genitalia (hypospadias), cryptorchidism, and testicular injury. In addition, permanent changes in sexually dimorphic structures, such as AGD and aereola/nipple retention, are observed. Although human data linking EDC exposure to reproductive malformations are limited, a recent epidemiological study demonstrated that certain phthalates are associated with a reduced AGD in human male infancies exposed prenatally to phthalates *(41)*.

2.2.1.1. Testicular Injury and Malformations of the Epididymis and Vas Deferens. Disturbance of normal fetal testicular Leydig function and/or development is one of the earliest effects observed following in utero phthalate exposure *(42–45)*. In the developing testes, fetal Leydig cell hyperplasia or the formation of large aggregates of fetal Leydig cells is commonly observed after in utero phthalate exposure (Fig. 4). A significant reduction in fetal testicular testosterone production precedes these morphological changes. Studies demonstrate that DBP at doses ≥50 mg/kg/day reduces mean fetal testicular testosterone levels to as low as 10 % of control levels at gestational day 19 at the peak of fetal testicular testosterone synthesis *(43,44,47)*.

A failure of normal Wolffian duct development into the vas deferens, epididymis, and seminal vesicles is also observed and is likely due to the reduction in fetal testicular testosterone production during the critical time of male reproductive tract development *(48)*. Following exposure to the anti-androgen DBP, epididymal lesions are not apparent until after development of the testicular lesions (Fig. 5). This implicates a direct effect of DBP on the fetal testes, and it is likely that the epididymal lesions are secondary to decreased testicular testosterone synthesis *(48)*. However, the adult testicular manifestations of in utero phthalate exposure are also indirectly induced because of the high incidence of epididymal effects and the induction of pressure atrophy in the testis, consequently the fluid produced has nowhere to go. Although testosterone levels were sufficient to initiate early differentiation of the Wolffian ducts, at late, gestation testosterone levels were insufficient to complete differentiation and development of the reproductive tract *(48)*.

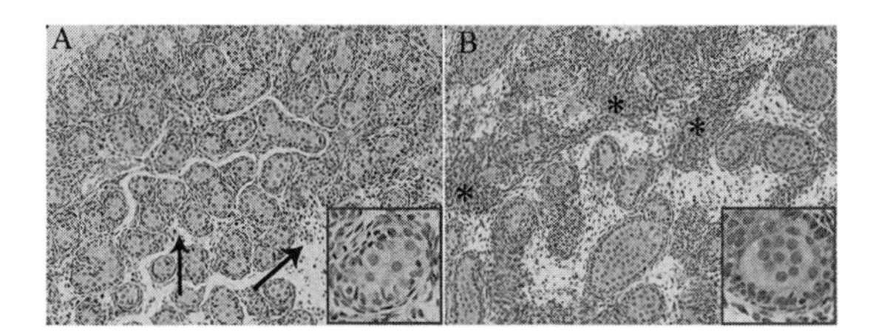

Fig. 4. Photomicrographs of testes obtained on GD21 of fetuses from control (**A**) or di-*n*-butyl phthalate (DBP)-treated dam (**B**). Normal small clusters of Leydig cells are observed in control testes (arrows), whereas large aggregates of Leydig cells are observed in the DBP offspring (∗). Insets: (**A**) Normal seminiferous cord from control animal compared with (**B**) a DBP-treated animal with a multi-nucleate gonocyte. Modified from *(46)*.

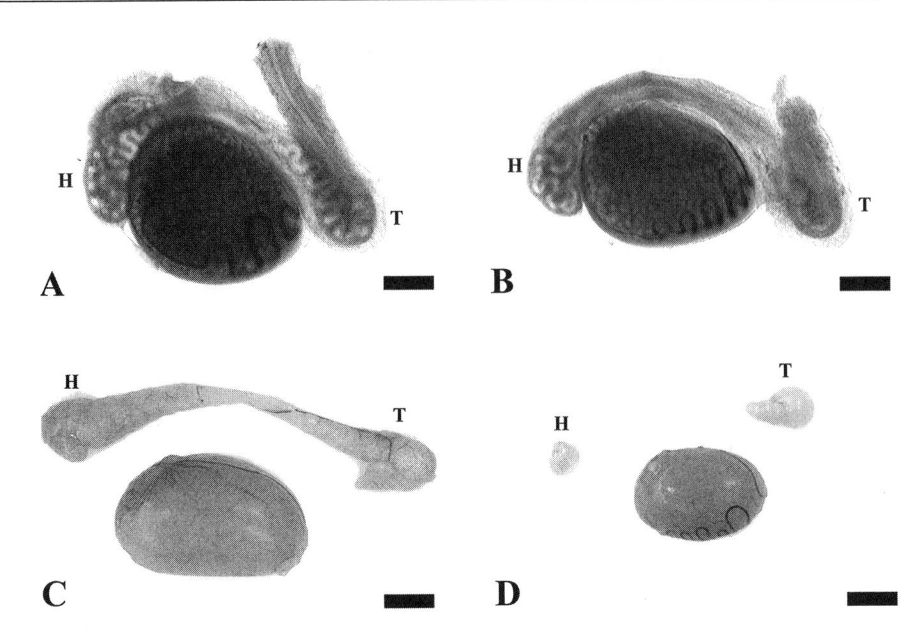

Fig. 5. Gross photographs of GD21 and PND 70 testes and epididymis from fetuses or pups of dams dosed by oral gavage with corn oil or 500 mg/kg/day DBP from GD12–21. (**A**) Control testis and epididymis on GD21. Note prominent coiling in the head and tail regions of the epididymis. (**B**) DBP-exposed testis and epididymis on GD21. Note decreased coiling of the epididymal duct in both the head and tail regions. The testis size is slightly decreased. Bar = 0.5 mm. (**C**) Control and (**D**) DBP-treated testis and epididymis on PND 70. Note the epididymal malformations and testicular atrophy in the DBP-exposed animal. Bar = 0.5 cm. H, epididymal head and T, epididymal tail. Modified from *(48)*.

2.2.1.2. Hypospadias and Nipple Development. In utero exposure to anti-androgens results in permanent phenotypic changes in external genitalia in addition to the internal male reproductive tract malformations. However, it is important to note that while the effects are similar, phthalates such as DBP are inhibitors of fetal testosterone biosynthesis, whereas other anti-androgens such as vinclozolin are competitive AR antagonists.

Testosterone is converted to the more potent androgen DHT, which acts to induce prostate development and also development of sexually dimorphic external genitalia structures. In utero exposure to AR antagonists induces alterations of DHT-mediated development with minimal affects on T-mediated development. Comparatively, phthalate in utero exposure induces malformations in T-dependent tissues, thus the incidence of hypospadias with phthalates is very low compared with the induction ability of AR antagonists. Thus, hypospadias and effects on the prostate are most prevalent following AR antagonist exposure.

Mammary gland development begins similarly in both male and female rodents; however, development of the rodent nipple is sexually dimorphic *(49,50)*. Female rodents have nipples but male rodents do not, because locally produced DHT causes regression of the nipple anlagen *(51–53)*. However, fetal exposure to anti-androgens blocks this process, and subsequently, these male offspring display nipples similar to female littermates (Fig. 6). DHT is also responsible for the growth of the perineum to produce the normal male AGD (approximately twice that of the female).

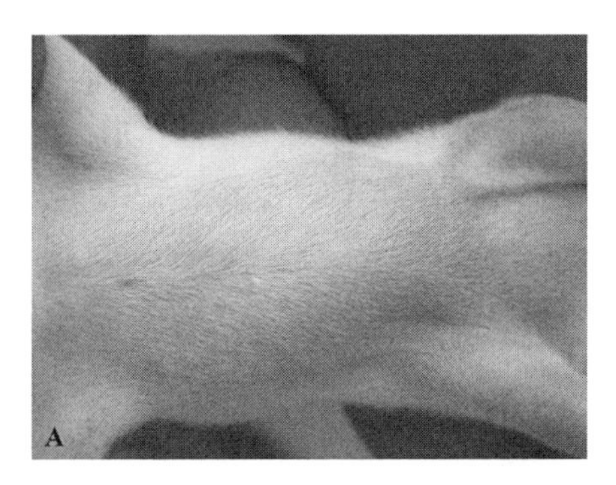

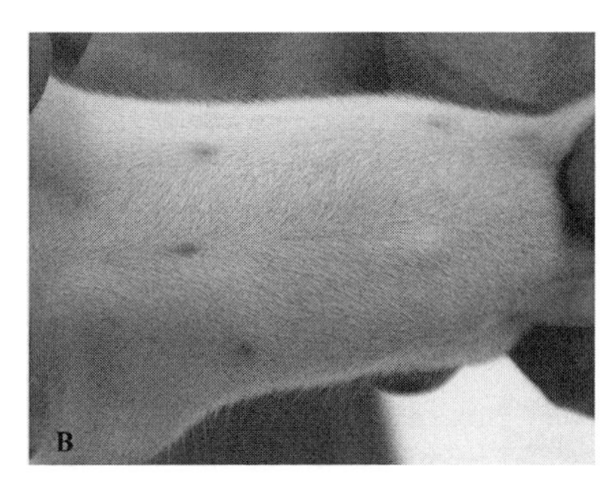

Fig. 6. A vehicle-exposed male rat on postnatal day 13 with no areolae nipples (**A**) and a male rat exposed to 100 mg/kg/day finasteride demonstrating prominent areolae-nipple retention (**B**). Adapted from *(54)*.

Demasculinization, or feminization, of the perineum resulting in a reduced AGD is observed following fetal exposure to anti-androgenic compounds *(29,31,36,39,40,51, 55,56)*.

2.2.1.3. Cryptorchidism. Normal androgen levels are also required for testicular descent into the scrotum; failure to descend results in cryptorchidism *(14,57)*. Cryptorchid or undescended testes are unable to produce mature sperm and are consequently infertile, presumably because of the higher temperatures in the abdominal cavity or inguinal canal. The initial stages of testicular descent require the actions of insulin-like factor 3 (*insl3*), a product of the Leydig cells. *Insl3* knockout mice demonstrate complete cryptorchidism. Following in utero DEHP, DBP, and BBP exposure, a significant reduction in *insl3* gene expression is noted *(58–60)*. The increased incidence of cryptorchidism following fetal phthalate exposure may therefore be related to the decreased expression of *insl3*, as well as a reduction in androgen levels.

Although the male reproductive tract is sensitive to altered androgen status, it is unclear how androgen-mediated gene disruption does in fact alter male reproductive tissue development. Preliminary studies are suggestive of altered paracrine interactions between ductal epithelial cells and the surrounding mesenchyme during differentiation due to lowered testosterone production, indicating critical growth factors and receptors that are involved in these interactions *(61,62)*.

3. PROSTATE GLAND AS A TARGET FOR EDCS

Although testosterone is necessary for maintaining the primordial structure of the male reproductive tract and directing its differentiation into adult reproductive structures *(63,64)*, androgens are also critical for specifying prostate development.

3.1. Prostate Gland Development

Growth and development of the prostate gland begins in fetal life and is complete at sexual maturity. The prostate, in both the rodent and human, arises from the urogenital sinus (UGS), a subdivision of the caudal terminus of the hindgut called the cloaca. Located just caudal to the neck of the developing bladder, the UGS is a midline structure comprising an endodermal-derived epithelial layer surrounded by a mesodermal-derived mesenchymal layer (Fig. 7). In humans, the UGS arises at around the 7th week of gestation *(65)* and approximately 13 days postcoitum in the rodent. The male and female UGS are morphologically indistinguishable until about 10–12 weeks in humans and 17.5 days postcoitum in rodents, at which time prostate morphogenesis begins. Prostate development begins with the outgrowth of five solid buds of urogenital sinus epithelium (UGE) into the surrounding urogenital sinus mesenchyme (UGM) *(66)*. The solid buds of UGE emanate from the urethral portion of the UGS, as bilateral ventral buds, above and below the entrance of the mesonephric (or Wolffian) ducts. The five pairs of epithelial buds (anterior, posterior, medial, and two lateral) undergo extensive branching to form a lobular arrangement of tubuloalveolar glands surrounded by developing stroma that encircle the developing urethra and ejaculatory ducts *(67,68)*. The precise spatial patterning of the epithelial outgrowths establishes the lobar subdivisions of the prostate *(66,69,70)*. The top pairs of buds composed of epithelia believed to be mesodermal in origin form the inner zones of the mature prostate; the lower buds, that form the outer zones of the mature prostate, are endodermally derived *(71)*.

The process by which these initial solid buds of UGE elongate, bifurcate at branch points, and form branches with terminal branch tips is collectively termed "branching morphogenesis," a process that also occurs in many other organs within the body, including kidney, lung, mammary gland, and pancreas *(72–75)*. Initiated in the solid epithelial buds, branching morphogenesis involves elongation of the UGE into the surrounding mesenchyme from the urethral terminus distally toward the ductal tips, as a result of intense proliferative activity at their tips *(76)* (Fig. 8). Concurrent with this process, epithelial and mesenchymal/stromal cytodifferentiation occurs. Basal epithelial cells become localized along the basement membrane forming a continuous layer of cells, whereas tall columnar luminal cells differentiate lining the ductal lumina. At the same time, ductal invagination and mesenchymal differentiation occur resulting in the formation of a glandular organ composed of tubuloalveolar glands surrounded by

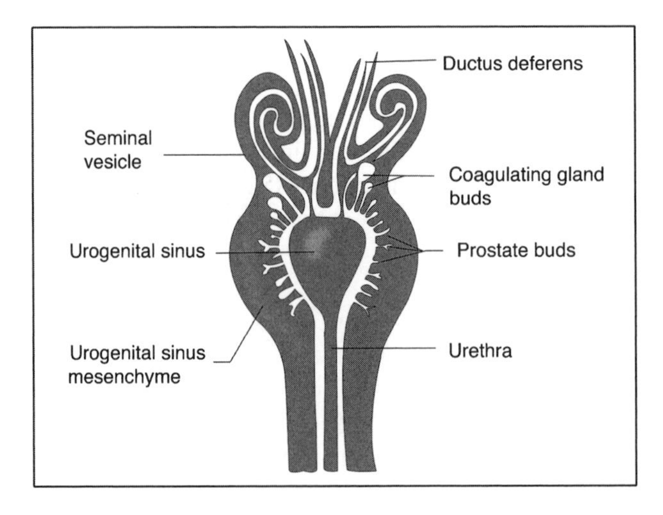

Fig. 7. Schematic illustration of a new-born rodent prostate. Note the growth of prostatic buds into the urogenital sinus mesenchyme. Adapted from *(67)*.

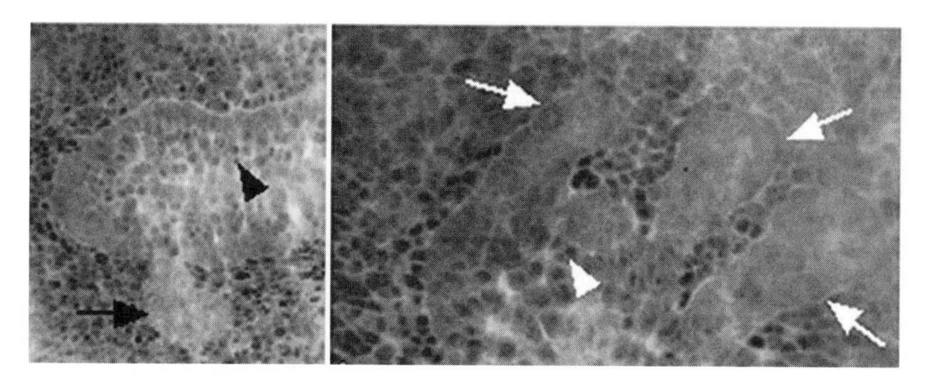

Fig. 8. Human male fetal reproductive tract demonstrating growth of prostatic epithelial buds (indicated by arrows) into the surrounding urogenital sinus mesenchyme. Modified from *(77)*.

mesenchyme containing fibrous tissue and differentiating smooth muscle *(78,79)*. At the organ's periphery, the mesenchyme thickens to encapsulate the gland *(80)*. These branching events are dependent on, and directed by, testicular androgens. The initial outgrowth of the epithelial buds is an androgen-driven process requiring AR expression in the surrounding UGM to facilitate the reciprocal interactions between the epithelia and the mesenchyme *(37,81)*. In the prostate, branching morphogenesis occurs early in development, during gestation or early neonatal life; this process determines the final volume, length, and morphological patterns of the arborized prostatic epithelial ducts that are present *(82)*.

At mid-gestational age, the glands consist of small ducts lined by undifferentiated epithelial cells, and as well as androgens these glands are exposed to increasing levels of maternal estrogens. Maternal estrogens cause squamous metaplasia of the epithelium—that is, multi-layering of the epithelial cells—and at birth, the epithelial cells lining the immature glands vary in both the incidence and extent of squamous metaplasia. However, at birth when the influence of maternal estrogens is no longer evident, this histology is reversed within 4 weeks *(83,84)*, and the neonatal gland consists of differentiated pseudostratified epithelia. Thus, during gestation, there is substantive evidence in human and rodent species that prostate development is responsive to androgens as well as estrogens.

The prostate in humans is well differentiated by the 4th month of fetal growth, and ongoing smooth muscle–epithelial cell interactions play a homeostatic role in maintaining prostatic structure and function *(85)*. Secretory activity detected during fetal life *(86)* is thought to be due to the actions of fetal testicular androgens.

At puberty, prostate maturation and growth proceed. Growth of the pubertal prostate gland is regulated by androgens, but more complex regulatory mechanisms also contribute to growth at this, and other, time of a man's life. The prostate is composed of increasingly complex tubuloaveolar glands arranged in lobules and surrounded by stroma containing fibroblasts, smooth muscle cells, vasculature, nerves, and lymphatics.

Prostate growth is exponential at puberty, in contrast to the mature human prostate gland that is a relatively growth-quiescent organ. At sexual maturity, the human prostate is a compact organ commonly described in three zones: central, transition, and peripheral, each reflecting a distinct ductal organization *(87)* (Fig. 9A). The different origins may be of significance, because the zones are susceptible to different disease

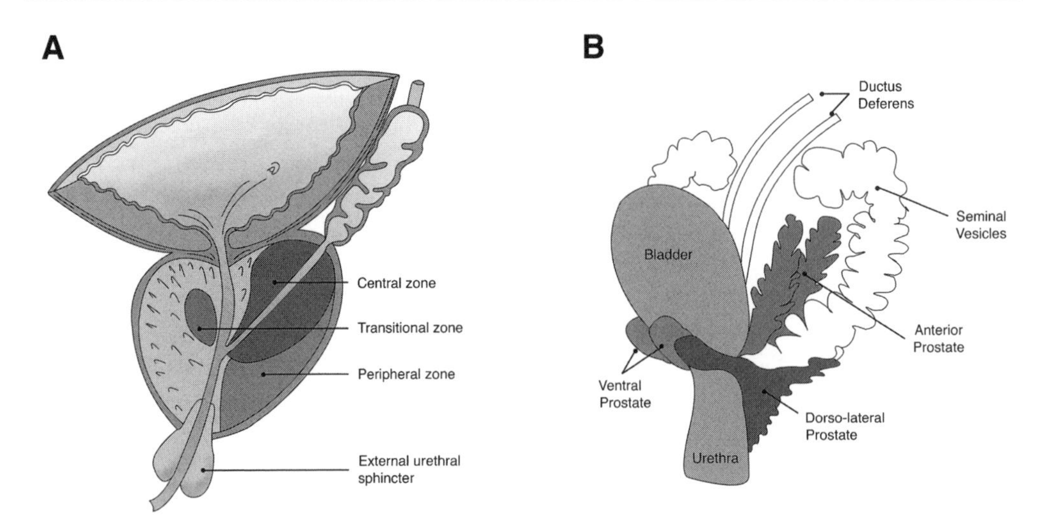

Fig. 9. Anatomy of the human and rodent prostate glands. (**A**) Sagittal view of the adult human prostate that is divided into zones including central zone, peripheral zone, and transitional zone. (**B**) Lateral view of the multi-lobular rodent prostate. Lobes include ventral prostate (VP), anterior prostate (AP), dorsal prostate (DP), lateral prostate (LP), and seminal vesicles (SV). Part **A** modified from *(88)*; part **B** modified from *(67)*.

processes; benign prostatic hyperplasia (BPH) disease arises mainly in the inner zone and prostate cancer in the outer zone *(87)*. By comparison, the rodent prostate is multi-lobular (ventral, anterior, and dorso-lateral), with each lobe exhibiting distinct branching and histological features (Fig. 9B). Although the ventral prostate lobe is commonly used in rodent studies, the consensus opinion from the Bar Harbor Meeting of the Mouse Models of Human Cancer Consortium Prostate Pathology Committee *(89)* is that there is no existing supporting evidence for a direct relationship between the specific rodent prostate lobes and the specific zones in the human prostate. No data currently exist that permit the assumption that one lobe of the rodent prostate is more relevant to human prostate disease than any other lobe *(89)*.

3.2. Essential Role of Sex Steroid Hormones on Normal Prostate Development Across the Life Cycle

3.2.1. Androgens

Testosterone is necessary for specifying development of the UGS along the prostatic lineage. This absolute requirement of androgens for prostate development is evident from the observations that the homologous embryonic region of the female UGS, which normally forms the vagina, undergoes differentiation into prostate if appropriately exposed to testosterone *(90,91)*. Thus, androgens over-ride the influence of genetic sex during development of the prostate gland.

Testosterone may exert its effects on androgen-responsive tissues by binding directly to ARs or alternatively, it may be enzymatically converted by the microsomal enzyme 5α-reductase into the more potent androgen, 5α-DHT *(92)*. For normal masculinization of the UGS, mesenchymal tissue associated with the UGS must express 5α-reductase, as a deficiency in the capacity to normally produce 5α-DHT prevents normal development of the prostate and male external genitalia *(93)*. Furthermore, rodents or humans that lack functional ARs due to inactivating mutations do not develop a prostate gland *(94–98)*, emphasizing the need for androgens in initiating and establishing prostate identity in the UGS.

In the prostate, 5α-DHT is the biologically active androgen with a higher affinity for ARs relative to testosterone. Prostate development and the process of branching morphogenesis are controlled by androgenic effects mediated through ARs located solely in the UGM prior to, and during, prostatic bud formation. Under the influence of androgens, mesenchymally derived paracrine signals induce epithelial budding, proliferation, and cell differentiation *(99)*; direct androgen binding to epithelial ARs is not required for initial epithelial development *(91,100)*.

Large transient surges of serum levels of androgens and estrogens normally occur very early in life in both human and rodent males. During postnatal life, androgen levels in the human male aged 2–3 months reach levels that are within the adult range and are 60-fold higher than normal prepubertal levels *(101,102)*. Known as the postnatal surge of testosterone, many rodent studies demonstrate the importance of postnatal hormone imprinting, as it is an important determinant of the long-term growth regulation of the gland, because the effects are permanent and long ranging *(103–105)*.

At puberty, the prostate undergoes a rapid phase of exponential growth corresponding to the rise in serum testosterone to adult levels. As a result of the increased testicular androgen secretion, the prostate gland grows to full size with prostatic glandular ducts forming patent lumens within the terminal acini and the epithelial lining becoming highly differentiated and beginning secretory activity *(68)*.

At maturity, androgen levels are maintained, prostate size remains fairly constant, and the organ is considered to be relatively growth quiescent until the fourth decade of life when growth is reinitiated. As rising serum androgens levels stimulate prostate growth in the pubertal male, how and in what way is the response to the same level of serum androgens modified in the mature adult male, so that no further growth occurs and prostate size is maintained? This is a critical question for which there is currently no clear explanation.

The majority of men over 50 years of age experience a decline in androgen levels. This reduction is referred to as "androgen decline in ageing males" (ADAM) or "andropause." Unlike menopause in women, this hormonal change is gradual and results in a decrease in androgen levels of approximately 35% *(106)*. Although the testicular and adrenal production of androgens declines with aging, levels of total plasma estradiol do not decline and may even increase with increased adiposity, resulting in a reduction in the testosterone : estrogen ratio. These age-dependent changes in the ratio of serum steroid concentrations may play a role in the onset of benign BPH, and it is well known that the incidence of BPH increases with aging *(107)*. The critical question that remains unanswered is whether and how the gradual decline in androgen levels upon aging causes reactivation of the growth processes resulting in prostatic enlargement and often disease?

3.2.2. Estrogens

In addition to androgens, prostate development is very sensitive to estrogens. Estrogens have dual actions in the prostate gland because of direct and indirect effects on epithelial cell differentiation and proliferation *(108–110)*. Maintaining the

appropriate androgen–estrogen balance is required for normal regulation of the structure and function of male reproductive tract development, including the prostate gland *(111)*.

Indirectly, estrogens act systemically on the HPG axis to suppress release of pituitary luteinizing hormone (LH), reducing testicular androgen synthesis and consequently lowering systemic androgen levels that induces apoptosis and prostatic epithelial atrophy, in addition to reducing secretory activity *(112–115)*. Direct action of estrogens is evident from the presence of two estrogen receptor (ER) subtypes (ER-α and ER-β) predominantly expressed within the prostate stroma and epithelia, respectively *(116)*. Acting locally through ER-α in the prostatic stroma, estrogens stimulate aberrant epithelial cell differentiation and proliferation leading to squamous metaplasia *(112,117)*. An anti-proliferative action of estrogen is suggested to be mediated through activation of epithelial ER-β *(118–120)* and was recently proven using selective estrogen receptor modulators (SERMS) *(121)*.

Prostate tissue itself also has the ability to locally synthesize estrogens. The prostate gland expresses the aromatase enzyme in the prostatic stroma, with endogenously synthesized estrogens acting directly on the prostate *(122)*. The aromatase knockout (ArKO) mouse has been a useful model to investigate the effects of endogenous estrogens. The ArKO mouse was generated by targeted disruption of the Cyp19 gene, and ArKO mice do not synthesize endogenous estrogens, yet remain responsive to the action of estrogen through both ER subtypes *(123,124)*. Aromatase deficiency results in hypertrophy and hyperplasia of the prostate gland at maturity, making it a useful model to investigate the effects of SERMS *(125)*. SERMs are estrogenic-like compounds displaying estrogen agonistic and antagonistic effects in different tissues *(126,127)*. SERMs have the ability to bind both ERs and also compete with and block estradiol (and other estrogenic) activity.

Systemic (serum) or locally (tissue) derived estrogens may target the prostate. Although serum estradiol levels are very high at birth in males, they fall to very low levels in the first few days after birth. In adult male serum, low levels of estrogens are detected and rise upon aging. Serum estrogens in the male are predominantly the products of peripheral aromatization of testicular and adrenal androgens. Estrogen levels in the ageing male are maintained due to the common increase in fat mass with aging (the source of peripheral aromatization), thus there is increased aromatase activity and conversion of testosterone to estrogen with aging *(128,129)*.

Therefore, the relationship between serum and tissue hormone levels is important. The serum hormone balance does not accurately reflect what occurs at the tissue level. This is particularly important in the prostate, because testicular androgens reach the prostate tissue through the serum but are metabolized into the more potent reduced androgen (5α-DHT) or estrogens through the actions of 5α-reductase and aromatase enzymes, respectively.

Despite systemic androgens being essential for coordinated growth the prostate, local hormones, and importantly estrogen, activity is equally essential to regulate the proliferative and anti-proliferative changes that occur during normal prostate development and differentiation. There is, therefore, considerable speculation regarding the potential for environmental chemicals with estrogenic or anti-androgenic activity to alter male reproductive tract development and function, specifically the prostate gland.

3.2.3. Steroid Hormones and Reciprocal Epithelial–Mesenchymal Interactions During Development

Epithelial–mesenchymal cell–cell communication is critical for appropriate action of androgens during prostate development and normal function. The classic work of Franks *(130)*, together with tissue recombination studies of Cunha and colleagues *(131, 132)*, emphasizes the important role played by the mesenchymal elements in controlling prostate growth and differentiation during the developmental period. These studies showed that prostatic morphogenesis and differentiation are the result of the localized instructive potential of the mesenchyme and the receptive epithelia in the UGE *(67, 81,100)*.

Briefly, UGM specifies prostatic epithelial identity, induces epithelial bud formation, elicits prostatic bud growth and regulates ductal branching, promotes differentiation of a secretory epithelium, and specifies the types of secretory proteins expressed *(132–134)*. In turn, differentiation of the UGM relies on interactions and reciprocal signaling from the adjacent epithelia to specify the morphological and spatial patterning of the smooth muscle *(132)*. For example, when human prostate epithelium was recombined with rat mesenchyme from UGM, thick sheaths of smooth muscle were formed, characteristic of human smooth muscle *(135)*. Therefore, in addition to the stroma controlling epithelial differentiation, the epithelium is critical in directing spatial patterning of the stroma, and these data highlighted the intimate relationship between the epithelium and surrounding stromal cells. The bidirectional nature of the relationship between epithelium and stroma is not restricted to development and continues, albeit in different ways, during normal and abnormal function of the gland *(85)*.

3.3. Evidence that Environmental Endocrine Disruptors Alter Prostate Developmental Processes

The synergistic actions of androgens and estrogens are essential in regulating normal proliferative and anti-proliferative changes that occur during prostate development and differentiation. However, during development, the fetal prostate is very sensitive to altered hormone levels and compounds that exhibit estrogenic or anti-androgenic properties. Environmental pollutants and industrial chemicals have the capacity to disrupt either by binding to endogenous hormone receptors, interfering with enzyme activity, or through other mechanisms including interfering with plasma transport of hormones *(136,137)*.

During development, the prostate undergoes DHT-mediated development and as such is sensitive to anti-androgens with AR antagonist actions, such as vinclozolin and flutamide. The fetal prostate lacks the necessary compensatory homeostatic mechanisms to prevent adverse effects of EDCs, and disruption of normal hormone levels ultimately alters the inductive and instructive properties of the neonatal prostatic stroma leading to perturbation of stromal–epithelial cell signaling and aberrant prostatic growth.

3.3.1. Endocrine Disruptive Chemicals with Estrogenic Activity

In recent years, there has been growing concern that the increases in human male reproductive disorders (testicular cancer, cryptorchidism, hypospadias, and low sperm counts) may arise from increased fetal (or neonatal) exposure of the developing male to estrogens and EDCs with estrogenic activity *(138,139)*. Exposure to man-made estrogen-mimicking chemicals can occur through many sources including industrial chemical components of plasticizers [bisphenol A (BPA)] and oral contraceptives (ethinylestradiol). It has been shown that numerous, ubiquitous environmental chemicals possess weak estrogenic activity when measured in both in vivo and in vitro test systems *(139)*. Despite these chemicals being only very weakly estrogenic (10,000-fold → 100,000-fold less potent than estradiol itself) *(139)*, studies have demonstrated that extremely low concentrations of weak environmental estrogens, such as BPA, are able to induce significant increases in prostate size in adulthood *(140,141)*. This, understandably, raises concern about the possible effects in humans.

During prenatal development, the Wolffian duct and UGS express ERs *(137,142)*, and thus, chemicals such as BPA that have the capacity to bind to ERs may directly affect the development and function of these organs. Estrogen exposure during the developmental neonatal period elicits both acute and long-term effects *(104,143–146)*, has been termed "developmental estrogenization" or "neonatal imprinting" *(147)*, and has become a popular model in rodents to investigate the actions of estrogens in the prostate gland. Along with direct actions on prostatic tissues, estrogens in males exert indirect effects through the suppression of pituitary gonadotropin levels and hence testicular testosterone levels. This centrally mediated effect of estrogen action causes changes in prostatic development.

Pharmacological doses of diethylstil bestrol (DES) given during neonatal imprinting lead to permanent, irreversible aberrations in the prostate gland, characterized by a reduction in prostate size and permanent alterations in prostatic growth, morphology, cellular organization, and secretory functions, leading to an increased incidence of prostatic lesions upon aging, including hyperplasia, inflammation, and dysplasia similar to prostatic intraepithelial neoplasia *(103,104,143,146,148–153)*. Similar observations have been reported in the mouse prostate *(105,154)*. Developmental estrogenization permanently reduces the levels of AR protein leading to reduced responsiveness to androgens *(103,104,145,147,151,155)*. Interestingly, the decreased expression of AR protein following neonatal estrogenization was not the result of transcriptional alterations but proteolytic degradation of the AR protein itself *(156)*.

It appears that the induction of reproductive tract abnormalities coincides with, and is completely dependent on, the suppression of androgen action through suppression of expression of AR. The observed estrogen-induced reproductive abnormalities cannot be reproduced by administration of an anti-androgen (such as flutamide) *(157)* or by suppressing androgen production through administration of a GnRH antagonist. Thus, to induce the major reproductive abnormalities observed following DES and other estrogenic EDC administration, there must not only be a suppression of androgen action (through a reduction in AR expression) but a coincident elevation of estrogen action *(157)*.

3.3.2. Endocrine-Disruptive Chemicals with Anti-Androgenic Activity

Environmental EDCs with anti-androgenic activity, such as the agricultural fungicides vinclozolin and procymidone, have the capacity to inhibit androgen-mediated male sexual development. Exposure to such substances during sexual differentiation may induce atrophy of the male reproductive organs while inducing hyperplasia of Leydig cells due to disruption of pituitary-testicular feedback mechanisms. Although the exposure may be transient, the effects are permanent and irreversible, with the absence of testosterone exposure resulting in the expression of the female phenotype, independently of the presence of the ovaries and male pseudohermaphroditism (incomplete masculinization of the male fetus). These functional alterations are not, however, discovered until puberty.

Not only is neonatal exposure to such chemicals is of concern but so too is maternal transfer of EDCs and the consequential transgenerational effects (Section 4.3.4.). The fungicide vinclozolin, commonly used on fruits, vegetables, ornamental plants, and vines, was one of the first reported chemicals to possess anti-androgenic activity *(35, 158)*. However, since then, other pesticides and fungicides have been shown to exhibit anti-androgen effects in vitro *(31,159,160)* and in vivo *(27,31,161–163)*.

Most commonly, anti-androgenic EDCs act to alter sexual differentiation in male rats by inhibition of AR-mediated gene activation *(158,164)*. Administration of vinclozolin to adult male rats induces Leydig cell hyperplasia and atrophy of the prostate and seminal vesicles, whereas administration to pregnant rats resulted in incomplete development of the male reproductive tract in male offspring *(35)*. Following developmental exposure (gestational day 14 to postnatal day 3), concentrations of vinclozolin as low as 3 mg/kg/day permanently reduce weights of androgen-dependent tissues, including the prostate *(27)*, emphasizing the sensitivity of the developing fetus to the activity of anti-androgenic chemicals.

3.3.3. EDCs with Estrogenic and Anti-Androgenic Activity

To further complicate the investigation of the impact of estrogenic and anti-androgenic toxic compounds on reproductive and gonadal development is the existence of EDCs with mixed estrogenic and anti-androgenic activity. One such compound is methoxychlor, a chlorinated hydrocarbon pesticide *(165)* currently used as a replacement for DDT in the USA. Methoxychlor is metabolized in the liver into two demethylated compounds: estrogenic metabolites that stimulate expression of ERs and metabolites that have anti-androgenic activity *(161,166–169)*. However, it has recently been demonstrated that the estrogenic metabolite possesses differential effects on ER-α and ER-β, namely, agonistic and antagonistic activity, respectively *(170,171)*. Thus, in examining the effects of such a compound, consideration must be given to the differential estrogenic and anti-androgenic activities.

3.4. Transgenerational Epigenetic Effects of EDCs

Transgenerational toxicological studies have been widely conducted in rodents to assess the adverse effects of chemicals, such as pesticides and fungicides, on reproductive function. Generally, these assessments are concerned with the effects on mating and fertility; however, it is becoming increasingly apparent that other endpoints are required to monitor additional effects *(172)*. Reproductive alterations in wildlife, decreases in human sperm counts, morphological anomalies in genitalia, and an increasing trend in testicular and prostatic cancers are all putative effects in response to EDC exposure.

Many transgenerational studies of EDCs observe effects on gestating mothers and subsequent actions on the offspring associated with the first filial (F1) generation *(173–175)*. However, the transgenerational effects on subsequent generations, whereby transmission occurs minimally to the F2 generation through the germline, have not been as thoroughly studied. There is the potential for EDCs to induce transgenerational effects through epigenetic alterations involving DNA methylations or stable chromosomal alterations *(176–178)*. The transgenerational effects of irradiation were the first to be identified by transmission of DNA mutations in the germline *(176,177,179)*.

Epigenetic alterations leading to transgenerational transmission of specific genetic traits or molecular events (imprinting) have been recently identified *(180–182)*. These observations are of considerable importance for human health as reprogramming through an altered methylation state of the germline is responsible *(183)*. Thus, EDCs have the potential to induce either a chromosomal or epigenetic alteration in the germline, resulting in permanent reprogramming and potentially a transgenerational phenotype that may promote or predispose a disease state. Further details are provided in Chapter 7 by Guerrero-Bosagna and Valladares.

There are numerous reports that the EDC vinclozin, which has anti-androgenic activities, affects subsequent generations of males *(31,158,162,184,185)*. Most recently, a study identified a transgenerational effect of vinclozolin and methoxychlor on rodent testis development and adult spermatogenic cells *(186)*. Despite the importance of this study, it is also highly controversial the administration method but and its mechanism of action. In this study, vinclozolin was administered through intraperitoneal injection as opposed to oral gavage. This is an important detail, because vinclozolin is metabolized to several metabolites, two of which (butenoic acid and enanilide, termed M1 and M2, respectively) have a far greater affinity for the AR (10–15 times) than the parent compound *(158,187)*. These metabolites can competitively inhibit the AR in vitro *(158,164)* and AR-dependent gene expression in vivo *(161)*. Despite this, the fact that the germline can be reprogrammed and induce transgenerational effects is critical and may have significant implications on human biology and disease etiology.

Notably, the effects observed in this study were the opposite of what might be expected for an AR antagonist with regard to sensitive exposure windows during pregnancy. In this study, exposure late in gestation, when peak androgen levels are required for normal reproductive tract development, did not produce a phenotype in the F1 despite the fact that the dose level was high enough to induce 100 % hypospadias as observed in previous studies. However, exposure early in gestation, during a period when it is highly debatable that a functioning AR is present, a distinct phenotype was observed to be transmitted to the fourth (F4) generation *(186)*. The critical question is if these effects are real then they may have nothing to do with AR biology and be the result of altered DNA methylation of several identified genes.

3.5. Additional Controversies

3.5.1. Limitations to the Use of Animal Models to Study the Effects of EDCs on Target Organs

A great deal of the work that has suggested an effect of EDCs on prostate health outcomes arose from studies using rodents. There are, however, a number of limitations to the utilization of animal models that create controversy that remains unresolved. Firstly, there are distinct anatomical differences between rodent and human prostate glands. As outlined in Section 3 above, compared anatomically to the human prostate that is a unilobular zonal structure *(66)*, rodent prostates are composed of several lobes that have a specific three-dimensional (3D) ductal network *(76,188)*. Yet the different origins of the human prostate may be of significance, because each zone is susceptible to different disease processes; benign BPH disease arises mainly in the inner zone and prostate cancer in the peripheral zone *(87)*. The rodent prostate, however, does not spontaneously develop prostate carcinogenesis or benign disease.

The variety of interspecies differences observed in the structure of the adult prostate gland emphasizes the difficulty associated with locating suitable animal models for the study of human prostate disease *(68)*. However, advances in computer technology now allow organ structure to be visualized by several 3D reconstruction techniques, providing a powerful tool for examining prostate anatomy.

Using a computer-assisted 3D approach to visualize the microanatomy of prostate development, Timms and colleagues compared the ductal budding patterns during prostate morphogenesis of the rat, mouse, and human *(66)*. This 3D reconstruction involves computer-assisted analysis of histological serial sections, requiring tracing, digitizing, and axial alignments of anatomical structures within each section. Albeit a labor-intensive process, this 3D reconstruction technique provides a basis for making quantitative comparisons of the developing glands *(66,189)*.

Despite the temporal and anatomical differences between the rodent and human, the prostate gland arises as a result of the same branching process. Recently, a newer computer-based method to detect and quantify early changes in prostate branching morphogenesis in normal and transgenic mice was developed *(82)*. This image analysis process uses binary images of confocal stacks *(190)* and involves fully automated measurements of the length of individual epithelial branches, ducts, and lobes and also the accumulative surface areas of the individual ducts. This allows temporal and spatial alterations in branching morphogenesis as a result of experimental manipulations.

These computer-assisted techniques have been particularly useful in understanding descriptive anatomy as well as the complex patterning of ductal morphogenesis—something that has previously been difficult to grasp when viewing 2D histological sections down a microscope *(66,68)*.

Using this method of prostate skeletonization and the ArKO mouse, the effects of removing estrogen on neonatal prostate development were examined (Almahbobi et al., 2006, unpublished data). It is predicted that estrogen has an important role in the regulation of prostate branching morphogenesis in rodent during the neonatal period and that alterations in branching morphogenesis may be identifiable well before the emergence of the phenotype in older animals. As such, this methodology may be suitable to investigate the role of estrogen or EDCs in the etiology of prostate disease well before it is evident in adulthood.

Another limitation of rodent studies relates to reproducibility, with conflicting and often opposing results being reported because of strain differences in mice and apparent differences between rats and mice. There is the question of low-dose effects and no effect level *(191)*, especially during pregnancy. This issue is particularly important as it is not known whether the human prostate has the same degree of sensitivity to EDCs as the rodent prostate.

3.5.2. Dose–Response Relationships

The dose–response relationship issue is one of the most controversial areas with regards to EDCs and reproductive health. As EDCs act directly on the endocrine system to mimic or antagonize naturally occurring hormones, the dose–response relationship is generally different than for other environmental chemicals, which do not directly act on the system. However, these relationships are also variable for different EDCs as well as different species endocrine mechanisms.

More controversial, however, are the numerous studies demonstrating the low-dose effects of EDCs. The prostate gland is also sensitive to low doses of estrogens and EDCs during development, with recent studies demonstrating that the differentiating male UGS is sensitive to low doses of EDCs *(192)*. vom Saal and coworkers proposed that an increase in reproductive organ disorders was linked to in utero exposure to EDCs in the environment *(138)*. Investigating the effects of fetal exposure to a full range of doses, vom Saal showed an inverted-U dose–response relationship, whereby low doses of EDCs were as effective as very high doses *(189)*. The implications of these studies are significant, because most toxicological testing does not reach into the lower dose range and will not detect these adverse effects of estrogenic compounds, natural or synthetic. However, these conclusions remain particularly controversial. The pioneering work of vom Saal was confirmed by Nagel and coworkers who showed similar effects, including prostate enlargement in 6-month-old CF1 mice after in utero exposure to low doses of BPA *(189)*. In contrast, at least two independent groups of investigators *(193,194)* were unable to confirm the effects of prenatal exposure to DES and BPA *(189)* on male CF1 mice. A recent study conducted in F344 and Sprague–Dawley rats, administering low and high doses of EB (β-estradiol-3-benzoate), demonstrated a similar inverted U-shaped response curve for prostate sizes during puberty, as previously described *(189)*, but this effect was not permanent and was absent from adult animals, suggesting that a "transient" effect was induced by low-dose estrogen treatment *(195)*.

The conflicting data regarding low-dose estrogens and estrogenic EDCs remain unresolved, but these reports generate a great deal of interest in the prostate biology field as it seeks to evaluate the potential effects of low-dose estrogen exposure that alter androgen action, particularly by EDCs. As hormones play a role in imprinting of the prostate gland during postnatal development, it is important to resolve these controversies to understand the permanent nature of early life events that may influence the onset of late life disease. Thus, it is imperative that dose–response issues should be considered during study design for risk evaluation for health effects. Ideally, doses should identify both toxic and mechanistic endpoints.

3.5.3. Long Latency and Environmental Effects During Lifetime Before Diseases Emerge upon Aging

One of the main complications associated with studying the effects of fetal EDC exposure and late life outcome in humans is the long latency of disease and consequently a long lead time to the identification of the effects of prenatal exposure on such reproductive indices as prostate disease, infertility, decreased sperm counts, and testicular cancer. In men, the process from fetal or neonatal development to late life, when adverse outcomes become evident, takes decades: 50–60 years or more. Thus, there is significant variation in environmental conditions during the intervening period when additional factors may impact on prostate health. This time line does not permit in vivo studies to be completed or causal relationships to be drawn and provides further justification for the development of more appropriate systems modeling human disease.

The critical question remains as to whether or not a pathological condition with long latency can be predicted by assessing earlier outcomes. A classic example of the delayed manifestation of the perinatal effect of an environmental agent is Young's syndrome, in which there is an obstructive lesion in the epididymis leading to azoospermia with significant bronchiectasis. This syndrome is associated with Pink's disease, a condition linked to the presence of mercury in teething powders. Removal of mercury from teething powders in the early 1960s led to the disappearance of Pink's disease and, 30–40 years later, Young's syndrome is no longer seen. Thus, the accurate diagnosis of earlier outcomes may represent a useful early sentinel marker of reproductive health in men. However, it will require serious consideration of the threat of exposure of the developing reproductive system to damaging environmental agents.

4. CONCLUSIONS AND THE WAY FORWARD

Analysis of human data is limited, and to date, there is no firm evidence demonstrating a direct causal association between EDC exposure and adverse reproductive health outcomes. Currently, the majority of evidence is correlative and associative with the incidence of late life disease. Nevertheless, there is general agreement that the critical role of hormones on reproductive development renders this process susceptible to disruption by factors that interfere with hormone production, bioavailability, metabolism, or action.

Animal models have been useful at highlighting important biological issues, including the identification of environmental agents with activities that may impact on the developing reproductive tract and potential mechanisms of action. Currently, the plausibility of the responses in animals versus the likelihood of occurrence in man and the relative sensitivity of our models versus human disease are still being assessed and determined. These studies are useful in directing where appropriate epidemiological studies need to be undertaken. This may ultimately mean that we can minimize exposures (by bans or reductions) to reduce risks.

A useful adjunct to rodent models is human cell and tissue models. However, human fetal and normal human prostate tissue from young adult men is difficult to obtain. Animal models have limited utility when extrapolating finding to humans because of their anatomical and developmental differences. Thus, it is imperative to establish a robust source of human prostate that accurately models normal prostate development, because this will enable researchers to study the effects of EDCs during development and their impact on normal maturation. One approach has been to develop an in vivo technique for the differentiation of human embryonic stem cells into immature and mature human prostate tissue by tissue recombination *(196)*. This is a novel, reliable, and reproducible model system that can be used to study human prostate development and maturation over 8–12 weeks and is comparable to the process that takes decades in the human male (Fig. 10). As the in vivo system uses rodent mesenchyme and grafts are hosted in mice, both the mesenchyme and host environment may be manipulated to identify critical systemic or local (stromal–epithelial) factors that influence prostate development and maturation. This model would be very useful in establishing that EDCs affect normal fetal and mature prostate tissue in a controlled experimental system.

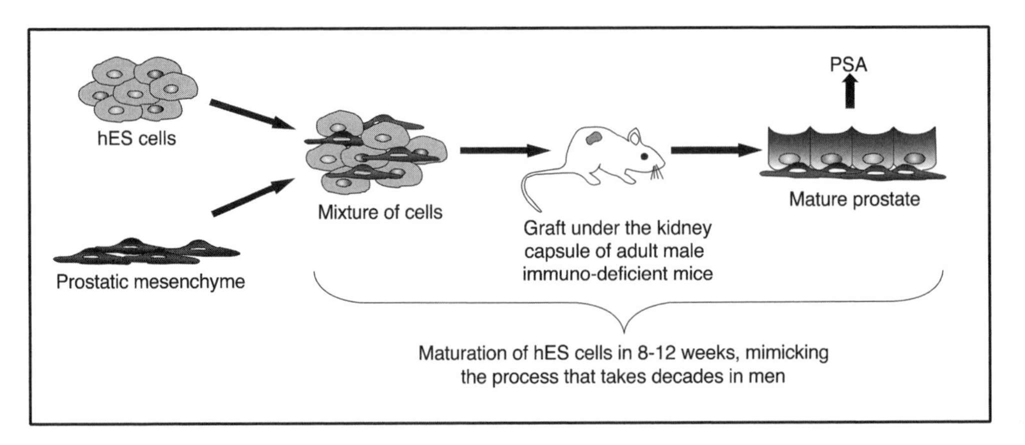

Fig. 10. Schematic illustration of the human prostate stem cell model. Neonatal prostatic rodent mesenchyme is recombined with human embryonic stem cells (hES cells). The resultant tissue graft is placed under the kidney capsule of adult male immuno-deficient mice and left to grow for up to 12 weeks. Maturation is evident by secretion of prostate-specific antigen (PSA). This process takes decades in the human males.

REFERENCES

1. Moore KL. *The Developing Human: Clinically Oriented Embryology*. 3rd ed. Philadelphia, PA: W. B. Saunders Company; 1982.
2. Pelliniemi LJFK, Parank J. *Cell Biology of Testicular Development*. CA: Academic Press Inc; 1993.
3. Byskov A. Differentiation of mammalian embryonic gonad. *Physiol Rev* 1986;66:71–117.
4. Pelliniemi L. Development of embryonic gonad. *Ultrastructural and Cytogenetic Study on the Pig* [Doctorate], University of Turku; 1975.
5. Eddy EMCJ, Gong D, Fenderson A. Origin and migration of primoridal germ cells in mammals. *Gamete Res* 1981;4:333–62.
6. Maitland P, Ullmann S. Gonadal development in the opossum, *Monodelphis domestica*: the rete ovarii does not contribute to the steroidogenic tissues. *J Anat* 1993;183:43–56.
7. Paranko J, Pelliniemi LJ, Vaher A, Foidart J-M, Lakkala-Paranko T. Morphogenesis and fibronectin in sexual differentiation of rat embryonic gonads. *Differentiation* 1983;23(Suppl):S72–8.
8. Koopman P, Munsterberg A, Capel B, Vivian N, Lovell-Badge R. Expression of a candidate sex-determining gene during mouse testis differentiation. *Nature* 1990;348:450–2.
9. Agelopoulou R, Magre S, Patsavoudi E, Jost A. Initial phases of the rat testis differentiation in vitro. *J Embryol Exp Morphol* 1984;83:15–31.
10. Pelliniemi LJDM. *The Fetal Gonad and Sexual Differentiation*. Philadelphia, PA: WB Saunders Company; 1980.
11. Pelliniemi LJ, Niei M. Fine structure of the human foetal testis. I. The interstitial tissue. *Z Zellforsch Mikrosk Anat* 1969;99(4):507–22.
12. Berman DM, Tian H, Russell DW. Expression and regulation of steroid 5 alpha-reductase in the urogenital tract of the fetal rat. *Mol Endocrinol* 1995;9(11):1561–70.
13. Clark RL, Anderson CA, Prahalada S, et al. Critical developmental periods for effects on male rat genitalia induced by finasteride, a 5 alpha-reductase inhibitor. *Toxicol Appl Pharmacol* 1993;119(1):34–40.
14. Imperato-McGinley J, Sanchez RS, Spencer JR, Yee B, Vaughan ED. Comparison of the effects of the 5 alpha-reductase inhibitor finasteride and the antiandrogen flutamide on prostate and genital differentiation: dose-response studies. *Endocrinology* 1992;131(3):1149–56.
15. Kassim NM, McDonald SW, Reid O, Bennett NK, Gilmore DP, Payne AP. The effects of pre- and postnatal exposure to the nonsteroidal antiandrogen flutamide on testis descent and morphology in the Albino Swiss rat. *J Anat* 1997;190(Pt 4):577–88.
16. Roy AK, Chatterjee B. Androgen action. *Crit Rev Eukaryot Gene Expr* 1995;5(2):157–76.

17. Silversides DW, Price CA, Cooke GM. Effects of short-term exposure to hydroxyflutamide in utero on the development of the reproductive tract in male mice. *Can J Physiol Pharmacol* 1995;73(11):1582–8.
18. Veyssiere G, Berger M, Jean-Faucher C, de Turckheim M, Jean C. Testosterone and dihydrotestosterone in sexual ducts and genital tubercle of rabbit fetuses during sexual organogenesis: effects of fetal decapitation. *J Steroid Biochem* 1982;17(2):149–54.
19. Wilson JD, Lasnitzki I. Dihydrotestosterone formation in fetal tissues of the rabbit and rat. *Endocrinology* 1971;89(3):659–8.
20. Nieschlag E, Behre HM, Van Ahlen H. *Andrology. Male Reproductive Health and Dysfunction.* 2nd ed. Heidelberg, Germany: Springer; 2000.
21. Klonisch T, Fowler PA, Hombach-Klonisch S. Molecular and genetic regulation of testis descent and external genitalia development. *Dev Biol* 2004;270(1):1–18.
22. Boisen KA, Kaleva M, Main KM, et al. Difference in prevalence of congenital cryptorchidism in infants between two Nordic countries. *Lancet* 2004;363(9417):1264–9.
23. Adami HO, Bergstrom R, Mohner M, et al. Testicular cancer in nine northern European countries. *Int J Cancer* 1994;59(1):33–8.
24. Toppari J, Kaleva M, Virtanen HE. Trends in the incidence of cryptorchidism and hypospadias, and methodological limitations of registry-based data. *Hum Reprod Update* 2001;7(3):282–6.
25. Sharpe RM. Pathways of endocrine disruption during male sexual differentiation and masculinization. *Best Pract Res Clin Endocrinol Metab* 2006;20(1):91–110.
26. Kelce WR, Stone CR, Laws SC, Gray LE, Kemppainen JA, Wilson EM. Persistent DDT metabolite p,p'-DDE is a potent androgen receptor antagonist. *Nature* 1995;375(6532):581–5.
27. Gray LE Jr, Wolf C, Lambright C, et al. Administration of potentially antiandrogenic pesticides (procymidone, linuron, iprodione, chlozolinate, p,p'-DDE, and ketoconazole) and toxic substances (dibutyl- and diethylhexyl phthalate, PCB 169, and ethane dimethane sulphonate) during sexual differentiation produces diverse profiles of reproductive malformations in the male rat. *Toxicol Ind Health* 1999;15(1–2):94–118.
28. McIntyre BS, Barlow NJ, Foster PM. Male rats exposed to linuron in utero exhibit permanent changes in anogenital distance, nipple retention, and epididymal malformations that result in subsequent testicular atrophy. *Toxicol Sci* 2002;65(1):62–70.
29. McIntyre BS, Barlow NJ, Wallace DG, Maness SC, Gaido KW, Foster PM. Effects of in utero exposure to linuron on androgen-dependent reproductive development in the male Crl:CD(SD)BR rat. *Toxicol Appl Pharmacol* 2000;167(2):87–99.
30. Hosokawa S, Murakami M, Ineyama M, et al. The affinity of procymidone to androgen receptor in rats and mice. *J Toxicol Sci* 1993;18(2):83–93.
31. Ostby J, Kelce WR, Lambright C, Wolf CJ, Mann P, Gray LE Jr. The fungicide procymidone alters sexual differentiation in the male rat by acting as an androgen-receptor antagonist in vivo and in vitro. *Toxicol Ind Health* 1999;15(1–2):80–93.
32. Noriega NC, Ostby J, Lambright C, Wilson VS, Gray LE Jr. Late gestational exposure to the fungicide prochloraz delays the onset of parturition and causes reproductive malformations in male but not female rat offspring. *Biol Reprod* 2005;72(6):1324–35.
33. Vinggaard AM, Christiansen S, Laier P, et al. Perinatal exposure to the fungicide prochloraz feminizes the male rat offspring. *Toxicol Sci* 2005;85(2):886–97.
34. Foster PM, McIntyre BS. Endocrine active agents: implications of adverse and non-adverse changes. *Toxicol Pathol* 2002;30(1):59–65.
35. Gray LE Jr, Ostby JS, Kelce WR. Developmental effects of an environmental antiandrogen: the fungicide vinclozolin alters sex differentiation of the male rat. *Toxicol Appl Pharmacol* 1994;129(1):46–52.
36. Gray LE Jr, Ostby J, Monosson E, Kelce WR. Environmental antiandrogens: low doses of the fungicide vinclozolin alter sexual differentiation of the male rat. *Toxicol Ind Health* 1999;15(1–2):48–64.
37. Cunha AR. Role of mesenchymal-epithelial interactions in normal and abnormal development of male urogenital glands. In: Riva A, Testa Riva F, Motta PM, eds. *Ultrastructure of Male Urogenital Glands: Prostate, Seminal Vesicles, Urethral, and Bulbourethral Glands*. Kluwer Academic Publishers; 1994:15–34.
38. Mylchreest E, Cattley RC, Foster PM. Male reproductive tract malformations in rats following gestational and lactational exposure to Di(n-butyl) phthalate: an antiandrogenic mechanism. *Toxicol Sci* 1998;43(1):47–60.

39. Mylchreest E, Sar M, Cattley RC, Foster PM. Disruption of androgen-regulated male reproductive development by di(n-butyl) phthalate during late gestation in rats is different from flutamide. *Toxicol Appl Pharmacol* 1999;156(2):81–95.
40. You L, Casanova M, Archibeque-Engle S, Sar M, Fan LQ, Heck HA. Impaired male sexual development in perinatal Sprague-Dawley and Long-Evans hooded rats exposed in utero and lactationally to p,p′-DDE. *Toxicol Sci* 1998;45(2):162–73.
41. Swan SH, Main KM, Liu F, et al. Decrease in anogenital distance among male infants with prenatal phthalate exposure. *Environ Health Perspect* 2005;113(8):1056–61.
42. Parks LG, Ostby JS, Lambright CR, et al. The plasticizer diethylhexyl phthalate induces malformations by decreasing fetal testosterone synthesis during sexual differentiation in the male rat. *Toxicol Sci* 2000;58(2):339–49.
43. Shultz VD, Phillips S, Sar M, Foster PM, Gaido KW. Altered gene profiles in fetal rat testes after in utero exposure to di(n-butyl) phthalate. *Toxicol Sci* 2001;64(2):233–42.
44. Mylchreest E, Sar M, Wallace DG, Foster PM. Fetal testosterone insufficiency and abnormal proliferation of Leydig cells and gonocytes in rats exposed to di(n-butyl) phthalate. *Reprod Toxicol* 2002;16(1):19–28.
45. Barlow NJ, Phillips SL, Wallace DG, Sar M, Gaido KW, Foster PM. Quantitative changes in gene expression in fetal rat testes following exposure to di(n-butyl) phthalate. *Toxicol Sci* 2003;73(2):431–1.
46. Foster PM. Disruption of reproductive development in male rat offspring following in utero exposure to phthalate esters. *Int J Androl* 2006;29(1):140–7.
47. Lehmann KP, Phillips S, Sar M, Foster PM, Gaido KW. Dose-dependent alterations in gene expression and testosterone synthesis in the fetal testes of male rats exposed to di(n-butyl) phthalate. *Toxicol Sci* 2004;81(1):60–8.
48. Barlow NJ, Foster PM. Pathogenesis of male reproductive tract lesions from gestation through adulthood following in utero exposure to di(n-butyl) phthalate. *Toxicol Pathol* 2003;31(4):397–410.
49. Kratochwil K. In vitro analysis of the hormonal basis for the sexual dimorphism in the embryonic development of the mouse mammary gland. *J Embryol Exp Morphol* 1971;25(1):141–53.
50. Kratochwil K, Schwartz P. Tissue interaction in androgen response of embryonic mammary rudiment of mouse: identification of target tissue for testosterone. *Proc Natl Acad Sci USA* 1976;73(11):4041–4.
51. Imperato-McGinley J, Binienda Z, Gedney J, Vaughan ED Jr. Nipple differentiation in fetal male rats treated with an inhibitor of the enzyme 5 alpha-reductase: definition of a selective role for dihydrotestosterone. *Endocrinology* 1986;118(1):132–7.
52. Kratochwil K. Development and loss of androgen responsiveness in the embryonic rudiment of the mouse mammary gland. *Dev Biol* 1977;61(2):358–65.
53. Kratochwil K. Tissue combination and organ culture studies in the development of the embryonic mammary gland. *Dev Biol (N Y 1985)* 1986;4:315–3.
54. Bowman CJ, Barlow NJ, Turner KJ, Wallace DG, Foster PM. Effects of in utero exposure to finasteride on androgen-dependent reproductive development in the male rat. *Toxicol Sci* 2003;74(2):393–406.
55. Imperato-McGinley J, Binienda Z, Arthur A, Mininberg DT, Vaughan ED Jr, Quimby FW. The development of a male pseudohermaphroditic rat using an inhibitor of the enzyme 5 alpha-reductase. *Endocrinology* 1985;116(2):807–12.
56. Hecker A, Hasan SH, Neumann F. Disturbances in sexual differentiation of rat foetuses following spironolactone treatment. *Acta Endocrinol (Copenh)* 1980;95(4):540–5.
57. George FW. Developmental pattern of 5 alpha-reductase activity in the rat gubernaculum. *Endocrinology* 1989;124(2):727–32.
58. Nef S, Parada LF. Cryptorchidism in mice mutant for Insl3. *Nat Genet* 1999;22(3):295–9.
59. Adham IM, Emmen JM, Engel W. The role of the testicular factor INSL3 in establishing the gonadal position. *Mol Cell Endocrinol* 2000;160(1–2):11–6.
60. Nef S, Shipman T, Parada LF. A molecular basis for estrogen-induced cryptorchidism. *Dev Biol* 2000;224(2):354–61.
61. Bowman CJ, Turner KJ, Sar M, Barlow NJ, Gaido KW, Foster PM. Altered gene expression during rat Wolffian duct development following di(n-butyl) phthalate exposure. *Toxicol Sci* 2005;86(1):161–74.

62. Turner KJ, McIntyre BS, Phillips SL, Barlow NJ, Bowman CJ, Foster PM. Altered gene expression during rat Wolffian duct development in response to in utero exposure to the antiandrogen linuron. *Toxicol Sci* 2003;74(1):114–28.
63. Cunha GR, Cooke P, Bigsby RM, Brody J. Ontogeny of sex steroid receptors in mammals. In: Parker M, ed. *The Structure and Function of Nuclear Hormone Receptors*. New York: Academic Press; 1991:235–68.
64. Shima H, Tsuji M, Young P, Cunha GR. Postnatal growth of mouse seminal vesicle is dependent on 5 alpha-dihydrotestosterone. *Endocrinology* 1990;127(6):3222–3.
65. Hamilton WJ, Boyd JD, Mossman HW. *Human Embryology*. Baltimore, MD: The Williams and Wilkins Company; 1959.
66. Timms BG, Mohs TJ, Didio LJ. Ductal budding and branching patterns in the developing prostate. *J Urol* 1994;151(5):1427–32.
67. Cunha GR, Donjacour AA, Cooke PS, et al. The endocrinology and developmental biology of the prostate. *Endocr Rev* 1987;8(3):338–62.
68. Timms BG. Anatomical perspectives in prostate development. In: Naz R, ed. *Prostate: Basic and Clinical Aspects*. New York: CRC Press; 1997:29–51.
69. Kellokumpu-Lehtinen P. Development of sexual dimorphism in human urogenital sinus complex. *Biol Neonate* 1985;48(3):157–67.
70. Lowsley O. The development of the human prostate gland with reference to the development of other structures at the base of neck of the urinary bladder. *Am J Anat* 1912;13:299–349.
71. Partin AW, Coffey DS. The molecular biology, endocrinology, and physiology of the prostate and seminal vesicles. In: Walsh PC, Retik AB, Vaughan EJD, Wein AJ, eds. *Campbell's Urology*. 7th ed. Philadelphia, PA: W.B. Saunders Company; 1998:1381–428.
72. Pohl M, Stuart RO, Sakurai H, Nigam SK. Branching morphogenesis during kidney development. *Annu Rev Physiol* 2000;62:595–620.
73. Hartmann D, Miura T. Modelling in vitro lung branching morphogenesis during development. *J Theor Biol* 2006;242:862–72.
74. Sternlicht MD. Key stages in mammary gland development: the cues that regulate ductal branching morphogenesis. *Breast Cancer Res* 2006;8(1):201.
75. Hisaoka M, Haratake J, Hashimoto H. Pancreatic morphogenesis and extracellular matrix organization during rat development. *Differentiation* 1993;53(3):163–72.
76. Sugimura Y, Cunha GR, Donjacour AA. Morphogenesis of ductal networks in the mouse prostate. *Biol Reprod* 1986;34(5):961–71.
77. Hayward SW, Baskin LS, Haughney PC, Cunha AR, Foster BA, Dahiya R, Prins GS, Cunha GR. Epithelial development in the rat ventral prostate, anterior prostate and seminal vesicle. *Acta Anatomica (Basel)* 1996;155(2):81–93.
78. Hayward SW, Cunha GR. The prostate: development and physiology. *Radiol Clin North Am* 2000;38(1):1–14.
79. Coffey DS. The molecular biology, endocrinology, and physiology of the prostate and seminal vesicles. In: Walsh PC, Retik AB, Stamey TA, Vaughan ED, eds. *Campbell's Urology*. 6th ed. Philadelphia, PA: WB Saunders; 1992:221–301.
80. Ayala AG, Ro JY, Babaian R, Troncoso P, Grignon DJ. The prostatic capsule: does it exist? Its importance in the staging and treatment of prostatic carcinoma. *Am J Surg Pathol* 1989;13(1): 21–7.
81. Cunha G, Alarid E, Turner T, Donjacour A, Boutin E, Foster B. Normal and abnormal development of the male urogenital tract. Role of androgens, mesenchymal-epithelial interactions, and growth factors. *J Androl* 1992;13(6):465–75.
82. Almahbobi G, Hedwards S, Fricout G, Jeulin D, Bertram JF, Risbridger GP. Computer-based detection of neonatal changes to branching morphogenesis reveals different mechanisms of and predicts prostate enlargement in mice haplo-insufficient for bone morphogenetic protein 4. *J Pathol* 2005;206(1):52–61.
83. Andrews G. The histology of the human foetal and prepubertal prostates. *J Anat* 1951;85: 44–54.
84. Driscoll SG, Taylor SH. Effects of prenatal maternal estrogen on the male urogenital system. *Obstet Gynecol* 1980;56(5):537–42.
85. Hayward SW, Rosen MA, Cunha GR. Stromal-epithelial interactions in the normal and neoplastic prostate. *Br J Urol* 1997;79(Suppl 2):18–26.

86. Xia TG, Blackburn WR, Gardner WA Jr. Fetal prostate growth and development. *Pediatr Pathol* 1990;10(4):527–37.
87. McNeal JE. An Atlas of Prostatic Diseases. Relationship of the origin of benign prostatic hypertrophy to prostatic structure of man and other mammals. In: Hinman F, ed. New York: Springer-Verlag; 1983:152–66.
88. Kirby RS. *An Atlas of Prostatic Diseases*. 3rd ed. New York: The Parthenon Publishing Group; 2003.
89. Shappell SB, Thomas GV, Roberts RL, et al. Prostate pathology of genetically engineered mice: definitions and classification. The consensus report from the Bar Harbor meeting of the Mouse Models of Human Cancer Consortium Prostate Pathology Committee. *Cancer Res* 2004;64(6):2270–305.
90. Cunha GR, Chung LW, Shannon JM, Reese BA. Stromal-epithelial interactions in sex differentiation. *Biol Reprod* 1980;22(1):19–42.
91. Takeda H, Lasnitzki I, Mizuno T. Analysis of prostatic bud induction by brief androgen treatment in the fetal rat urogenital sinus. *J Endocrinol* 1986;110:467–70.
92. Foster B, Cunha G. Efficacy of various natural and synthetic androgens to induce ductal branching morphogenesis in the developing anterior rat prostate. *Endocrinology* 1999;140:318–28.
93. Bardin CW, Catterall JF. Testosterone: a major determinant of extragenital sexual dimorphism. *Science* 1981;211(4488):1285–94.
94. Brown TR, Lubahn DB, Wilson EM, Joseph DR, French FS, Migeon CJ. Deletion of the steroid-binding domain of the human androgen receptor gene in one family with complete androgen insensitivity syndrome: evidence for further genetic heterogeneity in this syndrome. *Proc Natl Acad Sci USA* 1988;85(21):8151–5.
95. Charest N, Zhou ZX, Lubahn D, Olsen K, Wilson E, French F. A frameshift mutation destabilizes androgen receptor messenger RNA in the Tfm mouse. *Mol Endocrinol* 1991;5:573–81.
96. Gaspar ML, Meo T, Bourgarel P, Guenet JL, Tosi M. A single base deletion in the Tfm androgen receptor gene creates a short-lived messenger RNA that directs internal translation initiation. *Proc Natl Acad Sci USA* 1991;88(19):8606–10.
97. He WW, Kumar MV, Tindall DJ. A frame-shift mutation in the androgen receptor gene causes complete androgen insensitivity in the testicular-feminized mouse. *Nucleic Acids Res* 1991;19(9):2373–8.
98. Lubahn DB, Brown TR, Simental JA, et al. Sequence of the intron/exon junctions of the coding region of the human androgen receptor gene and identification of a point mutation in a family with complete androgen insensitivity. *Proc Natl Acad Sci USA* 1989;86(23):9534–8.
99. Shima H, Tsuji M, Elfman F, Cunha G. Development of male urogenital epithelia elicited by soluble mesenchymal factors. *J Androl* 1995;16(3):233–41.
100. Cunha GR, Chung LW, Shannon JM, Taguchi O, Fujii H. Hormone-induced morphogenesis and growth: role of mesenchymal-epithelial interactions. *Recent Prog Horm Res* 1983;39:559–98.
101. Pang SF, Chow PH, Wong TM. The role of the seminal vesicles, coagulating glands and prostate glands on the fertility and fecundity of mice. *J Reprod Fertil* 1979;56(1):129–32.
102. Forest MG. Plasma androgens (testosterone and 4-androstenedione) and 17-hydroxyprogesterone in the neonatal, prepubertal and peripubertal periods in the human and the rat: differences between species. *J Steroid Biochem* 1979;11(1B):543–8.
103. Prins GS. Neonatal estrogen exposure induces lobe-specific alterations in adult rat prostate androgen receptor expression. *Endocrinology* 1992;130(6):3703–14.
104. Prins GS, Birch L. Neonatal estrogen exposure up-regulates estrogen receptor expression in the developing and adult rat prostate lobes. *Endocrinology* 1997;138:1801–9.
105. Singh J, Handelsman DJ. Morphometric studies of neonatal estrogen imprinting in the mature mouse prostate. *J Endocrinol* 1999;162:39–48.
106. Baulieu EE. Androgens and aging men. *Mol Cell Endocrinol* 2002;198(1–2):41–9.
107. Shibata Y, Ito K, Suzuki K, et al. Changes in the endocrine environment of the human prostate transition zone with aging: simultaneous quantitative analysis of prostatic sex steroids and comparison with human prostatic histological composition. *Prostate* 2000;42(1):45–55.
108. Cunha GR, Wang YZ, Hayward SW, Risbridger GP. Estrogenic effects on prostatic differentiation and carcinogenesis. *Reprod Fertil Dev* 2001;13(4):285–96.
109. Ho SM. Estrogens and anti-estrogens: key mediators of prostate carcinogenesis and new therapeutic candidates. *J Cell Biochem* 2004;91(3):491–503.
110. Harkonen PL, Makela SI. Role of estrogens in development of prostate cancer. *J Steroid Biochem Mol Biol* 2004;92(4):297–305.

111. Oliveira CA, Mahecha GA, Carnes K, et al. Differential hormonal regulation of estrogen receptors ERalpha and ERbeta and androgen receptor expression in rat efferent ductules. *Reproduction* 2004;128(1):73–86.
112. Risbridger GP, Wang H, Frydenberg M, Cunha G. The metaplastic effects of estrogen on mouse prostate epithelium: proliferation of cells with basal cell phenotype. *Endocrinology* 2001;142(6):2443–50.
113. Wright AS, Thomas LN, Douglas RC, Lazier CB, Rittmaster RS. Relative potency of testosterone and dihydrotestosterone in preventing atrophy and apoptosis in the prostate of the castrated rat. *J Clin Invest* 1996;98(11):2558–63.
114. Neubauer B, Blume C, Cricco R, Greiner J, Mawhinney M. Comparative effects and mechanisms of castration, estrogen anti-androgen, and anti-estrogen-induced regression of accessory sex organ epithelium and muscle. *Invest Urol* 1981;18(4):229–34.
115. Thompson SA, Rowley DR, Heidger PM Jr. Effects of estrogen upon the fine structure of epithelium and stroma in the rat ventral prostate gland. *Invest Urol* 1979;17(1):83–9.
116. Jarred R, Cancilla B, Prins G, Thayer K, Cunha G, Risbridger G. Evidence that estrogens directly alter androgen regulated prostate development. *Endocrinology* 2000;141(9):3471–7.
117. Risbridger G, Wang H, Young P, et al. Evidence that epithelial and mesenchymal estrogen receptor-alpha mediates effects of estrogen on prostatic epithelium. *Dev Biol* 2001;229(2):432–2.
118. Imamov O, Morani A, Shim GJ, Omoto Y, Warner M, Gustafsson JA. Estrogen receptor-beta regulates epithelial cell differentiation in the mouse ventral prostate. *Horm Res* 2004;62 (Suppl 3):115.
119. Weihua Z, Makela S, Andersson LC, et al. A role for estrogen receptor beta in the regulation of growth of the ventral prostate. *Proc Natl Acad Sci USA* 2001;98(11):6330–5.
120. Weihua Z, Warner M, Gustafsson J. Estrogen receptor beta in the prostate. *Mol Cell Endocrinol* 2002;193(1–2):1.
121. McPherson SJ, Ellem SI, Simpson ER, Patchev V, Fritzemeier KH, Risbridger GP. Essential role for estrogen receptor β in stromal-epithelial regulation of prostatic hyperplasia. *Endocrinology* 2007;148:566–74.
122. Ellem SJ, Schmitt JF, Pedersen JS, Frydenberg M, Risbridger GP. Local aromatase expression in human prostate is altered in malignancy. *J Clin Endocrinol Metab* 2004;89(5):2434–41.
123. Fisher CR, Graves KH, Parlow AF, Simpson ER. Characterization of mice deficient in aromatase (ArKO) because of targeted disruption of the cyp19 gene. *Proc Natl Acad Sci USA* 1998;95(12):6965–70.
124. Murata Y, Robertson KM, Jones ME, Simpson ER. Effect of estrogen deficiency in the male: the ArKO mouse model. *Mol Cell Endocrinol* 2002;193(1–2):7–12.
125. McPherson SJ, Wang H, Jones ME, et al. Elevated androgens and prolactin in aromatase-deficient mice cause enlargement, but not malignancy, of the prostate gland. *Endocrinology* 2001;142(6):2458–67.
126. Burger HG. Selective oestrogen receptor modulators. *Horm Res* 2000;53(Suppl 3):25–9.
127. Dutertre M, Smith CL. Molecular mechanisms of selective estrogen receptor modulator (SERM) action. *J Pharmacol Exp Ther* 2000;295(2):431–7.
128. Gooren LJ, Toorians AW. Significance of oestrogens in male (patho)physiology. *Ann Endocrinol (Paris)* 2003;64(2):126–35.
129. Vermeulen A, Kaufman JM, Goemaere S, van Pottelberg I. Estradiol in elderly men. *Aging Male* 2002;5(2):98–102.
130. Franks LM, Riddle PN, Carbonell AW, Gey GO. A comparative study of the ultrastructure and lack of growth capacity of adult human prostate epithelium mechanically separated from its stroma. *J Pathol* 1970;100(2):113–9.
131. Chung LW, Cunha GR. Stromal epithelial interactions: II. Regulation of prostatic growth by embryonic urogenital sinus mesenchyme. *Prostate* 1983;4(5):503–11.
132. Cunha GR, Fujii H, Neubauer BL, Shannon JM, Sawyer L, Reese BA. Epithelial-mesenchymal interactions in prostatic development. I. Morphological observations of prostatic induction by urogenital sinus mesenchyme in epithelium of the adult rodent urinary bladder. *J Cell Biol* 1983;96(6):1662–70.
133. Hayashi N, Cunha GR, Parker M. Permissive and instructive induction of adult rodent prostatic epithelium by heterotypic urogenital sinus mesenchyme. *Epithelial Cell Biol* 1993;2(2):66–78.
134. Marker PC, Donjacour AA, Dahiya R, Cunha GR. Hormonal, cellular, and molecular control of prostatic development. *Dev Biol* 2003;253(2):165–74.

135. Hayward SW, Haughney PC, Rosen MA, et al. Interactions between adult human prostatic epithelium and rat urogenital sinus mesenchyme in a tissue recombination model. *Differentiation* 1998;63(3):131–40.
136. Colborn T, vom Saal FS, Soto AM. Developmental effects of endocrine-disrupting chemicals in wildlife and humans. *Environ Health Perspect* 1993;101(5):378–84.
137. Kavlock R, Daston G, DeRosa C, et al. Research needs for the risk assessment of health and environmental effects of endocrine disruptors: a report of the U.S. EPA-sponsored workshop. *Environ Health Perspect Suppl* 1996;104(Suppl 4):1–26.
138. Sharpe RM, Skakkebaek NE. Are oestrogens involved in falling sperm counts and disorders of the male reproductive tract. *Lancet* 1993;341(8857):1392–5.
139. Toppari J, Larsen JC, Christiansen P, et al. Male reproductive health and environmental xenoestrogens. *Environ Health Perspect* 1996;104(Suppl 4):741–803.
140. Nagel SC, vom Saal FS, Thayer KA, Dhar MG, Boechler M, Welshons WV. Relative binding affinity-serum modified access (RBA-SMA) assay predicts the relative in vivo bioactivity of the xenoestrogens bisphenol A and octylphenol. *Environ Health Perspect* 1997;105(1):70–6.
141. vom Saal FS, Cooke PS, Buchanan DL, et al. A physiologically based approach to the study of bisphenol A and other estrogenic chemicals on the size of reproductive organs, daily sperm production, and behavior. 1998;14(1–2):239–60.
142. Cooke PS, Young P, Hess RA, Cunha GR. Estrogen receptor expression in developing epididymis, efferent ductules and other male reproductive organs. *Endocrinology* 1991;128:2874–9.
143. Naslund MJ, Coffey DS. The differential effects of neonatal androgen, estrogen and progesterone on adult rat prostate growth. *J Urol* 1986;136:1136.
144. Piacsek BE, Hostetter MW. Neonatal androgenization in the male rat: evidence for central and peripheral defects. *Biol Reprod* 1984;30(2):344–51.
145. Prins G, Birch L. The developmental pattern of androgen receptor expression in rat prostate lobes is altered after neonatal exposure to estrogen. *Endocrinology* 1995;136(3):1303–4.
146. Rajfer J, Coffey DS. Effects of neonatal steroids on male sex tissues. *Invest Urol* 1979;17(1):3–8.
147. Prins GS, Birch L, Habermann H, et al. Influence of neonatal estrogens on rat prostate development. *Reprod Fertil Dev* 2001;13(4):241–52.
148. Cunha GR, Lung B, Reese B. Glandular epithelial induction by embryonic mesenchyme in adult bladder epithelium of BALB/c mice. *Invest Urol* 1980;17(4):302–4.
149. Gaytan F, Bellido C, Aguilar R, Lucena MC. Morphometric analysis of the rat ventral prostate and seminal vesicles during prepubertal development: effects of neonatal treatment with estrogen. *Biol Reprod* 1986;35(1):219–5.
150. Higgins SJ, Brooks DE, Fuller FM, Jackson PJ, Smith SE. Functional development of sex accessory organs of the male rat. Use of oestradiol benzoate to identify the neonatal period as critical for development of normal protein-synthetic and secretory capabilities. *Biochem J* 1981;194(3):895–905.
151. Prins G, Woodham C, Lepinske M, Birch L. Effects of neonatal estrogen exposure on prostatic secretory genes and their correlation with androgen receptor expression in the separate prostate lobes of the adult rat. *Endocrinology* 1993;132(6):2387–98.
152. Pylkkanen L, Makela S, Valve EPH, Toikkanen S, Santti R. Prostatic dysplasia associated with increased expression of c-myc in neonatally estrogenized mice. *J Urol* 1993;149:1593–601.
153. Rajfer J, Coffey DS. Sex steroid imprinting of the immature prostate. Long-term effects. *Invest Urol* 1978;16(3):186–90.
154. Singh J, Zhu Q, Handelsman DJ. Stereological evaluation of mouse prostate development. *J Androl* 1999;20(2):251–8.
155. Prins G. Developmental estrogenization of the prostate gland. In: Naz R, ed. *Prostate: Basic and Clinical Aspects*. New York: CRC Press; 1997:245–63.
156. Woodham C, Birch L, Prins GS. Neonatal estrogen down-regulates prostatic androgen receptor through a proteosome-mediated protein degradation pathway. *Endocrinology* 2003;144(11):4841–50.
157. McKinnell C, Atanassova N, Williams K, et al. Suppression of androgen action and the induction of gross abnormalities of the reproductive tract in male rats treated neonatally with diethylstilbestrol. *J Androl* 2001;22(2):323–8.
158. Kelce WR, Monosson E, Gamcsik MP, Laws SC, Gray LE Jr. Environmental hormone disruptors: evidence that vinclozolin developmental toxicity is mediated by antiandrogenic metabolites. *Toxicol Appl Pharmacol* 1994;126(2):276–85.

159. Kelce WR, Monosson E, Gray LE Jr. An environmental antiandrogen. *Recent Prog Horm Res* 1995;50:449–53.
160. Maness SC, McDonnell DP, Gaido KW. Inhibition of androgen receptor-dependent transcriptional activity by DDT isomers and methoxychlor in HepG2 human hepatoma cells. *Toxicol Appl Pharmacol* 1998;151(1):135–42.
161. Kelce WR, Lambright CR, Gray LE Jr, Roberts KP. Vinclozolin and p,p'-DDE alter androgen-dependent gene expression: in vivo confirmation of an androgen receptor-mediated mechanism. *Toxicol Appl Pharmacol* 1997;142(1):192–200.
162. Monosson E, Kelce W, Lambright C, Ostby J, Gray LJ. Peripubertal exposure to the antiandrogenic fungicide, vinclozolin, delays puberty, inhibits the development of androgen-dependent tissues, and alters androgen receptor function in the male rat. *Toxicol Ind Health* 1999;15:65–79.
163. Tamura H, Maness SC, Reischmann K, Dorman DC, Gray LE, Gaido KW. Androgen receptor antagonism by the organophosphate insecticide fenitrothion. *Toxicol Sci* 2001;60(1):56–62.
164. Wong C, Kelce WR, Sar M, Wilson EM. Androgen receptor antagonist versus agonist activities of the fungicide vinclozolin relative to hydroxyflutamide. *J Biol Chem* 1995;270(34):19998–20003.
165. Kapoor IP, Metcalf RL, Nystrom RF, Sangha GK. Comparative metabolism of methoxychlor, methiochlor, and DDT in mouse, insects, and in a model ecosystem. *J Agric Food Chem* 1970;18(6):1145–52.
166. Dehal SS, Kupfer D. Metabolism of the proestrogenic pesticide methoxychlor by hepatic P450 monooxygenases in rats and humans. Dual pathways involving novel ortho ring-hydroxylation by CYP2B. *Drug Metab Dispos* 1994;22(6):937–46.
167. Kupfer D, Bulger WH, Theoharides AD. Metabolism of methoxychlor by hepatic P-450 monooxygenases in rat and human. 1. Characterization of a novel catechol metabolite. *Chem Res Toxicol* 1990;3(1):8–16.
168. West P, Chaudhary S, Branton G, Mitchell R. High performance liquid chromatographic analysis of impurities and degradation products of methoxychlor. *J Assoc Off Anal Chem* 1982;65(6):1457–70.
169. Eroschenko VP, Rourke AW, Sims WF. Estradiol or methoxychlor stimulates estrogen receptor (ER) expression in uteri. *Reprod Toxicol* 1996;10(4):265–71.
170. Gaido KW, Leonard LS, Maness SC, et al. Differential interaction of the methoxychlor metabolite 2,2-bis-(p-hydroxyphenyl)-1,1,1-trichloroethane with estrogen receptors alpha and beta. *Endocrinology* 1999;140(12):5746–3.
171. Gaido KW, Maness SC, McDonnell DP, Dehal SS, Kupfer D, Safe S. Interaction of methoxychlor and related compounds with estrogen receptor alpha and beta, and androgen receptor: structure-activity studies. *Mol Pharmacol* 2000;58(4):852–8.
172. Takayama S, Akaike M, Kawashima K, Takahashi M, Kurokawa Y. A collaborative study in Japan on optimal treatment period and parameters for detection of male fertility disorders induced by drugs in rats. *J Am Coll Toxicol* 1995;14:266–92.
173. Tsui MT, Wang WX. Maternal transfer efficiency and transgenerational toxicity of methylmercury in Daphnia magna. *Environ Toxicol Chem* 2004;23(6):1504–1.
174. Shimada A, Shima A. Transgenerational genomic instability as revealed by a somatic mutation assay using the medaka fish. *Mutat Res* 2004;552(1–2):119–24.
175. Nomura T, Nakajima H, Ryo H, et al. Transgenerational transmission of radiation- and chemically induced tumors and congenital anomalies in mice: studies of their possible relationship to induced chromosomal and molecular changes. *Cytogenet Genome Res* 2004;104(1–4):252–60.
176. Hoyes KP, Lord BI, McCann C, Hendry JH, Morris ID. Transgenerational effects of preconception paternal contamination with (55)Fe. *Radiat Res* 2001;156(5 Pt 1):488–94.
177. Mohr U, Dasenbrock C, Tillmann T, et al. Possible carcinogenic effects of X-rays in a transgenerational study with CBA mice. *Carcinogenesis* 1999;20(2):325–32.
178. Dubrova YE. Radiation-induced transgenerational instability. *Oncogene* 2003;22(45):7087–93.
179. Barber R, Plumb MA, Boulton E, Roux I, Dubrova YE. Elevated mutation rates in the germ line of first- and second-generation offspring of irradiated male mice. *Proc Natl Acad Sci USA* 2002;99(10):6877–82.
180. Rakyan VK, Chong S, Champ ME, et al. Transgenerational inheritance of epigenetic states at the murine Axin(Fu) allele occurs after maternal and paternal transmission. *Proc Natl Acad Sci USA* 2003;100(5):2538–43.
181. Rakyan V, Whitelaw E. Transgenerational epigenetic inheritance. *Curr Biol* 2003;13(1):R6.
182. Roemer I, Reik W, Dean W, Klose J. Epigenetic inheritance in the mouse. *Curr Biol* 1997;7(4): 277–80.

183. Skinner MK, Anway MD. Seminiferous cord formation and germ-cell programming: epigenetic transgenerational actions of endocrine disruptors. *Ann N Y Acad Sci* 2005;1061:18–32.
184. Gray LE Jr, Ostby JS. The effects of prenatal administration of azo dyes on testicular development in the mouse: a structure activity profile of dyes derived from benzidine, dimethylbenzidine, or dimethoxybenzidine. *Fundam Appl Toxicol* 1993;20(2):177–83.
185. Wolf CJ, LeBlanc GA, Ostby JS, Gray LE Jr. Characterization of the period of sensitivity of fetal male sexual development to vinclozolin. *Toxicol Sci* 2000;55(1):152–61.
186. Anway MD, Cupp AS, Uzumcu M, Skinner MK. Epigenetic transgenerational actions of endocrine disruptors and male fertility. *Science* 2005;308(5727):1466–9.
187. Pothuluri JV, Freeman JP, Heinze TM, Beger RD, Cerniglia CE. Biotransformation of vinclozolin by the fungus Cunninghamella elegans. *J Agric Food Chem* 2000;48(12):6138–48.
188. Hayashi N, Sugimura Y, Kawamura J, Donjacour AA, Cunha GR. Morphological and functional heterogeneity in the rat prostatic gland. *Biol Reprod* 1991;45(2):308–21.
189. vom Saal FS, Timms BG, Montano MM, et al. Prostate enlargement in mice due to fetal exposure to low doses of estradiol or diethylstilbestrol and opposite effects at high doses. *Proc Natl Acad Sci USA* 1997;94:2056–61.
190. Fricout G, Cullen-McEwen L, Harper IS, Jeulin D, Bertram JF. A quantitative method for analysing 3D branching in embryonic kidneys: development of a technique and preliminary data. *Image Anal Stereol* 2001;20(Suppl 1):36–41.
191. Noel A, Hajitou A, L'Hoir C, et al. Inhibition of stromal matrix metalloproteases: effects on breast-tumor promotion by fibroblasts. *Int J Cancer* 1998;76(2):267–73.
192. Timms BG, Howdeshell KL, Barton L, Bradley S, Richter CA, Vom Saal FS. Estrogenic chemicals in plastic and oral contraceptives disrupt development of the fetal mouse prostate and urethra. *Proc Natl Acad Sci USA* 2005;102(19):7014–9.
193. Ashby J, Tinwell H, Haseman J. Lack of effects for low dose levels of bisphenol A and diethylstilbestrol on the prostate gland of CF1 mice exposed in utero. *Regul Toxicol Pharmacol* 1999;30 (2 Pt 1):156–66.
194. Cagen SZ, Waechter JM Jr, Dimond SS, et al. Normal reproductive organ development in CF-1 mice following prenatal exposure to bisphenol A. *Toxicol Sci* 1999;50(1):36–44.
195. Putz O, Schwartz CB, Kim S, LeBlanc GA, Cooper RL, Prins GS. Neonatal low- and high-dose exposure to estradiol benzoate in the male rat: I. Effects on the prostate gland. *Biol Reprod* 2001;65(5):1496–505.
196. Cowin PA, Taylor RA, Cunha GR, et al. Formation of human prostate tissue from embryonic stem cells. *Nat Methods* 2006;3(3):179–81.

4 Endocrine-Disrupting Chemicals and the Brain

Deena M. Walker, BS,
and Andrea C. Gore, PhD

Contents

1. INTRODUCTION

Vertebrates have two systems responsible for communication in the body: the nervous system and the endocrine system. The brain is not only the major regulatory element for both but also a mediator between the two systems. The brain coordinates inputs from the environment with hormonal outputs from the endocrine system through the autonomic nervous system. Endocrine glands in the body are innervated by the autonomic nervous system, which controls glandular function in two ways: (i) by regulating blood flow into the gland and (ii) by regulating the release of hormones from the gland. The hypothalamus, located at the base of the brain, is especially important for regulation of endocrine function as it serves as an interface between the nervous system and endocrine systems, and these endocrine functions that receive a driving input from the brain are called "neuroendocrine" (Fig. 1). There are two types of neuroendocrine outputs that are involved in the control of homeostatic processes. One of these outputs involves the release of peptide hormones, specifically vasopressin or oxytocin, from neurons that originate in the hypothalamus and terminate at a bloodstream located in the posterior pituitary gland. The second neuroendocrine output

From: *Endocrine-Disrupting Chemicals: From Basic Research to Clinical Practice*
Edited by: A. C. Gore © Humana Press Inc., Totowa, NJ

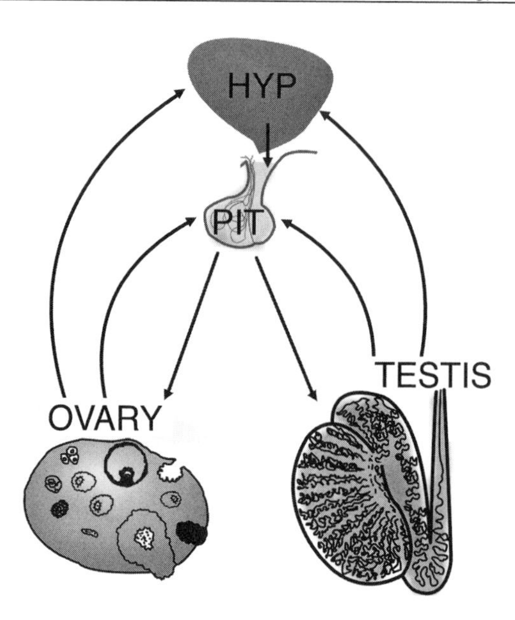

Fig. 1. Diagram of the reproductive neuroendocrine axis, consisting of the hypothalamus (HYP), pituitary (PIT) and gonads, the ovary in females, and the testis in males. Arrows indicate feedforward and feedback regulation by hormones.

involves the release of peptide hormones collectively called hypothalamic releasing or inhibiting hormones into a blood system at the base of the hypothalamus. This vasculature, called the portal capillary system, transports the hypothalamic peptides to the anterior pituitary gland, where they bind to specific pituitary receptors, a process that in turn results in the synthesis and release of secondary hormones. The anterior pituitary hormones then act at peripheral endocrine target glands to cause the release of tertiary hormones that affect various functions throughout the body. Along with these actions, tertiary endocrine hormones can exert negative feedback onto hypothalamic neurons or their inputs to suppress or modify the release of the initial hypothalamic hormones (Fig. 1). Although all neuroendocrine systems exhibit negative feedback regulation, the reproductive hypothalamic–pituitary–gonadal (HPG) system of females has the unique ability to exert a positive feedback signal just prior to the time of ovulation *(1)*.

Endocrine hormones are essential for the normal development and function of the central nervous system (CNS). The prenatal and early postnatal period is particularly important in this regard as exposure to hormones during early development results in the normal organization and facilitation of many behaviors in the adult organism. For example, glucocorticoids released from the neonatal adrenals are important for proliferation of neurons in the hippocampus during development and maturation, whereas removal of the adrenals leads to apoptosis of hippocampal dentate gyrus granule cells, a decrease in synaptic plasticity, and deficits in learning in rats *(2,3)*. In the mammalian brain, thyroid hormone is essential for neuronal proliferation in the adult hippocampus and dentate gyrus *(4)* as well as normal cerebellar development and differentiation *(5)*. It is also important for cellular metabolism. Disruptions to the thyroid gland that negatively affect the release or production of thyroid hormone have been implicated in causing cognitive deficits [reviewed in *(6)*]. In the case of the HPG axis, neonatal exposure to the sex steroid hormones, estradiol (E_2) and testosterone, is crucial for normal sex-typical sexual differentiation of the brain and behavior *(7)*.

The role of the brain as the primary regulator of neuroendocrine functions, together with the brain's responsiveness to hormonal feedback, makes it an important organ to consider when studying endocrine-disrupting chemicals (EDCs).

EDCs are compounds in the environment that imitate, block, alter, or otherwise modify endogenous hormonal activity. Alterations of hormone levels, receptors, or actions in the brain through exposure to EDCs are not only disruptive to the structure and function of the brain itself but also to the regulation of hormones downstream from the brain. The mechanisms of endocrine disruption are diverse and complex; in the nervous system, they probably involve actions through hormone receptors, the enzymes involved in the biosynthesis and degradation of hormones, and other aspects of hormone regulation. In addition, EDCs can be overtly neurotoxic through actions independent (or in addition to) hormone receptors, such as through actions on neurotransmitter receptors, transporters, and biosynthesis.

The discovery that the synthetic estrogen diethylstilbesterol (DES) can cause gynecological cancers and reproductive tract abnormalities in the daughters of mothers who took the drug during pregnancy was a landmark finding that demonstrated endocrine disruption in humans. It is now recognized that along with effects on the reproductive system, EDCs can disrupt numerous hormonal systems in the body, including stress, thyroid, growth, and other metabolic hormones. This chapter will focus primarily on estrogenic EDCs and their effects on the mammalian brain in vivo. For information regarding other hormonal systems refer to the following reviews on thyroid [*(8)* and Chapter 12 of this book by David Armstrong]; glucocorticoids [*(6,9)*]; and androgens [*(10)* and Chapter 3 of this book by Prue Cowin et al.] As discussed in greater detail throughout this chapter, the mediation of estrogenic effects of EDCs occurs through the two nuclear estrogen receptors, estrogen receptor-alpha (ER-α) and ER-β, both of which are abundant in brain but which may be differentially affected by EDCs. There are also non-nuclear estrogen receptors that may be affected by EDCs. For illustrative purposes, this review will discuss three families of extensively studied estrogenic endocrine disruptors: phytoestrogens, organochlorine pesticides, and polychlorinated biphenyls (PCBs).

2. OVERVIEW OF THREE TYPES OF EDCS AND THEIR ACTIONS IN THE BRAIN

2.1. Phytoestrogens

Phytoestrogens are non-steroidal plant compounds that act as weak estrogens or antiestrogens in the body. They are found in many plant species, including soy (isoflavones genistein and daidzen), alfafa (coumestrol), and grapes (resveratrol) (Fig. 2). These compounds have been touted in the popular press for their potential usage in the prevention and/or treatment of cancers (especially cancers in reproductive tissues) as well as for non-pharmacologic postmenopausal hormone treatment and hence are drawing attention from the scientific community. However, careful clinical studies on phytoestrogens are lacking. Because of their potential activity as estrogens, they may have much broader effects on target organs, including the CNS in which the estrogen receptor is abundantly expressed.

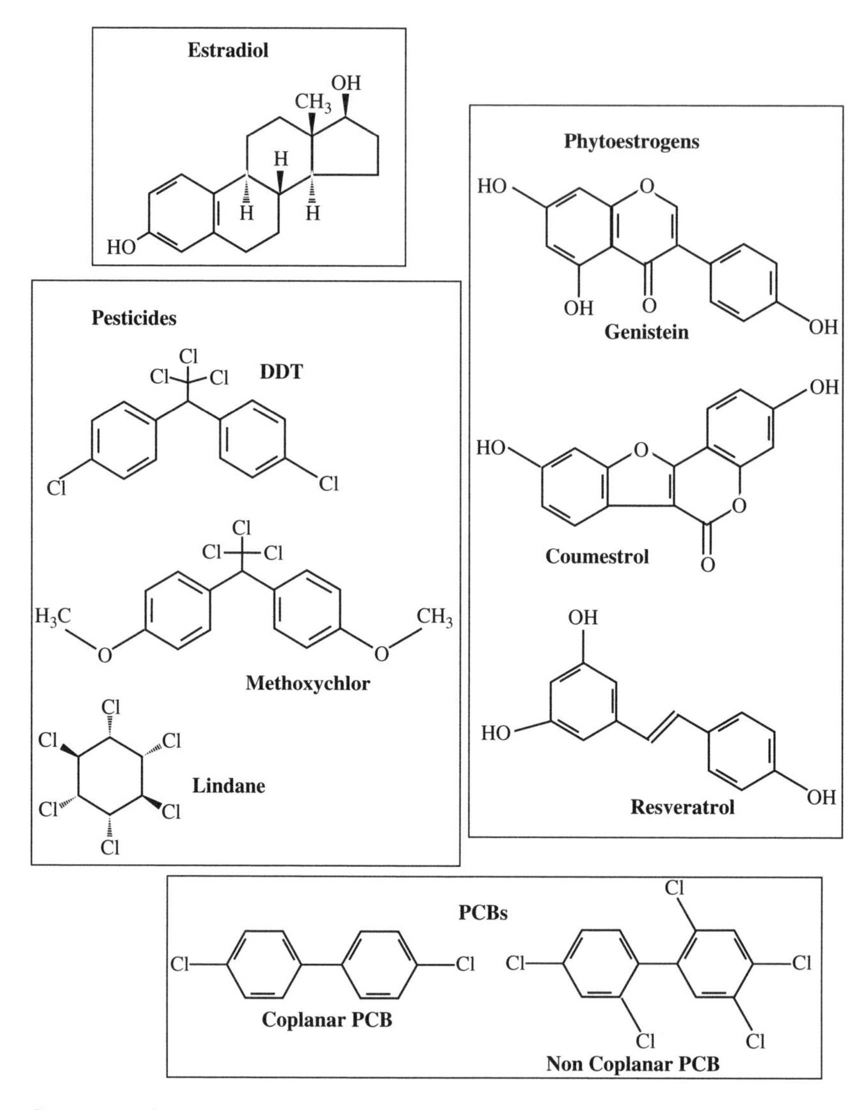

Fig. 2. Structures of representative endocrine-disrupting chemicals.

The actions of phytoestrogens appear to be limited to those mediated by estrogen receptors. This is an important point because other EDCs tend to act through multiple hormonal and neurotransmitter systems, not just estrogenic systems. This relative specificity of phytoestrogens provides valuable information regarding the estrogenic actions of EDCs as opposed to the endocrine effects that may be the result of system-wide alterations. Recent evidence suggests that phytoestrogens may have greater affinity and more pronounced actions on the ER-β than the ER-α *(11)*, and this highlights the point that different classes of EDCs, while estrogenic, may have differential actions because of diverse effects upon target receptors.

2.2. Organochlorine Pesticides

Organochlorine pesticides are synthetic compounds, such as dichlorodiphenyltrichloroethane (DDT) (Fig. 2), DDE (a metabolite of DDT), dieldrin, endosulfan, lindane, chlopyrifos, and methoxychlor (Fig. 2). They are persistent in the environment and owing to their hydrophobic/lipophilic properties have the potential to bioaccumulate in fatty tissues. Although many of these compounds (e.g., DDT, heptachlor, and dieldren) have been banned for decades or more recently (e.g., chlopyrifos) in the USA and Western Europe, they are still widely used in other parts of the world, and human contamination continues through food consumption. Most organochlorine pesticides are weakly estrogenic; some alter thyroid function, interfere with immune function, or act upon neurotransmitter synthesis, degradation, and/or receptors. It is not surprising that organochlorine pesticides affect the mammalian CNS as studies focusing on their mechanism of action have found that the pesticidic nature of these compounds is through neurotoxic, specifically neurostimulatory, effects *(12)*. Regarding their actions on neurotransmitter receptors and transporters, there is evidence for disruption of cholinergic, dopaminergic, and GABAergic systems. For example, some of these EDCs are gamma-aminobutyrate acid receptor subtype A ($GABA_A$) antagonists and disruptors of dopamine transporters (DATs), DAT and vesicular monoamine transporter 2 (VMAT2) *(13,14)*. Although the relationship between these neurotransmitters and neuroendocrine functions may not be immediately obvious, it is important to point out that acetylcholine, dopamine, GABA, and other neurotransmitters can influence hypothalamic neurons through CNS connections. Thus, by disrupting central neurotransmission, these pesticides can exert severe *indirect* effects on an organism's neuroendocrine function *(15)*.

2.3. Polychlorinated Biphenyls

PCBs are synthetic compounds that were manufactured in the USA from the 1930s through 1970s. They are categorized into two groups, coplanar and non-coplanar, as per their molecular structure, and they exist in various states of chlorination (Fig. 2). These structural features affect the ability of PCBs to bind to various hormone and neurotransmitter receptors and to act as agonists, antagonists, or mixed agonists/antagonists. Although banned in the 1970s, PCBs are environmentally persistent and widely dispersed throughout the globe, including in regions where PCBs were never used (e.g., the Arctic) due to populations of animals exposed to PCBs in industrial regions that migrate to non-industrial regions and enter the food chain. Additionally, PCBs are promiscuous compounds in that they exert actions on multiple classes of neurotransmitter and hormonal targets that are not always predictable. Some non-coplanar PCBs affect neurotransmitter systems (dopamine, serotonin, and acetylcholine). In this way, they may indirectly affect neuroendocrine functions. Other PCBs (usually coplanar) interact directly with endocrine systems to exert effects on thyroid hormone, glucocorticoids [reviewed in *(16,17)*], and sex steroid hormone levels. To make matters even more complicated, PCBs are both estrogenic and antiestrogenic as well antiandrogenic, an effect that is usually based on the structure of the specific congener [reviewed in *(17)*]. Finally, some "dioxin-like" PCBs may interact with orphan receptors such as the aryl hydrocarbon receptor (AhR) [reviewed in *(18)*]; this makes an understanding of PCB actions even more difficult as the endogenous ligand of the AhR is not currently known. It is this diversity of interactions that makes the PCB compounds such an interesting subject for endocrine studies and a potentially dangerous compound for human exposure.

3. EDC ACTIONS IN THE BRAIN DURING FETAL AND EARLY POSTNATAL DEVELOPMENT

The development of the brain begins in embryonic life and continues through puberty and into adulthood. Neural development is a carefully regulated sequence of events that is controlled by several factors including genes, neurotransmitters, growth factors, hormones, and their interactions. Disruption of any one of these events by EDCs can have permanent effects on brain morphology and function, resulting in changes to physiology and behavior. To follow is a brief summary of the importance of normal prenatal and early postnatal exposures to steroid hormones to be followed by how early exposures to EDCs can disrupt these processes.

3.1. Brain Development, Sexual Differentiation, and the Role of Steroid Hormones

The brain is sexually dimorphic, meaning that there are distinct differences in structure and morphology of regions between males and females. These differences are permanent and determined during a critical period of sexual differentiation, largely through exposures to sex steroid hormones in fetal development or shortly after birth. In male mammals, much of sexual differentiation of the brain occurs through activation of the fetal testis to release the testicular hormone testosterone. Testosterone and its metabolite, E_2, are responsible for sexual differentiation of the male brain. This is surprising to many people who mistakenly believe that E_2 is a "female" hormone. In fact, the male testis produces appreciable levels of E_2, the major estrogen in mammals, because of the aromatization of testosterone in the testis. In addition, testicular androgens may be aromatized to E_2 at target tissues that contain the aromatase enzyme. The brain is an organ that has particularly high levels of aromatase, and together with the brain's high expression of both androgen and estrogen receptors, both endogenous and exogenous hormones may significantly impact neural function. Under normal developmental circumstances, early exposure to gonadal steroids lays the groundwork for the male physiology and behavior in the organism; the absence of such exposure results in a female phenotype [reviewed in *(19)*; see Chapter 3 by Cowin et al. for additional details].

The mechanisms for these effects of steroids in the nervous system are beginning to be understood. Steroid hormones, acting as transcription factors, determine which genes will be expressed in certain areas of the brain. Additionally, hormones may affect sexual differentiation at the cellular level by determining the brain's capacity to express sex steroid hormone receptors in a region-specific and sex-specific manner *(20)*. They may also cause other neurobiological changes such as dendrite outgrowth and synaptogenesis [reviewed in *(21,22)*]. In addition, structural effects of steroids, through programmed cell death (apoptosis), have been observed in some areas of the brain. This is known to be a factor in the development of several sexually dimorphic nuclei in which hormones play an important role in both cell survival and apoptosis. The balance of these processes determines whether a brain region is masculinized, de-masculinized, feminized, or de-feminized depending upon the sex and the type of exposure. For example, testosterone decreases cell death in three sexually dimorphic regions: the spinal nucleus of the bulbocavernosus (SNB) in male rats involved in penile erection *(23)*; the sexually dimorphic nucleus of the preoptic area (SDN-POA) thought to be involved in masculine sexual behavior in rats *(24)*; and the bed nucleus of the stria terminalis (BNST) involved in reproductive physiology and behavior and in affective behavior pathways [reviewed in *(20)*]. On the other hand, testosterone *increases* cell death in the anteroventral periventricular nucleus (AVPV) [reviewed in *(25)*] thought to regulate preovulatory gonadotropin-releasing hormone (GnRH) release *(26)*. Several of these aforementioned processes are not mediated by testosterone acting on its androgen receptor but rather by the metabolism of testosterone to E_2 and subsequent actions on estrogen receptors. Several relevant points need to be addressed in this regard. First, these early "organizational" effects of sex steroid hormones are necessary for the appropriate expression of sex-typical dimorphic behaviors later in life including reproductive behaviors such as the lordosis reflex in female rodents and mounting behavior in males. These latter behaviors require appropriate exposure to steroid hormones not only prenatally in the organizational period, but also pubertally, and the manifestation of these behaviors after puberty is referred to as the "activational" effects of steroid hormones. Second, early sex hormone exposure organizes other dimorphic non-reproductive behaviors such as infant play, aggression, learning, exploration, and activity level [reviewed in *(21)*]. Finally, recent evidence suggests that the perinatal period is not the only one during which sexually dimorphic circuits are organized. For example, the pubertal period is associated with dramatic changes in sex steroid hormones that may exert novel organizational effects upon the brain that may be sexually dimorphic in nature [reviewed in *(27)*].

The fact that sex steroid hormones play a large role in the normal sexual differentiation of the brain makes early exposure to estrogenic/antiestrogenic and androgenic EDCs a potent dysregulator of brain function from genes to behavior. Notably, hormonal levels involved in controlling such events are extremely low (1 part per trillion), and the fetus is exceptionally sensitive to even the slightest shift in the hormonal milieu. Therefore, even the slightest imbalance can have exponential effects *(28)*. Because the events of early development are relatively well understood and the developing fetus/neonate is especially sensitive to alterations in hormonal concentrations, it is not surprising that many of the studies focusing on effects of EDCs on the brain examine their role in early development. To follow is such a discussion of effects of EDC exposure in this period, a field often referred to as the "fetal (or developmental) basis of adult disease" [reviewed in *(29)*].

3.2. Phytoestrogens and Early Brain Development

The finding that plant compounds could act like estrogens in animals was first published in 1946 *(28)*. Since this early discovery, phytoestrogens have been studied as potential treatments for symptoms of menopause and reproductive cancers. However, it should not be overlooked that while they may have therapeutic potential they may also be disruptive to developing organisms. The literature on phytoestrogens' effects on the brain is not as expansive as that of other EDCs, such as PCBs, discussed in section 3.4. Nevertheless, there is increasing evidence that the brain may be an important target of these compounds, and therefore early exposure, particularly through consumption of soy formula, may be relevant to neurodevelopment. To follow is a review of the literature examining the effects of phytoestrogens on the neonate and adult when exposure is limited to early development. For specific information regarding age of treatment, age of examination, mode of exposure, and dosage, refer to Table 1A.

Table 1

Effects of perinatal (fetal and early postnatal) exposures to (A) phytoestrogens, (B) Organochlorine pesticides, and (C) PCBs on brain morphology and function

	Sex	*Compound*	*Age of treatment*	*Age of testing*	*Method of exposure*	*Dose*	*Organism*	*Reference*
A. Phytoestrogens								
Hypothalamus/pituitary								
↓ ERβ gene expression	M	Daidzein	E85–E114	P1	Maternal diet	8 mg/kg	Pig	*30*
↑ Vasopressin protein levels	M/F	Genistein	E7–P77	P77	Maternal and personal diet	1250 ppm	Rat	*36*
↑ TH (ir cells)	M	Genistein	P1–P2	P19	SC injections	250 μg	Rat	*19*
- TH (ir cells)	F							
↓ TH + ERα cells (ir cells)	F							
↓ BAD protein expression	M	Soy isoflavones	E0–P120	P120	Maternal and personal diet	600 ppm	Rat	*34*
↑ Neuron-specific beta III tubulin protein expression								
↑ SDN-POA volume	M	Genistein	E0–P140	P140	Maternal and personal diet	5, 100, and 500 ppm	Rat	*36*
- SDN-POA volume	F							
- SDN-POA volume	M/F	Genistein	P1–P21	P22	SC injection (P1–P7) and daily gavage until P21	4 and 40 mg/kg	Rat	*35*
↓ SDN-POA volume	F	Genistein	P1–P10	P42	SC injection	100 μg 500 and 1000 μg	Rat	*41,37*
-SDN-POA volume	M/F	Resveratrol	Gestation to lactation	~P100	Maternal drinking water	5 mg/l	Rat	*39*

↓ SDN-POA volume ↑ AVPV volume (5 and 50 μM only)	M	Resveratrol	P1–P22	P132–P140	Maternal drinking water	5, 50, and 100 μM	Rat	*38*
Reproductive function and behavior								
↓ Intromissions - Mounting	M	Resveratrol	Gestation to Lactation	~P100	Maternal drinking water	5 mg/l	Rat	*39*
↑ Mounting	M	Resveratrol	P1–P22	P132–P140	Maternal drinking water	5, 50, and 100 μM	Rat	*38*
↓ Mounting ↓ Ejaculations	M	Coumestrol	P1–P21	P132	Maternal diet	100 μg/g	Rat	*42*
↑ Irregular cycles ↓ Lordosis	F	Genistein	P1–P5	P60	SC injection	1 mg	Rat	*43*
↓ Lordosis response ↓ Sexual receptivity ↑ Irregular cycles ↑ Time to pubertal onset	F	Resveratrol	Gestation to lactation	~P100	Maternal drinking water	5 mg/l	Rat	*39*
- Lordosis response - Sexual receptivity - Irregular cycles - Time to pubertal onset	F	Resveratrol	P1–P22	P132–P140	Maternal drinking water	5, 50, and 100 μM	Rat	*38*
↑ Irregular cycles	F	Coumestrol	P1–P21	P132	Maternal diet	100 μg/g	Rat	*42*

(Continued)

Table 1
(*Continued*)

	Sex	*Compound*	*Age of treatment*	*Age of testing*	*Method of exposure*	*Dose*	*Organism*	*Reference*
- Irregular cycles	F	Coumestrol	P1–P10	P132	Maternal diet	100 μg/g	Rat	*42*
↓ Intromissions	M	Resveratrol	Gestation to Lactation	~P100	Maternal drinking water	5 mg/l	Rat	*39*
- Mounting ↑ Mounting intromissions (N/A)	M	Resveratrol	P1–P22	P132–P40	Maternal drinking water	5, 50, and 100 μM	Rat	*38*
↓ Mounting ↓ Ejaculations	M	Coumestrol	P1–P21	P132	Maternal diet	100 μg/g	Rat	*42*
Hippocampus								
↑ Neuron-specific beta III tubulin protein expression	M	Soy isoflavones	E0–P120	P120	Maternal and personal diet	600 ppm	Rat	*34*
Frontal Cortex								
↓ BAD protein expression ↑ Neuron-specific beta III tubulin protein expression	M	Soy isoflavones	E0–P120	P120	Maternal and personal diet	600 ppm	Rat	*34*
B. Organochlorine pesticides								
Hypothalamus/pituitary								
↓ PR gene expression	M	MXC	E15–P10	P10	Maternal diet	240 ppm	Rat	*44*
↑ LH protein (ir cells) ↑ FSH protein (ir cells) ↑ PRL protein (ir cells)	M	MXC	E15–P10	~P21	Maternal diet	1200 ppm	Rat	*49*

↑ PR gene expression ↓ ERβ gene expression ↓ SRC-1 gene expression	F	MXC	E15–P10	P10	Maternal diet	1200 ppm 240 ppm	Rat	*44*
↑ LH protein (ir cells) ↑ FSH protein (ir cells) ↑ PRL protein (ir cells)	F	MXC	E15–P10	~P21 ~P77 ~P77	Maternal diet	1200 ppm 240 ppm	Rat	*49*
Reproductive function and behavior								
Delays pubertal onset	M	MXC	E14–P42	P42	Maternal and personal gavage	50 and 150 mg/kg/day	Rat	*56*
↓ Serum LH (1200 ppm only) ↓ Serum FSH (all doses) - Serum testosterone - Copulatory behavior	M	MXC	E15–P10	8–15 weeks	Maternal diet	24, 240, and 1200 ppm	Rat	*57*
↓ Sexual motivation ↓ Sexual arousal ↓ Serum testosterone	M	MXC	E5–E7	4 months	Maternal SC injections	33 mg/kg	Mice	*58*
Advances pubertal onset (all doses) ↑ Persistant vaginal cornification (50 and 150) ↑ Irregular cycles (50 and 150) ↑ Serum estradiol/progesterone ratio (50 and 150)	F	MXC	E14–P42	P42	Maternal and personal gavage	5, 50, and 150 mg/kg/day	Rat	*56*

(continued)

Table 1
(*Continued*)

	Sex	*Compound*	*Age of treatment*	*Age of testing*	*Method of exposure*	*Dose*	*Organism*	*Reference*
↓ Serum FSH during estrus (all doses)								
Advances pubertal onset	F	MXC	E0–P110	P55	Maternal and personal diet	800 ppm	Rat	*60*
↑ Persistant vaginal cornification ↑ Irregular cycles ↑ Persistant vaginal cornification (1200 ppm) ↑ Irregular cycles (240 and 1200 ppm) ↓ Preovulatory LH surge (all doses) ↓ Lordosis response (all doses)	F	MXC	E15–P10	8–15 weeks	Maternal diet	24, 240, and 1200 ppm	Rat	*57*
Advances puberty Advances reproductive senescence	F	DDT	Various	Various	Various	Various	Rodents	*61–64*
↑ Pulsatile GnRH release	F	DDT	P5–P10	Not specified	Not specified	Not specified	Rats	*64*
Aggression								
↑ Territorial marking - Intrasex aggression	M	DDT	E11–E17	P60–P90	Maternal diet	20 and 200 μg/kg/day	Mice	*67*
↑ Territorial marking - Intrasex Aggression	M M/F	MXC	E11–E17	P60–P90	Maternal diet	20 and 200 μg/kg/day	Mice	*67*

Delays onset of intrasex aggression	M		E11–E17	P60–P90	Maternal ingestion	0.02, 0.2, and 2 μ/kg/day	Mice	*67*
↓ Aggression toward stranger	F	MXC	E1 to Weaning	P60	Maternal ingestion	2000 μg/kg	Vole	*71*
Pair bonding								
↓ Time spent with mate	F	MXC	E1 to Weaning	P60	Maternal ingestion	2000 μg/kg	Vole	*71*
Basal Ganglia								
↑ DAT protein expression	M	Heptachlor	E1–P21	P28	Maternal ingestion	3 mg/kg	Mice	*13*
↑ VMAT2 protein expression								
↑ TH protein expression								
Cortex								
↑ MAChR density	M	DDT	P10	4 months	Ingestion	0.5 mg/kg	Mice	*54*
Motor activity								
↑ Spontaneous activity	M	DDT	P10	4 months	Ingestion	0.5 mg/kg	Mice	*54*
- Wheel running	F	MXC	E8–E 15	P77	Maternal diet	10, 100 and 1000 ppm	Rat	*66*
↑ Locomotor activity	F	MXC	E 14–P21	P 31–P67	Maternal and personal gavage	50 mg/kg/day	Rat	*56*

(*continued*)

Table 1
(*Continued*)

	Sex	*Compound*	*Age of treatment*	*Age of testing*	*Method of exposure*	*Dose*	*Organism*	*Reference*
- Locomotor activity or amphetamine-induced locomotor activity	F	MXC	E0–P110	P64–P65	Maternal and personal diet	800 ppm	Rat	*60*
- Neuromuscular reflexes	M and F	DDT	E11–E17	P2/P5	Maternal ingestion	20 μg/kg/day	Mice	*70*
↓ Neuromuscular reflexes	M and F	MXC	E11–E17	P2	Maternal ingestion	20 μg/kg/day	Mice	*70*
↑ Reactivity to novel environments	M	MXC	E7–P21	P31–P67	Maternal and personal gavage	150 mg/kg	Rat	*56*
↑ Reactivity (150 mg/kg)								
Brain stem								
↓ $GABA_A$ receptor subunit α 1,β 3, and γ 1	M	Dieldrin	E12–E17	E17	IP injection	2 mg/kg/day	Rat	*14*
C. PCBs								
Hypothalamus/pituitary								
↓ ERβ protein expression (AVPV)	F	A1221	E16, P1 and P4	P42	IP injection	0.34 mg/kg	Rat	*74*
↓ Aromatase activity	M	PCBs (RM)	E0–P0	P0	Maternal diet	40 mg/kg	Rat	*77*
- Aromatase activity	M	A1254	E0–P0	P0			Rat	

Reproductive function and behavior								
↓ Prepubertal serum LH delays pubertal onset	F	PCB 153	E60–P91	9 months	Maternal ingestion	98 μg/kg bw	Goat	*85*
↑ Progesterone during ovulation								
↓ Serum estradiol	F	PCB (RM)	E0–P21	P21	Maternal diet	20 and 40 mg/kg diet	Rat	*89*
↑ Serum testosterone	M	PCBs (RM)	E0–P21	P170	Maternal diet	40 mg/kg diet	Rat	*77*
↓ Serum testosterone	M	PCBs (RM)	E0–P21	P310	Maternal diet	40 mg/kg diet	Rat	*89*
↓ Sexual receptivity (lordosis)	F	A1221	E14, P1, and P10	P70–P105	IP injection	5 and 15 mg	Rat	*87*
↑ Lordosis quotient	F	A1221	E16 and E18	P60	IP injection	1 and 10 mg/kg	Rat	*88*
↓ Other Receptive Behaviors								
↓ Sexual receptivity (lordosis)	F	A1254	P1–P7	P60	IP injection	2.5 and 5 mg	Rat	*86*
↓ Sexual motivation	F	A1254	E14, P1, and P10	P70–P105	IP injection	5 and 15 mg	Rat	*87*
↑ Sweet preference	M	PCBs (RM)	E0–P21	P180	Maternal diet	40 mg/kg diet	Rat	*77,89*
↑ Preference for a testosterone-paired compartment	M	PCB (RM)	E0–P21	P110	Maternal diet	40 mg/kg diet	Rat	*89*
Hippocampus								
↓ ChAT activity	M/F	A1254	E0–P15	P15	Maternal diet	62.5, 125, and 250 ppm	Rat	*78*

(*Continued*)

Table 1
(*Continued*)

	Sex	Compound	*Age of treatment*	*Age of testing*	*Method of exposure*	*Dose*	*Organism*	*Reference*
- CHAT activity	M/F	A1254	E0–P28	P90	Maternal diet	125 and 250 ppm	Rat	*79*
↑ Serotonin	M/F	A1254	E10–P16	P90	Maternal gavage	5 and 25 mg	Rat	*81*
↑ 5-HIAA (serotonin metabolite)								
↓ Mossy fiber growth	M/F	A1254	E0–P16, P30, and P60	P16, P30, and P60	Maternal diet	125 ppm	Rat	*83*
Cerebellum								
↓ Cerebellar mass	M	A1254	E11–P21	P21	Maternal ingestion	10 mg/kg/day	Rat	*84*
Cortex								
↑ Dopamine concentration (PFC)	M and F	TCB and PtCB	E6–P21	P60	Maternal ingestion	100 μl/ 200 g BW	Rat	*76*
↓ Dopamine concentration (PFC)		o' TCB	E6–P21	P60	Maternal ingestion	100 μl/ 200 g BW	Rat	
↑ Serotonin	M and F	A1254	E10–E16	P90	Maternal gavage	5 and 25 mg	Rat	*81*
↑ 5-HIAA (serotonin metabolite)								
Basal Ganglia								
↑ Dopamine concentration	M and F	A1016	E 8–E21	P 25, P35, and P60	Maternal diet	100 and 300 ppm	Rat	*147*

Motor Activity								
↑ Hyperactivity	M	PCB118	E6	P70–P74	Matemal gavage	375 μg	Rat	*90*
Delays reflex development	M (F)	A1254	E11–P21		Matemal ingestion	10 mg/kg/day	Rat	*84*
Righting response				P3–P6				
Negative geotaxis				P5–P7				
Rotorod								
Delays reflex development	M	A1254	E0–P21	P12	Matemal diet	26 ppm		*91*
Auditor y startle								
Negative geotaxis								
Air righting								
↑ Hyperactivity	M	A1254	E6–P21	P34	Matemal gavage	10 mg/kg/day	Mouse	*92*
Cognition								
↓ Spatial learning and memory	M/F	Various	Perinatal	Adult	Various	Various	NHP rodents	Reviewed in *(95)*
↓ Spatial alternation tasks	F	A1254	Perinatal	Adult	Various	Various	Rats	Reviewed in *(6)*
↓ Working reference memory	M	A1254						

↑, increase; ↓, decrease; -, no change; A, Aroclor; AVPV, anteroventral periventricular nucleus; BW, body weight; CHAT, choline acetyltransferase; DAT, dopamine transporter; DDT, dichlorodiphenyltrichloroethane; E, embryonic day of age; ER, estrogen receptor; F, female; FSH, follicle-stimulating hormone; $GABA_A$, gamma-aminobutyrate acid receptor subtype A; IP, intraperitoneal; ir cells, immunoreactive cells; LH, luteinizing hormone; M, male; MAChR, muscarinic acetylcholine receptor; MXC, methoxychlor; NHP, non-human primates; P, postnatal day of age; PCB, polychlorinated biphenyl; PFC, prefrontal cortex; PR, progesterone receptor; PtCB, 3,4,5,3′,4′-pentachlorobiphenyl; RM, reconstituted mixture of PCBs; SC, subcutaneous; SDN-POA, sexually dimorphic nucleus of the preoptic area; SRC-1, steroid receptor co-activator; TCB, 3,4,3′4-tetrachlorobiphenyl; TH, tyrosine hydroxylase; and VMAT2, vesicular monoamine transporter 2.

3.2.1. Gene Expression, Protein Expression, and Brain Morphology

Phytoestrogens may act through both ER-α and ER-β. As discussed in section 3.1, activation of the estrogen receptor (or other steroid hormone receptors) can cause alterations at various levels of cellular structure and function from genes through behavior. However, there are few experiments that have sought to identify specific genes that are transcriptionally activated or repressed by phytoestrogens. A study by Ren et al. *(30)* reported that ER-β gene expression was downregulated in the hypothalamus of neonate male piglets that had been exposed to daidzen in utero. Many other studies have focused on the effects of *neonatal* phytoestrogen exposure on *adult* protein levels in the brain. When rats were treated with genistein throughout embryonic and early development, increased levels of the peptide vasopressin were detected in the hypothalamus *(31)*. Vasopressin is a neuropeptide that influences physiological events, such as water absorption in the kidneys *(32)*, and it is also a neuromodulator in the CNS *(33)*. Recently, Patisaul et al. *(19)* showed that neonatal exposure to genistein resulted in long-term effects on both tyrosine hydroxylase (TH), the enzyme important for dopamine production, and ER-α expression in the AVPV. Sex differences were observed: specifically, in genistein-treated males, a significant increase above controls was observed in the number of TH immunoreactive (ir) cells, whereas no effect was detected in treated females. However, when genistein-treated females were compared with estrogen-treated females, a significant increase in TH-ir cells was seen. Additionally, a significant decrease in cells co-expressing ER-α and TH in the medial AVPV was observed. These results suggest that ER-α is important for the masculinization of the brain and that genistein acts as an antiestrogen in the AVPV *(19)*. In addition to the regulation of gene and protein expression, phytoestrogens are implicated in altering brain structure by decreasing or increasing apoptosis in some areas of the brain. Male rats fed with a high phytoestrogen diet (mainly soy isoflavones) throughout their lifetime display an increase in BAD, a proapototic protein, in the amygdala and a significant decrease in the frontal cortex and medial basal hypothalamus (MBH) when compared with their control counterparts. Bu and Lephart *(34)* also observed an increase in the expression of neuron-specific beta III tubulin, a protein important for neuronal diffentiation and survival, in the amygdala, frontal cortex, hippocampus, and MBH. Taken together, these data suggest that a high phytoestrogen diet may be neuroprotective *(34)*.

Apoptosis is believed to be one pathway that leads to the development of morphologic sexual dimorphisms in the size of brain nuclei, as discussed in section 3.1. As estrogens play an important role in the induction of these dimorphisms, it has been hypothesized that estrogen-like compounds may affect the development and maintenance of these nuclei. Several studies have sought to determine the role of phytoestrogens in this process. However, these studies have produced inconsistent results. For a detailed review of the effect of phytoestrogens on the SDN-POA volume, refer to Table 1A. Briefly, perinatal exposure to various phytoestrogens can increase or decrease the volume of the SDN-POA depending on timing of exposure, dose, and compound *(35–39)*.

Studies focusing on resveratrol show similar effects on sexually dimorphic nuclei. In untreated rats, the locus coeruleus (LC), a key noradrenergic releasing region located in the brainstem, is larger in female than male rats, and this sex difference is abolished when female rats are exposed to resveratrol in utero. The LC is important for the stress response and expresses both ER-α and ER-β *(39)*. In addition, projections from the LC to the hypothalamus are believed to be involved in other neuroendocrine functions including the regulation of HPG function *(40)*. However, there were no effects of resveratrol noted in the SDN-POA in either sex, demonstrating region specificity of the effect to the LC *(39)*.

3.2.2. Reproductive Function and Behavior

The size of sexually dimorphic areas of the brain correlates strongly with dimorphic brain function and behavior, suggesting that perturbations of these nuclei by phytoestrogens have functional consequences. Several studies have looked at how developmental exposure to phytoestrogens can affect these endpoints. For example, Faber and Hughes demonstrated that neonatal exposure to genistein alters adult pituitary function [through luteinizing hormone (LH) release] in a dose-dependent manner. High levels of genistein caused a decrease in pituitary response to GnRH and a larger SDN-POA in females, whereas exposure to a low level of genistein caused a hypersensitivity of the pituitary with no effect on the volume of the SDN-POA. In males, the effect on pituitary responsiveness was identical; however, no effect on SDN-POA volume was observed *(37,41)*.

In addition to altering adult brain morphology and reproductive physiology, developmental exposure to phytoestrogens can modify adult reproductive behavior, including masculine sexual behavior (Table 1A). Male-typical sexual behavior in rats is characterized by a series of behaviors beginning with the male mounting the female, followed by mounts with intromissions of the penis, and finally ejaculation. Limited exposure during the postnatal period to phytoestrogens reduces mounting behavior *(38,42)*; however, when exposure includes prenatal exposure, a significant decrease in the number of intromissions is observed with no effect on mounting behavior *(39)*. Seemingly, these differences are due to timing of exposure although one cannot rule out the effects of other variables in these experiments. Early phytoestrogen exposure can disrupt feminine reproductive behavior as well. Briefly, perinatal exposure to phytoestrogens can advance or delay puberty *(39,43)*, result in irregular estrous cycles *(42,43)*, and may reduce lordosis behavior (a marker for sexual receptivity) when exposure is limited to the postnatal period *(38,39,43)*. However, it should be noted that these results seem to depend greatly on timing of exposure, dosage, experimental compound, and so on, and further studies are needed to verify the effects of phytoestrogen on adult female reproductive behavior.

3.2.3. Conclusions

Experiments focusing on the effects of early exposure to phytoestrogens are intriguing in that these compounds alter the adult brain at several morphologic and behavioral levels. Any inconsistencies in these studies are more than likely because of differences in experimental design, and further studies are needed to determine the mechanism by which these compounds are exerting their effects.

3.3. *Organochlorine Pesticides and Early Brain Development*

DDT, probably the most well known organochlorine pesticide, was first found to be weakly estrogenic in chickens in 1960 *(28)*. Since then, DDT and other organochlorine pesticides have been extensively studied as endocrine disruptors in several species. Although results of experiments analyzing neonatal exposure to these substances have sometimes varied in the details, as a whole, the literature supports the likelihood that exposure during the critical period of brain sexual differentiation to organochlorine pesticides can affect all aspects of brain physiology. To follow is a general review of such literature. For details of methodology used in these experiments, including time of exposure, age of examination, mode of exposure and dosage, see Table 1B.

3.3.1. Gene Expression, Protein Expression, and Brain Morphology

As with other EDCs, organochlorine pesticide exposure during the critical period of brain sexual differentiation can alter gene expression later in life. Methoxychlor, a substitute for DDT, was shown to alter gene expression of several estrogen-responsive genes in the juvenile rat. When mothers were fed high doses of methoxychlor, progesterone receptor (PR) expression in the medial preoptic area (MPOA) decreased in males and increased in females when compared with controls. Normally, this expression is sexually dimorphic with expression being higher in the male MPOA, suggesting that high doses of methoxychlor can have demasculinizing effects in males and masculinizing effects on females. Interestingly, lower doses had no effect on PR expression but resulted in a decrease in ER-β and steroid receptor co-activator 1 (SRC-1) gene expression in females *(44)*. SRC-1 plays an important role in the hypothalamus during development by aiding in the determination of sexual-specific behavior in females *(45,46)*. Additionally, reducing SRC-1 protein levels in the hypothalamus results in dysfunctional male reproductive behavior [reviewed in *(47)*]. The lowest dose of methoxychlor had no effect on gene expression in either male or the female pups *(44)*. These data suggest that varying doses of methoxychlor can alter the transcription of estrogen-responsive genes differently.

Another mechanism for the effects of organochlorine pesticides is through actions on the transcription of genes in an estrogen receptor-independent manner. For example, $GABA_A$ receptor subunit mRNA expression is decreased by dieldrin exposure in the brainstem of embryonic mice *(14)*. Although these effects were observed in the brainstem, it is worth investigating whether dieldrin and other organochlorine pesticides have similar effects in other brain regions, as $GABA_A$ receptors are important for many functions including the regulation of the hypothalamic GnRH system that drives HPG function *(48)*.

By altering the expression of numerous neuronal genes, organochlorine pesticides can alter protein levels in the brain thereby profoundly impacting cellular function. For example, rats exposed to methoxychlor in utero and lactationally showed significant alterations in pituitary protein levels. Specifically, in 3-week old males, a high dose of methoxychlor increased LH, FSH, and prolactin but only increased LH in comparable females. At 11 weeks of age, females treated with low levels of methoxychlor had an increase in the number of prolactin ir cells, whereas those exposed to a high dose showed only an increase in FSH *(49)*. These data suggest that perinatal exposure to methoxychlor can have significant long-term effects on protein translation in the pituitary.

In another study, Caudle et al. *(13)* found that gestational and lactational exposure to heptachlor increased DAT, VMAT2, and TH in the striatum of mice. The transporters, DAT and VMAT2, are important for normal dopamine neurotransmission.

DAT is responsible for the reuptake of dopamine into the presynaptic terminal, whereas VMAT2 in the presynaptic terminal is responsible for transporting dopamine into vesicles (both small and dense core) thereby preparing dopamine for release into the synapse *(50,51)*. Additionally, the ratio of DAT : VMAT2 was increased by 29 % by heptachlor. The striatum is an area of the brain that controls movement, balance, and walking, requiring dopamine to function and is compromised in Parkinson's disease. The authors speculate that the ratio of DAT : VMAT2 is important for predicting susceptibility to Parkinson's disease *(13)*. Interestingly, E_2 can modulate both DAT and VMAT2 in the rat striatum *(52)* and can be neuroprotective in adult animals *(53)*. Together, these results suggest that early exposure to the weakly estrogenic compound heptachlor may increase susceptibility to neurodegeneration and potentially play a role in Parkinson's disease.

In addition to these previously discussed actions on the GABAergic and dopaminergic systems, organochlorine pesticides are strongly implicated in modulating the synthesis and degradation of neurotransmitter acetylcholine. Specifically, DDT decreases expression of muscarinic cholinergic receptors, which are inhibitory acetylcholine receptors found throughout the body. Eriksson et al. found that a single dose of DDT on P10 decreased muscarinic acetylcholine receptor density in the cerebral cortex of 4-month-old male mice. A decrease in receptor density appears to translate into an increase in spontaneous locomotor activity, a behavior that is sexually dimorphic. This effect is seemingly time dependent as it was not observed when mice were treated on P3 or P19 *(54)*.

As discussed previously (Section 3.2.1.), gene expression and protein levels directly affect cellular differentiation, cell survival, and brain morphology and function. To our knowledge, the only study examining effects of early methoxychlor exposure on the volume of the SDN-POA did not detect any effect *(55)*.

3.3.2. Reproductive Function and Behavior

Although there is little evidence that organochlorine pesticides change brain morphology, there is substantial evidence that exposure during development to these compounds alters both brain function and behavior. In males, organochlorine pesticides are detrimental to several reproductive parameters (Table 1B). First, methoxychlor delays the onset of puberty *(56)* and decreases serum LH and FSH, but not testosterone levels or copulatory behavior *(57)*. However, when exposure is limited to the time of blastocyst implantation, a profound effect on male reproductive behavior is observed. Specifically, male sexual motivation and arousal is virtually ablated. Amstislavsky et al. *(58)* demonstrated that treated males avoid a receptive female and show no change in basal testosterone levels when placed in close proximity to them. Normally, serum testosterone levels will rise in response to a receptive female *(58,59)*. Organochlorine pesticides also have significant affects on the reproductive function and behavior of adult females. Several studies have demonstrated that perinatal exposure to methoxychlor and DDT advances puberty *(56,60–63)*, increases persistent vaginal cornification, increases the percentage of animals with irregular estrous cycles *(56,57,60)*, accelerates the loss of fertility *(61–63)*, dampens the preovulatory LH surge, and suppresses the lordosis reflex *(57)*. Recently, Rasier et al. suggested that the possible mechanism for the induction of precocious puberty by demonstrating that infantile exposure to DDT not only resulted in precocious puberty but also accelerated pulsatile GnRH release in the female hypothalamus [review in *(64)*].

3.3.3. Non-Reproductive Sexually Dimorphic Behavior

Recently, evidence has been published that embryonic exposure to methoxychlor can alter the reward circuitry in the female rat. This circuitry is known to be sexually dimorphic. For example, female mice are more sensitive to amphetamine-induced place conditioning than males. However, when treated with methoxychlor, females did not display preference for the amphetamine-paired chamber. These results suggest that embryonic exposure to methoxychlor impairs the female brain reward pathways. Interestingly, methoxychlor-treated males did not display this impairment suggesting a potential link to the estrogenic properties of methoxychlor. Additionally, the dopamine system, the main system associated with the reward system, is modulated by estrogen, suggesting that these data may be the result of an altered dopaminergic system *(65)*.

Several studies have observed altered wheel running and open field exploration behaviors, both of which are sexually dimorphic, in response organochlorine pesticides. Eriksson et al. demonstrated that spontaneous locomotor activity and rearing were significantly increased in males treated with 0.5 mg/kg DDT on P10. This increase in activity is thought to be the result of a long-term decrease in muscarinic acetylcholine receptors in the cerebral cortex (Section 3.3.1.) *(54)*. In females, running wheel activity was not altered when exposure was limited to gestation *(66)*, but locomotor activity was increased when treatment of a low dose of methoxychlor was extended through weaning *(56)*. Interestingly, continuous exposure to high doses of methoxychlor did not alter other basal or amphetamine-induced locomotor activity in females or males *(60)*.

Still another sexually dimorphic behavior that is altered by organochlorine pesticides is that of sodium chloride solution intake. Under control conditions, females drink more than males; however, treatment with high levels of methoxychlor increased sodium chloride solution intake in both females and males *(66)*. These data are interesting because intake was increased in both sexes suggesting males to be feminized and the females to be hyperfeminized.

Exposure to organochlorine pesticides during development affects other dimorphic behaviors such as aggression and territorial marking. Palanza et al. showed that prenatal exposure to low doses of *o, p′*-DDT and methoxychlor increased the rate of territorial urine marking but had no significant effects on intrasex aggression in males regarding the number of attacks observed and the latency to attack. However, the intensity of attacks by males treated with *o, p′*-DDT decreased significantly (bite frequency, tail rattling, and total number of attacks) *(67)* [reviewed in *(21)*]. Additionally, the onset of intrasex aggression is significantly delayed in pubertal male mice exposed to low doses of methoxychlor but did not alter adult aggressive behavior *(68)*. Interestingly, *o, p′*-DDT had no effect on intrasex aggression in females *(67,69)* [reviewed in *(21)*].

Finally, although it is beyond the scope of this article to discuss non-sexually dimorphic behaviors, readers are referred to published studies in this area *(70,71)*.

3.3.4. Conclusions

As a whole, studies focusing on early developmental exposure to organochlorine pesticides suggest that their effects can be observed throughout the brain. Notably, these compounds seem to have their most profound (and reproducible) effects on the reproductive system. For example, methoxychlor and DDT have repeatedly been shown to advance puberty in female rodents and delay puberty in males. Additionally, they disrupt estrous cycles and seem to have detrimental effects on reproductive behavior. Studies focusing on other areas of the brain are more difficult to interpret as there are inconsistencies in the data or some parameters simply have not been examined. It should be noted that the inconsistencies observed are most likely due to differences in experimental design (i.e., dose, length and age of exposure, and compound), and further research is required to determine the effects of early exposure to organochlorine pesticides on the development of the brain.

3.4. PCB Exposure During Early Brain Development

The estrogenic nature of PCBs has been known for decades. Exposure to these compounds during critical periods in development is especially detrimental to wildlife *(28,72)*. Additionally, PCBs are lipophilic and so have the potential to bioaccumulate in adipose tissue. This is of particular concern when PCBs accumulate in tissues of prereproductive or reproductive females, because these compounds can be transferred to the developing offspring either through the placenta or postnatally through lactation. Therefore, it is not surprising that many of the laboratory experiments examining the role of PCBs as endocrine disruptors have focused on developmental exposure and their long-term effects on estrogen-responsive genes, cellular processes, and structural alterations and behaviors. A summary of details on the literature about PCBs and CNS function is provided in Table 1C.

3.4.1. Gene Expression, Protein Expression, and Brain Morphology

A study from our laboratory showed that young adult female rats exposed to Arochlor 1221 (A1221) (a mixture of lightly chlorinated PCB congeners) during development had significantly lower ER-β protein expression in the AVPV, a sexually dimorphic brain region involved in the control of female ovulation. Because ER-β expression in this brain region is also sexually dimorphic, with more ER-β-positive cells in females than in males *(73)*, the reduction in ER-β-positive cells in adult females exposed to A1221 in our study *(74)* suggests that the brain region was masculinized during development by PCBs. This effect was not observed in the supraoptic nucleus (SON), an area of the brain important for osmoregulation, regulation of blood pressure, parturition, and lactation, indicating regional specificity of the PCB treatment *(74)*. PCBs may also act on neurotransmitter receptors in the brain, with strongest evidence for the dopaminergic system. Moreover, findings for interactions between dopamine and estrogen signaling *(75)* suggest a dual mechanism for effects of early PCB exposures. For example, Seegal et al. *(76)* found that gestational exposure to 3,4,3′4-tetrachlorobiphenyl (TCB) and 3,4,5,3′,4′-pentachlorobiphenyl (PtCB) (both coplanar congeners) significantly increases dopamine levels in the prefrontal cortex of adult rats. However, when an *ortho*-substituted compound (*o*′-TCB) was used, a significant decrease was observed in the prefrontal cortex of adult rats. These results suggest that structure of the compound can have substantially different effects in the adult organism when exposure is limited to the perinatal period. Additionally, as dopamine release in the prefrontal cortex is modulated by estrogen, these data suggest that the estrogenic effects of coplanar PCBs may play a role in these effects *(76)*.

In addition to protein expression, enzymatic activity is also affected by early exposure to PCBs. Gestational exposure to a reconstituted mixture of PCBs decreased aromatase activity in the hypothalamus/preoptic area of male rat pups *(77)*. This enzyme converts testosterone to estrogen in the hypothalamus and preoptic area, a reaction that is essential for the masculinization of the brain. Another PCB mixture, Arochlor 1254, did not affect aromatase activity in male pups *(77)*. However, Arochlor 1254 may alter the activity of other enzymes. For example, exposure during gestation to A1254 reduces choline acetyltransferase (ChAT) activity, the biosynthetic enzyme for acetylcholine, in the basal forebrain and hippocampus *(78)*. Notably, that same laboratory showed that A1254 had no effect on ChAT activity when using a slightly different experimental paradigm (Table 1C) *(79)*. Although this does not directly alter the endocrine system, both the basal forebrain and hippocampus communicate with the neuroendocrine system through neural connections, suggesting that alterations in other brain regions can potentially alter endocrine function *(80)*. In addition to the dopaminergic and acetylcholine systems, serotonin levels are also affected by prenatal exposure to PCBs. Adult female and male rats exposed during gestation to A1254 showed a marked increase in serotonin and its metabolite 5-hydroxyindole acetic acid in the lateral olfactory tract, prefrontal cortex, and hippocampus *(81)*. Taken together, these data suggest that PCBs are affecting several neurotransmitter systems ranging from the synthesis, storage, release, and metabolism.

The consequences of EDCs altering protein expression and enzymatic activity during development include potentially profound neurodevelopmental effects. Although there is considerable evidence for phytoestrogens in this regard (Section 3.2.1. and Table 1A), PCBs specifically have not been implicated in the morphological alterations of sexually dimorphic areas of the brain. However, some structural (non-sexually dimorphic) changes have been observed. For example, exposure to A1254 results in a decrease in brain weight *(82)*, reduces the growth of the mossy fibers in the hippocampus *(83)*, and results in a male-specific reduction in cerebellar mass *(84)* (see Table 1C for experimental details).

3.4.2. Reproductive Function and Behavior

It is not surprising that PCBs alter hormonal levels, as they are known to affect almost every aspect of the endocrine system. Recently, Lyche et al. *(85)* determined that PCB153, but not PCB126, decreased prepubertal LH concentrations, delayed puberty, and resulted in higher progesterone concentrations during ovulation (luteal phase) in 9-month-old female goats exposed during gestation and lactation *(85)*. Additionally, Hany et al. *(77)* found that treatment with a reconstituted mixture of PCBs (Section 3.4.1.) throughout gestation resulted in a decrease in serum testosterone levels in adult male rats. These have strong implications for the effects of PCBs on pubertal onset and sexual development.

Exposure to PCBs during development alters adult reproductive and other sexually dimorphic behaviors. For example, *neonatal* exposure to A1254 reduces sexual receptivity (lordosis quotient), whereas A1221 has no effect *(86)*. However, when treatment includes prenatal *and* postnatal exposure, A1254 decreases sexual motivation and A1221 decreases sexual receptivity *(87)*. Our laboratory has recent preliminary evidence that limited prenatal exposure to A1221 significantly alters specific receptive behaviors in female rats *(88)*. These data suggest that timing of exposure is an extremely important parameter to consider when analyzing the effects of EDCs on sexual behavior.

3.4.3. Sexually Dimorphic Non-Reproductive Behaviors

As with organochlorine pesticide exposure, PCBs alter sexually dimorphic behaviors. For example, when rats are exposed to a reconstituted mixture that is similar to that found in human breast milk, an increase in sweet preference in males was observed. This sexually dimorphic behavior is usually increased in females, suggesting that gestational and lactational exposure can feminize sexually dimorphic behaviors in adult rats *(77,89)*.

3.4.4. Cerebellar Function and Locomotion

Developmental exposure to PCBs has profound effects on cerebellar function. Briefly, PCBs cause delays in reflex development in rodents *(84,90,91)*, alter locomotor activity and fine motor control *(82,84)*, and result in hyperactivity *(90,92)*. However, there are several inconsistencies in these studies that are most likely due to differences in timing of exposure, dose of exposure, type or mix of PCB used, and methodology used to access locomotion. For a more comprehensive review of the literature on PCBs effects on motor function, see Roegge and Schantz *(93)*. Interestingly, although cerebellar function is affected by perinatal exposure to PCBs, cerebellar cell structure does not seem to be affected, as dendritic branching, dendritic length, number and structure of Purkinje cells, and parmedian lobule volume were unaltered by exposure to A1254 *(94)*.

3.4.5. Cognition

Finally, early exposure to PCB mixtures or *ortho*-substituted congeners has detrimental effects on cognition. Monkeys exposed to a complex mixture of PCBs during gestation and lactation displayed deficits in spatial learning and memory as adults. These deficits suggest that PCB exposure alters the function of the dorsolateral region of the prefrontal cortex, an area important for spatial learning *(95)*. Rodent studies have demonstrated similar results. Interestingly, there were several sex differences found in the cognitive deficits in rats. For example, only females were affected by A1254 treatment when tested on a spatial alternation task, whereas only males showed deficits in working and reference memory. These sex differences suggest that some of cognitive deficits observed in rodents may be due to the endocrine-disrupting actions of PCBs *(6,96)*.

3.4.6. Conclusions

Taken together, these data suggest that perinatal exposure to PCBs have consequences throughout the brain, and these effects are, in many cases, persistent. However, it is difficult to interpret these results further as the studies discussed differ in so many experimental parameters (dose, age and duration of exposure, nature of compound, etc.). The few conclusions that can be made suggest that PCBs have effects on neurotransmitter release and metabolism, reproductive function, and alter normal development (both somatic and cognitive). Further studies are needed to conclusively determine how early exposure to PCBs affect brain development and function.

4. EFFECTS OF EDC EXPOSURE DURING LATER POSTNATAL AND PUBERTAL DEVELOPMENT

4.1. Background on Hormonal Changes During Postnatal/Pubertal Life Stages

The pubertal process is a protracted event that is characterized by the maturation of the reproductive system, an increase in gonadal steroid hormone synthesis and release, and the eventual attainment of adult reproductive capacity. Puberty is initiated by an activation of hypothalamic GnRH release *(97)*, followed by a stimulation of pituitary gonadotropins, and finally the activation of the ovary/testis. At this time, folliculogenesis, ovulation (females), and spermatogenesis (males) are stimulated. In addition, the increases in sex steroid hormones trigger the development secondary sex characteristics. Increases in sex steroid hormones during puberty also exert important actions on the brain called "activational effects," whereby those organizational actions of neonatal steroid hormones (Section 3.1.) are triggered. For example, early steroid hormone exposure organizes male or female-typical brain morphology and neurochemistry (Section 3.1.), but these effects are not manifested until the pubertal process when activational effects of steroids occur. In addition, there is mounting evidence that they exert organizational effects in the CNS leading to differences in adult behavior [reviewed in *(27)*]. Regardless of the mechanism, pubertal steroid hormones exert profound morphological changes in the brain such as changes in neurogenesis, synaptogenesis, apoptosis, and dendritic growth or pruning. Thus, exposure to environmental toxicants during this sensitive developmental phase may have pathological outcomes [reviewed in *(27)*].

Along with hormonal changes during puberty, several neurotransmitter systems are also maturing at this time in development. For example, there is dopaminergic maturation during puberty in non-human primates *(98)*. As discussed previously, EDCs such as organochlorine pesticides and PCBs can cause significant alterations to neurotransmitter systems. Because these systems are still maturing during the pubertal period, it is worth noting that exposure to EDCs could potentially alter normal maturation of neurotransmitter systems and therefore have long-term detrimental effects.

Cognition may also be a potent target of EDCs during pubertal development as the frontal cortex is one of the last areas of the brain to mature, and it is known that maturation of this process can continue through adolescence *(99)*. Additionally, the prefrontal cortex is thought to be an estrogen-responsive target tissue, at least in regard to cognitive function. For example, girls with Turner syndrome (XO genotype) show an immature pattern of frontal cortex activation. These studies indicate that the pubertal period may be a sensitive period of development and that exposure to EDCs during this period has the potential to have long-term effects on cognition *(99)*. Although publications in this area are few in number, to follow is a brief summary of this literature. For details regarding dosage, length of exposure, age of exposure, and method of exposure, refer to Table 2.

4.2. Pubertal Exposure to EDCs

4.2.1. Phytoestrogens

The few studies that have looked at pubertal exposure to phytoestrogens have focused on isoflavones found in soy products. In rats exposed to genistein at a time of

Table 2
Effects of pubertal exposures to (A) phytoestrogens, (B) organochlorine pesticides, and (C) PCBs on brain morphology and function

	Sex	Compound	*Age of treatment*	*Age of testing*	*Method of exposure*	*Dose*	*Organism*	*Reference*
A. Pyhtoestrogens								
Hypothalamus/pituitary								
↓ ERβ gene expression	F	Genistein	P21 to adult	Adult	Diet	500 ppm	Rat	*100*
↑ VA1 Binding by vasopressin in lateral hypothalamus	M	PhytoE diet	P28 to P75	P75	Diet	460 ppm	Syrian Hamster	*101*
↓ VA1 binding in lateral septum								
↑ Serum testosterone levels								
B. Organochlorine pesticides								
Reproductive Function and behavior								
• Advances puberty	F	MXC/ Lindane	Various	Various	Various	Various	Rat	*102–104*
• Advances first estrus								
• Alters estrous cyclicity								
↑ Size of pituitary	F	Lindane	Chronic	P28	Injection	5, 10,20 & 40 mg/kg	Rat	*104*
↑ Serum FSH levels								
↓ Serum LH levels								
• Delays puberty	M	MXC	P21–P300		Gavage	200, 300 & 400 mg/kg	Rat	*107*
↓ Fertility								
↑ GnRH protein	M	MXC	P21–P77	P77	Gavage	50 mg/kg	Rat	*108*

(***Continued***)

Table 2
(*Continued*)

	Sex	Compound	*Age of treatment*	*Age of testing*	*Method of exposure*	*Dose*	*Organism*	*Reference*
Miscellaneous areas								
• Impairs blood brain barrier	M	Lindane	P10–P13 P10–P17	P13 P17	Oral Oral	2 mg/kg	Rat	*109*
Cognition								
• Impairs cognition	M & F	MXC	20–32 mo	Up to 46 mo	Oral	50 mg/kg	Rhesus Monkeys	*98*
↓ Performance on spatial learning tasks								
C. PCBs								
Hypothalamus/pituitary								
↓ α-Synuclein protein expression	M	A1254	P30	P36/P51	Gavage	100 µg/kg/day	Rat	*110*
Cortex, hippocampus and cerebellum								
↓ α-Synuclein/ ↓ Parkin protein expression	M	A1254	P30	P36	Gavage	100 µg/kg/day	Rat	*110*

↑, increase; ↓, decrease; -, no change; A = Aroclor; E = embryonic day of age; ER, estrogen receptor; F, female; FSH, follicle-stimulating hormone; GnRH, gonadotropin-releasing hormone; LH, luteinizing hormone; M, male; MXC, methoxychlor; P, postnatal day of age; PCB, polychlorinated biphenyl; PhytoE diet, high phytoestrogen diet; and VA = vasopressin 1 receptor.

development that corresponds to early puberty, an increase in ER-β mRNA expression was observed in the paraventricular nucleus (PVN) of the hypothalamus, whereas E_2 treatment decreased ER-β expression (Table 2A). This suggests that genistein has antiestrogenic actions in the PVN during puberty. In addition, there was no effect on oxytocin receptor (OTR) mRNA expression (known to be regulated by ER-α) in the ventromedial nucleus (VMN) of the hypothalamus. Taken together, these findings suggest that genistein is acting through ER-β and its effects are region specific *(100)*. In male Syrian hamsters, a high-phytoestrogen diet results in altered binding of vasopressin to its receptor (V1A) in animals exposed during puberty and adulthood. An increase in V1A binding was observed in the lateral hypothalamus with decreased binding in the lateral septum. Additionally, the authors found an increase in serum testosterone levels (Table 2A). The authors hypothesize that changes in V1A binding in these regions, which are known regulators of aggressive behavior, in concert with increased serum testosterone levels, indicate that phytoestrogen exposure during puberty may affect male-typical behavior *(101)*.

4.2.2. Organochlorine Pesticides

The few studies in this arena have focused on brain function and behavior as opposed to gene expression or structural changes that may occur in the pubertal brain. To follow is a brief overview of these data, for experimental details regarding dose, method of exposure, and so on, refer to Table 2B. A majority of studies in this field have investigated organochlorine pesticide exposure during puberty and the effects on reproductive parameters. For example, Gray et al. and Laws et al. both reported that methoxychlor significantly advanced the onset of puberty (vaginal opening) and first estrus when weanlings were treated extensively, through P110 *(102)* or for three days from P21–P23 *(103)*. This effect was also observed with treatment with lindane from P21–P110 *(104)* (Table 2B). Additionally, lindane increased the size of the pituitary as well as FSH concentration and decreased LH *(104)*. Both Gray and Cooper noted that pubertal exposure to methoxychlor and lindane altered estrous cyclicity *(102,104–106)*. In males exposed to high doses of methoxychlor for 10 months, beginning on the day of weaning, fertility was substantially impaired, as evidenced by a long time required for males to impregnate dams, a smaller number of dams impregnated, and smaller litter sizes. However, it significantly stimulated their libido as both inter-intromission interval and latency to ejaculate were significantly decreased in all treatment groups. Additionally, as with neonatal exposure, the timing of puberty was delayed *(107)*. Males exposed throughout puberty to high levels of methoxychlor displayed a significant increase in GnRH protein levels in the mediobasal hypothalamus (MBH), and in vitro slices of the MBH released more GnRH when stimulated with a 60 mM KCl pulse in vitro. However, no effects were observed on serum LH, FSH, or testosterone levels in vivo *(108)*.

In addition to reproductive parameters, pesticides may alter blood-brain barrier permeability *(109)*. Finally, pubertal exposure to organochlorine pesticides can also alter cognitive function. Female rhesus monkeys exposed throughout puberty for 1 year (6 months before predicted menarche and 6 months after) were subject to several cognitive tests after a 6-month recovery. Methoxychlor-treated females performed poorly on a battery of cognitive tests when compared with their control counterparts. Additionally, those treated with the highest dose of methoxychlor (Table 2B) failed to improve on a spatial learning task unlike all other groups (control and treatment). These results suggest that pubertal exposure to methoxychlor can have detrimental and persistent effects on cognition *(98)*.

4.2.3. PCBs

To our knowledge, only a few studies have observed the effects of PCB exposure during postnatal development and puberty in vivo (see Table 2C for experimental details). One study found that exposure to A1254 on P30 for 6 days resulted in alteration of protein levels that are important for synaptic transmission *(110)*. Specifically, the authors found a decrease in α-synuclein, a protein important for release of neurotransmitter [reviewed in *(111)*], in the cortex, hippocampus, hypothalamus, and cerebellum 2 days after the treatment had ended and in the hypothalamus 21 days after the treatment had ended *(110)*. Parkin protein levels, a protein thought be involved with the degradation of α-synuclein [reviewed in *(111)*], were also affected, but an increase was observed in cortex, hippocampus, and cerebellum 2 days but not 21 days after treatment. These results suggest that PCB exposure for a limited time during puberty may have transient effects on synaptic transmission *(110)*.

4.2.4. Conclusions

As a whole, these data on effects of phytoestrogens, organochlorine pesticides, and PCBs suggest that puberty is a sensitive period for effects of EDCs to be exerted, and some of these alterations are enduring. More studies are required to determine other long-lasting effects and to decipher the underlying mechanisms.

5. EFFECTS OF EDC EXPOSURE IN ADULTHOOD ON THE NERVOUS SYSTEM

5.1. Estrogen's Role in Adulthood and Aging

In adulthood, estrogens continue to modify the brain by driving gene transcription, altering dendritic spine density, causing neurogenesis, changing brain morphology, and even affecting brain function. The major estrogen E_2 is well established as a regulator of ovulatory cycles in females. During menstrual or estrous cycles, E_2 exerts primarily negative feedback onto hypothalamic and pituitary components of the HPG axis. However, shortly before the period of ovulation, the feedback regulation of E_2 onto GnRH neurons becomes positive, resulting in a "GnRH surge" that in turn causes the preovulatory LH surge. Along with fluctuations during reproductive cycles, ovarian steroids also undergo changes during the life cycle, with increased overall concentrations of E_2 occurring at puberty and decreased levels later in life at the time of reproductive failure (menopause in women) *(112)*. Aging-related alterations in E_2 are also responsible for or contribute to changes in temperature regulation, fertililty, fecundity, cardiovascular function, and cognition. Although it is known that E_2 plays an important role in both ovulation and menopause, the mechanism by which it exerts these effects are still under investigation, and further, these differ among species *(112)*.

Estrogens in the adult brain are not only important for reproductive function, but they also play non-reproductive roles in areas of the brain beyond the hypothalamus. These include the hippocampus, amygdala, cingulate cortex, LC, and midbrain raphe nuclei, which are abundant in their expression of estrogen receptors. In each of these areas, estrogen has been shown to regulate the release of neurotransmitters such as serotonin, acetylcholine, and glutamate. This question is particularly pertinent in the context of reproductive aging, as lower estrogen levels can lead to many common CNS symptoms of menopause such as hot flashes, headaches, and depression *(113)*. For further review of estrogen's effects in the adult brain. See McEwen *(114)*.

5.2. *Phytoestrogens*

There is substantial evidence that phytoestrogens have profound effects on the adult brain, ranging from altered neural gene expression to manifestations at the behavioral level. To follow is a general review of the literature, for more detailed information regarding age, duration of treatment, dosage, and method of exposure, refer to Table 3A.

5.2.1. Gene Expression, Protein Expression, and Morphological Changes

Gene expression studies show that phytoestrogens can act as antagonists on both ER-α and ER-β in the brain. In rats, given a commercially developed phytoestrogen supplement (mainly daidzen and genistein) in addition to estrogen, a marked reduction in the induction of oxytocin receptor (OTR) mRNA was observed in the VMN. In addition, the supplement increased ER-β gene expression, a process normally *repressed* by ER activation, in another brain region, the PVN. These results suggest that phytoestrogens are acting as antiestrogens in two areas of the brain known to be important for female reproductive function and behavior *(115)*. Coumestrol is also thought to exert antiestrogenic effects on both ER-α and ER-β in the brain. Dietary exposure to coumestrol has similar effects on ER-β mRNA expression, the commercial supplement mentioned above *(116)*, and attenuates the effects of E_2 on protein expression. Jacob et al. *(117)* showed that coumestrol significantly decreased the number of PR-ir in mice that were concomitantly treated with estrogen versus mice treated with estrogen alone. Additionally, when ER-α knockout mice were treated with both E_2 and coumestrol, no effect was observed *(117)*.

When considering the effects of phytoestrogens on gene expression, it is also important to account for morphological changes that might arise from exogenous hormonal activation. Few studies have examined this question. One group found that male rats fed with a phytoestrogen-rich diet during development and puberty, and then switched to a phytoestrogen-free diet in adulthood, had smaller SDN-POAs than males fed with phytoestrogens throughout their entire life *(118)*. Although these results are interesting, it should be noted that no negative controls were used, that is, there were no animals that were continuously kept on a phytoestrogen-free diet, making these results difficult to interpret but worth pursuing.

5.2.2. Reproductive Behavior and Function

A number of groups have examined functional/physiological changes in the CNS in response to phytoestrogens in adulthood *(119,120)*. For example, coumestrol has effects on GnRH/LH pulsatile release. Ovariectomized rats treated with coumestrol showed a significant decrease in LH pulse amplitude and frequency, and this appears to be mediated through hypothalamic (presumably GnRH) activity *(120)*. Interestingly, a high concentration of coumestrol and exogenous GnRH was unable to elicit LH

Table 3

Effects of adult exposures to (A) phytoestrogens, (B) organochlorine pesticides, and (C) PCBs on brain morphology and function

	Sex	*Compound*	*Length of treatment*	*Method of exposure*	*Dose*	*Organism*	*Reference*
A. Phytoestrogens							
Hypothalamus/pituitary							
↑ Oxytocin receptor gene expression	F	PhytoE sup	weeks	Diet	~13 ppm G/33 ppm D	Rat	*115*
↑ ERβ gene expression							
↓ ERβ gene expression	F	Coumestrol	1 week	Diet	100 mg/kg diet	Rat	*116*
↓ PR protein (ir cells)	F	Coumestrol	10 days	Diet	0.02 % of diet	Mouse	*117*
Reproductive function and behavior							
↓ LH pulse amplitude and frequency	F	Coumestrol	8.5 h	Injection	1.6 mg Bolus + 2.4 mg/h	Rat	*120*
↓ Lordosis response	F	PhytoE sup	5 days	Diet	~13 ppm G/33 ppm D	Rat	*122*
↑ Anxiety	M	PhytoE diet	18 days	Diet	150 μg/g	Rat	*123*
↑ Serum vasopressin and corticosterone							
• Altered aggressive behavior	M and F	Soy protein	15 months	Diet	0.94 and 1.88 mg/g	NHP	*124*
Maternal behavior							
-Nursing behavior	F	Genistein	Gestation	Diet	5, 100, and 500 ppm	Rat	*126*

Other Effects							
↓ Core body temperature	F	PhytoE diet	6 weeks	Diet	600 ppm	Rat	*121*
↑ Neuroprotective	M and F	Various	Various	Various	Various	Mouse/rat/NHP	*119,126–133*
Cognition							
↑ Working memory	F	PhytoE diet	10 months	Diet	600 ppm	Rat	*137*
↓ Drug-induced amnesia	F	Daidzein	4 weeks	Diet	4.5 mg/kg BW	Mice	*138*
Other effects							
↓ Core body temperature	F	PhytoE diet	6 weeks	Diet	600 ppm	Rat	*121*
↑ Neuroprotective	M and F	Various	Various	Various	Various	Mouse/rat/NHP	*119,126–133*
B. Organochlorine pesticides							
Hypothalamus/pituitary							
↑ ERE–luciferase induction	M	DDT		Injection	5 mg/kg	Rat	*139*
↑ VEGF protein Expression	F	MXC	1 and 3 weeks	Gavage	50, 100, and 200 mg/kg/day	Rat	*140*
↑ Vascularization							
Basal ganglia							
↑ DAT activity	M	Heptachlor	2 weeks	IP injection	6 and 12 mg/kg	Mouse	*141*
↓ O_2 consumption							
↓ Mitochodrial respiration	F	MXC	20 days	IP injection	10 μg/ml	Mouse	*142*
Cognition							
↓ Visual discrimination-reversal learning	N/A	Dieldren	Not specified	Not specified	Not specified	Sheep/squirrel monkeys	Reviewed in *(6)*

(*Continued*)

Table 3
(***Continued***)

	Sex	*Compound*	*Length of treatment*	*Method of exposure*	*Dose*	*Organism*	*Reference*
↓ Retention of passive avoidance tests	M	Lindane	Not specified	Not specified	Not specified	Rat	Reviewed in *(6)*
↓ Acquisition of active avoidance	M	Lindane/ endosulfan	Not specified	Not specified	Not specified	Rat	Reviewed in *(6)*
Maternal behavior							
↓ Time spent nursing	F	MXC	8 days	Oral	0.02, 0.2, and 2 μg/g BW	Mice	*(68)*
Other Effects							
↑ Neurotoxicity	M and F	Dieldren	Various	Various	Various	Rodents/ NHP	Reviewed in *(143)*
C. PCBs							
Neurotransmitter							
↓ Dopamine concentrations	M	A1016 and 1260	Various	Various	Various	Rat/NHP	*144,145, 147–149*
↓ Norepinephrin concentrations	M	A1260 and A1254	1,3,7, and 14 days	Gavage	500 and 1000 mg/kg	Rat	
↓ Serotonin concentrations	M						
Miscellaneous Proteins							
↓ Anti-oxidant proteins	M	A1254	30 days	IP Injection	2 mg/kg BW/day	Rat	*150*

Reproductive function and behvaior							
• Disruption of estrous cyclicity	F	A1245	1 month	Oral	30 mg/kg	Rat	*151*
↓ Sexual receptivity							
• Delayed copulation							
Maternal behavior							
↓ High crouch nursing time	F	PCB 77	12 days	SC injections	2 mg/kg	Rat	*152*
↑ Grooming							
↑ Amount of time on nest							
↓ High crouch nursing time	F	PCB 77	12 days	SC injections	2 and 4 mg/kg	Rat	*153*
↑ Licking and grooming					4 mg/kg		
↑ Amount of time on nest					4 mg/kg		
Motor function							
↓ Spontaneous locomotor activity	M	A1254	Various	Various	Various	Rodents	Reviewed in *(154)*

↑, increase; ↓, decrease; -, no change; A, aroclor; BW, body weight; D, daidzein; DAT, dopamine transporter; ER, estrogen receptor; ERE, estrogen response element; F, female; G, genistein; IP, intraperitoneal; ir cells, immuno-reactive cells; LH, luteinizing hormone; M, male; MXC, methoxychlor; NHP, non-human primates; phytoe diet, high phytoestrogen diet; phytoe sup, phytoestrogen supplement; PR, progesterone receptor; SC, subcutaneous; VEGF, vascular endothelial growth factor.

release from the pituitary. Genistein, however, did not have this effect *(120)*. One study focusing on a different hypothalamic function, that of regulation of core body temperature, found that phytoestrogens cause a reduction in this endpoint in female rats throughout the estrous cycle *(121)*. Functional changes often translate to behavioral endpoints, and several studies have demonstrated that phytoestrogens affect a variety of behaviors. Patisaul et al. *(122)* reported that in addition to inhibition of gene expression (Section 5.2.1.), rats treated with a phytoestrogen supplement showed an inhibition of lordosis, an estrogen-dependent behavior *(122)*.

5.2.3. Sexually Dimorphic Behaviors

In adult male rats fed with a high phytoestrogen diet, serum vasopressin and corticosterone levels were elevated. These changes in hormone levels correlated with an increased display of anxiety *(123)*. Phytoestrogen diets also increase intense aggression and intense submission in non-human primates fed with soy protein for 15 months, meaning there was an enhancement of dominant and submissive behaviors. Additionally, monkeys on the diet displayed reduction in the amount of time spent in physical contact with others and an increased amount of time the monkeys spent alone by 30 % *(124)*.

5.2.4. Maternal Behavior

One behavior that may be affected by EDCs that is extremely understudied is that of maternal behavior. This is an important endpoint to be observed, because there is extensive evidence indicating that alterations in maternal behavior can have profound enduring (even transgenerational) effects on the developing offspring [reviewed in *(125)*]. Altering maternal behavior may be one of the mechanisms responsible for the long-term effects observed in offspring exposed to EDCs during the perinatal period. One study focusing on genistein found that it had no effect on nursing behavior in the dams *(126)*. However, no other maternal behaviors were investigated, such as licking and grooming, and it is important to note that this endpoint deserves further investigation.

5.2.5. Neuroprotective Effects

As estrogens are thought to be neuroprotective in nature, this trait has also been ascribed to phytoestrogens. There is substantial evidence supporting this. Briefly, both soy extract and a high dose of genistein reduced neurotoxicity in response to kainic acid in the dentate gyrus (hippocampus) in ovariectomized female rats *(126)*. In vivo studies have shown that soy phytoestrogens protect cholinergic neurons and reduce age-related cognitive decline in male rats *(127)*. Low doses of genistein reduce apoptosis, *(119)* oxidative stress *(128,129)*, and glutamate excitotoxicity *(130)*. Additionally, resveratrol has neuroprotective effects throughout the brain *(131–133)* and prevents the reuptake of both serotonin and norepinephrine in the adult rat brain *(134)*. However, chronic doses of genistein at high concentrations are cytotoxic *(135)*.

5.2.6. Cognitive Effects

Soy isoflavones can improve cognitive function in female rats; however, it is unclear whether soy isoflavones have an effect on male cognition *(136)*. Ovariectomized females treated for 10 months with soy isoflavones displayed an improvement in working memory (radial arm maze). Additionally, lifelong exposure to a high phytoestrogen diet resulted in faster acquisition on a radial arm maze. Interestingly, when those females were taken off of the high phytoestrogen diet, they performed more poorly on the radial arm maze test than those females who remained on the diet *(136,137)*. Finally, daidzein reduces drug-induced amnesia and improves performance and a Y-maze test after scopolamine (used to induces amnesia) injections *(138)*.

5.2.7. Conclusions

Although there is less data implicating endocrine disruption of phytoestrogens in the adult organism when compared with the effects on developing mammals, there is ample evidence suggesting that these compounds can cause significant alterations to the adult brain. Phytoestrogens affect the reproductive system more than other systems in the organism. Superficially, this seems reasonable, as these compounds interact with the estrogen receptor and should produce the most profound effects on estrogen-regulated processes. However, it is known that E_2 can regulate other neurotransmitter systems (see section 5.1), and more studies are needed to determine how phytoestrogens may be interfering in these processes as well.

5.3. Organochlorine Pesticides

Because most of the organizational effects of hormones on the nervous system are believed to have already taken place by adulthood, it has been postulated that exposure to EDCs during adulthood is less toxic to the CNS than exposure at earlier developmental time points. However, there are several studies that have looked at the consequences of organochlorine pesticide exposure to the adult organism and demonstrated significant endocrine-disrupting effects. To follow is a summary of the results of these studies, for experimental details regarding dosage, duration of exposure, method of exposure, and so on, refer to Table 3B.

5.3.1. Gene Expression, Protein Expression, and Brain Morphology

p, *p*′-DDT can activate the estrogen receptor in the brain of estrogen response element (ERE)-luciferase (Luc; a reporter for activation of the gene) transgenic male mice. The ERE is the estrogen-response element, a region of a gene promoter that is responsive to binding of estrogen or agonists to the estrogen receptor. Interestingly, the induction of ERE-Luc by *p*, *p*′-DDT has different kinetics than that of 17-β-E_2. Maximum induction of the ERE-Luc protein by *p*, *p*′-DDT was observed 16 h after treatment, whereas 17-β-E_2 induced expression after 6 h. These data suggest that this DDT metabolite has high estrogenic activity *(139)*. In a different study, methoxychlor treatment increases vascularization in the pituitary in a dose-dependent manner. Goldman et al. determined that methoxychlor treatment for 1–3 weeks increases both vascular endothelial growth factor (VEGF) protein expression and the number of vessels (after 3 weeks of treatment only) in ovariectomized rats. Because VEGF is estrogen responsive, these data suggest that methoxychlor (MXC) treatment may be initiating angiogenesis in a similar manner as estrogen *(140)*.

5.3.2. Neurotranmitter Systems

Organochlorine pesticides affect cellular function in several areas of the brain. For example, heptachlor increases DAT activity in the mouse striatum and decreases O_2 consumption *(141)*. Another study found that chronic treatment with methoxychlor resulted in a decrease in brain mitochondrial respiration in female mice *(142)*. Dieldren has also been implicated in several facets of neurotoxicity in adults including oxidative stress, apoptosis, and dopamine depletion throughout the brain [reviewed in *(143)*]. Although these results do not directly evaluate the estrogenic effects of organochlorine pesticides, it is important to reiterate the point that the endocrine system is significantly affected by neurotransmitter/neuromodulator systems and is constantly integrating information from the rest of the brain to respond appropriately to the environment. If neuromodulator/neurotransmitter systems throughout the brain are impaired by EDCs, this can have profound effects on the brain's ability to regulate the endocrine system.

5.3.3. Maternal Behavior

As with phytoestrogens, few studies have observed how organochlorine pesticides can alter maternal behavior. However, one study found that mice treated with methoxychlor spent less time nursing than their control counterparts *(68)*. As mentioned before, even small changes in maternal behavior can have effects on the pups, and this effect provides a possible mechanism for the effects of organochloride pesticides observed in the developing organism.

5.3.4. Cognition

Many of the studies exploring organochlorine exposure during adulthood have focused on cognition. Dieldrin exposure caused deficits in visual discrimination-reversal learning in sheep and squirrel monkeys and caused rats to make more mistakes on a zig-zag maze [reviewed in *(6)*]. Lindane exposure did not cause deficits in the acquisition of a passive avoidance test but did impair retention as rats performed poorly when retested after 7 days. Additionally, both lindane and endosulfan exposure resulted in deficits in rats ability to learn an active avoidance test [reviewed in *(6)*]. In summary, organochlorine pesticides appear to have detrimental effects on an animal's ability to acquire new information. However, it should be noted that the literature exploring cognition in exposed adults is sparse and further research is needed to conclusively determine organochlorine pesticides effects on cognition in the adult organism.

5.3.5. Conclusions

Organochlorine pesticides are implicated in the alteration of the dopamine system, and these effects have even been linked to neurodegenerative diseases such as Parkinson's. Additionally, they have been shown to be estrogenic in the adult brain and have been linked to deficiencies in cognitive function. Taken together, organochlorine pesticides provide a very interesting model for studying endocrine disruption as many of the functional and behavioral deficits studies are also regulated by estrogen. More studies examining the role of the estrogenic properties of these compounds and the mechanism by which they exert their effects would be of great benefit to the field.

5.4. PCBs

PCBs have been known for decades to be toxic to the adult organism (Table 3C). This was first brought into the spotlight in 1968 and again in 1978 when thousands of people in Yusho, Japan, and Yu-Cheng Taiwan suffered from PCB poisoning after consuming rice oil that was contaminated during the manufacturing process. Those exposed individuals exhibited symptoms including increased skin pigmentation, severe chloracne, thickening of the nailbeds, and numbness in the extremities. Other long-term effects included elevated levels of serum thyroid hormone levels (both T3 and T4), thereby providing some of the first evidence implicating PCBs as endocrine disruptors [reviewed in *(95)*].

5.4.1. Neurotransmitter Systems

In adult mammals, many studies have focused on the neurotoxic effects of PCBs on neurotransmitter systems, with the greatest focus on the dopaminergic system. Seegal et al. have repeatedly demonstrated that exposure to Aroclor (A)1016 and A1260 (*ortho*-substituted PCB congeners) suppresses dopamine concentration in the brains of non-human primates and rats in several regions of the brain including the caudate nucleus, substantia nigra, and the hypothalamus *(144–146)*. Interestingly, coplanar/non-*ortho*-substituted congeners do not affect the adult dopaminergic system but rather elevate dopamine concentrations when administered during development *(76,147)*. Norepinephrine and serotonin concentrations are reduced in the prefrontal cortex and hippocampus lateral olfactory tract (serotonin only) of adult male rats treated with A1254 and A1260 *(148,149)*. Notably, these results are not seen when animals are exposed to much lower concentrations during development, as norepinephine is not affected and serotonin is increased in the prefrontal cortex, hippocampus, and the lateral olfactory tract (Section 3.4.1.) *(81)*. It should be mentioned that the former study did not look at long-term alterations of neurotransmitter systems, and these differences may be due to the elapsed time between treatment and euthanasia.

PCBs not only alter protein concentrations in the adult brain, but they also affect protein function. For example, treatment with A1254 for 30 days resulted in decrease of several antioxidant proteins (enzymatic and non-enzymatic) in the rat hypothalamus. These include superoxide dismutase, catalase, glutathione peroxidase, glutathione reductase, and acetylcholine esterase. Reducing the activity of these enzymes may be one way in which PCBs exert oxidative stress in the hypothalamus *(150)*.

5.4.2. Reproductive Function and Behavior

To reiterate, altered protein concentration and function in brain regions can cause profound effects on brain function and behavior. Through these alterations, PCBs can affect reproductive function and behavior in adult organisms. Brezner et al. *(151)* reported that treating mature female rats for 30 days resulted in disruption of estrous cycles (prolonging them by 1–3 days by increasing the number of days in diestrus), decreased sexual receptivity, and delayed copulation *(151)*.

5.4.3. Maternal Behavior

Two recent reports have found that PCBs have significant effects on maternal behavior. Both studies observed that PCB77 reduces high crouch nursing time, increases licking and grooming, and increases the amount of time the mother spends on the nest *(152,153)*. These results are interesting because maternal behavior can result in long-term alterations in the offspring, and further studies are needed to determine whether this is a mechanism for the developmental effects observed above (Section 3.4.).

5.4.4. Motor Behavior

PCB exposure during adulthood significantly decreases spontaneous motor activity in rats and mice after one treatment with A1254 [reviewed in *(154)*]. This finding is in contrast to reports that exposure to PCBs during development results in an increase in spontaneous motor activity (Section 3.4.3.). Again, these data provide evidence for the importance of timing of exposure and resulting effects.

5.4.5. Cognition

PCBs have long been implicated in impairment of cognitive function. However, to our knowledge, the studies focusing on their effects in adults have concentrated mainly on humans and do not apply to the scope of this review. For more details on adult exposure to PCBs and cognitive function, see Schantz et al., Faroon et al., and Altenkirch et al. *(6,154,155)*.

5.4.6. Conclusions

Because of the relative lack of material on adult exposure to PCBs in animal models, it is difficult to draw strong conclusions. Nevertheless, PCBs exert actions throughout the brain. They disrupt neurotransmitter systems, have effects on the reproductive system, and are implicated in decreasing the motor function in the adult organism. However, further studies are needed to examine the mechanism behind many of these studies. PCBs are well known disruptors of thyroid function, but little is known regarding the estrogenic actions of these compounds. Mechanistic studies focused on teasing out the effects of thyroid hormone disruption and estrogenic disruption would be an asset to this field of study.

6. CONCLUDING REMARKS

Taken together, these data suggest that phytoestrogens, organochlorine pesticides, and PCBs affect the CNS. These effects range from altering gene expression, protein expression, enzyme function, brain morphology, and behavior. Whereas there are many inconsistencies in the data, it can be said that the developing mammal is highly sensitive to even trace amounts of these compounds, and exposure at inappropriate times can have enduring effects. Inconsistencies observed are more than likely due to differences in experimental design, and further studies are needed to determine the mechanisms by which these compounds are exerting their effects.

Exposure during the pubertal phase is a burgeoning field that deserves much more attention. If this period is indeed another stage of organization for the brain, exposure to EDCs during this developmental period could have profound effects in the pubertal and adult brain. Further studies are needed to follow up on these ideas, and it is also important to point out that these studies would not only help to identify possible alterations in pubertal organism but could also provide important information regarding the mechanisms of puberty. Finally, another important future direction is to better understand the long-term effect of chronic exposures throughout the life cycle, particularly their relevance to reproductive aging and menopause.

REFERENCES

1. Gore AC, Roberts JL. Neuroendocrine systems. *Fundamental Neuroscience*, 2nd ed. New York: Academic Press; 2003.
2. Mirescu C, Gould E. From neurotoxin to neurotrophin. *Nat Neurosci* 2004;7(9):899–900.
3. Mirescu C, Peters JD, Gould E. Early life experience alters response of adult neurogenesis to stress. *Nat Neurosci* 2004;7(8):841–6.
4. Ambrogini P, Cuppini R, Ferri P, Mancini C, Ciaroni S, Voci A, Gerdoni E, Gallo G. Thyroid hormones affect neurogenesis in the dentate gyrus of adult rat. *Neuroendocrinology* 2005;81(4): 244–53.
5. Koibuchi N, Fukuda H, Chin WW. Promoter-specific regulation of the brain-derived neurotropic factor gene by thyroid hormone in the developing rat cerebellum. *Endocrinology* 1999;140(9): 3955–61.
6. Schantz SL, Widholm JJ. Cognitive effects of endocrine-disrupting chemicals in animals. *Environ Health Perspect* 2001;109(12):1197–206.
7. Cooke PS, Buchanan DL, Lubahn DB, Cunha GR. Mechanism of estrogen action: lessons from the estrogen receptor-alpha knockout mouse. *Biol Reprod* 1998;59(3):470–5.
8. Zoeller RT. Environmental chemicals as thyroid hormone analogues: new studies indicate that thyroid hormone receptors are targets of industrial chemicals. *Mol Cell Endocrinol* 2005;242 (1–2):10–5.
9. Moritz KM, Boon WM, Wintour EM. Glucocorticoid programming of adult disease. *Cell Tissue Res* 2005;322(1):81–8.
10. Kelce WR, Wilson EM. Environmental antiandrogens: developmental effects, molecular mechanisms, and clinical implications. *J Mol Med* 1997;75(3):198–207.
11. Kuiper GG, Lemmen JG, Carlsson B, Corton JC, Safe SH, van der Saag PT, van der Burg B, Gustafsson JA. Interaction of estrogenic chemicals and phytoestrogens with estrogen receptor beta. *Endocrinology* 1998;139(10):4252–63.
12. Colosio C, Tiramani M, Maroni M. Neurobehavioral effects of pesticides: state of the art. *Neurotoxicology* 2003;24(4–5):577–91.
13. Caudle WM, Richardson JR, Wang M, Miller GW. Perinatal heptachlor exposure increases expression of presynaptic dopaminergic markers in mouse striatum. *Neurotoxicology* 2005;26(4):721–8.
14. Liu J, Brannen KC, Grayson DR, Morrow AL, Devaud LL, Lauder JM. Prenatal exposure to the pesticide dieldrin or the GABA(A) receptor antagonist bicuculline differentially alters expression of GABA(A) receptor subunit mRNAs in fetal rat brainstem. *Dev Neurosci* 1998;20(1):83–92.
15. Belsham DD, Lovejoy DA. Gonadotropin-releasing hormone: gene evolution, expression, and regulation. *Vitam Horm* 2005;71:59–94.
16. Meserve LA, Murray BA, Landis JA. Influence of maternal ingestion of aroclor 1254 (PCB) or FireMaster BP-6 (PBB) on unstimulated and stimulated corticosterone levels in young rats. *Bull Environ Contam Toxicol* 1992;48(5):715–20.
17. Schantz SL, Gasior DM, Polverejan E, McCaffrey RJ, Sweeney AM, Humphrey HE, Gardiner JC. Impairments of memory and learning in older adults exposed to polychlorinated biphenyls via consumption of Great Lakes fish. *Environ Health Perspect* 2001;109(6):605–11.
18. Denison MS, Nagy SR. Activation of the aryl hydrocarbon receptor by structurally diverse exogenous and endogenous chemicals. *Annu Rev Pharmacol Toxicol* 2003;43:309–34.
19. Patisaul HB. Neonatal genistein or bisphenol – A exposure alters sexual differentiation of the AVPV. *Neurotoxicol Teratol* 2006;28(1):111–118.
20. Simerly RB. Organization and regulation of sexually dimorphic neuroendocrine pathways. *Behav Brain Res* 1998;92(2):195–203.
21. Palanza P, Morellini F, Parmigiani S, vom Saal FS. Prenatal exposure to endocrine disrupting chemicals: effects on behavioral development. *Neurosci Biobehav Rev* 1999;23(7):1011–27.
22. Matsumoto A. Synaptogenic action of sex steroids in developing and adult neuroendocrine brain. *Psychoneuroendocrinology* 1991;16(1–3):25–40.
23. Johansen JA, Jordan CL, Breedlove SM. Steroid hormone masculinization of neural structure in rats: a tale of two nuclei. *Physiol Behav* 2004;83(2):271–7.
24. Davis EC, Popper P, Gorski RA. The role of apoptosis in sexual differentiation of the rat sexually dimorphic nucleus of the preoptic area. *Brain Res* 1996;734(1–2):10–8.

25. Forger NG. Cell death and sexual differentiation of the nervous system. *Neuroscience* 2006;138(3):929–38.
26. Wiegand SJ, Terasawa E. Discrete lesions reveal functional heterogeneity of suprachiasmatic structures in regulation of gonadotropin secretion in the female rat. *Neuroendocrinology* 1982;34(6): 395–404.
27. Sisk CL, Zehr JL. Pubertal hormones organize the adolescent brain and behavior. *Front Neuroendocrinol* 2005;26(3–4):163–74.
28. Colbert T, Dumanoski D, Peterson Myers J. *Our Stolen Future*. New York: Penguin Group; 1997.
29. Gore AC, Heindel JJ, Zoeller RT. Endocrine disruption for endocrinologists (and others). *Endocrinology* 2006;147(6 Suppl):S1–3.
30. Ren MQ, Kuhn G, Wegner J, Nurnberg G, Chen J, Ender K. Feeding daidzein to late pregnant sows influences the estrogen receptor beta and type 1 insulin-like growth factor receptor mRNA expression in newborn piglets. *J Endocrinol* 2001;170(1):129–35.
31. Scallet AC, Wofford M, Meredith JC, Allaben WT, Ferguson SA. Dietary exposure to genistein increases vasopressin but does not alter beta-endorphin in the rat hypothalamus. *Toxicol Sci* 2003;72(2):296–300.
32. Marunaka Y. Hormonal and osmotic regulation of NaCl transport in renal distal nephron epithelium. *Jpn J Physiol* 1997;47(6):499–511.
33. de Vries GJ, Miller MA. Anatomy and function of extrahypothalamic vasopressin systems in the brain. *Prog Brain Res* 1998;119:3–20.
34. Bu L, Lephart ED. Soy isoflavones modulate the expression of BAD and neuron-specific beta III tubulin in male rat brain. *Neurosci Lett* 2005;385(2):153–7.
35. Lewis RW, Brooks N, Milburn GM, Soames A, Stone S, Hall M, Ashby J. The effects of the phytoestrogen genistein on the postnatal development of the rat. *Toxicol Sci* 2003;71(1):74–83.
36. Scallet AC. Dietary exposure to genistein increases vasopressin, but does not alter beta-endorphin in the rat hypathalamus. *Toxicol Sci* 2003;72(2):296–300.
37. Faber KA, Hughes CL Jr. Dose-response characteristics of neonatal exposure to genistein on pituitary responsiveness to gonadotropin releasing hormone and volume of the sexually dimorphic nucleus of the preoptic area (SDN-POA) in postpubertal castrated female rats. *Reprod Toxicol* 1993;7(1):35–9.
38. Henry LA, Witt DM. Effects of neonatal resveratrol exposure on adult male and female reproductive physiology and behavior. *Dev Neurosci* 2006;28(3):186–95.
39. Kubo K, Arai O, Omura M, Watanabe R, Ogata R, Aou S. Low dose effects of bisphenol A on sexual differentiation of the brain and behavior in rats. *Neurosci Res* 2003;45(3):345–56.
40. Anselmo-Franci JA, Franci CR, Krulich L, Antunes-Rodrigues J, McCann SM. Locus coeruleus lesions decrease norepinephrine input into the medial preoptic area and medial basal hypothalamus and block the LH, FSH and prolactin preovulatory surge. *Brain Res* 1997;767(2):289–96.
41. Faber KA, Hughes CL Jr. The effect of neonatal exposure to diethylstilbestrol, genistein, and zearalenone on pituitary responsiveness and sexually dimorphic nucleus volume in the castrated adult rat. *Biol Reprod* 1991;45(4):649–53.
42. Whitten PL, Lewis C, Russell E, Naftolin F. Phytoestrogen influences on the development of behavior and gonadotropin function. *Proc Soc Exp Biol Med* 1995;208(1):82–6.
43. Kouki T, Kishitake M, Okamoto M, Oosuka I, Takebe M, Yamanouchi K. Effects of neonatal treatment with phytoestrogens, genistein and daidzein, on sex difference in female rat brain function: estrous cycle and lordosis. *Horm Behav* 2003;44(2):140–5.
44. Takagi H, Shibutani M, Lee KY, Masutomi N, Fujita H, Inoue K, Mitsumori K, Hirose M. Impact of maternal dietary exposure to endocrine-acting chemicals on progesterone receptor expression in microdissected hypothalamic medial preoptic areas of rat offspring. *Toxicol Appl Pharmacol* 2005;208(2):127–36.
45. Apostolakis EM, Ramamurphy M, Zhou D, Onate S, O'Malley BW. Acute disruption of select steroid receptor coactivators prevents reproductive behavior in rats and unmasks genetic adaptation in knockout mice. *Mol Endocrinol* 2002;16(7):1511–23.
46. Molenda HA, Griffin AL, Auger AP, McCarthy MM, Tetel MJ. Nuclear receptor coactivators modulate hormone-dependent gene expression in brain and female reproductive behavior in rats. *Endocrinology* 2002;143(2):436–4.
47. Charlier TD, Balthazart J. Modulation of hormonal signaling in the brain by steroid receptor coactivators. *Rev Neurosci* 2005;16(4):339–57.

48. Compere V, Li S, Leprince J, Tonon MC, Vaudry H, Pelletier G. In vivo action of a new octadecaneuropeptide (ODN) antagonist on gonadotropin-releasing hormone gene expression in the male rat brain. *Neuroscience* 2004;125(2):411–5.
49. Masutomi N, Shibutani M, Takagi H, Uneyama C, Lee KY, Hirose M. Alteration of pituitary hormone-immunoreactive cell populations in rat offspring after maternal dietary exposure to endocrine-active chemicals. *Arch Toxicol* 2004;78(4):232–40.
50. Giros B, Caron MG. Molecular characterization of the dopamine transporter. *Trends Pharmacol Sci* 1993;14(2):43–9.
51. Miller GW, Gainetdinov RR, Levey AI, Caron MG. Dopamine transporters and neuronal injury. *Trends Pharmacol Sci* 1999;20(10):424–9.
52. Le Saux M, Di Paolo T. Influence of oestrogenic compounds on monoamine transporters in rat striatum. *J Neuroendocrinol* 2006;18(1):25–32.
53. Jourdain S, Morissette M, Morin N, Di Paolo T. Oestrogens prevent loss of dopamine transporter (DAT) and vesicular monoamine transporter (VMAT2) in substantia nigra of 1-methyl-4-phenyl-1,2,3,6-tetrahydropyridine mice. *J Neuroendocrinol* 2005;17(8):509–17.
54. Eriksson P, Ahlbom J, Fredriksson A. Exposure to DDT during a defined period in neonatal life induces permanent changes in brain muscarinic receptors and behaviour in adult mice. *Brain Res* 1992;582(2):277–81.
55. Masutomi N, Shibutani M, Takagi H, Uneyama C, Takahashi N, Hirose M. Impact of dietary exposure to methoxychlor, genistein, or diisononyl phthalate during the perinatal period on the development of the rat endocrine/reproductive systems in later life. *Toxicology* 2003;192(2–3):149–70.
56. Chapin RE, Harris MW, Davis BJ, Ward SM, Wilson RE, Mauney MA, Lockhart AC, Smialowicz RJ, Moser VC, Burka LT, Collins BJ. The effects of perinatal/juvenile methoxychlor exposure on adult rat nervous, immune, and reproductive system function. *Fundam Appl Toxicol* 1997;40(1):138–57.
57. Suzuki M, Lee HC, Chiba S, Yonezawa T, Nishihara M. Effects of methoxychlor exposure during perinatal period on reproductive function after maturation in rats. *J Reprod Dev* 2004;50(4):455–61.
58. Amstislavsky SY, Amstislavskaya TG, Eroschenko VP. Methoxychlor given in the periimplantation period blocks sexual arousal in male mice. *Reprod Toxicol* 1999;13(5):405–11.
59. Eroschenko VP, Amstislavsky SY, Schwabel H, Ingermann RL. Altered behaviors in male mice, male quail, and salamander larvae following early exposures to the estrogenic pesticide methoxychlor. *Neurotoxicol Teratol* 2002;24(1):29–36.
60. You L, Casanova M, Bartolucci EJ, Fryczynski MW, Dorman DC, Everitt JI, Gaido KW, Ross SM, Heck Hd H. Combined effects of dietary phytoestrogen and synthetic endocrine-active compound on reproductive development in Sprague-Dawley rats: genistein and methoxychlor. *Toxicol Sci* 2002;66(1):91–104.
61. Bulger WH, Kupfer D. Estrogenic action of DDT analogs. *Am J Ind Med* 1983;4(1–2):163–73.
62. Heinrichs WL, Gellert RJ, Bakke JL, Lawrence NL. DDT administered to neonatal rats induces persistent estrus syndrome. *Science* 1971;173(997):642–3.
63. Welch RM, Levin W, Conney AH. Estrogenic action of DDT and its analogs. *Toxicol Appl Pharmacol* 1969;14(2):358–67.
64. Parent AS, Rasier G, Gerard A, Heger S, Roth C, Mastronardi C, Jung H, Ojeda SR, Bourguignon JP. Early onset of puberty: tracking genetic and environmental factors. *Horm Res* 2005;64 (Suppl 2):41–7.
65. Laviola G, Gioiosa L, Adriani W, Palanza P. D-amphetamine-related reinforcing effects are reduced in mice exposed prenatally to estrogenic endocrine disruptors. *Brain Res Bull* 2005;65(3):235–40.
66. Flynn KM, Delclos KB, Newbold RR, Ferguson SA. Long term dietary methoxychlor exposure in rats increases sodium solution consumption but has few effects on other sexually dimorphic behaviors. *Food Chem Toxicol* 2005;43(9):1345–54.
67. vom Saal FS, Nagel SC, Palanza P, Boechler M, Parmigiani S, Welshons WV. Estrogenic pesticides: binding relative to estradiol in MCF-7 cells and effects of exposure during fetal life on subsequent territorial behaviour in male mice. *Toxicol Lett* 1995;77(1–3):343–50.
68. Palanza P, Morellini F, Parmigiani S, vom Saal FS. Ethological methods to study the effects of maternal exposure to estrogenic endocrine disrupters: a study with methoxychlor. *Neurotoxicol Teratol* 2002;24(1):55–69.

69. Palanza P, Parmigiani S, Liu H, vom Saal FS. Prenatal exposure to low doses of the estrogenic chemicals diethylstilbestrol and o,p′-DDT alters aggressive behavior of male and female house mice. *Pharmacol Biochem Behav* 1999;64(4):665–72.
70. Palanza P, Parmigiani S, vom Saal FS. Effects of prenatal exposure to low doses of diethylstilbestrol, o,p′DDT, and methoxychlor on postnatal growth and neurobehavioral development in male and female mice. *Horm Behav* 2001;40(2):252–65.
71. Engell MD, Godwin J, Young LJ, Vandenbergh JG. Perinatal exposure to endocrine disrupting compounds alters behavior and brain in the female pine vole. *Neurotoxicol Teratol* 2006;28(1): 103–10.
72. Ferguson SA, Scallet AC, Flynn KM, Meredith JM, Schwetz BA. Developmental neurotoxicity of endocrine disrupters: focus on estrogens. *Neurotoxicology* 2000;21(6):947–56.
73. Orikasa C, Kondo Y, Hayashi S, McEwen BS, Sakuma Y. Sexually dimorphic expression of estrogen receptor beta in the anteroventral periventricular nucleus of the rat preoptic area: implication in luteinizing hormone surge. *Proc Natl Acad Sci USA* 2002;99(5):3306–11.
74. Salama J, Chakraborty TR, Ng L, Gore AC. Effects of polychlorinated biphenyls on estrogen receptor-beta expression in the anteroventral periventricular nucleus. *Environ Health Perspect* 2003;111(10):1278–82.
75. Smith CL, Conneely OM, O'Malley BW. Modulation of the ligand-independent activation of the human estrogen receptor by hormone and antihormone. *Proc Natl Acad Sci USA* 1993;90(13): 6120–4.
76. Seegal RF, Brosch KO, Okoniewski RJ. Coplanar PCB congeners increase uterine weight and frontal cortical dopamine in the developing rat: implications for developmental neurotoxicity. *Toxicol Sci* 2005;86(1):125–31.
77. Hany J, Lilienthal H, Sarasin A, Roth-Harer A, Fastabend A, Dunemann L, Lichtensteiger W, Winneke G. Developmental exposure of rats to a reconstituted PCB mixture or aroclor 1254: effects on organ weights, aromatase activity, sex hormone levels, and sweet preference behavior. *Toxicol Appl Pharmacol* 1999;158(3):231–43.
78. Juarez de Ku LM, Sharma-Stokkermans M, Meserve LA. Thyroxine normalizes polychlorinated biphenyl (PCB) dose-related depression of choline acetyltransferase (ChAT) activity in hippocampus and basal forebrain of 15-day-old rats. *Toxicology* 1994;94(1–3):19–30.
79. Corey DA, Juarez de Ku LM, Bingman VP, Meserve LA. Effects of exposure to polychlorinated biphenyl (PCB) from conception on growth, and development of endocrine, neurochemical, and cognitive measures in 60 day old rats. *Growth Dev Aging* 1996;60(3–4):131–43.
80. Sapolsky RM, Krey LC, McEwen BS. The adrenocortical axis in the aged rat: impaired sensitivity to both fast and delayed feedback inhibition. *Neurobiol Aging* 1986;7(5):331–5.
81. Morse DC, Seegal RF, Borsch KO, Brouwer A. Long-term alterations in regional brain serotonin metabolism following maternal polychlorinated biphenyl exposure in the rat. *Neurotoxicology* 1996;17(3–4):631–8.
82. Roegge CS, Wang VC, Powers BE, Klintsova AY, Villareal S, Greenough WT, Schantz SL. Motor impairment in rats exposed to PCBs and methylmercury during early development. *Toxicol Sci* 2004;77(2):315–24.
83. Pruitt DL, Meserve LA, Bingman VP. Reduced growth of intra- and infra-pyramidal mossy fibers is produced by continuous exposure to polychlorinated biphenyl. *Toxicology* 1999;138(1):11–7.
84. Nguon K, Baxter MG, Sajdel-Sulkowska EM. Perinatal exposure to polychlorinated biphenyls differentially affects cerebellar development and motor functions in male and female rat neonates. *Cerebellum* 2005;4(2):112–22.
85. Lyche JL, Oskam IC, Skaare JU, Reksen O, Sweeney T, Dahl E, Farstad W, Ropstad E. Effects of gestational and lactational exposure to low doses of PCBs 126 and 153 on anterior pituitary and gonadal hormones and on puberty in female goats. *Reprod Toxicol* 2004;19(1):87–95.
86. Chung YW, Nunez AA, Clemens LG. Effects of neonatal polychlorinated biphenyl exposure on female sexual behavior. *Physiol Behav* 2001;74(3):363–70.
87. Chung YW, Clemens LG. Effects of perinatal exposure to polychlorinated biphenyls on development of female sexual behavior. *Bull Environ Contam Toxicol* 1999;62(6):664–70.
88. Steinberg R, Walker D, Gore A. *Prenatal PCB Exposure Results in Altered Development and Sexual Behaviors in Female Rats*. Endocrine Society Forum on Endocrine-Disrupting Chemicals Abst, San Diego, CA, 2005.
89. Kaya H, Hany J, Fastabend A, Roth-Harer A, Winneke G, Lilienthal H. Effects of maternal exposure to a reconstituted mixture of polychlorinated biphenyls on sex-dependent behaviors

and steroid hormone concentrations in rats: dose-response relationship. *Toxicol Appl Pharmacol* 2002;178(2):71–81.
90. Kuriyama SN, Chahoud I. In utero exposure to low-dose 2,3′,4,4′,5-pentachlorobiphenyl (PCB 118) impairs male fertility and alters neurobehavior in rat offspring. *Toxicology* 2004;202(3):185–97.
91. Overmann SR, Kostas J, Wilson LR, Shain W, Bush B. Neurobehavioral and somatic effects of perinatal PCB exposure in rats. *Environ Res* 1987;44(1):56–70.
92. Branchi I, Capone F, Vitalone A, Madia F, Santucci D, Alleva E, Costa LG. Early developmental exposure to BDE 99 or Aroclor 1254 affects neurobehavioural profile: interference from the administration route. *Neurotoxicology* 2005;26(2):183–92.
93. Roegge CS, Schantz SL. Motor function following developmental exposure to PCBS and/or MEHG. *Neurotoxicol Teratol* 2006;28(2):260–77.
94. Roegge CS, Morris JR, Villareal S, Wang VC, Powers BE, Klintsova AY, Greenough WT, Pessah IN, Schantz SL. Purkinje cell and cerebellar effects following developmental exposure to PCBs and/or MeHg. *Neurotoxicol Teratol* 2006;28(1):74–85.
95. Seegal RF, Schantz SL. *Neurochemical and Behavioral Sequelae of Exposure to Dioxins and PCBs. In Dioxins and Health.* New york: Plenum Press; 1994;409–447.
96. Weiss B. Sexually dimorphic nonreproductive behaviors as indicators of endocrine disruption. *Environ Health Perspect* 2002;110(Suppl 3):387–91.
97. Gore A. Modulation of the GnRH gene and onset of puberty. *Control of the Onset of Puberty.* Amsterdam: Elsevier; 2000;25–35.
98. Golub MS, Germann SL, Hogrefe CE. Endocrine disruption and cognitive function in adolescent female rhesus monkeys. *Neurotoxicol Teratol* 2004;26(6):799–809.
99. Golub MS. Adolescent health and the environment. *Environ Health Perspect* 2000;108(4):355–62.
100. Patisaul HB, Melby M, Whitten PL, Young LJ. Genistein affects ER beta- but not ER alpha-dependent gene expression in the hypothalamus. *Endocrinology* 2002;143(6):2189–97.
101. Moore TO, Karom M, O'Farrell L. The neurobehavioral effects of phytoestrogens in male Syrian hamsters. *Brain Res* 2004;1016(1):102–10.
102. Gray LE Jr, Ostby J, Ferrell J, Rehnberg G, Linder R, Cooper R, Goldman J, Slott V, Laskey J. A dose-response analysis of methoxychlor-induced alterations of reproductive development and function in the rat. *Fundam Appl Toxicol* 1989;12(1):92–108.
103. Laws SC, Carey SA, Ferrell JM, Bodman GJ, Cooper RL. Estrogenic activity of octylphenol, nonylphenol, bisphenol A and methoxychlor in rats. *Toxicol Sci* 2000;54(1):154–67.
104. Cooper RL, Chadwick RW, Rehnberg GL, Goldman JM, Booth KC, Hein JF, McElroy WK. Effect of lindane on hormonal control of reproductive function in the female rat. *Toxicol Appl Pharmacol* 1989;99(3):384–94.
105. Gray LE Jr, Ostby JS, Ferrell JM, Sigmon ER, Goldman JM. Methoxychlor induces estrogen-like alterations of behavior and the reproductive tract in the female rat and hamster: effects on sex behavior, running wheel activity, and uterine morphology. *Toxicol Appl Pharmacol* 1988;96(3): 525–40.
106. Rasier G, Toppari J, Parent AS, Bourguignon JP. Female sexual maturation and reproduction after prepubertal exposure to estrogens and endocrine disrupting chemicals: a review of rodent and human data. *Mol Cell Endocrinol* 2006;254–255:187–201.
107. Gray LE Jr, Ostby J, Cooper RL, Kelce WR. The estrogenic and antiandrogenic pesticide methoxychlor alters the reproductive tract and behavior without affecting pituitary size or LH and prolactin secretion in male rats. *Toxicol Ind Health* 1999;15(1–2):37–47.
108. Goldman JM, Cooper RL, Rehnberg GL, Hein JF, McElroy WK, Gray LE Jr. Effects of low subchronic doses of methoxychlor on the rat hypothalamic-pituitary reproductive axis. *Toxicol Appl Pharmacol* 1986;86(3):474–83.
109. Gupta A, Agarwal R, Shukla GS. Functional impairment of blood-brain barrier following pesticide exposure during early development in rats. *Hum Exp Toxicol* 1999;18(3):174–9.
110. Malkiewicz K, Mohammed R, Folkesson R, Winblad B, Szutowski M, Benedikz E. Polychlorinated biphenyls alter expression of alpha-synuclein, synaptophysin and parkin in the rat brain. *Toxicol Lett* 2006;161(2):152–8.
111. Abeliovich A, Beal MF. Parkinsonism genes: culprits and clues. *J Neurochem* 2006;99(4):1062–72.
112. Maffucci J, Gore A. Age-related changes in hormones and their receptors in animal models of female reproductive senescence. *Handbook of Models for Human Aging*. Amsterdam Academic Press/ Elsevier; 2006.

113. Perz JM. Development of the menopause symptom list: a factor analytic study of menopause associated symptoms. *Women Health* 1997;25(1):53–69.
114. McEwen BS. Invited review: estrogens effects on the brain: multiple sites and molecular mechanisms. *J Appl Physiol* 2001;91(6):2785–801.
115. Patisaul HB, Dindo M, Whitten PL, Young LJ. Soy isoflavone supplements antagonize reproductive behavior and estrogen receptor alpha- and beta-dependent gene expression in the brain. *Endocrinology* 2001;142(7):2946–52.
116. Patisaul HB, Whitten PL, Young LJ. Regulation of estrogen receptor beta mRNA in the brain: opposite effects of 17beta-estradiol and the phytoestrogen, coumestrol. *Brain Res Mol Brain Res* 1999;67(1):165–71.
117. Jacob DA, Temple JL, Patisaul HB, Young LJ, Rissman EF. Coumestrol antagonizes neuroendocrine actions of estrogen via the estrogen receptor alpha. *Exp Biol Med (Maywood)* 2001;226(4):301–6.
118. Lund TD, Rhees RW, Setchell KD, Lephart ED. Altered sexually dimorphic nucleus of the preoptic area (SDN-POA) volume in adult Long-Evans rats by dietary soy phytoestrogens. *Brain Res* 2001;914(1–2):92–9.
119. Linford NJ, Dorsa DM. 17beta-estradiol and the phytoestrogen genistein attenuate neuronal apoptosis induced by the endoplasmic reticulum calcium-ATPase inhibitor thapsigargin. *Steroids* 2002;67(13–14):1029–40.
120. McGarvey C, Cates PA, Brooks A, Swanson IA, Milligan SR, Coen CW, O'Byrne KT. Phytoestrogens and gonadotropin-releasing hormone pulse generator activity and pituitary luteinizing hormone release in the rat. *Endocrinology* 2001;142(3):1202–8.
121. Bu LH, Lephart ED. Effects of dietary phytoestrogens on core body temperature during the estrous cycle and pregnancy. *Brain Res Bull* 2005;65(3):219–3.
122. Patisaul HB, Luskin JR, Wilson ME. A soy supplement and tamoxifen inhibit sexual behavior in female rats. *Horm Behav* 2004;45(4):270–7.
123. Hartley DE, Edwards JE, Spiller CE, Alom N, Tucci S, Seth P, Forsling ML, File SE. The soya isoflavone content of rat diet can increase anxiety and stress hormone release in the male rat. *Psychopharmacology (Berl)* 2003;167(1):46–53.
124. Simon NG, Kaplan JR, Hu S, Register TC, Adams MR. Increased aggressive behavior and decreased affiliative behavior in adult male monkeys after long-term consumption of diets rich in soy protein and isoflavones. *Horm Behav* 2004;45(4):278–84.
125. Szyf M, Weaver IC, Champagne FA, Diorio J, Meaney MJ. Maternal programming of steroid receptor expression and phenotype through DNA methylation in the rat. *Front Neuroendocrinol* 2005;26(3–4):139–62.
126. Flynn KM, Ferguson SA, Delclos KB, Newbold RR. Multigenerational exposure to dietary genistein has no severe effects on nursing behavior in rats. *Neurotoxicology* 2000;21(6):997–1001.
127. Lee YB, Lee HJ, Won MH, Hwang IK, Kang TC, Lee JY, Nam SY, Kim KS, Kim E, Cheon SH, Sohn HS. Soy isoflavones improve spatial delayed matching-to-place performance and reduce cholinergic neuron loss in elderly male rats. *J Nutr* 2004;134(7):1827–31.
128. Ho KP, Li L, Zhao L, Qian ZM. Genistein protects primary cortical neurons from iron-induced lipid peroxidation. *Mol Cell Biochem* 2003;247(1–2):219–2.
129. Sonee M, Sum T, Wang C, Mukherjee SK. The soy isoflavone, genistein, protects human cortical neuronal cells from oxidative stress. *Neurotoxicology* 2004;25(5):885–91.
130. Zhao L, Chen Q, Diaz Brinton R. Neuroprotective and neurotrophic efficacy of phytoestrogens in cultured hippocampal neurons. *Exp Biol Med (Maywood)* 2002;227(7):509–19.
131. Gao ZB, Hu GY. Trans-resveratrol, a red wine ingredient, inhibits voltage-activated potassium currents in rat hippocampal neurons. *Brain Res* 2005;1056(1):68–75.
132. Kasdallah-Grissa A, Mornagui B, Aouani E, Hammami M, Gharbi N, Kamoun A, El-Fazaa S. Protective effect of resveratrol on ethanol-induced lipid peroxidation in rats. *Alcohol Alcohol* 2006;41(3):236–9.
133. Marambaud P, Zhao H, Davies P. Resveratrol promotes clearance of Alzheimer's disease amyloid-beta peptides. *J Biol Chem* 2005;280(45):37377–82.
134. Yanez M, Fraiz N, Cano E, Orallo F. Inhibitory effects of cis- and trans-resveratrol on noradrenaline and 5-hydroxytryptamine uptake and on monoamine oxidase activity. *Biochem Biophys Res Commun* 2006;344(2):688–95.
135. Choi EJ, Lee BH. Evidence for genistein mediated cytotoxicity and apoptosis in rat brain. *Life Sci* 2004;75(4):499–509.

136. Lee YB, Lee HJ, Sohn HS. Soy isoflavones and cognitive function. *J Nutr Biochem* 2005; 16(11):641–9.
137. Lund TD, West TW, Tian LY, Bu LH, Simmons DL, Setchell KD, Adlercreutz H, Lephart ED. Visual spatial memory is enhanced in female rats (but inhibited in males) by dietary soy phytoestrogens. *BMC Neurosci* 2001;2:20.
138. Heo HJ, Suh YM, Kim MJ, Choi SJ, Mun NS, Kim HK, Kim E, Kim CJ, Cho HY, Kim YJ, Shin DH. Daidzein activates choline acetyltransferase from MC-IXC cells and improves drug-induced amnesia. *Biosci Biotechnol Biochem* 2006;70(1):107–11.
139. Mussi P, Ciana P, Raviscioni M, Villa R, Regondi S, Agradi E, Maggi A, Di Lorenzo D. Activation of brain estrogen receptors in mice lactating from mothers exposed to DDT. *Brain Res Bull* 2005;65(3):241–7.
140. Goldman JM, Murr AS, Buckalew AR, Schmid JE, Abbott BD. Methoxychlor-induced alterations in the histological expression of angiogenic factors in pituitary and uterus. *J Mol Histol* 2004;35(4):363–75.
141. Kirby ML, Barlow RL, Bloomquist JR. Neurotoxicity of the organochlorine insecticide heptachlor to murine striatal dopaminergic pathways. *Toxicol Sci* 2001;61(1):100–6.
142. Schuh RA, Kristian T, Gupta RK, Flaws JA, Fiskum G. Methoxychlor inhibits brain mitochondrial respiration and increases hydrogen peroxide production and CREB phosphorylation. *Toxicol Sci* 2005;88(2):495–504.
143. Kanthasamy AG, Kitazawa M, Kanthasamy A, Anantharam V. Dieldrin-induced neurotoxicity: relevance to Parkinson's disease pathogenesis. *Neurotoxicology* 2005;26(4):701–19.
144. Seegal RF, Bush B, Brosch KO. Comparison of effects of Aroclors 1016 and 1260 on non-human primate catecholamine function. *Toxicology* 1991;66(2):145–63.
145. Seegal RF, Bush B, Brosch KO. Sub-chronic exposure of the adult rat to Aroclor 1254 yields regionally-specific changes in central dopaminergic function. *Neurotoxicology* 1991;12(1):55–65.
146. Seegal RF, Bush B, Brosch KO. Decreases in dopamine concentrations in adult, non-human primate brain persist following removal from polychlorinated biphenyls. *Toxicology* 1994;86(1–2):71–87.
147. Seegal RF. The neurochemical effects of PCB exposure are age-dependent. *Arch Toxicol Suppl* 1994;16:128–37.
148. Seegal RF, Brosch KO, Bush B. Regional alterations in serotonin metabolism induced by oral exposure of rats to polychlorinated biphenyls. *Neurotoxicology* 1986;7(1):155–65.
149. Seegal RF, Bush B, Brosch KO. Polychlorinated biphenyls induce regional changes in brain norepinephrine concentrations in adult rats. *Neurotoxicology* 1985;6(3):13–23.
150. Muthuvel R, Venkataraman P, Krishnamoorthy G, Gunadharini DN, Kanagaraj P, Jone Stanley A, Srinivasan N, Balasubramanian K, Aruldhas MM, Arunakaran J. Antioxidant effect of ascorbic acid on PCB (Aroclor 1254) induced oxidative stress in hypothalamus of albino rats. *Clin Chim Acta* 2006;365(1–2):297–303.
151. Brezner E, Terkel J, Perry AS. The effect of Aroclor 1254 (PCB) on the physiology of reproduction in the female rat–I. *Comp Biochem Physiol C* 1984;77(1):65–70.
152. Cummings JA, Nunez AA, Clemens LG. A cross-fostering analysis of the effects of PCB 77 on the maternal behavior of rats. *Physiol Behav* 2005;85(2):83–91.
153. Simmons SL, Cummings JA, Clemens LG, Nunez AA. Exposure to PCB 77 affects the maternal behavior of rats. *Physiol Behav* 2005;84(1):81–6.
154. Faroon O, Jones D, de Rosa C. Effects of polychlorinated biphenyls on the nervous system. *Toxicol Ind Health* 2001;16(7–8):305–3.
155. Altenkirch H, Stoltenburg G, Haller D, Hopmann D, Walter G. Clinical data on three cases of occupationally induced PCB-intoxication. *Neurotoxicology* 1996;17(3–4):639–43.

5 Heavy Metals as Endocrine-Disrupting Chemicals

Cheryl A. Dyer, PhD

CONTENTS

1. INTRODUCTION

Heavy metals are present in our environment as they formed during the earth's birth. Their increased dispersal is a function of their usefulness during our growing dependence on industrial modification and manipulation of our environment *(1,2)*. There is no consensus chemical definition of a heavy metal. Within the periodic table, they comprise a block of all the metals in Groups 3–16 that are in periods 4 and greater. These elements acquired the name heavy metals because they all have high densities, >5 g/cm^3 *(2)*. Their role as putative endocrine-disrupting chemicals is due to their chemistry and not their density. Their popular use in our industrial world is due to their physical, chemical, or in the case of uranium, radioactive properties. Because of the reactivity of heavy metals, small or trace amounts of elements such as iron, copper, manganese, and zinc are important in biologic processes, but at higher concentrations they often are toxic.

Previous studies have demonstrated that some organic molecules, predominantly those containing phenolic or ring structures, may exhibit estrogenic mimicry through actions on the estrogen receptor. These xenoestrogens typically are non-steroidal organic chemicals released into the environment through agricultural spraying, industrial activities, urban waste and/or consumer products that include organochlorine pesticides, polychlorinated biphenyls, bisphenol A, phthalates, alkylphenols, and parabens *(1)*. This definition of xenoestrogens needs to be extended, as recent investigations have yielded the paradoxical observation that heavy metals mimic the biologic

From: *Endocrine-Disrupting Chemicals: From Basic Research to Clinical Practice*
Edited by: A. C. Gore © Humana Press Inc., Totowa, NJ

activity of steroid hormones, including androgens, estrogens, and glucocorticoids. Early studies demonstrated that inorganic metals bind the estrogen receptor. Zn(II), Ni(II), and Co(II) bind the estrogen receptor, most likely in the steroid-binding domain, but in this study neither Fe(II) nor Cd(II) bound the receptor *(3)*. Certain metals bind the zinc fingers of the estrogen receptor and could alter the receptor's interaction with DNA *(4)*. Several metals can displace or compete with estradiol binding to its receptor in human Michigan Cancer Foundation MCF-7 breast cancer cells *(5–7)*. Recently, cadmium has been shown to act like estrogen *in vivo* affecting estrogen-responsive tissues such as uterus and mammary glands *(8)*. Metals that mimic estrogen are called metalloestrogens *(9,10)*.

Five heavy metals have been sufficiently investigated to provide insight into the means of their impact on mammalian reproductive systems. Arsenic, a metalloid and borderline heavy metal, is included because it is often found in the earth associated with other heavy metals, such as uranium. Additional heavy metals to be discussed are cadmium, lead, mercury, and uranium—the heaviest naturally occurring element. In this chapter, for each heavy metal, descriptions will be provided for the environmental exposure, history of its use, and thus potential for increased dispersal in our environment, targeted reproductive organs, and specific effects or means of action, usually as a function of low versus high concentration. An important tenet is that earlier (developmental) ages of exposure increase the impact of the endocrine-disrupting chemical or heavy metal on the normal development of reproductive organs, which may be permanently affected. Thus, where known, I will describe the direct action of a heavy metal on a growing embryo, perhaps through epigenetic changes, to set the stage for increased chance of disease later in life when the individual is challenged by another environmental insult. In the case of uranium, I will describe my laboratory's research that supports the conclusion that uranium is a potent estrogen mimic at concentrations at or below the United States Environmental Protection Agency (USEPA) safe drinking water level.

2. ARSENIC

The abundance of arsenic (As) in the Earth's crust is 1.5–3.0 mg/kg, making it the 20th most abundant element in the earth's crust *(11)*. Arsenic has been in use by man for thousands of years. It is infamous as a favored form of intentional poisoning and famous for being developed by Paul Erlich as the first drug to cure syphilis *(12)*. Today, arsenic is used in semiconductor manufacture and pesticides *(13)*. It serves as a wood preservative in chromated copper arsenate (CCA). CCA-treated lumber products are being removed voluntarily from consumer use as of 2002 and were banned as of January 1, 2004. CCA-treated lumber is a potential risk of exposure of children to arsenic in play-structures *(14)*. Another source of environmental arsenic is from glass and copper smelters, coal combustion, and uranium mining. The most extensive environmental exposure is in drinking water. For instance, since the 1980s, the provision of arsenic-contaminated Artesian well water in Bangladesh has exposed an estimated 50–75 million people to very high levels of arsenic *(11)*. Given the latency of 30–50 years for arsenic-related carcinogenesis, epidemiological data on arsenic-induced cancers including skin, lung, urinary bladder, kidney, and liver are only now becoming available *(15)*.

Inorganic arsenic in the forms +3 (arsenite) or +5 charges are the most often encountered forms of arsenic and are most readily absorbed from the gastrointestinal tract; therefore, these forms cause the greatest number of health problems. A new USEPA limit of arsenic standard for drinking water has recently gone into effect, lowering the limit from 50 to 10 μg/L. Compliance of water systems with this standard became enforceable as of January 23, 2006 *(16)*. However, achieving this limit will be problematic for many smaller water municipalities because of the expense of installing equipment to reduce arsenic to <10 μg/L *(17)*.

2.1. Arsenic as an Endocrine-Disrupting Chemical in Reproductive Systems

Arsenic-mediated endocrine disruption has been reported in research animals and potentially in humans. For instance, adult rats that consume drinking water with arsenite at 5 mg/kg of body weight per day 6 days a week for 4 weeks have reproductive tract abnormalities such as suppression of gonadotrophins and testicular androgen, and germ cell degeneration—all effects similar to those induced by estrogen agonists *(18)*. In this study, it was concluded that arsenite may exhibit estrogenic activity. However, there was no evidence presented to indicate estrogen receptor specificity by demonstrating that an antiestrogen such as ICI 182,780 prevented the arsenic-induced changes. Thus, the degenerative problems could have resulted from arsenic chemical toxicity. Similar to this study are those conducted by Waalkes' research group. In mice that were injected with sodium arsenate at 0.5 mg/kg i.v. once a week for 20 weeks, males had testicular interstitial cell hyperplasia and tubular degeneration that probably resulted from the interstitial cell hyperplasia *(19)*. Arsenate injections in female mice caused cystic hyperplasia of the uterus, which is often related to abnormally high, prolonged estrogenic stimulation. Again, as these changes were unexpected, there was no attempt to determine the dependence on the estrogen receptor by using an antiestrogen to block the changes in the male and female reproductive tissues *(19)*. This same research group went onto to show that *in utero* exposure to arsenic leads to changes in the male and female offspring that indicate they have been exposed to an estrogenic influence *(20)*. In addition, *in utero* arsenic-exposed mice are much more prone to urogenital carcinogenesis, urinary bladder, and liver carcinogenesis when they are exposed postnatally to the potent synthetic estrogen, diethylstilbestrol (DES) or tamoxifen *(21,22)*. The altered estrogen signaling may cause over expression of estrogen receptor-α through promoter region hypomethylation, suggesting an epigenetic change was caused by *in utero* As exposure *(23)*. Together, the *in vivo* data support the hypothesis that arsenic can produce estrogenic-like effects by direct or indirect stimulation of estrogen receptor-α. The levels of As used in the *in vivo* studies are high, similar to the high levels in drinking water in Bangladesh, on average in the 0.1–1 mM range, and thus, these studies are environmentally relevant for people living with one of the worst scenarios of As environmental contamination. Arsenic levels at 0.4 ppm/day, 40 times more than current USEPA safe drinking water level, when given daily in drinking water to rats results in reduced gonadotrophins, plasma estradiol, and decreased activities of these steroidogenic enzymes, 3β hydroxysteroid dehydrogenase (HSD), and 17β HSD *(24)*. At the same time there was no change in body weight, but ovarian, uterine, and vaginal weights were significantly reduced, suggesting that As treatment caused organ toxicity but not general toxicity. For a full description of inorganic arsenic-mediated reproductive toxicity in animals and human see Golub and Macintosh *(25)*.

2.2. Relationship Between Arsenic and Diabetes

In those parts of the world with the most elevated levels of environmental As in drinking water, there is a proposed relationship to type 2 diabetes, as arsenic may cause insulin resistance and impaired pancreatic β-cell functions including insulin synthesis and secretion *(26)*. Blackfoot disease, which is associated with drinking As-contaminated drinking water, is endemic in southwestern Taiwan and also associated with the increased prevalence of type 2 diabetes *(27)*. Type 2 diabetes compromises fertility *(28)*, making As a potential endocrine-disrupting chemical on both the diabetes and reproductive systems. Another mechanism of heavy metals and As is through formation of reactive oxygen and nitrogen species that cause non-specific damage such as oxidative damage to DNA and lipid peroxidation that can contribute to reproductive problems *(29)*. For instance, there are low birth weight infants, more spontaneous abortions, and congenital malformations in female employees and women living close to copper smelters as reported in Sweden and Bulgaria *(30,31)*. However, this mechanism of As action is certainly due to its chemical toxicity rather than its mimicry of endocrine agents such as estrogen.

2.3. Mechanisms of Arsenic Actions on Endocrine Systems

There are limited *in vitro* based studies of the putative estrogenic activity of As. In MCF-7 breast cancer cells, which are often used to assess estrogenic activity of endocrine-disrupting chemicals *(32)*. In these cells, arsenite at low micromolar concentrations stimulated increased proliferation, steady state levels of progesterone receptor, pS2, and decreased estrogen receptor-α mRNA expression *(33)*. The antiestrogen ICI 182,780 or fluvestrant blocked the effects of arsenite indicating the dependence on the estrogen receptor. In addition, by using binding assays and receptor activation assays, it was determined that As interacts with the hormone-binding domain of the estrogen receptor *(33)*. Another group tested the estrogenicity of several heavy metals and arsenite treatment stimulated MCF-7 cell growth but relative to other metals was not very potent *(34)*. In contrast, arsenic trioxide, an approved treatment of acute promyelocytic leukemia, blocks MCF-7 cell proliferation without binding the ligand-binding domain of the estrogen receptor but does interfere with estrogen receptor-signaling pathway indicating that the chemical state of As is key in determining its biologic activity *(35)*.

Arsenite binds to the Zinc (Zn) finger region of the estrogen-binding region of estrogen receptor-α, and the binding affinity is influenced by the amino acid length between two cysteines *(36,37)*. However, these investigations are strictly cell-free assays; so, it is difficult to extrapolate to whole cell responses. Finally, arsenite at 100 μM binds the glucocorticoid receptor in the steroid-binding domain but does not compete for binding to progesterone, androgen, or estrogen receptors in MCF-7 cells at this high concentration *(38)*. On the contrary, arsenite from 0.3 to 3.3 μM, a non-toxic dose, interacts with the glucocorticoid receptor in human breast cancer cells and rat hepatoma cells to inhibit glucocorticoid receptor-mediated gene transcription *(39–41)*. In addition, glucocorticoid receptor binding of its ligand dexamethasone is blocked by low micromolar concentrations of arsenite but not arsenate. Arsenite interacts with the vicinal dithiols of the glucocorticoid receptor as is the case with its interaction with the estrogen receptor *(42,43)*. Arsenite at low micromolar concentrations binds to the estrogen receptor and glucocorticoid receptor to alter gene expression in rat and human cells. At concentrations >100 μM arsenic may act through chemical toxicity to non-specifically damage DNA or proteins through reactive oxidative species *(29)*. As a whole, these studies suggest influence of As on the stress neuroendocrine system.

In sum, there is suggestive evidence from *in vivo* studies that As may have estrogenic activity. Nevertheless, further proof that antiestrogens may block the responses elicited by As would allow a stronger connection between As and putative estrogenic activity to be drawn. Moreover, there could be indirect endocrine effects of As because of its causing insulin resistance and reducing insulin levels leading to type 2 diabetes that potentially would compromise reproductive tissue responses. The evidence in MCF-7 cell E-Screen bioassays strongly supports the conclusion that As can bind the estrogen receptor in the ligand-binding domain to activate the receptor and exert downstream signaling events that are blocked by the antiestrogen ICI 182,780. In addition, there is strong evidence to support the conclusion that arsenite binds the glucocorticoid receptor to either activate and/or inhibit gene transcription. Thus it appears that at low concentrations (<10 μM) there are observations of specific interaction with steroid receptors whereas at higher concentrations (>100 μM), As reactive chemistry prevails and non-specific interactions with DNA and protein causes toxicity and leads to cell death.

3. CADMIUM

Cadmium (Cd) is dispersed through out the environment primarily from mining, smelting, electroplating, and it is found in consumer products such as nickel/Cd batteries, pigments (Cd yellow) and plastics *(13)*. Tobacco smoke is one of the most common sources of Cd exposure because the tobacco plant concentrates Cd *(13)*. Smoking one pack of cigarettes a day results in a dose of about 1 mg Cd/year *(13)*. Cadmium is very slowly excreted from the body so it accumulates with time. Of all the heavy metals the most data has been collected both regarding Cd's biologic activity as well as in support of its being an endocrine-disrupting chemical *(44)*.

3.1. Cadmium Effects on Pregnancy and the Fetus

The greatest environmental Cd exposure is in the Jinzu River basin in Japan because of an effluent from an upstream mine. Maternal exposure to high levels of Cd has led to a significant increase in premature delivery *(45)*. This has led to investigation of the possible mechanisms for Cd-induced premature delivery, possibly by compromising placental function. There are enhanced concentrations of Cd in follicular fluid and placentae of smokers that are correlated with lower progesterone *(46,47)*. Cd at high concentrations inhibits placental progesterone synthesis and expression of the low-density lipoprotein receptor that is needed to bring cholesterol substrate into the cells *(48,49)*. Detailed analysis of the Cd-mediated reduction in progesterone production by cultured human trophoblast cells indicated that the decrease is not due to cell death or apoptosis. Rather, there is a specific block of P450 side chain cleavage expression and activity. This was shown by blocking P450 side chain cleavage activity with aminoglutethimide and adding pregnenolone, which was converted to progesterone by the unaffected activity of 3β HSD *(50)*.

Placental 11β HSD activity is critical to protect the fetus from maternal cortisol, which suppresses fetal growth, by converting it to inactive cortisone. Mutation or reduced expression of 11β HSD is associated with fetal growth restriction and is a significant risk factor for obesity, type 2 diabetes, and cardiovascular disease later in life *(51,52)*. There is an inverse relationship between birth weight and number of cigarettes smoked per day *(53)*. Cd accumulates in the placenta so significant amounts do not reach the growing fetus *(54)*. Of the thousands of toxic chemicals in cigarette smoke, Cd is one that has been linked to placental deficiencies *(54)*. A recent report describes that Cd at $<1\ \mu M$ reduces 11β HSD type 2 activity and expression in cultured human trophoblast cells *(50)*. Cd's effect was unique because it was not mimicked by other metal divalent cations such as Zn, Mg, or Mn *(51)*. Cadmium may downregulate 11β HSD by mimicking the ability of estrogen to attenuate the expression of this placental enzyme *(55)*. Thus, Cd environmental exposure in cigarette smoke, either first or second hand, could contribute to risk of major diseases later in life, particularly for the low birth weight fetus that was not protected from maternal cortisol.

The detrimental actions of Cd are seen at concentrations $>5\ \mu M$. For instance, in human granulosa cells collected during *in vitro* fertilization (IVF) procedures, Cd $> 16\ \mu M$ inhibited progesterone production *(56)*. However, at concentrations $<5\ \mu M$, Cd stimulates transcription of P450 side chain cleavage in porcine granulosa cells that results in greater progesterone production *(57)*. Cadmium may act to stimulate gene transcription by its high-affinity displacement of calcium from its binding to calmodulin and activation of protein kinase-C and second messenger pathways *(58)*. P450 side chain cleavage is the rate-limiting step for steroidogenesis. Thus, Cd's ability to either stimulate or suppress this enzyme could have a profound impact in all steroidogenic tissues.

In primary ovarian cell cultures from either cycling or pregnant rats, or human placental tissue, Cd at concentrations $>100\ \mu M$ suppressed progesterone and testosterone production *(59,60)*. In addition, *in vivo* Cd-treated rat ovaries exhibited suppressed progesterone, testosterone, and estradiol production in culture *(61)*. All these experiments used Cd concentrations that probably induced toxicity through one or more of numerous mechanisms such as inhibition of DNA repair, decreased antioxidants, activated signal transduction, or cell damage *(62)* rather than acting through a specific receptor or mechanism to inhibit steroidogenesis.

3.2. Cadmium and Testicular Toxicity

There are hundreds of articles describing toxic effects of Cd on the testes, as first reported in 1919 with the finding that testicular necrosis was induced by Cd *(63)*. As in the female, there is a causal relationship between cigarette Cd exposure and impaired male fertility *(64,65)*. In research models, such as rat Leydig cells, Cd is toxic to steroidogenesis but at concentrations $>10\ \mu M$ that coincide with cell death *(66)*. However, in another study, also using rat Leydig cells, $100\ \mu M$ Cd treatment doubled testosterone production with no change in cell viability *(67)*. Consistent with the *in vitro* observation of increased testosterone in the presence of Cd, chronic Cd oral exposure increased plasma testosterone in rats *(68)*. The increase in plasma testosterone was not evident until after more than 1-month exposure to Cd in the drinking water. At the same time, there was an increase in testicular weight *(68)*. In contrast, Cd given by subcutaneous injection to adult rats caused a decrease in plasma testosterone *(69)*.

Discrepancies between these studies suggest that the route of exposure to Cd affects whether it stimulates or inhibits testicular androgen production. Human Cd exposure through ingestion or occupationally also is associated with increased testosterone and estradiol *(70,71)*. Even postmenopausal women demonstrate a correlation between urinary cadmium and significantly elevated serum testosterone *(72)*. The mechanism for Cd-induced increase in human testosterone is unknown.

3.3. Cadmium as a Metalloestrogen

One of the most important studies indicating that Cd is an estrogen mimic, published in 2003 *(8)*, showed that female rats injected with Cd experienced earlier puberty onset, increased uterine weight, and enhanced mammary development. Cadmium treatment induced estrogen-regulated genes such as progesterone receptor and complement component C3. It also promoted mammary gland development with an increase in the formation of side branches and alveolar buds. In utero exposure of female offspring resulted in their reaching puberty earlier and an increase in epithelial area and number of terminal end buds in the mammary glands. Importantly the effect of Cd on uterine weight, mammary gland density, and progesterone receptor expression in uterus and mammary gland was blocked by coadministration of the antiestrogen ICI 182,780 *(8)*. Thus far, this *in vivo* study showing the reversibility of these Cd-induced effects by an antiestrogen is the most robust in supporting the conclusion that Cd is an endocrine-disrupting chemical and a putative metalloestrogen.

Evidence for Cd interaction with the estrogen receptor is the best characterized of all the heavy metals. Cd-treated MCF-7 human breast cancer cells demonstrate many responses to Cd that are the same as those elicited by estrogen. Cadmium treatment stimulates MCF-7 cell growth, downregulates the estrogen receptor, stimulates the expression of the progesterone receptor, and stimulates estrogen response element in transient transfection experiments *(73)*. In these studies, Zn treatment did not elicit these cellular responses demonstrating that Cd's effect was specific and not due to general effects of heavy metals. The specific nature of Cd's interaction with the estrogen receptor was examined in further detail *(74)*. Low concentrations of Cd activate the estrogen receptor-α by interacting non-competitively with the hormone-binding domain to block the binding of estradiol. It is notable that the ability of Cd to block estradiol binding occurs over 8 logs of concentration from 10^{-13} to 10^{-5} M but Zn at 10^{-5} M did not compete. Within the binding domain, the specific amino acids engaged by Cd are cysteines, glutamic acid, and histidine. These residues, particularly the cysteines, react with As through dithiol coordination suggesting that As and Cd share similar chemistry in interacting with the estrogen receptor. The same research group demonstrated that Cd at environmentally relevant concentrations also binds to the androgen receptor in human prostate cancer cells, LNCaP, to activate the receptor and stimulate cell growth *(75)*. As the same heavy metal Cd binds both the estrogen receptor and the androgen receptor, and in many tissues in the reproductive system expresses both types of receptors, it presents the scenario where the same metal exposure could lead to different responses depending on the relative localization and activation of the two steroid receptors in various tissues.

There are additional reports of Cd stimulating MCF-7 breast cancer cell gene transcription and increased cell growth. For instance, Cd-stimulated proliferation of MCF-7 cells is blocked by melatonin, the pineal gland indole hormone *(76)*. Cd treatment significantly activated both estrogen receptor-α and estrogen receptor-β, with a greater effect on the estrogen receptor-α. Additionally, Cd activated the transcription factor AP-1 through estrogen receptor-α similar to the response caused by estrogens *(77,78)*. To aid identification of estrogen mimetics the cell line, T47D-KBluc, derived from a human breast cancer cell line, has been genetically modified to be a specific, sensitive estrogen-responsive gene expression assay *(79)*. Cd treatment of these cells induced gene expression as indicated by reporter gene luciferase-mediated light generation. At concentrations as low as 0.01×10^{-9} M, Cd induced a significant increase in luciferase gene expression that was completely blocked by the antiestrogen ICI 182,780 *(78)*. Cd induces at least two types of genes: (i) genes for cytoprotective proteins, i.e. metallothioneins, heat-shock proteins and Zn transporter proteins and (ii) early proto-oncogenes related to cell proliferation, i.e. *c-fos (79)*. The first type of genes are induced by Cd at 10–30 μM whereas the stimulation of cell-proliferation related genes occurs at 0.1 μM leading to mitogen-activated protein kinase (MAPK) cascade activation *(80)*. But there is a fly in the Cd ointment. Recently, it was reported that Cd is neither estrogenic, as it does not induce increased MCF-7 cell proliferation, nor does it induce phosphorylation of MAPK *(81)*. Cd was able to interact with the estrogen receptor to prevent estrogen from binding, but these investigators did not observe Cd-mediated increased transcriptional activation as was previously reported by Stoica et al. *(74)*. Therefore, further investigation is needed to clarify the interaction of Cd with the estrogen receptor and downstream consequences.

4. LEAD

Lead (Pb) is a ubiquitous environmental contaminant. In the 1940s, dietary intake of Pb was approximately 500 μg/day in the US population, but now, that intake is <20 μg/day as a result of removing or reducing the primary sources of Pb: leaded-gasoline, lead-based paints, lead-soldered food cans, and lead plumbing pipes *(13)*. Thanks, in particular to the ban on leaded gasoline in 1979, the US population Pb blood level dropped precipitously from 13 μg/dL in the 1980s to <5 μg/dL *(82)*. Lead is similar to calcium in its disposition in the body. Its half-life in the blood is 1–2 months, but depending on exposure, it can accumulate in bone where its half-life is 20–30 years *(13)*. Lead-based paint remains the most common source of Pb exposure for children <6 years old. However, acute lead poisoning as well as chronic low-level Pb exposure can come from handling and/or swallowing metallic charms *(83)*. Pb poisoning is most dangerous to children as it causes mental impairment. There is a 2–4 point IQ deficit for each μg/dL increase in blood Pb within the range of 5–35 μg/dL. Thus, the CDC has set blood lead concentrations of 10 μg/dL or greater to indicate excessive absorption in children and triggers the need for environmental assessment and remediation *(13)*. Recent data suggest that even Pb < 10 μg/dL is associated with impaired intellectual performance in children *(84)*.

4.1. Lead as an Endocrine-Disrupting Chemical in Humans and Animals

Of the five heavy metals discussed here, Pb has the strongest evidence to connect its exposure to endocrine disruption in human populations. Three independent studies indicate that environmental exposure to Pb leads to delay in growth and pubertal development in girls. The first study found that a blood Pb concentration of 3 μg/dL was associated with delayed puberty after adjustment for body size and other confounders *(85)*. The second study found a similar relationship between blood Pb concentration and delayed attainment of menarche even after adjusting for race/ethnicity, age, family size, residence in metropolitan area, poverty income ratio, and body mass index *(86)*. The most recent study associated blood Pb with later menarche controlling for other toxicants, age, and socioeconomic status in Akwesasne Mohawk girls *(87)*. In all these studies, the relationship between blood Pb and puberty was significant even after adjusting for body size. This indicates that Pb's effect was probably direct through its impact on the hypothalamic–pituitary–ovarian axis rather than secondary to Pb-related decreased body size, which can be associated with the timing of the onset of puberty *(88)*. In the same population, exposure to polychlorinated biphenyls resulted in reduced size at birth but advanced sexual maturation, indicating that different pollutants exert effects through different physiology or endocrinology pathways *(89)*.

How does Pb exposure in children lead to delayed puberty? Research results from experiments with rats show that growth and sex hormones are altered from prenatal, lactational, and prepubertal exposure to Pb. These treatments delayed the age of vaginal opening, first estrus, and disrupted estrous cycling associated with suppressed serum levels of insulin-like growth factor-1 (IGF-1), a liver hormone involved in growth and reproduction *(90–92)*. Moreover, Pb affected hormones and responsiveness of all levels of the hypothalamic–pituitary–ovarian axis *(93,94)*. Dietary Pb may delay the onset of puberty in female mice, as observed in rats *(95)*, although by contrast, very low levels of dietary Pb, 0.02 ppm, were associated with a marked and significant acceleration of puberty in mice *(95,96)*, indicating an effect of dose on the pubertal outcome. In the last decade, puberty onset has advanced in the USA even in children migrating from foreign countries in Western European countries. It has been suggested that environmental endocrine-disrupting chemicals may be contributing to the earlier onset of puberty *(97)*.

4.2. Lead Effects on the Ovary and on Steroidogenesis

There are few studies of the direct effect of Pb on ovarian steroidogenesis. Lead exposure *in vivo* in cynomolgus monkey suppressed circulating luteinizing hormone (LH), follicle-stimulating hormone (FSH), and estradiol without affecting progesterone or causing overt signs of menstrual irregularity *(98)*. Prenatal and neonatal Pb exposure resulted in suppressed rat ovarian homogenate Δ4 androgen production whereas 5α-reduced androgens were increased *(99)*. There are more studies that examine changes in follicle populations after Pb exposure. For instance, mice exposed to Pb *in utero* experienced a significant reduction in the number of ovarian primordial follicles *(100)*. Adult mice given Pb by gavage for 60 days had significant changes in ovarian small, medium, and large follicle populations *(101)*. But, in another study, Pb was given by injection, and there was no change in either antral follicles but decreased primordial follicles and increased growing and atretic follicles *(102)*. Accelerated elimination of ovarian follicles, common to the above-cited reports, will ultimately lead to premature ovarian failure if the reduced follicle pool is the non-regenerating primordial follicles or disrupted cycles if the growing follicles are targeted.

Lead treatment of culture human ovarian granulosa cells retrieved during IVF reduces mRNA and protein levels of both P450 aromatase and estrogen receptor-β *(103)*. However, the mechanism responsible for Pb suppression of these two targets is unknown. Another molecular target of Pb is steroidogenic acute regulatory protein (StAR) that mediates the transfer of cholesterol into mitochondria *(104)*. Female rats exposed to Pb *in utero* have decreased basal ovarian StAR mRNA and protein but this is reversed by stimulating with gonadotrophins before collecting the ovaries. On the basis of these results, it was concluded that Pb acts at the hypothalamic–pituitary level of the reproductive axis.

4.3. Effects of Lead on Testicular Function and Steroidogenesis

Substantial research has been conducted regarding the specific effects of Pb on testicular and male reproductive function *(105)*. Pb accumulates in male reproductive organs. Exposure to Pb causes altered and delayed spermatogenesis accompanied by decreased fertility. For instance, in one IVF clinic, >40 % of the males who were not exposed to Pb occupationally and did not smoke cigarettes had blood and seminal plasma Pb concentrations greater than the permissible limit in men who are exposed to Pb in the work place *(105)*. In fact, there was an inverse relationship between Pb concentration in blood and seminal plasma and rate of fertilization that was due to altered sperm function *(105)*. The progesterone-dependent acrosome reaction was the sperm function most affected by the presence of Pb.

Lead inhibits both Sertoli and Leydig cell steroid production at every step of synthesis. Expression and/or activity of gonadotrophin receptors, StAR, p450 side chain cleavage, 3β HSD, and P450c17, the enzyme that converts progesterone into testosterone, are significantly if not dramatically suppressed by Pb *in vivo, ex vivo*, or *in vitro (106–110)*. Given the high concentrations of Pb used in all of these studies and often more severe inhibition with greater length of exposure, it is likely that the suppression of steroidogenesis at these step is due to toxicity rather than specific action or mimicry of an endocrine hormone.

4.4. Lead and Sex Ratios

Lead has been implicated in shifting the sex ratio to fewer boys born and maybe related to low testosterone at the time of conception *(111)*. Professional drivers exposed to excessive petroleum products father fewer sons *(112)*. Consistent with the occupational exposure to Pb leading to fewer boys born also is observed in filling station workers *(113)*. Both these reports suggest a relationship between occupational exposure to Pb and reduced male sex ratio, although this area is still controversial *(114)*, in part because of separating the consequences of exposure of one or both parents *(112)*. The most recent brief report revealed a dose–response with Pb such that workers with higher blood Pb levels had significantly reduced odds of having male offspring *(115)*. These data are intriguing, but much more investigation is needed to support or reject the hypothesis that Pb exposure impacts paternal-related sex ratio.

Lead can activate estrogen receptor-dependent transcriptional expression assay and stimulate MCF-7 breast cancer cell growth. In both these assays, Pb was not very effective and far less efficient than Cd *(34)*. Specific interaction of Pb with estrogen receptor was not determined in these studies, leaving me to conclude that there is insufficient evidence to conclude that Pb is an estrogen mimetic. However, the effects of subtoxic doses of Pb on reproductive functions described above support Pb as an endocrine-disrupting chemical whose specific mechanisms need to be determined.

5. MERCURY

Depending on its chemical form mercury (Hg) can be very toxic. Occupational exposure leads to neurodegeneration, behavioral changes, and death. Over 400 years ago, mercuric nitrate used in the felt hat industry gave rise to the phrase "mad as a hatter" depicted by the Mad Hatter in Lewis Carroll's *Alice's Adventures in Wonderland* *(13)*. Tragically, a recent accidental dimethylmercury poisoning reaffirmed the dangers associated with working with Hg *(116,117)*. Large-scale Hg poisonings have occurred in Minamata, Japan, and Iraq by industrial or inadvertent introduction of Hg into the food chain *(118)*. Industrial uses of Hg range from laboratory, dental, thermometers, paints, electrical equipment, and chloralkali. For the general public, there are three major sources of environmental Hg: fish consumption, dental amalgams, and vaccines *(119)*.

Fish contaminated with Hg poses a serious health risk to pregnant women and their babies. Methyl mercury bioaccumulates and biomagnifies in muscles of predatory fish that are at the top of the food chain, such as albacore *(116)*. Most concern is centered on neurological and behavioral problems that occur following exposure of the fetus or newborn, as Hg easily crossing the placenta and passes through the undeveloped infant blood–brain barrier. Thimerosal is an ethyl mercury-based vaccine preservative used since the 1930s *(116)*. It has received much attention of late because of its possible link to autism, but this is unproven and very controversial.

5.1. Mercury and Reproduction

Women exposed to Hg at work have been reported to experience reproductive dysfunction. Occupational exposure to mercury either in mercury vapor lamp factor or in dentistry is associated with menstrual disorders, subfecundity, and adverse pregnancy outcomes *(120,121)*. Interestingly, at very low Hg exposure, women working in dentistry were more fertile, suggesting a U-shaped dose–response curve *(122)*. Much more research is needed to establish a causal relationship between occupational Hg exposure and compromised fertility *(123)*.

Animal studies provide some insight into the impact of Hg exposure on female reproduction. In hamster, subcutaneous mercuric chloride treatment disrupts estrous cycles, suppresses follicular maturation, reduces plasma and luteal progesterone levels, and may disrupt hypothalamus-pituitary gonadotrophin secretion *(124)*. Similar changes were observed in female rats exposed to Hg vapor. The estrous cycle was lengthened, and morphological changes were evident in the corpora lutea, but ovulation, implantation, or maintenance of first pregnancy was unchanged *(125)*.

In wildlife, it has been proposed that Hg exposure has been responsible for increased cryptorchidism in the Florida panther as a result of exposure through bioaccumulation *(126)*. The authors report no significant difference between serum estradiol levels in male and female panthers, suggesting demasculinization and feminization of males. However, it is important to consider that the reproductive impairment and cryptorchidism may be genetically rather than environmentally based because there is limited genetic variation in the remaining Florida panther population *(127)*. Nevertheless, based on analysis of panther hair from museum collections, there is no doubt that current Florida panther Hg levels have increased since the 1890s *(128)*.

Investigation of the impact of either intraperitoneal injection of mercury or oral dosing of male rats or mice reveals consistent changes in the reproductive system. For instance, in rats exposed for 90 days to Hg by i.p. injection, testicular steroidogenesis was suppressed at the 3β HSD synthetic step with a significant decrease in serum testosterone and LH *(129–131)*. Oral exposure of rats to mercuric chloride for 45 days resulted in suppressed testosterone and increased testicular cholesterol *(132)*. The authors suggest that the increased cholesterol is due to the block of its biosynthetic conversion to sex steroid hormones such as testosterone. Another possibility is that Hg mimics the effect of estrogen on the testes which is to both inhibit androgen production and cause accumulation of cholesterol probably because of upregulation of the high-density lipoprotein (HDL) receptor, scavenger receptor class B, type I (SR-BI) *(133)*.

5.2. Estrogenic Mechanisms of Mercury

The estrogenicity of Hg was examined in MCF-7 cells *(34)*. Mercuric chloride stimulated both estrogen receptor-dependent transcription and increased proliferation of MCF-7 cells *(34)*. A more detailed study of the methyl mercury impact on MCF-7 cells was performed by Sukocheva et al. *(134)*. In this study, instead of measuring increased number of MCF-7 cells, the number of postconfluent foci that formed with estrogen treatment was counted. Multicellular foci form in response to estrogen agonists and are proportional to hormone dose or concentration *(135)*. A very narrow concentration range, 0.5×10^{-7} to 1×10^{-6} M, of methyl mercury stimulated MCF-7 cell foci formation but did not reach the maximum response elicited by estradiol, indicating that Hg is a weak estrogen mimic. Hg exhibited estrogen receptor agonist–antagonist properties depending on concentration. Its stimulation of foci formation is blocked by the antiestrogen ICI 182,780. However, Hg is poor at competing for ^{3}H estradiol binding to recombinant estrogen receptor as displacement was only observed and 10^{-4} M. As is the case with estradiol, Hg stimulated Erk1/2 activation that was dependent on mobilization of intracellular Ca^{+2}, suggesting similar signaling mechanisms. Methyl mercury reacts with sulfhydryls and could interact with protein thiol groups such as those located in the ligand-binding domain of the estrogen receptor to stimulate MCF-7 cell proliferation. The Erk1/2 activation pathway is involved in cell proliferation and gene expression. A number of extra cellular signals have been shown to induce MAPK Erk1/2 including many other well-known endocrine-disrupting chemicals such as bisphenol A *(136)* (see Chapter 2 by Soto, Rubin, and Sonnenschein).

6. URANIUM

Uranium (U) is the heaviest naturally occurring element, certainly qualifying it as a heavy metal. Similar to Cd, U was used initially for its pigmentation in bright orange-red Fiestaware. Uranium was the first element to be identified as fissile. Uranium supports nuclear chain reactions leading to the development of atomic weapons and later as a fuel for nuclear reactors. Its current use is as depleted uranium (DU) in armed conflict first in the Balkans and now Iraq for both munitions and armor. Environmental sources and thus risk of contact with U are mining, production of nuclear weapons, nuclear reactor industry and disasters, and US-facilitated armed conflict.

Nuclear accidents at Three Mile Island and Chernobyl quashed hopes of developing nuclear energy to replace dependence on fossil fuels for decades. With today's dwindling petroleum supplies and significant green house gas emissions from coal-produced energy, interest in nuclear energy is renewed. Australia has the world's largest U reserves, but Canada is the largest U exporter. In the USA, the greatest U reserves are in the southwest, specifically the states of the Four Corners, AZ, CO, NM, UT, where the Navajo Reservation Nation is located.

6.1. Uranium, the Navajo Reservation, and Human Exposures

The Navajo Reservation is the largest Native American reservation in the USA. It covers over 27,000 square miles. The Navajo Nation comprises 110 Chapters, the political/social units. According to the 2000 census there are 250,000 members of the Navajo tribe, making it the largest Native American tribe in the USA. This group continues to live its traditional lifestyle and rely on its native language to sustain and nurture the "Navajo way." Over 170,000 people live on Reservation and the majority of the remaining tribal members live in "border towns" such as Farmington, NM, and Flagstaff, AZ. Fifty four percent of households have no indoor plumbing or running water necessitating half of the Reservation households to haul water from the nearest available source. The Navajo Reservation evokes images of a peaceful and healthy environment, but the reality of environmental contamination presents a different picture.

From the 1945 through 1988 there was intensive U mining and milling *(137)*. In 1980, the market price for U crashed and effectively ended the mining. However, the market value of U ore has quintupled in the last four years because of renewed interest in nuclear energy and the ongoing need for depleted U for armed conflict. Thus, U mining is on the upsurge, with 700 claims filed 2005 in the northern part of AZ. At the writing of this chapter, U is poised to climb to a record high of $50/lb in the next 6 months stimulating revived interest in mining on the Navajo Reservation.

Well-documented health problems that arise from U mining and milling are lung cancer and respiratory diseases *(137,138)*. These health problems result from radiation exposure after inhalation of radon-rich dust released during U mining. Hundreds of Navajo and Mormon miners and ore truck drivers have developed these diseases, and if they can document their work experience with pay stubs or other evidence they may qualify for compensation from the Federal Radioactive Exposure Compensation Act (RECA). However, many Navajo miners do not have the documentation to prove their employment history and die before they can be compensated *(137)*.

It is estimated that there are over 1500 abandoned mines on the Navajo Reservation that have not been properly closed, allowing U to be dispersed by the elements over the last 62 years throughout the natural environment. The US Army Corps of Engineers performed a survey of water sources in 30 Chapters on the Navajo Nation, and the results are posted by USEPA on this website http://yosemite.epa.gov/r9/sfund/r9sfdocw.nsf/vwsoalphabetic/Abandoned+Uranium+Mines+On+The+Navajo+Nation?OpenDocument. In every chapter surveyed, there was at least one water source with U water levels that exceeds the USEPA safe drinking water level of 30 μg/L. Because half of the households on the Reservation haul water, it is certain that many people are exposed to unsafe levels of U in their drinking and household water.

Health problems that result from drinking U-contaminated water include kidney disease, kidney cancer, and possibly stomach cancer *(138)*. Kidney disease results from U's heavy metal poisoning of the proximal tubules and interferes with glucose uptake *(139)*. Cancer in the kidney, stomach and bone marrow results from U-derived alpha radiation causing DNA damage leading to cell transformation and cancer. Owing to its chemical properties, U homes to bone and bioaccumulates, leading to increased risk of leukemia. Other routes of exposure are inhalation of small particles blown from tailings or dust and dirt brought into the home on the miner's clothing often washed with the clothes of other family members *(137)*. Depending on size, the particles can enter the bloodstream leading to the above-mentioned diseases. Allowable exposure limits vary for U. The USEPA safe drinking water limit of 30 μg/L is based on economic feasibility while the World Health Organization standard of 2 μg/L is based on the fractional source of U assuming that an adult drinks 2 L of water a day. However, a recent study suggests that even low U concentrations in drinking water can cause nephrotoxic effects, and after long-term ingestion, elevated concentrations of U in urine can be detected up to 10 months after the exposure has stopped *(140)*.

6.2. Uranium as an Endocrine-Disrupting Chemical on Reproductive Systems

Numerous studies have investigated the reproductive toxicology of both U and DU in experimental mice and rats *(141–143)*. The original study conducted by Maynard and Dodge *(141)* showed that breeder rats consuming chow containing 2 % uranyl nitrate hexahydrate (UN) for 7 months gave birth to half as many litters as the control chow-fed rats. The litter size of the U-consuming rats was significantly less as well. For the next 5 months all rats ate control chow; however, the female rats that had been exposed to UN still only produced less than half as many litters as the rats that were on control chow for the entire 12 months. Rats eating U-containing chow also had irregular estrous cycles compared with the control chow-fed rats. Even though the U-fed rats lost weight, which could have contributed to their reduced fertility, the fact that after the rats were returned to control chow and regained weight, but they still produced half the number of litters indicates that the impact of U was permanent.

Several studies have documented U's toxicity in the male reproductive system. General features reported are degenerative changes in the testes such as aspermia in the testes and epididymis, testicular atrophy, interstitial cell alterations, Leydig cell vacuolization, and reduced successful female mouse impregnation *(142,143)*. DU exposure in human has been a result of warfare. Gulf war 1990–1991 UK veterans may have impaired fertility, and pregnancies took longer to conceive *(144)*. Also, there was a 40 % increased risk of miscarriage among pregnancies fathered by men who served in the first Gulf war *(145)*. Another cohort of US Gulf war veterans struck by friendly fire had DU shrapnel embedded in muscle and soft tissue and were still excreting greater than background levels of U 9–11 years after sustaining their injuries *(146)*.

A follow-up of 30 friendly fire cohort members found evidence of subtle perturbations in the reproductive system indicated by significantly elevated prolactin levels *(147)*. In another DU-exposed friendly fire cohort, there were significantly elevated sperm counts and a higher percentage of progressive spermatozoa among veterans excreting high levels of U as compared with veterans excreting low levels of U in their urine *(148)*. In a different population, Czechoslovakian uranium miners fathered significantly more girls than boys, suggesting a shift in the sex ratio *(149)*. It is important to consider that these reports are preliminary, and there is no evidence to suggest that altered male fertility is caused by exposure to U or DU.

U exposure of pregnant rodents produces maternal toxicity, fetal toxicity, and developmental defects *(150)*. There are more absorptions, dead fetuses, fewer live-born fetuses, and pup body weight and length were significantly reduced. In addition, there is a higher incidence of cleft palate and dorsal and facial hematomas *(150)*. In humans, there is sufficient epidemiological evidence to suggest an increased risk of birth defects in offspring of persons exposed to DU *(151)*. On the Navajo Reservation exposure to environmental U was statistically associated with uranium operations and unfavorable birth outcome, including cleft palate and craniofacial developmental defects, if the mother lived near mine tailings or mine dumps *(152)*. Again, the possible association of environmental U exposure and adverse human health outcomes is correlative, and more research is needed to draw any casual connections.

6.3. Uranium as an Environmental Estrogen

Probably because of U's radioactive nature, it has been studied intensely with regard to harmful effects from its ionizing radiation and chemical toxicity but had not previously been tested for its potential estrogenic activity as a heavy metal. Recently, we have discovered that U is estrogenic in female reproductive tissues *(153,154)*. The source of U in our *in vivo* and *in vitro* studies is UN and is DU. It should be noted that natural U and DU have the same chemistry, and therefore, both will cause the same changes that are dependent on chemical properties.

Although there are many publications describing the effects of U exposure on reproduction in female, none of the reports mentioned the impact of U on ovarian follicle populations *(141–143)*. We wanted to determine whether U exposure would target a specific ovarian follicle population or more likely cause a non-specific general ovarian toxicity. In our original studies, high levels of UN at mg/L in the drinking water were consumed for 30 days by intact immature female mice. UN exposure specifically targeted primary follicles whose number was reduced. Unexpectedly, in these mice, there was a trend, although not statistically significant, of increased uterine weight. In this experiment at these high doses of U, kidney weight was reduced, consistent with U's well-known nephrotoxicity *(139)*. Intrigued by the possible uterotrophic effect of U at mg/L concentrations, we next tested whether μg/L levels would cause uterine weight to increase in ovariectomized mice *(155)*. Ovariectomized mice that drank UN-containing water for 10 days at concentrations starting at the USEPA safe drinking water level of 30 μg/L down to 0.3 μg/L had significantly increased uterine weights *(153)*. If the mice were treated with the antiestrogen ICI 182,780 while drinking the U-contaminated water, the increase in uterine weight did not occur. We concluded from this experiment that U at low, environmentally relevant concentrations, acted like estrogen. Another biological response elicited by estrogen in ovariectomized immature mice is accelerated vaginal opening. This response was also observed in mice drinking U-contaminated water that was again prevented by the coincident treatment with ICI 182,780 *(153)*. Finally, we observed the persistent presence of vaginal cornified epithelial cells from immature ovariectomized mice drinking U-contaminated water, and as before, the antiestrogen ICI 182,780 ablated this response. Cornified epithelial cells from vaginal smears are an indication that there is an ongoing estrogenic stimulation of the reproductive tract *(156)*. In all these experiments, we compared the responses induced by U to those caused by the potent synthetic estrogen DES. The magnitude of the responses were comparable at equivalent molarity for U and DES, indicating that U is a potent estrogen mimic *in vivo*.

Our preliminary results analyzing the putative estrogenic activity of U *in vitro* using MCF-7 breast cancer cells indicate that UN at nanomolar to low micromolar concentrations stimulates cell proliferation *(154)*. The estrogen receptor was involved because the antiestrogen ICI 182,780 blocked both 17β-estradiol and UN-stimulated MCF-7 cell proliferation. Cell proliferation is a response that takes days to occur through the classic estrogen receptor-dependent genomic pathway. Recent reports indicate that estrogen-mediated responses can occur in just a few minutes *(157)*. To detect whether rapid responses occurred in DES- or UN-treated MCF-7 cells, we used scanning electron microscopy to visualize cell surface morphological changes within minutes of exposure. Both DES and UN treatment caused a significant increase in number and branching of MCF-7 cell surface microvilli, and the cell surface changes were blocked by ICI 182,780 *(154)*. The rapid morphological changes suggest that, similar to estrogen, UN causes cell responses independent of genomic responses *(158)*.

As with the other heavy metals, low concentrations ($<10\,\mu M$) mediate specific responses for instance through the estrogen receptor or other steroid receptors, whereas higher concentrations ($>100\,\mu M$) cause non-specific toxic responses. For instance, in Chinese hamster ovary cells, uranyl nitrate killed the cells and induced chromosome aberrations and sister-chromatid exchanges *(159)*. Uranyl acetate at 200 μM killed and mutagenized Chinese hamster ovary cells perhaps by causing DNA strand breaks and forming uranium-DNA adducts *(160)*. Thus, depending on the environmental level of U and exposure or contact, the biologic responses can range from subtle through steroid receptor interaction and activation to striking through cytotoxic and genotoxic mechanisms. The concentrations of U and oral route of exposure at which we see estrogenic responses *in vivo* and *in vitro* are environmentally relevant to exposures on the Navajo Reservation as well as other communities in the USA where U is in drinking water at levels at or below the USEPA safe drinking water limit.

7. CONCLUSIONS

Historically, heavy metal toxicity targeting the kidney or nervous system has been the focus of most research efforts. However, accumulating evidence, particularly when studies are performed with low micromolar concentrations, indicates that heavy metals can act as endocrine-disrupting substances through specific, high-affinity pathways. Thus far, heavy metals primarily have been described to interact with the estrogen receptor giving rise to the term metalloestrogens. This is just the beginning of our understanding of the subtle means by which heavy metals disrupt endocrine function—many more pathways need scrutiny *(161)*.

REFERENCES

1. Colborn T, vom Saal FS, Soto AM. Developmental effects of endocrine-disrupting chemicals in wildlife and humans. *Environ Health Perspect* 1993; 101:378–84.
2. Jarup L. Hazards of heavy metal contamination. *Br Med Bull* 2003; 68:167–82.
3. Medici N, Minucci S, Nigro V, Abbondanza C, Armetta I, Molinari AM, Puca GA. Metal binding sites of the estradiol receptor from calf uterus and their possible role in the regulation of receptor function. *Biochemistry* 1989; 28:212–19.
4. Predki PF, Sarkar B. Effect of replacement of zinc finger zinc on estrogen receptor DNA interactions. *J Biol Chem* 1992; 267:5842–46.
5. Martin MB, Reiter R, Pham T, Avellanet YR, Camara J, Lahm M, Pentecost E, Pratap K, Gilmore BA, Divekar S, Dagata RS, Bull JL, Stoica A. Estrogen-like activity of metals in Mcf-7 breast cancer cells. *Endocrinology* 2003; 144:2425–36.
6. Nesatyy VJ, Rutishauser BV, Eggen RIL, Suter JF. Identification of the estrogen receptor Cd-binding sites by chemical modification. *Analyst* 2005; 130:1087–97.
7. Nesatyy VI, Ammann AA, Rutishauser BV, Suter MJF. Effect of cadmium on the interaction of 17β-estradiol with the rainbow trout estrogen receptor. *Environ Sci Technol* 2006; 40:1358–63.
8. Johnson MD, Kenney N, Stoica A, Hilakivi-Clarke L, Singh B, Chepko G, Clarke R, Sholler PF, Lirio AA, Foss C, Reiter R, Trock B, Paik S, Martin MB. Cadmium mimics the *in vivo* effects of estrogen in the uterus and mammary gland. *Nat Med* 2003; 9:1081–84.
9. Safe S. Cadmium's disguise dupes the estrogen receptor. *Nat Med* 2003; 9:1000–1.
10. Darbre PD. Metalloestrogens: an emerging class of inorganic xenoestrogens with potential to add to the oestrogenic burden of the human breast. *J Appl Toxicol* 2006; 26:191–7.
11. Mandal BK, Suzuki KT. Arsenic round the world: a review. *Talanta* 2002; 58:201–35.
12. Waxman S, Anderson KC. History of the development of arsenic derivatives in cancer therapy. *Oncologist* 2001; 6(Suppl 2):3–10.
13. Klaassen CD. Heavy metals and heavy-metal antagonists. In: Brunton LL, ed. *Goodman & Gilman the Pharmacological Basis of Therapeutics*, 11th ed. New York: McGraw-Hill, 2006:1753–75.
14. Kwon E, Zhang H, Wang Z, Jhangri GS, Lu X, Fok N, Gabos S, Li XF, Le XC. Arsenic on the hands of children after playing in playgrounds. *Environ Health Perspect* 2004; 112:1375–80.
15. Tapio S, Grosche B. Arsenic in the aetiology of cancer. *Mutat Res* 2006; 612:215–46.
16. U.S. Environmental Protection Agency. Arsenic in drinking water. http://www.epa.gov/safewater/arsenic/index.html
17. Frost FJ, Muller T, Petersen HV, Thomson B, Tollestrup K. Identifying US populations for the study of health effects related to drinking water arsenic. *J Expo Anal Environ Epidemiol* 2003; 13:231–39.
18. Jana K, Jana S, Samanta PK. Effects of chronic exposure to sodium arsenite on hypothalamo-pituitary-testicular activities in adult rats: possible an estrogenic mode of action. *Reprod Biol Endocrinol* 2006; 4:9.
19. Waalkes MP, Keefer LK, Diwan BA. Induction of proliferative lesions of the uterus, testes, and liver in Swiss mice given repeated injections of sodium arsenate: possible estrogenic mode of action. *Toxicol Appl Pharmacol* 2000; 166:24–35.
20. Waalkes MP, Ward JM, Liu J, Diwan BA. Transplacental carcinogenicity of inorganic arsenic in the drinking water: induction of hepatic, ovarian, pulmonary, and adrenal tumors in mice. *Toxicol Appl Pharmacol* 2003; 186:7–17.
21. Waalkes MP, Liu J, Ward JM, Powell DA, Diwan BA. Urogenital carcinogenesis in female CD1 mice induced by *in utero* arsenic exposure is exacerbated by postnatal diethylstilbestrol. *Cancer Res* 2006; 66:1337–45.
22. Waalkes MP, Liu J, Ward JM, Diwan B. Enhanced urinary bladder and liver carcinogenesis in male CD1 mice exposed to transplancental inorganic arsenic and postnatal diethylstilbestrol or tamoxifen. *Toxicol Appl Pharmacol* 2006; May 16, Epub ahead of print.
23. Waalkes MP, Liu J, Chen H, Xie Y, Achanzar WE, Zhou YS, Cheng ML, Diwan BA. Estrogen signaling in livers of male mice with hepatocellular carcinoma induced by exposure to arsenic *in utero*. *J Natl Cancer Inst* 2004; 96:466–74.
24. Chattopadhyay S, Ghosh S, Chaki S, Debnath J, Ghosh D. Effect of sodium arsenite on plasma levels of gonadotrophins and ovarian steroidogenesis in mature albino rats: duration-dependent response. *J Toxicol Sci* 1999; 24:425–31.

25. Golub MS, Macintosh MS. Developmental and reproductive toxicity on inorganic arsenic: animal studies and human concerns. *J Toxicol Environ Health B Crit Rev* 1998; 1:199–241.
26. Tseng GH. The potential biological mechanisms of arsenic – induced diabetes mellitus. *Toxicol Appl Pharmacol* 2004; 197:67–83.
27. Tseng GH. Blackfoot disease and arsenic: a never-ending story. *J Environ Sci Health C Environ Carcinog Ecotoxicol Rev* 2005; 23:55–74.
28. Livingstone C, Collison M. Sex steroids and insulin resistance. *Clin Sci (Lond)* 2002; 102:151–66.
29. Valko M, Morris H, Cronin MTD. Metals, toxicity and oxidative stress. *Curr Med Chem* 2005; 12:1161–208.
30. Nordstrom S, Beckman L, Nordenson I. Occupational and environmental risks in and around a smelter in northern Sweden. V. Spontaneous abortion among female employees and decreased birth weight in their offspring. *Hereditas* 1979; 90:291–6.
31. Tabacova S, Baird DD, Balabaeva Lolova LD, Petrov I. Placental arsenic and cadmium in relation to lipid peroxides and glutathione levels in maternal-infant pairs from a copper smelter area. *Placenta* 1994; 15:873–1.
32. Soto AM, Maffini MV, Schaeberle CM, Sonnenschein C. Strengths and weaknesses of in vitro assays for estrogenic and androgenic activity. *Best Pract Res Clin Endocrinol Metab* 2006; 20:15–33.
33. Stoica A, Pentecost E, Martin MB. Effects of arsenite on estrogen receptor-α expression and activity in MCF-7 breast cancer cells. *Endocrinology* 2000; 141:3595–602.
34. Choe SY, Kim SJ, Kim HG, Lee JH, Choi Y, Lee H, Kim Y. Evaluation of estrogenicity of major heavy metals. *Sci Total Environ* 2003; 312:15–21.
35. Chow SKY, Chan JYW, Fung KP. Suppression of cell proliferation and regulation of estrogen receptor α signal pathway by arsenic trioxide on human breast cancer MCF-7 cells. *J Endocrinol* 2004; 182:325–7.
36. Kitchin KT, Wallace K. Arsenite binding to synthetic peptides based on the Zn finger region and the estrogen binding region of the human estrogen receptor. *Toxicol Appl Pharmacol* 2005; 206:66–72.
37. Kitchin KT, Wallace K. Arsenite binding to synthetic peptides: the effect of increasing length between two cysteines. *J Biochem Mol Toxicol* 2006; 20:35–8.
38. Lopez S, Miyashita Y, Simons SS. Structurally based selective interaction of arsenite with steroid receptors. *J Biol Chem* 1990; 265:16039–42.
39. Hamilton JW, Kaltreider RC, Bajenova OV, Ihnat MA, McCaffrey J, Turpie TW, Rowell EE, Oh J, Nemeth MJ, Pesce CA, Lariviere JP. Molecular basis for effects of carcinogenic heavy metals on inducible gene expression. *Environ Health Perspect* 1998; 106:1005–5.
40. Kaltreider RC, Pesce CA, Ihnat MA, Lariviere JP, Hamilton JW. Differential effects of arsenic(III) and chromium(VI) on nuclear transcription factor binding. *Mol Carcinog* 1999; 25:219–9.
41. Kaltreider RC, Davis AM, Lariviere JP, Hamilton JW. Arsenic alters the function of the glucocorticoid receptor as a transcription factor. *Environ Health Perspect* 2001; 109:245–51.
42. Simons SS, Chakraborti PK, Cavanaugh AH. Arsenite and cadmium(II) as probes of glucocorticoid receptor structure and function. *J Biol Chem* 1990; 265:1938–45.
43. Kitchin KT, Wallace K. Dissociation of arsenite-peptide complexes: triphasic nature, rate constants, half-lives and biological importance. *J Biochem Mol Toxicol* 2006; 20:48–56.
44. Henson MC, Chedrese PJ. Endocrine disruption by cadmium, a common environmental toxicant with paradoxical effects on reproduction. *Exp Biol Med (Maywood)* 2004; 229:383–92.
45. Nishijo M, Nakagawa H, Honda R, Tanebe K, Saito S, Teranishi H, Tawara K. Effects of maternal exposure to cadmium on pregnancy outcome and breast milk. *Occup Environ Med* 2002; 59:394–97.
46. Zenzes MT, Krishnan S, Krishnan B, Zhang H, Casper RF. Cadmium accumulation in follicular fluid of women in *in vitro* fertilization-embryo transfer is higher in smokers. *Fertil Steril* 1995; 64:599–603.
47. Piasek M, Blanusa M, Kostial K, Laskey JW. Placental cadmium and progesterone concentrations in cigarette smokers. *Reprod Toxicol* 2001; 15:673–81.
48. Jolibois LS, Shi W, George WJ, Henson MC, Anderson MB. Cadmium accumulation and effects on progesterone release by cultured human trophoblast cells. *Reprod Toxicol* 1999; 13:215–1.
49. Jolibois LS, Burow ME, Swan KF, George WJ, Anderson MB, Henson MC. Effects of cadmium on cell viability, trophoblastic development, and expression of low density lipoprotein receptor transcripts in cultured human placental cells. *Reprod Toxicol* 1999; 13:473–80.
50. Kawai M, Swan KF, Green AE, Edwards DE, Anderson MB, Henson MC. Placental endocrine disruption induced by cadmium: effects on P450 cholesterol side-chain cleavage and

3β-hydroxysteroid dehydrogenase enzymes in cultured human trophoblasts. *Biol Reprod* 2002; 67:178–83.

51. Yang K, Julan L, Rubio F, Sharma A, Guan H. Cadmium reduces 11β-hydroxysteroid dehydrogenase type 2 activity and expression in human placental trophoblast cells. *Am J Physiol Endocrinol Metab* 2006; 290:E135–42.
52. Osmond C, Barker DJ. Fetal, infant, and childhood growth are predictors of coronary heart disease, diabetes, and hypertension in adult men and women. *Environ Health Perspect* 2000; 108(Suppl 3):545–3.
53. Cnattingius S. The epidemiology of smoking during pregnancy: smoking prevalence, maternal characteristics, and pregnancy outcomes. *Nicotine Tob Res* 2004; 6(Suppl 2):S125–140.
54. Bush PG, Mayhew TM, Abramovich DR, Aggett PJ, Burke MD, Page KR. A quantitative study on the effects of maternal smoking on placental morphology and cadmium concentration. *Placenta* 2000; 21:247–56.
55. Sun K, Yang K, Challis JRG. Regulation of 11β-hydroxysteroid dehydrogenase type 2 by progesterone, estrogen, and the cyclic adenosine 5'-monophosphate pathway in cultured human placental and chorionic trophoblasts. *Biol Reprod* 1998; 58:1379–84.
56. Paksy K, Rajczy K, Forgacs Z, Lazar P, Bernard A, Gati I, Kaali G. Effect of cadmium on morphology and steroidogenesis of cultured human ovarian granulosa cells. *J Appl Toxicol* 1997; 17:321–27.
57. Smida AD, Valderrame XP, Agostini MC, Furlan MA, Chedrese J. Cadmium stimulates transcription of the cytochrome P450 side chain cleavage gene in genetically modified stable porcine granulosa cells. *Biol Reprod* 2004; 70:25–31.
58. Long GJ. The effect of cadmium on cytosolic free calcium, protein kinase C, and collagen synthesis in rat osteosarcoma (ROA 17/2.8) cells. *Toxicol Appl Pharmacol* 1997; 143:189–95.
59. Piasek M, Laskey JW. Effect of *in vitro* cadmium exposure on ovarian steroidogenesis in rats. *J Appl Toxicol* 1999; 19:211–17.
60. Piasek M, Laskey JW, Kostial K, Blanusa M. Assessment of steroid disruption using cultures of whole ovary and/or placenta in rat and in human placental tissue. *Int Arch Occup Environ Health* 2002; 75(Suppl):S36–44.
61. Piasek M, Laskey JW. Acute cadmium exposure and ovarian steroidogenesis in cycling and pregnant rats. *Reprod Toxicol* 1994; 8:495–507.
62. Waisberg M, Joseph P, Hale B, Beyersmann. Molecular and cellular mechanisms of cadmium carcinogenesis. *Toxicology* 2003; 192:95–117.
63. Alsberg CL, Schwartze EW. Pharmacological action of Cd. *Pharmacology* 1919; 13:504–9.
64. Chia SE, Xu B, Ong CN, Tsakok FM, Lee ST. Effect of cadmium and cigarette smoking on human semen quality. *Int J Fertil Menopausal Stud* 1994; 39:292–8.
65. Al-Bader A, Omu AE, Dashti H. Chronic cadmium toxicity to sperm of heavy cigarette smokers: immunomodulation by zinc. *Arch Androl* 1999; 43:135–40.
66. Yang JM, Arnush M, Chen QY, Wu XD, Pang B, Jiang XZ. Cadmium-induced damage to primary cultures of rat Leydig cells. *Reprod Toxicol* 2003; 17:553–60.
67. Laskey JW, Phelps PV. Effect of cadmium and other metal cations on *in vitro* Leydig cell testosterone production. *Toxicol Appl Pharmacol* 1991; 108:296–306.
68. Zeng X, Jin T, Zhou Y, Nordberg GF. Changes of serum sex hormone levels and MT mRNA expression in rats orally exposed to cadmium. *Toxicol* 2003; 186:109–8.
69. Lafuente A, Marquez N, Perez-Lorenzo M, Pazo D, Esquifino AI. Pubertal and postpubertal cadmium exposure differentially affects the hypothalamic-pituitary-testicular axis function in the rat. *Food Chem Toxicol* 2000; 38:913–23.
70. Zeng X, Jin T, Zhou Y. Alterations of serum hormone levels in male workers occupationally exposed to cadmium. *J Toxicol Environ Health A* 2002; 65:513–21.
71. Jurasovic J, Cvitkovic P, Pizent A, Colak B, Telisman S. Semen quality and reproductive endocrine function with regard to blood cadmium in Croation male subjects. *Biometals* 2004; 17:735–43.
72. Nagata C, Nagao Y, Shibuya C, Kashiki Y, Shimizu H. Urinary cadmium and serum levels of estrogens and androgens in postmenopausal Japanese women. *Cancer Epidemiol Biomarkers Prev* 2005; 14:705–8.
73. Garcia-Morales P, Saceda M, Kenney N, Kim N, Salomon DS, Gottardis MM, Solomon HB, Sholler PF, Jordon VC, Martin MB. Effect of cadmium on estrogen receptor levels and estrogen-induced responses in human breast cancer cells. *J Biol Chem* 1994; 269:16896–901.

74. Stoica A, Katzenellenbogen BS, Martin MB. Activation of estrogen receptor-α by the heavy metal cadmium. *Mol Endocrinol* 2000; 14:545–3.
75. Martin MB, Voeller HJ, Gelmann Lu EPJ, Stoica EG, Hebert EJ, Reiter R, Singh B, Danielsen M, Pentecost E, Stoica A. Role of cadmium in the regulation of AR gene expression and activity. *Endocrinology* 2002; 143:263–75.
76. Martinez-Campa C, Alonso-Gonzalez CA, Mediavilla MD, Cos S, Gonzalez A, Ramons S, Sanchez-Barcelo EJ. Melatonin inhibits both ERα activation and breast cancer cell proliferation induced by a metalloestrogen, cadmium. *J Pineal Res* 2006; 40:291–96.
77. Beyersmann D, Hechtenberg S. Cadmium, gene regulation, and cellular signaling in mammalian cells. *Toxicol Appl Pharmacol* 1997; 144:247–61.
78. Wilson VS, Bobseine K, Gray LE. Development and characterization of a cell line that stably expresses an estrogen-responsive luciferase reporter for the detection of estrogen receptor agonist and antagonists. *Toxicol Sci* 2004; 81:69–77.
79. Beyersmann D. Effects of carcinogenic metals on gene expression. *Toxicol Lett* 2002; 127:63–8.
80. Ding W, Templeton DM. Activation of parallel mitogen-activated protein kinase cascades and induction of *c-fos* by cadmium. *Toxicol Appl Pharmacol* 2000; 162:93–9.
81. Silva E, Lopez-Espinosa MJ, Molina-Molina JM, Fernandez M, Olea N, Kortenkamp A. Lack of activity of cadmium in *in vitro* estrogenicity assays. *Toxicol Appl Pharmacol* 2006; 216:20–8.
82. Pirkle JL, Kaufmann RB, Brody DJ, Hickman T, Gunter EW. Paschal DC. Exposure of the DCUS population to lead, 1991-1994. *Environ Health Perspect* 1998; 106:745–50.
83. Centers for Disease Control (CDC). Death of a child after ingestion of a metallic charm—Minnesota, 2006. *MMWR Morb Mortal Wkly Rep* 2006; 55:340–1.
84. Canfield RL, Henderson CR, Cory-Slechta DA, Cox C, Jusko TA, Lanphear BP. Intellectual impairment in children with blood lead concentrations below 10 μg per deciliter. *N Engl J Med* 2003; 348:1517–26.
85. Selevan SG, Rice DC, Hogan KA, Euling SY, Pfahles-Hutchens A, Bethel J. Blood concentration and delayed puberty in girls. *N Engl J Med* 2003; 348:1527–36.
86. Wu T, Buck GM, Mendola P. Blood lead levels and sexual maturation in U.S. girls: the third Nation Health and Nutrition Examination Survey, 1988-1994. *Environ Health Perspect* 2003; 111:737–41.
87. Denham M, Schell LM, Deane G, Gallo MV, Ravenscroft J, DeCaprio AP, Akwesasne Task Force on the Environment. Relationship of lead, mercury, mirex, dichlorodiphenyldichloroethylene, hexachlorobenzene, and polychlorinated biphenyls to timing of menarche among Akwesasne Mohawk girls. *Pediatrics* 2005; 115:e127–34.
88. Kaplowitz PB, Slora EJ, Wasserman RC, Pedlow SE, Herman-Giddens ME. Earlier onset of puberty in girls: relation to increased body mass index and race. *Pediatrics* 2001; 108:347–53.
89. Schell LM, Gallo MV, Denham M, Ravenscroft J. Effects of pollution on human growth and development: an introduction. *J Physiol Anthropol* 2006; 25:103–2.
90. Ronis MJJ, Badger TM, Shema SJ, Roberson PK, Shaikh F. Effects on pubertal growth and reproduction in rats exposed to lead perinatally or continuously throughout development. *J Toxicol Environ Health A* 1998; 53:327–41.
91. Ronis MJJ, Gandy J, Badger T. Endocrine mechanisms underlying reproductive toxicity in the developing rat chronically exposed to dietary lead. *J Toxicol Environ Health A* 1998; 54:77–99.
92. Dearth RK, Hiney JK, Srivastava V, Burdick SB, Bratton GR, Dees WL. Effects of lead (Pb) exposure during gestation and lactation on female pubertal development in the rat. *Reprod Toxicol* 2002; 16:343–52.
93. McGivern RF, Sokol RZ, Berman NG. Prenatal lead exposure in the rat during the third week of gestation: long-term behavioral, physiological and anatomical effects associated with reproduction. *Toxicol Appl Pharmacol* 1991; 110:206–15.
94. Pine MD, Hiney JK, Dearth RK, Bratton GR, Dees WL. IGF-1 administration to prepubertal female rats can overcome delayed puberty caused by maternal Pb exposure. *Reprod Toxicol* 2006; 21:104–9.
95. Iavicoli I, Carelli G, Stanek EJ, Castellino N, Calabrese EJ. Effects of low doses of dietary lead on puberty onset in female mice. *Reprod Toxicol* 2004; 19:35–41.
96. Iavicoli I, Carelli G, Stanek EJ, Castellino N, Li Z, Calabrese EJ. Low doses of dietary lead are associated with a profound reduction in the time to the onset of puberty in female mice. *Reprod Toxicol* 2006; 22:586–90.

97. Parent AS, Teilmann G, Juul A, Skakkebaek NE, Toppari J, Bourguignon JP. The timing of normal puberty and the age limits of sexual precocity: variations around the world, secular trends, and changes after migration. *Endocr Rev* 2003; 24:668–93.
98. Foster WG. Reproductive toxicity of chronic lead exposure in the female cynomolgus monkey. *Reprod Toxicol* 1992; 6:123–31.
99. Wiebe JP, Barr KJ, Buckingham KD. Effect of prenatal and neonatal exposure to lead on gonadotropin receptors and steroidogenesis in rat ovaries. *J Toxicol Environ Health* 1988; 24:461–76.
100. Wide M. Lead exposure on critical days of fetal life affects fertility in the female mouse. *Teratology* 1985; 32:375–80.
101. Junaid M, Chowdhuri DK, Narayan R, Shanker R, Saxena DK. Lead-induced changes in ovarian follicular development and maturation in mice. *J Toxicol Environ Health* 1997; 50:31–40.
102. Taupeau C, Poupon J, Nome F, Lefevre B. Lead accumulation in the mouse ovary after treatment-induced follicular atresia. *Reprod Toxicol* 2001; 15:385–91.
103. Taupeau C, Poupon J, Treton D, Brosse A, Richard Y, Machelon V. Lead reduces messenger RNA and protein levels of cytochrome P450 aromatase and estrogen receptor β in human ovarian granulosa cells. *Biol Reprod* 2003; 68:1982–88.
104. Srivastava V, Dearth RK, Hiney JK, Ramirez LM, Bratton GR, Dees WL. The effects of low-level Pb on steroidogenic acute regulatory protein (StAR) in the prepubertal rat ovary. *Toxicol Lett* 2004; 77:35–40.
105. Benoff S, Jacob A, Hurley IR. Male infertility and environmental exposure to lead and cadmium. *Human Reprod Update* 2000; 6:107–21.
106. Wiebe JP, Salhanick AI, Myer KI. On the mechanism of action of lead in the testis: in vitro suppression of FSH receptors, cyclic AMP and steroidogenesis. *Life Sci* 1983; 32:1997–2005.
107. Rodamilans M, Martinez-Osaba J, To-Figueras J, Rivera-Fillat F, Torra M, Perez P, Corbella J. Inhibition of intratesticular testosterone synthesis by inorganic lead. *Toxicol Lett* 1988; 42:285–90.
108. Thoreux-Manlay A, Le Goascogne C, Segretain D, Jegou B, Pinon-Lataillade P. Lead affects steroidogenesis in rat Leydig cells in vivo and in vitro. *Toxicology* 1995; 103:53–62.
109. Thoreux-Manlay A, Velez de la Calle JF, Olivier MF, Soufir JC, Masse R, Pino-Lataillade G. Impairment of testicular endocrine function after lead intoxication in the adult rat. *Toxicology* 1995; 100:101–9.
110. Liu MY, Leu SF, Yang HY, Huang BM. Inhibitory mechanisms of lead on steroidogenesis in MA-10 mouse Leydig tumor cells. *Arch Androl* 2003; 49:29–38.
111. James WH. Offspring sex ratios at birth as markers of paternal endocrine disruption. *Environ Res* 2006; 100:77–85.
112. Dickinson H, Parker L. Do alcohol and lead change the sex ratio. *J Theor Biol* 1994; 169:313–15.
113. Ansari-Lari M, Saadat M, Hadi N. Influence of GSTT1 null genotype on the offspring sex ratio of gasoline filling station workers. *J Epidemiol Commun Health* 2004; 58:393–4.
114. James WH. Further evidence that mammalian sex ratios at birth are partially controlled by parental hormone levels around the time of conception. *Human Reprod* 2004; 19:1250–56.
115. Simonsen CR, Roge R, Christiansen U, Larsen T, Bonde JP. Effects of paternal blood levels on offspring sex ratio. *Reprod Toxicol* 2006; 22:3–4.
116. Clarkson TW. The three modern faces of mercury. *Environ Health Perspect* 2002; 110(Suppl 1):11–23.
117. Nierenberg DW, Nordgren RE, Chang MB, Siegler RW, Blayney MB, Hochberg F, Toribara TY, Cernichiari E, Clarkson T. Delayed cerebellar disease and death after accidental exposure to dimethylmercury. *N Engl J Med* 1998; 338:1672–6.
118. Mottet NK, Shaw CM, Burbacher TM. Health risks from increases in methylmercury exposure. *Environ Health Perspect* 1985; 63:133–40.
119. Clarkson TW, Magos L, Myers GJ. The toxicology of mercury – current exposures and clinical manifestations. *N Engl J Med* 2003; 349:1731–7.
120. De Rosis F, Anastasio SP, Selvaggi L, Beltrame A, Moriani G. Female reproductive health in two lamp factories: effects of exposure to inorganic mercury vapour and stress factors. *Br J Ind Med* 1985; 42:488–94.
121. Sikorski R, Juszkiewicz T, Paszkowski T, Szprengier-Juszkiewcz T. Women in dental surgeries: reproductive hazards in occupational exposure to metallic mercury. *Int Arch Occup Environ Health* 1987; 59:551–7.

122. Rowland AS, Baird DD, Weinberg CR, Shore DL, Shy CM, Wilcox AJ. The effect of occupational exposure to mercury vapour on the fertility of female dental assistants. *Occup Environ Med* 1994; 51:28–34.
123. Schuurs AHB. Reproductive toxicology of occupational mercury. A review of the literature. *J Dent* 1999; 27:249–56.
124. Lamperti AA, Printz RH. Effects of mercuric chloride on the reproductive cycle of the female hamster. *Biol Reprod* 1973; 8:378–87.
125. Davis BJ, Price HC, O'Connor RW, Fernando R, Rowland AS, Morgan DL. Mercury vapor and female reproductive toxicology. *Toxicol Sci* 2001; 59:291–6.
126. Facemire CF, Gross TS, Guillette LJ Jr. Reproductive impairment in the Florida panther: nature or nurture. *Environ Health Perspect* 1995; 103(Suppl 4):79–86.
127. Mansfield KG, Land ED. Cryptorchidism in Florida panthers: prevalence, features, and influence of genetic restoration. *J Wildl Dis* 2002; 38:693–98.
128. Newman J, Zillioux E, Rich E, Liang L, Newman C. Historical and other patterns of monomethyl and inorganic mercury in the Florida panther (*Puma concolor coryi*). *Arch Environ Contam Toxicol* 2004; 48:75–80.
129. Vachhrajani KD, Chowdhury AR. Distribution of mercury and evaluation of testicular steroidogenesis in mercuric chloride and methylmercury administered rats. *Indian J Exp Biol* 1990; 28:746–51.
130. Nagar RN, Bhattacharya L. Effect of mercuric chloride on testicular activities in mice, Musculus albinus. *J Environ Biol* 2001; 22:15–8.
131. Ramalingam V, Vimaladevi V, Rajeswary S, Suryavathi V. Effect of mercuric chloride on circulating hormones in adult albino rats. *J Environ Biol* 2003; 24:401–4.
132. Rao MV, Sharma PSN. Protective effect of vitamin E against mercuric chloride reproductive toxicity in male mice. *Reprod Toxicol* 2001; 15:705–12.
133. Tong MH, Christenson LK, Song WC. Aberrant cholesterol transport and impaired steroidogenesis in Leydig cells lacking estrogen sulfotransferase. *Endocrinology* 2004; 145:2487–97.
134. Sukocheva OA, Yang Y, Gierthy JF, Seegal RF. Methyl mercury influences growth-related signaling in MCF-7 breast cancer cells. *Environ Toxicol* 2005; 20:32–44.
135. Gierthy JF, Lincoln DW, Roth KE, Bowser SS, Bennett JA, Bradley L, Dickerman HW. Estrogen-stimulation of postconfluent cell accumulation and foci formation of human MCF-7 breast cancer cells. *J Cell Biochem* 1991; 45:177–87.
136. Li X, Zhang S, Safe S. Activation of kinase pathways in MCF-7 cells by 17β-estradiol and structurally diverse estrogenic compounds. *J Steroid Biochem Mol Biol* 2006; 98:122–32.
137. Brugge D, Goble R. The history of uranium mining and the Navajo people. *Am J Public Health* 2002; 92:1410–9.
138. Brugge D, de Lemos JI, Oldmixon B. Exposure pathways and health effects associated with chemical and radiological toxicity of natural uranium: a review. *Rev Environ Health* 2005; 20:177–93.
139. Goldman M, Yaari A, Doshnitzki Z, Cohen-Luria R, Moran A. Nephrotoxicity of uranyl acetate: effect on rat kidney brush border membrane vesicles. *Arch Toxicol* 2006; 80:387–93.
140. Orloff KG, Mistry K, Charp P, Metcalf S, Marino R, Shelly T, Melaro E, Donohoe AM, Jones RL. Human exposure to uranium in groundwater. *Environ Res* 2004; 94:319–26.
141. Maynard E, Dodge H. Studies of the toxicity of various uranium compounds when fed to experimental animals. In: Voeglen C, ed. *Pharmacology and Toxicology of Uranium Compounds*, New York: McGraw-Hill, 1949:309–76.
142. Arfsten DP, Still KR, Ritchie GD. A review of the effects of uranium and depleted uranium exposure on reproduction and fetal development. *Toxicol Ind Health* 2001; 17:180–91.
143. Domingo JL. Reproductive and developmental toxicity of natural and depleted uranium: a review. *Reprod Toxicol* 2001; 15:603–9.
144. Maconochie N, Doyle P, Carson C. Infertility among male UK veterans of the 1990-1 Gulf war: reproductive cohort study. *BMJ* 2004; 329:196–201.
145. Doyle P, Maconochie N, Davies G, Maconochie I, Pelerin M, Prior S, Lewis S. Miscarriage, stillbirth and congenital malformation in the offspring of UK veterans of the first Gulf war. *Int J Epidemiol* 2004; 33:74–86.
146. McDiarmid MA, Engelhardt S, Oliver M, Guccer P, Wilson PD, Kane R, Kabat M, Kaup B, Anderson L, Hoover D, Brown L, Handwerger B, Albertini RJ, Jacobson-Kram D, Thorne CD, Squibb KS. Health effects of depleted uranium on exposed Gulf war veterans: a 10-year follow-up. *J Toxicol Environ Health A* 2004; 67:277–96.

147. McDiarmid MA, Keogh JP, Hooper FJ, McPhaul K, Squibb K, Kane R, DiPino R, Kabat M, Kaup B, Anderson L, Hoover D, Brown L, Hamilton M, Jacobson-Kram D, Burrows B, Walsh M. Health effects of depleted uranium on exposed Gulf war veterans. *Environ Res* 2000; 82:168–80.
148. McDiarmid MA, Squibb K, Engelhardt S, Oliver M, Gucer P, Wilson PD, Kane R, Kabat M, Kaup B, Anderson L, Hoover D, Brown L, Jacobson-Kram D. Depleted uranium follow-up program. *J Occup Environ Med* 2001; 43:991–1000.
149. Muller C, Ruzicka L, Bakstein J. The sex ratio in the offsprings of uranium miners. *Acta Univ Carol Med* 1967; 13:599–603.
150. Domingo JL, Paternain JL, Llobet JM, Corbella J. The developmental toxicity of uranium in mice. *Toxicology* 1989; 55:143–52.
151. Hindin R, Brugge D, Panikkar B. Teratogenicity of depleted uranium aerosols: a review from an epidemiological perspective. *Environ Health* 2005; 4:17.
152. Shields LM, Wiese WH, Skipper BJ, Charley B, Benally L. Navajo birth outcomes in the Shiprock uranium mining area. *Health Phys* 1992; 63:542–1.
153. Whish SR, Mayer LP, Robinson ZD, Layton TM, Hoyer PB, Dyer CA. Uranium is an estrogen mimic and causes changes in female mouse reproductive tissues. The Endocrine Society 87th Annual Meeting 2005; Abstract P2-261.
154. Getz JE, Sellers M, Whish SR, Dyer CA. Uranium mimics the effects of 17β-estradiol, mediating rapid cell surface morphological changes in MCF-7 human breast cancer cells. *Society for the Study of Reproduction 38th Annual Meeting* 2005; Abstract 64.
155. Clode SA. Assessment of in vivo assays for endocrine disruption. *Best Pract Res Clin Endocrinol Metab* 2006; 20:35–43.
156. Gordon MN, Osterburg HH, May PC, Finch CE. Effective oral administration of 17β-estradiol to female C57BL/6J mice through the drinking water. *Biol Reprod* 1986; 35:1088–95.
157. Zhang D, Trudeau VL. Integration of membrane and nuclear estrogen receptor signaling. *Comp Biochem Physiol A Mol Integr Physiol* 2006; 144:306–15.
158. Song RX, McPherson RA, Adam L, Bao Y, Shupnik M, Kumar R, Santen RJ. Linkage of rapid estrogen action to MAPK activation by ERalpha-Shc association and Shc pathway activation. *Mol Endocrinol* 2002; 16:116–27.
159. Lin RH, Wu LJ, Lee CH, Lin-Shiau SY. Cytogenetic toxicity of uranyl nitrate in Chinese hamster ovary cells. *Mutat Res* 1993; 319:197–203.
160. Stearns DM, Yazzie M, Bradley AS, Coryell VH, Shelley JT, Ashby A, Asplund CS, Lantz RC. Uranyl acetate induces hprt mutations and uranium-DNA adducts in Chinese hamster ovary EM9 cells. *Mutagenesis* 2005; 20:417–23.
161. Tabb MM, Blumberg B. New modes of action for endocrine-disrupting chemicals. *Mol Endocrinol* 2006; 20:475–82.

6 Cellular Mechanisms of Endocrine Disruption

Traditional and Novel Actions

Stuart R. Adler, MD, PhD

Contents

1. ENDOCRINE DISRUPTION AS A HEALTH ISSUE

It is now well recognized that chemicals and compounds used in our industrialized society may act as endocrine disrupters; yet, the implications on human health and the health of wildlife remain unclear and controversial. Perhaps, the first recognized possible effects of these compounds were their potential to act on reproductive and developmental pathways. Indeed, in the laboratory, many compounds such as detergents, insecticides, plasticizers, and dioxins can be classified as weak estrogens and as potential endocrine disrupters. Specific determination of the human health implications of these environmental estrogens are still being defined and are extremely controversial, and have been for over a decade *(1,2)*. Rises in the rate of birth defects, including cryptorchidism and hypospadias *(3)*, decreased sperm counts *(4)*, breast cancer *(5,6)*, as well as neurological defects *(7–10)* all have been proposed and investigated for relationships to exposures to chemical estrogens and to endocrine disrupters acting through other endocrine systems.

From: *Endocrine-Disrupting Chemicals: From Basic Research to Clinical Practice*
Edited by: A. C. Gore © Humana Press Inc., Totowa, NJ

Contributing to this persistent controversy is a paucity of information on the specific biological effects of chemicals alone or in combinations likely to be encountered from environmental exposures. Three major arguments originally raised 10 years ago still remain active areas of concern *(2)*; although, with the passage of time and increasing evidence from research, some progress has been made in addressing these issues. The first concern has been whether the impact of estrogenic endocrine disrupters on human health may have been overstated because many of these compounds bind to estrogen receptor (ER) only weakly. Second, humans may exhibit less sensitivity to these compounds than other species. The final concern is whether certain compounds may act as anti-estrogens and balance the effects of other compounds acting as estrogen agonists *(2)*. Additional issues that have been raised more recently include, on the one hand, the possibility of potential favorable adaptations to compensate for and mitigate the effects of endocrine disrupter compounds and, on the other hand, persistent and transgenerational effects continuing in the absence of ongoing exposure to endocrine disrupters *(11–14)*. All these considerations have important implications in the process of risk assessment for exposures to these "environmental estrogens." All of these concerns can be addressed by careful in vitro and in vivo studies that consider the biological mechanisms through which these agents affect human biology and appropriate comparisons to the biology of wildlife and laboratory animals.

2. WHICH HORMONES CAN BE MIMICKED OR DISRUPTED BY EDCs?

Although much of the initial concern regarding endocrine-disrupting chemicals (EDCs) focused on environmental estrogens, other hormonal systems are also susceptible to disruption. Androgens, through agonist or antagonist interactions with the androgen receptor, notably through metabolites of DDT *(15)* or phthalates *(4,16)*, or through metabolic effects on hormone synthesis *(17)* or metabolism, including aromatase *(18,19)*, also can be targets for endocrine disruption. Thyroid hormone can also be considered as a target for endocrine disruption *(10)*. Recently, the peroxisome proliferator activator receptor (PPAR)/retinoid X receptor (RXR) system has also been identified as a target for disruption through organotin agents that are used as marine-defouling agents *(20,21)*.

2.1. Environmental Estrogens and Endocrine Disruption

For the remainder of this chapter, the focus will be on environmental estrogens and disruption of estrogenic pathways. The spectrum of chemical compounds to which we may be exposed is broad. These chemicals, including organochlorine pesticides, polychlorinated biphenyls, alkyl phenols, dibenzo-*p*-dioxins, and dibenzofurans—although structurally unrelated to estradiol, are linked by their potential to influence estrogenic pathways. Furthermore, these chemicals can be widespread, long lasting, and accumulate in adipose tissue in our food chain and ultimately in humans *(7)*. The potential of many of these chemicals to persist in the environment and in bodies of women long after exposure, coupled with possible epigenetic effects lasting generations even in the absence of continued exposure are particularly threatening.

2.1.1. Estrogen Action in Health and Disease

The functions of estrogens in cellular metabolism and signaling have been increasingly expanded beyond the now classical and conventional functions mediated by a nuclear receptor and identified through molecular biology as ERα. Not only are we now aware of the ERβ *(22–27)*, but it is now generally accepted that ERs have, in addition to gene activation effects, effects on gene repression *(28)*, and effects through non-genomic mechanisms mediated by membrane ERs *(29–31)*, the possible participation of G-coupled receptors *(32–34)*, an increasing understanding of roles for ERs in the mitochondria *(29,35,36)*, and distinct metabolic pathways with the potential to affect the cellular status through oxidative metabolites *(37)*. Each one of these identified roles for the natural hormone estradiol and its metabolites or derivatives is a potential target for chemicals that act as mimics, antagonists, or affect the metabolic pathways by increasing or decreasing steady-state levels of the receptors themselves. Compounds may also induce enzymes that accelerate the metabolism of estradiol. This type of activity has been reported for dioxins *(38,39)*. A review of all of these functions is beyond the intended scope of this review. The remainder of the chapter will focus on actions of the two nuclear receptors, ERα and ERβ, and the genomic or nuclear activities that may be subject to disruption by EDCs.

3. NUCLEAR GENE REGULATION THROUGH ERS

The ERs (α and β) and the other steroid hormone receptors are part of a superfamily of nuclear transcription factors *(40)*. These intracellular receptors are the targets for a variety of systemic hormones that regulate a multitude of bodily functions including male and female sexual development, reproductive and daily cycles, maintenance of metabolic homeostasis, the physiologic responses to stress, fluid and electrolyte regulation, and maintenance of bone density *(41)*. These intracellular receptors directly participate in the cellular response to hormones, altering nuclear gene expression *(40)*. Our basic concept of the classical ER, ERα, as a transcription factor is well established and well reviewed *(40)*. Most of our knowledge of ERα function is based on the ability of ligand-activated ER to increase the transcription of target genes in target tissues *(40)*. The basic model of positive gene regulation is shared by all the steroid hormone receptors *(40)*. Although there is only limited data available on the physiology of ERβ, this receptor has been shown to activate gene expression in transfection assays using reporter genes containing a classical estrogen response elements (EREs) as determined from studies of ERα *(23–25)*.

Alternatively, steroid hormones and their receptors can decrease specific gene transcription *(28)*. This negative form of regulation may be equally important in understanding hormone action; yet, gene repression has been less well characterized *(28)*. Several possible mechanisms have been proposed to explain how genes may be reversibly negatively regulated by steroid hormone receptors, as well as by other factors *(28)*. The molecular details of these mechanisms differ significantly from positive regulation *(28)*. It is apparent that interaction with other transcription factors plays a dominant role in negative regulation and that target genes do not require the classic ERE-binding sites *(40,42)*. It is likely that these dissimilar mechanisms of gene regulation require participation of different receptor domains, features, or structures.

A third form of regulation, termed composite regulation, involves interactions between activated steroid receptor and members of the AP-1 complex, Fos and Jun *(43–48)*. Composite regulation can result in either gene activation or gene repression *(43)*. Although many details and potential interactions make this type of regulation difficult to study, at least one simple model system has been developed for examining composite regulation by AP-1 and ER *(49)*. Two types of regulatory effects were studies using model AP-1 and ERE reporter genes *(49)*. Tamoxifen had profound stimulatory effects on an AP-1-containing promoter in cell lines of uterine (Ishikawa) but not breast (MCF-7 or ZR75) origin *(49)*. This interaction required both ER and Jun, and was independent of an ERE-binding site *(49)*. This is in contrast to ERE-containing reporters, which were activated by estradiol (E2), but not by tamoxifen *(49)*. This functional linkage of ER to AP-1 provides a potential mechanism for estrogenic growth stimulation. Furthermore, the cell-type-specific effects of tamoxifen on model AP-1 reporters is similar to the cell-type-specific effects of tamoxifen on growth, that is growth stimulation in uterus, but growth repression in breast.

It is not clear how negative, positive, and composite regulation together participate in the complex cellular functions we associate with the actions of estrogens, such as differentiation, organ development, and growth. Although changes in gene expression must certainly participate in these complex responses, these changes likely include negative regulation of cellular gene inhibitors, *(50,51)* as well as increased expression of cellular gene activators.

3.1. Ligands can Selectively Influence Receptor Regulatory Competence

Surprisingly, ligand can play a selective role in distinguishing the competence of ERs to function in different mechanisms of gene regulation. If this were not the case, all ligands would perform the same functions, which has certainly not been observed. In humans, ERα is encoded by a unique, single-copy gene *(52)*, and all functions of ERα, therefore, are derived from the same gene. The ERα protein itself contains two activation domains, an N-terminal gene activation domain (AF-1, formerly TAF-1) and a C-terminal domain (AF-2, formerly TAF-2) *(53,54)*.

Studies have demonstrated that both AF-1 and AF-2 can participate in E2-dependent gene activation functions of ERα *(53,55)*. However, there is both cell-type specificity and promoter specificity *(53,55)*. Hydroxytamoxifen has been shown to allow gene activation—that is to act as an agonist—for genes activated by the AF-1 functional domain of ERα, but not AF-2 *(56)*. Promoters that require the AF-2 function for gene activation were not activated, and the effects of E2 were blocked by hydroxytamoxifen *(56)*. This was not observed with ERβ *(22)*. In contrast to this selective activity of tamoxifen, ICI 164,384, a pure antagonist compound, was not competent as a ligand for gene activation by ER from any promoter and always blocked the actions of E2 *(56)*.

Work in my laboratory has demonstrated that tamoxifen can act either as an agonist or an antagonist depending on the molecular function of ERα *(57)*. This was demonstrated for specific transcriptional effects on two distinct target genes *(57)*. These differential activities cannot be explained by the presence of different tamoxifen metabolites, or cell-type or species differences *(57)*. Squelching and non-receptor-mediated effects were also excluded as possible mechanisms *(57)*. Similar results with clomiphene also indicate that these results are not exclusive to tamoxifen *(57)*. Rather, different functional activities and conformational requirements distinguish positive and negative regulation by the ER. Our data show tamoxifen has gene-specific effects acting as either an agonist or an antagonist, depending on the molecular mechanism and the corresponding functional requirements for activated receptor *(57)*. Work from other investigators also illustrates differential effects comparing gene activation on an ERE compared to composite regulation by ERs on an AP-1 site *(58)*. It thus is important to evaluate the activity of potential estrogenic EDCs using a variety of assays that illustrate a range of mechanisms used by ER for gene regulation.

All these data are important because they support the notion that individual ligands for ER do not exhibit the same effects *(59)*. Responses may differ in various target tissues and for different functional roles that ER may play. Certainly, one must also consider whether compounds may exhibit different activities when evaluated using ERα versus ERβ. This is apparently the case for hydroxytamoxifen, which acts as an agonist with ERα in Cos-1 cells but does not with ERβ *(25)*. Evaluation of chemicals for biological effects must therefore consider regulation of more than just one "model" gene, and must take into account different mechanisms of gene regulation, as well as the distinct effects through ERα and ERβ.

3.2. What do we Know About ERβ

It has now been over a decade since a new form of ER, ERβ has been identified *(22–27)* and first raised the possibility that phytoestrogens and environmental estrogens may also influence human health by effects on pathways regulated by this form of the nuclear receptor. Clones for ERβ have been isolated from mice, rats, and humans, indicating that ERβ is evolutionarily conserved. The DNA-binding domain of ERβ is highly similar to that of ERα, and the same ERE identified for ERα will also serve to bind ERβ and to confer regulation by ERβ. In contrast to the high similarity of DNA-binding domains, other parts and domains of the two ERs are not as similar. This suggests a potential for different ligand specificities and different regulatory roles or effects, including different interactions with coregulators. Competitive binding studies on expressed ERβ indicate a slightly higher affinity for the soy isoflavone and phytoestrogen, genistein, than observed with ERα *(27,60)*, and efforts in many laboratories are directed toward identifying compounds exhibiting markedly selective activity as ligands and defining distinct biological roles for each receptor subtype *(61–66)*. Interestingly, a natural compound has recently been proposed to be an ERβ specific ligand *(67)* that may activate prostatic ERβ and play role in preventing progression to prostate cancer *(68)*. Also, ERα and ERβ exhibit different effects with non-steroidal ligands (e.g., raloxifene) in some mechanisms of gene regulation *(22,58)*. Expression of ERβ mRNA has been reported in the testis, ovary, prostate, and brain *(69,70)*. This pattern suggests potential roles for ERβ in both male and female reproduction and development, as well as in central neuroendocrine regulation. While initial reports indicated that ERβ is not expressed in breast, more recent studies indicate ERβ can be found in breast *(71–73)*.

Recently, several accessory factors and/or co-regulators that interact with members of the nuclear receptor family have been identified *(74,75)*. Some of these cellular factors participate with activated ERα in mediating its transcriptional effects *(74,75)*. The interaction between ERα and these factors can be ligand dependent, as successful association is observed in the presence of estradiol but not with antagonists *(74,75)*.

These data extend our molecular understanding of how estrogens and the ER work. Because these interactions distinguish agonists from antagonists, they can provide a novel molecular test for estrogenic ligands. The limited information available for interaction of ERβ with cellular coregulators shows that, like ERα, ERβ associates with steroid receptor co-activator (SRC)-1 and that SRC-1 enhances the transcriptional activation observed with ERβ *(25)*. As with ERα, this association is ligand dependent and is not observed with the antagonists hydroxytamoxifen and ICI 182,780 *(25)*. However, there may be distinct differences in the interactions between coregulators and the ERs beyond the p160 family. TRAP220 is reported to preferentially interact with ERβ *(76)*. This selectivity of coregulators could therefore contribute to cell-type and receptor specificity of responses, not only for E2 but possibly to environmental estrogens and to dietary phytoestrogens as well.

3.3. Additional Mechanistic Considerations for Endocrine Disrupters

Key to a mechanistic evaluation of environmental estrogens is an understanding of how non-steroidal compounds, such as environmental estrogens, can mediate estrogenic biological effects. In addition to actions through ERs, including agonist, antagonist, and mixed types of activities, actions based on other activities also must be considered.

One possibility is that alternative "estrogenic receptors" may exist and mediate parallel pathways of estrogenic regulation. Evaluation of this hypothesis could certainly include analyses of the orphan receptors called ER-related receptors (ERRs), ERR1 and ERR2, members of the steroid hormone receptor superfamily isolated based on a marked sequence similarity to the ER *(77)*. One might also consider actions through weaker estradiol-binding sites or the so-called anti-estrogen triphenyl-ethylene-binding sites classically identified by tamoxifen binding.

It has been suggested that certain compounds that can act as ligands for ER also function through non-receptor-mediated pathways. Effects of tamoxifen on calmodulin regulation have been invoked to explain some of the actions of tamoxifen, and the interaction with calmodulin is not ER mediated *(78)*. Tamoxifen has also been shown to inhibit protein kinase C directly through non-ER mechanisms *(79,80)*. It is therefore conceivable that some compounds could exert apparent estrogenic growth effects by interacting directly with other cellular factors. This could occur downstream from ER in an estrogenic regulatory cascade in the target cells, perhaps by direct interactions with proteins, with which activated ER interacts. Alternatively, growth-promoting targets might participate in completely different regulatory pathways.

Another attractive hypothesis is that a biotransformation takes place in vivo that converts the endocrine disrupters from inactive compounds to active, ER-binding estrogens. It is worthwhile to note that biotransformation has been described for tamoxifen *(81)*. 4-OH-tamoxifen, a metabolite of tamoxifen, can be isolated from animals and humans treated with tamoxifen *(81)*. 4-OH-tamoxifen is also a potent anti-estrogen, with an even higher affinity for ER than for its parent compound *(81)*. For tamoxifen, it is not known whether this conversion takes place in the target cells or only in the liver or other distinct organs in the intact animal *(81)*. It is worth noting that MCF-7 human breast cells are competent to hydroxylate, methylate, and conjugate estradiol, and that some chemicals, such as dioxins, can alter the level of these metabolic activities *(38,39)*. There are many EDCs that are metabolically converted to active agents, including ethylene glycol monomethyl ester (EGME) *(82)* and methoxychlor *(83)*. If a biotransformation is required for activity, the catalytic enzymes required for the activation of the compounds may be ubiquitous or the activated compounds may be available systemically. Alternatively, if these transforming enzymes are limited in cell-type distribution, the effects of compounds may be cell-type specific, observed primarily in target cells defined by their capacity to metabolically activate these compounds. Such a cell-type specificity has an added difficulty in terms of risk assessment and health implications as it is not only difficult to determine the identity of the final active compound but it may also be difficult to identify the effects of the parent compound in screening tests.

4. EVALUATING CANDIDATE EDCs

Many compounds have been identified as potential EDCs. Evaluation of several of these compounds will be presented to illustrate the range of actions these compounds exhibit in conventional activities of estrogenic gene regulation. These compounds include

1. Bisphenol A *(84,85)*, used as a component in many resins and polycarbonate plastics. These have uses in the food industry in canned foods and plastic containers, and in dentistry in plastic fillings and as a coating on teeth.
2. *p*-Nonyl phenol, *(86–89)* an alkylphenol detergent used not only as a cleaning agent but also in a variety of processes as a dispersant, and in plastics as a stabilizer.
3. Diethylhexylphthalate and dipentyl phthalate, plasticizers used to make plastics more flexible. Both are found in the food industry as a part of foil/plastic heat seal caps *(90)*.
4. Pesticides including lindane, *(91)* as well as chlordane, dieldrin, and p, p′-dichlorodiphenyldichloroethene (DDE) a metabolite of DDT.

4.1. Testing Compounds for Specific Binding to Human ER

One mechanism by which a compound may directly affect estrogenic pathways as an endocrine disrupter, whether as an agonist or an antagonist, is by binding to ER. Previously used methods for determining binding to ER rely on competition with radioactive or fluorescent ligands. While these approaches have been useful, certain compounds may not be accurately assayed using these techniques. This may apply if their affinity is much less than that of estradiol or if the kinetics of their binding differ from that of estradiol. In these cases, a direct binding assay might be more useful. Direct binding assays have their own problems, particularly if the ligand must be modified, by making either radioactive or fluorescent derivatives, as these derivatives may differ in binding affinity from the parent compound. Also, binding assays require a preparation of ER. Many assays use animal uteri as a source of ER or use tissue culture cells as a source of ER. For translation of assay results to humans, a source of human ER is desirable, as the effects of non-steroidal endocrine disrupters may be species specific.

Our studies have applied an assay for measuring the direct binding of a potential chemical ligand to ERs. This assay does not rely on competition with E2. It does not require chemical modifications of compounds, processes that may alter their affinity for receptor. It also does not require radioactive labeling which may be costly or also require chemical modifications. The assay uses in vitro translated ERs as a readily available source of human ERs. The use of human receptors represents an advantage over the use of tissues from other species for studies of effects in humans.

The assay is a ligand-dependent electromobility-gel shift assay *(92–96)*, and it is performed using in vitro translated ERs prepared in reticulocyte lysates *(92,95,96)*. Translated ER is readily inactivated by a heat treatment in the absence of ligand. In contrast, ligand, whether E2, tamoxifen, or even ICI 164,384, stabilizes the receptor, allowing it to bind to an endlabeled duplex ERE oligonucleotide in a subsequent binding reaction *(92–96)*. Active, DNA-bound receptor is resolved on a low percentage polyacrylamide gel, and the fraction of oligonucleotide shifted in the gel, quantitated by PhosphorImager analysis, reflects the amount of translated ER that has bound ligand.

We have used this assay on a variety of compounds, including known ligands and compounds with previously unknown binding properties. Experiments have included RU486, the progesterone receptor/glucocorticoid receptor antagonist, as well as dexamethasone and progesterone, steroidal compounds that are not ligands for ER. Known estrogens included clomiphene, nafoxidene, diethylstilbestrol, and estradiol. In all cases, ligands—whether agonists or antagonists—show activity, and non-ligands do not (data not shown). Even RU486, which is ordinarily not considered a ligand for ER, scores as a ligand. This, however, only serves to validate the assay further, because RU486 does have estrogenic effects and RU486 can interact with ER *(97)*. Our results indicate that this assay is concentration dependent for ligands and that it displays the appropriate specificity. We have demonstrated binding for both agonists and antagonists, and also have shown that compounds that do not bind to ER, including dexamethasone and progesterone, do not score in this assay (data not shown).

Figure 1 shows the results obtained for two environmental estrogens, *p*-nonyl phenol, and bisphenol A. Similar assays were performed with dioctyl phthalate and lindane. We detected no binding to ERα by lindane and dioctyl phthalate (not shown). Both *p*-nonyl phenol and bisphenol A bound to the ER with relative binding affinities of approximately 1/300 that of estradiol. Bisphenol A achieved binding comparable to estradiol, but required high concentrations to achieve complete binding. *p*-Nonyl phenol was not able to achieve full binding, achieving only 60% of full binding. The reason for this is not known. Some previous studies of *p*-nonyl phenol have also shown less than full estrogenic activity for *p*-nonyl phenol even at high concentrations. Our results, therefore, may either reflect a limitation of our binding assay or, perhaps, a limitation in the ability of *p*-nonyl phenol to interact with ERα.

4.2. Testing Compounds for Transcriptional Regulation of Specific Estrogen-Regulated Model Reporter Genes

Compounds may differ in their ability to activate ER to perform the different mechanisms of gene regulation. Therefore functional analysis of each compound must consider testing for activity in gene regulation for distinct mechanisms and for the possibility of agonist and antagonist activity.

4.2.1. Estrogenic Positive Regulation

Transient co-transfection assays for studies of positive gene regulation by steroid hormone receptors are widely used and well established. We have previously used this type of assay for structure-function studies of the human ER *(54)* to investigate

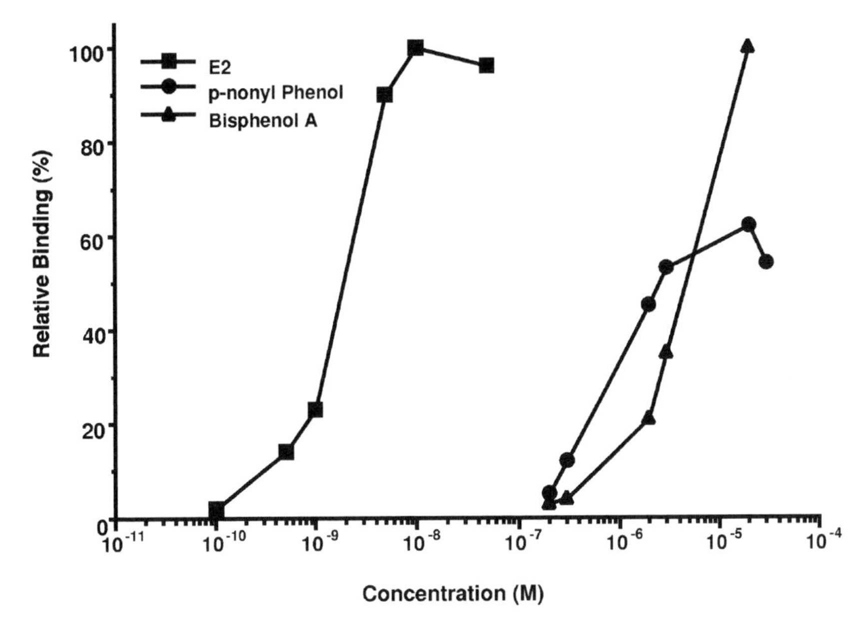

Fig. 1. Relative binding of environmental estrogens. Dose–response profiles for direct binding to human estrogen receptor α were obtained using a modified gel shift assay (section 4.1). No direct binding was detected for dioctyl phthalate and lindane (not shown). Results are included for estradiol. Relative binding affinity can be estimated from these data as approximately 2.5×10^{-3} M for *p*-nonyl phenol and approximately 5×10^{-4} M for bisphenol A.

estrogenic activity of RU486 *(97)* and tamoxifen *(57)*, and for studies of the unusual estrogens, the doisynolic acids *(95,96,98)*. Similar assays have been used in the field of environmental endocrine disrupters to analyze effects of DDT metabolites as androgen antagonists *(15)*. All assays are performed under estrogen-free conditions by using phenol red-free media and charcoal-stripped sera *(42,54,98)*.

Figure 2 shows the results of gene activation assays in HeLa cells for each of a panel of four environmental estrogens: *p*-nonyl phenol, bisphenol A, dioctyl phthalate, and lindane. Whereas *p*-nonyl phenol and bisphenol A are competent to mediate gene activation of an ERE-containing reporter gene with activities comparable to E2, dioctyl phthalate and lindane are minimally active, if at all.

For each compound, dose–response profiles were determined using a range of concentrations in these assays (Fig. 3). These dose–response profiles show that despite lower potency, *p*-nonyl phenol and bisphenol A are capable of full activity in assays of gene activation. Only slight activation by lindane and dioctyl phthalate is seen at the highest concentrations.

4.2.1.1. Activity Dependence on ER. Two types of experiments were performed to demonstrate the ER-dependence of activation by the environmental estrogens. First, the activities of the compounds were tested in the presence of competing concentrations of ICI 164,384 (Fig. 2, hatched bars). The steroidal estrogen antagonist ICI 164,384 has been shown to act as an antagonist for ERα, as well as ERβ for conventional gene activation, as demonstrated by its effects with 17β-estradiol in this assay. Antagonism

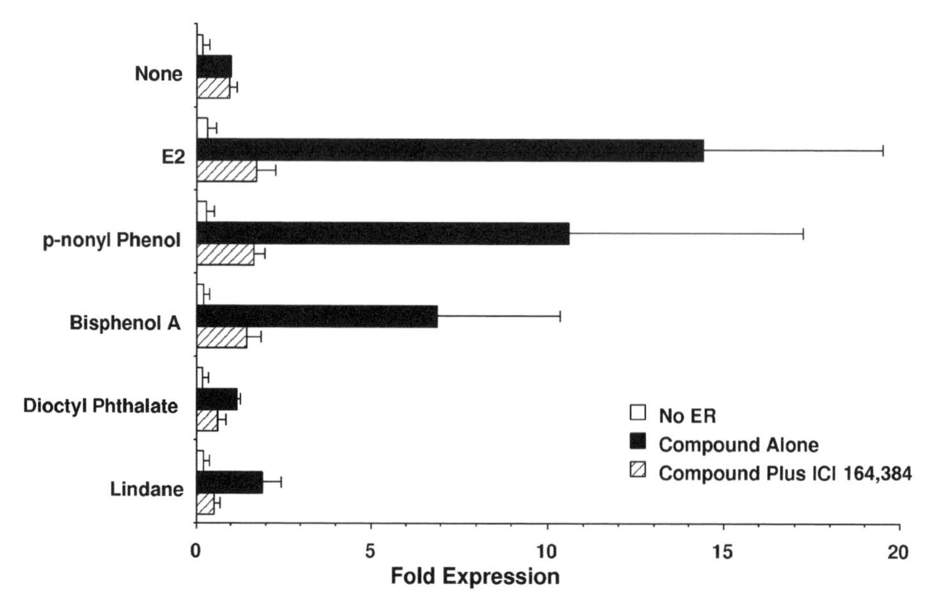

Fig. 2. Environmental compounds act as agonists for estrogenic gene activation. For gene activation, hela cells were cotransfected with VIT2P36L reporter plasmid plus RSV HEG0 for expression of estrogen receptor α (ERα), or RSV Neo, as a control. Concentrations for each compound were added as indicated. Light units are expressed as folds relative to the activity observed with expressed estrogen receptor (ER) in the absence of added compounds set to 1. Data are the mean and SEM from three experiments. Open bars, activity of each compound with expressed neomycin phosphotransferase II (Neo) control; filled bars, activity of each compound with expressed ERα; hatched bars, activity of each compound with expressed ERα in the presence of 100 nm ICI 164,384. None, no compound added; E2, estradiol at 3×10^{-10} M; *p*-nonyl phenol, 1×10^{-5} M; bisphenol A, 1×10^{-5} M; dioctyl phthalate, 1×10^{-5} M; lindane, 1×10^{-5} M.

of their activity by ICI 164,384 is one additional way to determine a receptor-dependent mechanism and may suggest interaction of the compound at the receptor ligand-binding site. As shown, ICI 164,384 blocks the specific gene activation effect of *p*-nonyl phenol and bisphenol A. This indicates that ER participates in the observed gene activation.

Second, we used an expression vector for neomycin phosphotransferase II (Neo), rather than ERα, to demonstrate the requirement for ER (Fig. 2, empty bars). In the absence of expressed ER, when the control gene *Neo* is expressed instead, neither estradiol nor the environmental estrogens can effectively activate the target gene. This type of experiment also demonstrates that gene activation of this ERE-containing model reporter gene by the environmental compounds requires functional ER.

Compounds showing only weak or no activity as agonists, or exhibiting atypical dose–response profiles, can be tested for antagonist activity. Each compound can be tested alone and in the presence of a concentration of estradiol that activates gene expression, typically 3×10^{-10} M. Controls of no hormone, estradiol alone, ICI 164,384 (an antagonist), and ICI 164,384 plus estradiol, are appropriately included. Typically, a single high concentration of each compound is tested and several independent experiments are combined. These experiments to determine whether the compounds have antagonist activity were performed with dioctyl phthalate and lindane (Fig. 4). The activity of estradiol was not affected by the presence of either compound; thus, neither

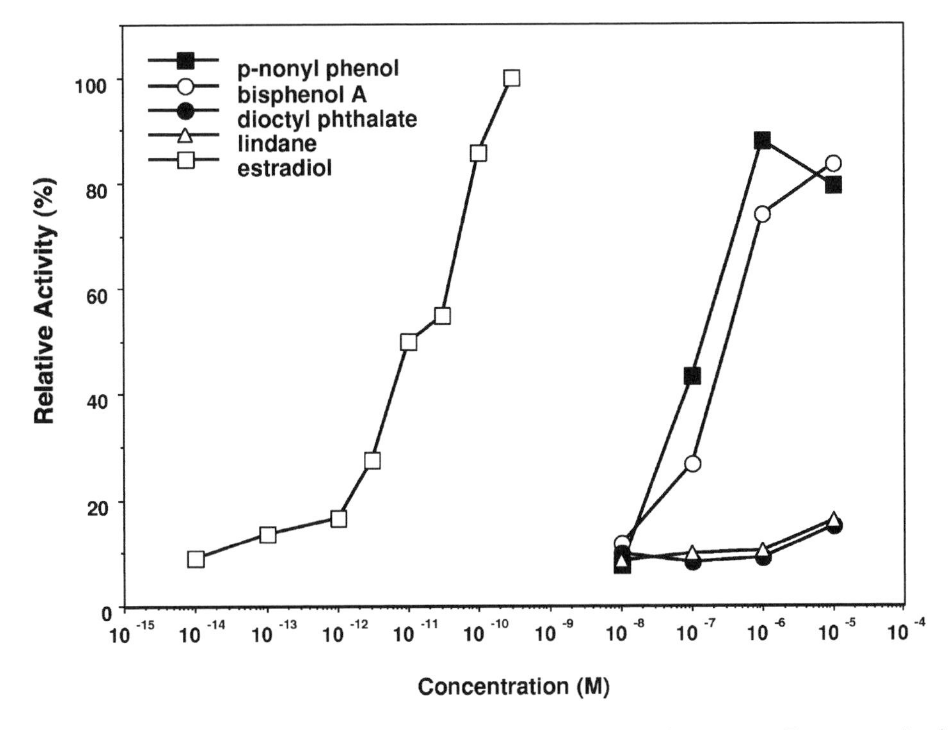

Fig. 3. Gene activation dose–response profiles for the environmental estrogens. For gene activation, hela cells were cotransfected with VIT2P36L reporter plasmid plus RSV HEG0 for expression of estrogen receptor α (ERα). Concentrations for each compound were added as indicated. Light units are expressed relative to the maximal activation achieved with estradiol (100%) and represent the means from several experiments. [Filled box], *p*-nonyl phenol ($n = 3$); [empty circle], bisphenol A ($n = 3$); [filled circle], dioctyl phthalate ($n = 3$); [triangle], lindane ($n = 3$); [empty box], estradiol ($n = 8$). Effective concentrations for half-maximal gene activation can be estimated from these data as approximately 1×10^{-7} M for *p*-nonyl phenol and approximately 3×10^{-7} M for bisphenol A. Only slight activity at high concentrations was detected for dioctyl phthalate and lindane.

compound showed antagonist activity. In contrast, ICI 164,384, a known antagonist clearly, shows a reduction in estradiol-mediated gene activation.

4.2.1.2. Cell-Type- or Species-Specific Effects. Some non-steroidal compounds have exhibited differing potencies or activities that are species-dependent or cell-type-dependent. We have recently investigated this type of activity difference for a model doisynolic acid derivative, (±)-*Z-bis*dehydrodoisynolic acid 3-methyl ether *(95)*. Limited clinical experience with (±)-*Z-bis*dehydrodoisynolic acid 3-methyl ether suggested that the estrogenic activity of this compound differs between humans and other species *(100)*. This type of differing activity is a concern in evaluating environmental estrogens, as it is not clear that effects in animals and wildlife necessarily will be indicative of health effects in humans *(2)*. Our evaluation of this compound used the same reporter gene for positive transcriptional regulation but used several different cell types. Hela human cervical carcinoma cells, MDA MB 231 human breast carcinoma cells, and CV-1 monkey kidney cells are widely used as model systems for studying gene regulation by nuclear receptors. They express no detectable endogenous ERα or ERβ. For these assays, the human ERα expression vector, RSV HEG0 *(97,101)*,

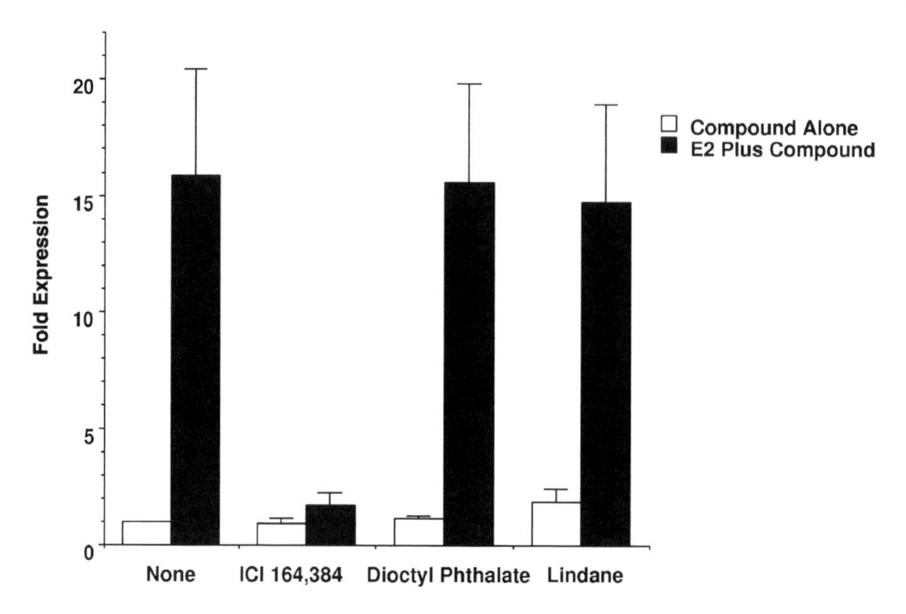

Fig. 4. Environmental compounds do not act as antagonists for estrogenic gene activation. Hela cells were cotransfected with VIT2P36L reporter plasmid plus RSV HEG0 for expression of estrogen receptor α (ERα). Concentrations for each compound were added as indicated. Light units are expressed as folds relative to the activity observed with expressed ER in the absence of added compounds set to 1. Data are the mean and SEM from two experiments. Open bars, activity of each compound alone; filled bars, activity of each compound with estradiol at 3×10^{-10} M. None, no compound added; ICI 164,386, 100 nM; dioctyl phthalate, 1×10^{-5} M; lindane, 1×10^{-5} M.

is co-transfected with a specific, ERE-containing luciferase reporter gene, Vit2P36 luciferase *(97)*, which contains two 26-bp ERE from the xenopus vitellogenin A2 gene, linked to a minimal 36-bp promoter. In each set of experiments, controls with no hormone and with 17β-estradiol are routinely included, to assess the extent to which estrogen-free conditions are achieved, and to determine the extent of gene activation achieved by an authentic physiologic ligand. Activity of estradiol was comparable in all three cell types *(95)*. However, for the compound (±)-*Z-bis*dehydrodoisynolic acid 3-methyl ether, activity was higher in the monkey CV-1 cells than in either human cell line *(95)*. That is, our previous studies of a doisynolic acid derivative exhibited cell-type-specific effects (with ERα) that were apparent in comparisons between human and non-human cell lines *(95)*.

The potential for an endocrine disrupter to exhibit similar differences in activities raises a potential problem for appropriate evaluation, whether dependent on cell-type-specific biotransformation events or due to interaction with other kinds of cell-type-specific factors or based on the different species of origin (human versus monkey) or organ- and cell-type-specific effects (breast, cervix, or kidney). Indeed there may well be additional mechanisms for different activities specific to intact animals or intact humans.

Based on concerns from our previous data, we have examined the activities of environmental estrogens in both Hela cells and CV-1 cells. Figure 5 shows results for these paired assays for dipentyl phthalate and dieldrin. In contrast to the results with (±)-*Z-bis*dehydrodoisynolic acid 3-methyl ether *(95)*, both these EDC compounds

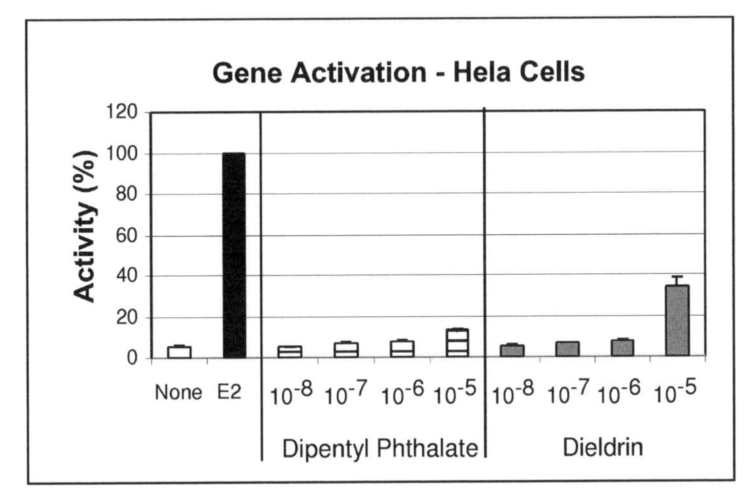

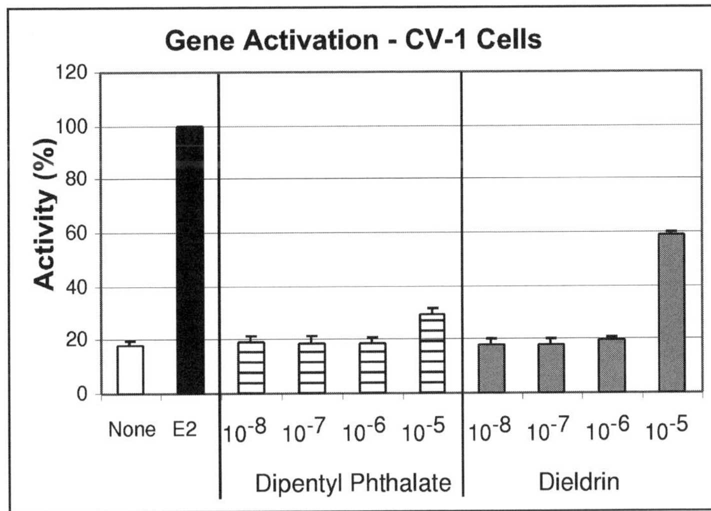

Fig. 5. Gene activation dose–response profiles for the environmental estrogens in both hela and CV-1 cells. For gene activation, hela cells (top) and CV-1 cells (bottom) were cotransfected with VIT2P36L luciferase reporter plasmid plus RSV HEG0 for expression of ERα. A no treatment control (empty bars) and dipentyl phthalate (hatched) and dieldrin (shaded) were added at concentrations as indicated. Light units are reported as relative expression compared to the maximal activation 100% *(100)* achieved with 1 nM estradiol (black bars), and all values represent the means ± SEM from three independent experiments.

showed similar activity in Hela cells and CV-1 cells. The inclusion of activity assays in these two cell lines is a part of our analysis of estrogenic activity of environmental estrogens to evaluate the potential for species-specific transcriptional effects.

Similar assays can be performed to evaluate the activity of EDC compounds through ERβ. Figure 6 shows the results obtained with two additional compounds, chlordane and DDE. As described above, activity with ERβ was determined, along with evaluation that observed activity could be clocked by an estrogen antagonist, ICI 164,384. The compounds were also tested together with E2 to evaluate potential antagonist activity of the EDC compounds themselves. Both compounds exhibit weak activity at

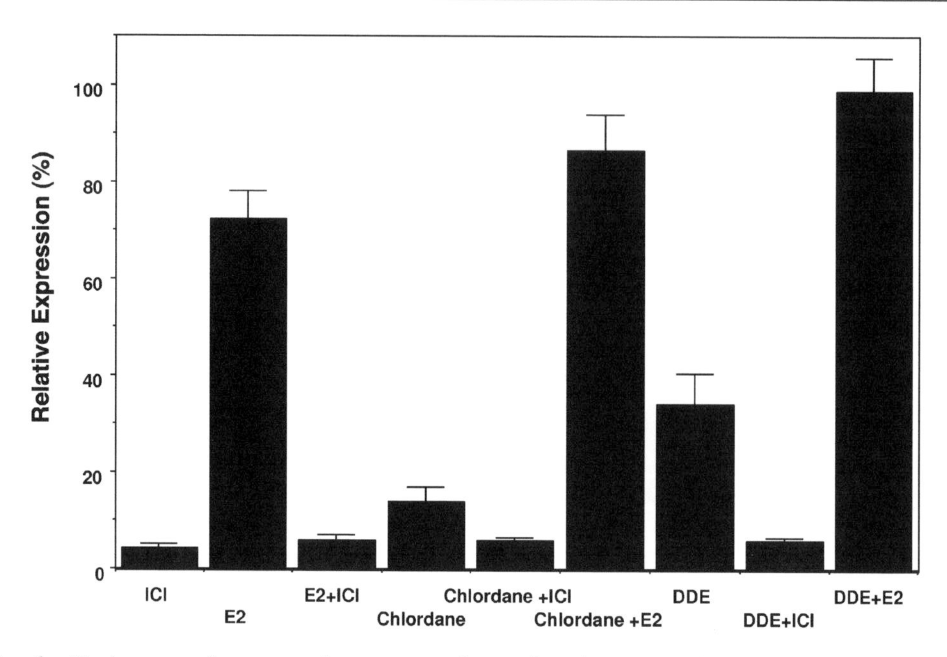

Fig. 6. Environmental compounds act as weak agonists for estrogenic gene activation through estrogen receptor β (ERβ). For gene activation, hela cells were cotransfected with VIT2P36L reporter plasmid plus RSV ERβ for expression of ERβ from rat. Concentrations of 10^{-7} M chlordane or 10^{-5} M DDE were added as indicated, alone, together with 10^{-7} M ICI 164,384, or together with 10^{-10} M E2. Light units are expressed as relative expression with maximal activity indicated as 100 %. Data are the mean ± SEM from three experiments.

high concentrations. The activity is inhibited by ICI 164,384. No antagonist activity is seen in combination with E2, and overall estrogenic activity of the mixture appears to be additive, rather than decreased as with the true antagonist, ICI.

4.2.2. Estrogenic Negative Regulation

Molecular studies of negative regulation by ER *(42)* have been used to evaluate estrogenic activity of a variety of potential estrogenic compounds. This system has been used to evaluate the unusual non-steroidal estrogens related to doisynolic and allenolic acids *(95)*, and the assay may be particularly useful for evaluation of selective estrogen receptor modulator (SERM) activity. In this assay, tamoxifen and hydroxytamoxifen act as agonists, while they act as antagonists in an ERE-dependent gene activation assay in the same cell line *(57)*. We recently have determined that, in contrast to tamoxifen, raloxifene acts as an antagonist in this assay (see Fig. 7, top). In addition to performing these assays with the native human ERα, we also performed these assays with a partially constitutive receptor containing the Y→N mutation described by Zhang et al. and Tremblay et al. *(25,102)* (see Fig. 7, bottom). In this assay, because the receptor is partially active in the absence of added hormone or compound, one can easily determine either agonist or antagonist activity from a single determination, without requiring two sets of assays—one testing compound alone, and a second series performed with compound in the presence of submaximal concentrations of E2. This assay has the potential to simplify screening of compounds, such as industrial chemicals.

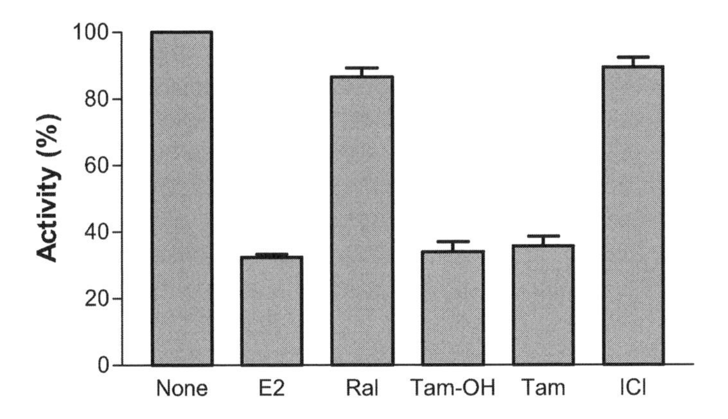

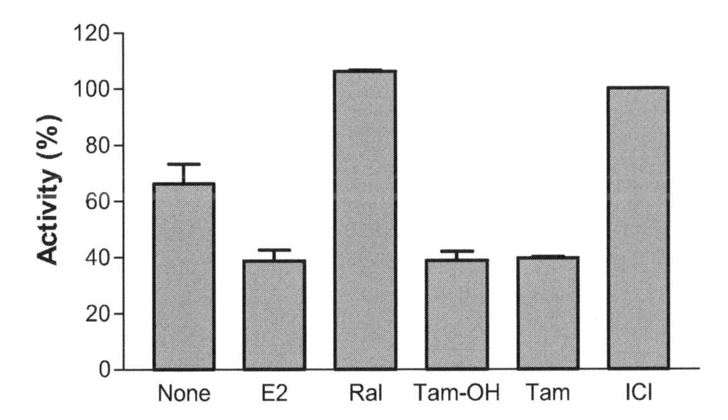

Fig. 7. A gene repression assay for selective estrogen receptor modulators (SERMs) and its modification to show either agonist or antagonist activity. For gene repression, GC cells were cotransfected with D/E pGL3 reporter plasmid plus RSV ER alpha (HEG0) (top) or RSV ER alpha Y→N (bottom) for expression of estrogen receptor (ER). Concentrations for each compound were added as indicated. Light units are expressed relative to the 100% value observed with no compound (top) or ICI (bottom) and represent the means ± SEM from at least three experiments. E2, 0.3 nM estradiol; ral, 1 μM raloxifene; tam-OH, 10 nM hydoxy-tamoxifen; tam, 1 μM tamoxifen; ICI, 0.1 μM ICI 164,384.

We have used this reporter gene assay to assess negative regulatory competence of environmental endocrine disrupters. Negative regulation mediated by human ERα (HEG0) *(101)* was assayed using the modified rat prolactin control region, PRL D/E *(42)*. Figure 8 shows the results of this type of assay used to generate dose–response profiles for the panel of environmental estrogens. In this assay, *p*-nonyl phenol and bisphenol A are agonists and, like estradiol, repress expression of the reporter gene. Although the effective concentrations required are very high compared with E2, these two compounds mediate negative regulation, repressing expression by approximately 75% and acting as an ER agonist. Dioctyl phthalate and lindane are only weakly active at the highest concentrations. Furthermore, as for positive regulation, submaximal concentrations of E2 plus dioctyl phthalate or lindane showed no antagonist effects on gene repression (Fig. 9).

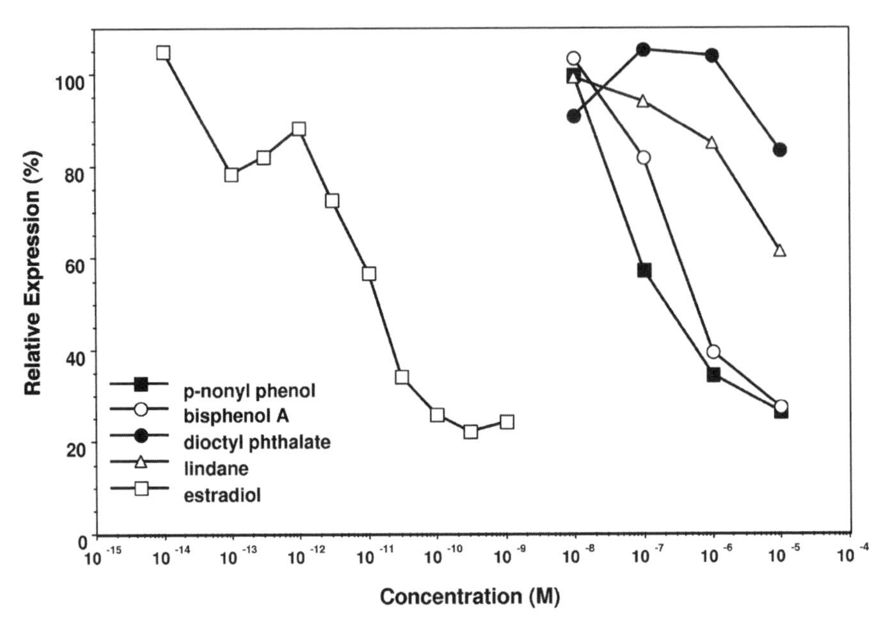

Fig. 8. Gene repression by environmental estrogens. Dose–response profiles for gene repression of a model estrogen responsive reporter gene by human estrogen receptor were obtained using a transient transfection assay (section 4.2.2). Results are included for estradiol. Effective concentrations for half-maximal gene repression can be estimated from these data as approximately 1×10^{-7} M for *p*-nonyl phenol and approximately 3×10^{-7} M for bisphenol A. Repression was detected for dioctyl phthalate and lindane only at higher concentrations.

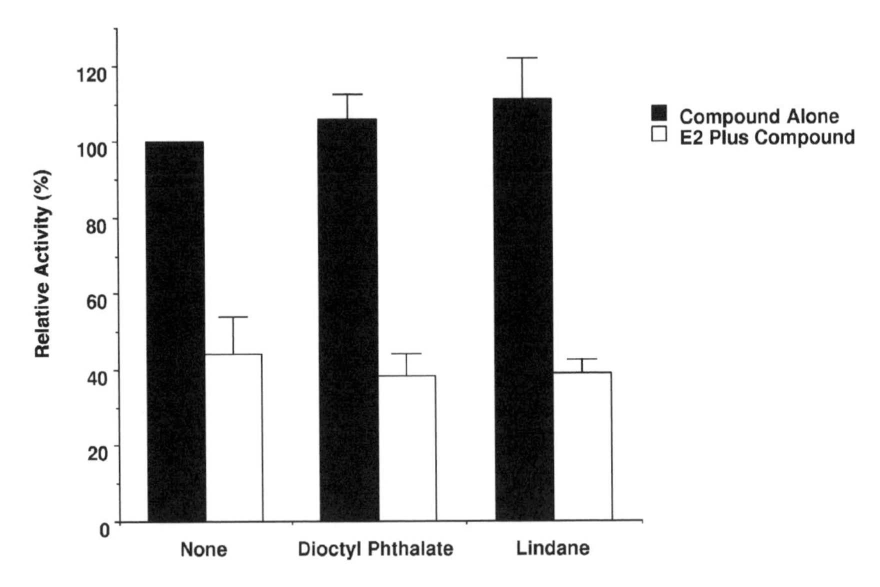

Fig. 9. Environmental compounds do not act as antagonists for estrogenic gene repression. GC cells were cotransfected as described. Concentrations for each compound were added as indicated. Light units are expressed as folds relative to the activity observed with expressed estrogen receptor (ER) in the absence of added compounds set to 100. Data are the mean and SEM from two experiments. Filled bars, activity of each compound alone; open bars, activity of each compound with estradiol at 3×10^{-10} M. None, no compound added; dioctyl phthalate, 1×10^{-6} M; lindane, 1×10^{-6} M.

4.2.2.1. Negative Regulatory Activity is Dependent on/and Mediated Through ER. Our previous analyses of gene repression by estradiol, tamoxifen, and doisynolic acid derivatives *(95)* have demonstrated that repression in this system by these compounds is receptor dependent *(42,57)*. We have used the same experimental design to determine whether any observed activity of the environmental chemicals is dependent on, and mediated through, the ER.

First, the repressive activity of the compounds was determined in the presence of competing concentrations of ICI 164,384. ICI 164,384 is a true steroidal ER ligand that acts as an antagonist of gene repression (prevents the decrease in gene expression by agonists) in this assay. Such an assay is illustrated in Fig. 10. Second, we used expression of a dominant negative mutant of human ER, ER 251 *(42,54,57)* rather than wild-type receptor, to demonstrate the requirement for ER. In the presence of this mutant receptor, rather than wild-type ER, neither estradiol, nor other agonists can fully repress the target gene (Fig. 11).

4.3. Traditional Activity of EDCs

Potential EDCs may exhibit positive and negative gene regulation through binding to the ERs and by traditional nuclear pathways, as illustrated by the panel of endocrine disrupter compounds. Data can be obtained for risk assessment, including full dose–response profiles by using the compounds. Controls, including dependence on ER for observed estrogenic activity and blocking of observed estrogenic effects by a true steroidal antagonist (ICI 164,384), have been used to confirm that the observed estrogenic effects are actually mediated through the ERs and that the compounds

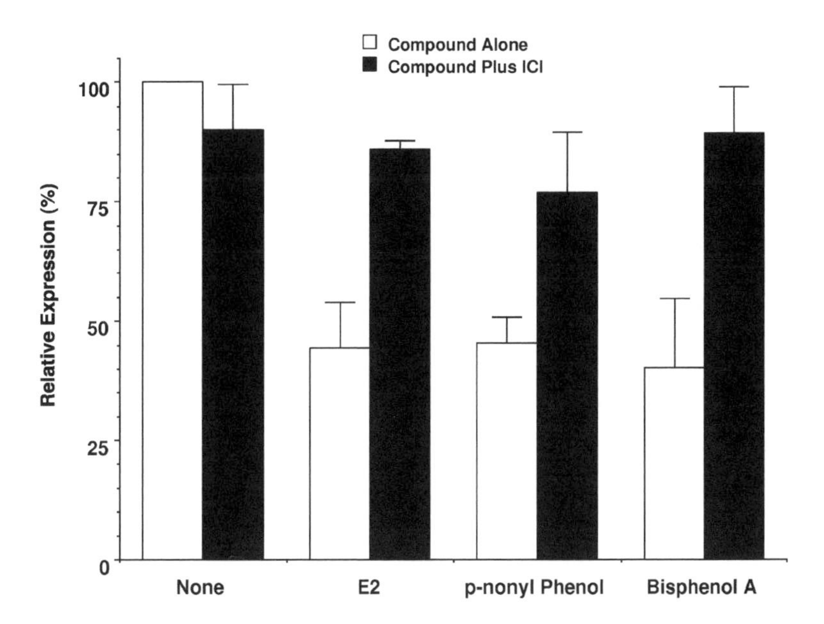

Fig. 10. Negative gene regulation is blocked by ICI 164,384. Assays of specific gene repression using the D/E PRL luciferase reporter are demonstrated for estradiol, *p*-nonyl phenol, and bisphenol A. The environmental compounds decrease gene expression similar to the effects of E2 (empty bars). This effect is blocked by 10^{-7} M ICI 164,384 (solid bars). Estradiol, 3×10^{-10} M; *p*-nonyl phenol, 1 μM; bisphenol A, 1 μM.

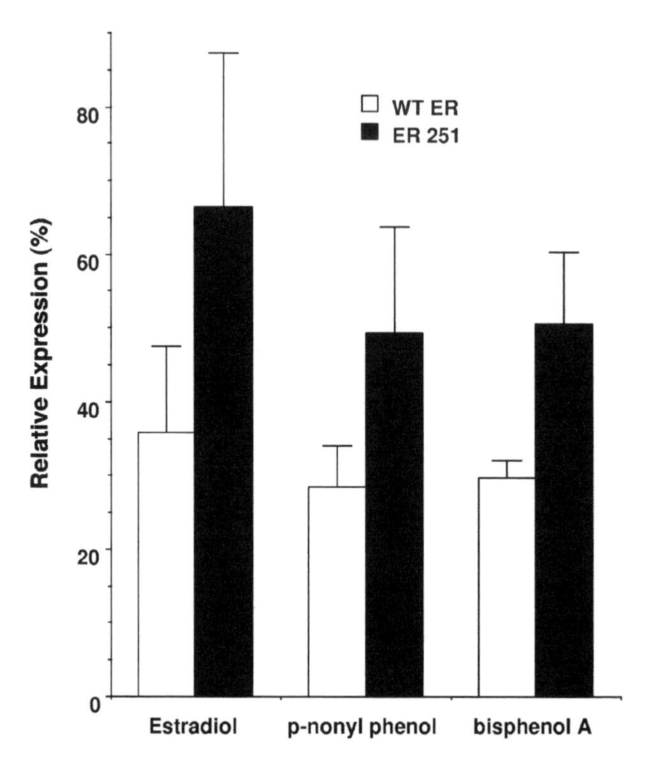

Fig. 11. Negative gene regulation for test compounds—dependence on functional estrogen receptor (ER). Assays of specific gene repression used the D/E PRL luciferase reporter with co-transfection of RSV HEG0 ER (empty bars) or the dominant negative receptor mutant, RSV 251 ER (solid bars). Full inhibition by E2 and the environmental compounds required functional ER. Estradiol, 3×10^{-10} M; *p*-nonyl phenol, 10 μM; bisphenol A, 10 μM.

are acting as ligands for the ER. For positive regulation, in contrast to our previous studies of model compounds, we found no evidence for a difference in activity of these endocrine disrupters between the human cell lines and the non-human cell lines. For negative regulation, similar estrogenic activities were observed for this different mechanism of ER action. Again, appropriate controls demonstrate that these compounds require ER for their action and that their actions are blocked by a steroidal antagonist. For both positive and negative regulation, *p*-nonyl phenol and bisphenol A were agonists. No significant activity was observed as agonists or antagonists for lindane and dioctyl phthalate, including two distinct mechanisms of gene regulation, several cell types, and an in vitro direct binding assay.

5. NON-TRADITIONAL AND NOVEL ACTIONS OF ESTROGENS AND EDCs

Many previous studies of estrogenic effects have either relied on gross responses [such as uterotropic responses *(81)*, vaginal cornification *(85)*, or cell proliferation *(88)*], or a few selected regulated genes, including progesterone receptor *(88)*, cathepsin D *(103)*, prolactin *(104)*, luteinizing hormone (LH) *(105)*, or follicle-stimulating hormone (FSH) *(105)*. These assays usually reflected only estrogenic increases in gene expression, and expression of these specific genes is often limited by cell-type specificity. It is perhaps more desirable and more meaningful to evaluate the expression of either unselected endogenous gene sets or collections of selected genes large enough to include genes that might show unpredicted regulatory responses. In addition, the ability to determine gene repression as well as gene activation may be important for a complete analysis of traditional and novel regulatory mechanisms. One of the first methods for this type of analysis was differential display, a method now largely replaced by analysis of gene arrays.

5.1. Differential Display Suggests Non-Traditional Regulation

Differential Display was one of the first methods to utilize polymerase chain reaction (PCR) to allow one to analyze specific regulated but unselected subsets of mRNAs in a cell. Treatment with hormones or other perturbants may change the level of genes expressed or may result in the addition or loss of expressed genes. Using a standard sequencing acrylamide gel, treatment-specific differences in the amount of mRNA for a particular gene, or the appearance or disappearance of an mRNA from this subset, can easily be seen by direct comparison of band patterns from each tested condition *(104,107)*. Differential display generates a characteristic gel pattern or "regulatory fingerprint," subsets of small, poly A-tailed cDNAs in a way that is both reproducible and sensitive to changes in the composition of the original RNA sample. By performing several reactions, a broad sampling of total cellular mRNAs can be achieved.

As an illustration of the ligand-specific effects of estrogenic compounds, Fig. 12 shows a single differential display (E. Fels and S. Adler, unpublished data) using Ishikawa endometrial cells, a human endometrial cell line *(108)*. The Ishikawa cell line has been used as a model system for studies of estrogen action, as well as studies of hydroxy-tamoxifen action *(108)*. Expression of progesterone receptor in these cells is activated by both E2 and hydroxytamoxifen (*108*; Y. Sadovsky, personal communication). It represents an important and useful cell model for estrogenic responses typical of uterine cells. Cells grown in estrogen-free conditions with phenol-red free media and charcoal stripped serum, were treated for 48 h with either no hormone, 1×10^{-8} M E2, or 0.1 μM tamoxifen. Some gene bands are not regulated (Fig. 12, "*") but others are repressed or activated by E2 and/or tamoxifen. One striking result, apparent even from this single regulatory fingerprint, is that the overall pattern of tamoxifen on gene expression is not similar to that of E2. E2 and tamoxifen generate distinct regulatory fingerprints. Even in this tissue, for which the profound clinical effects of tamoxifen on growth and differentiation are estrogenic, the effects of E2 and tamoxifen are markedly different at the level of gene expression. This provides evidence that suggests that, even in a single cell type, traditional gene regulatory mechanisms cannot fully account for all the observed regulatory changes and that complete evaluation of EDCs cannot rely on growth responses or on analysis of a few selected genes.

5.2. Are Phytoestrogens also EDCs?

Insecticides, plasticizers, and detergents can be classified as potential EDCs. At least some of these compounds bind to ERs as ligands and therefore may affect gene regulation in estrogenic pathways. Exposure to these chemicals in the environment,

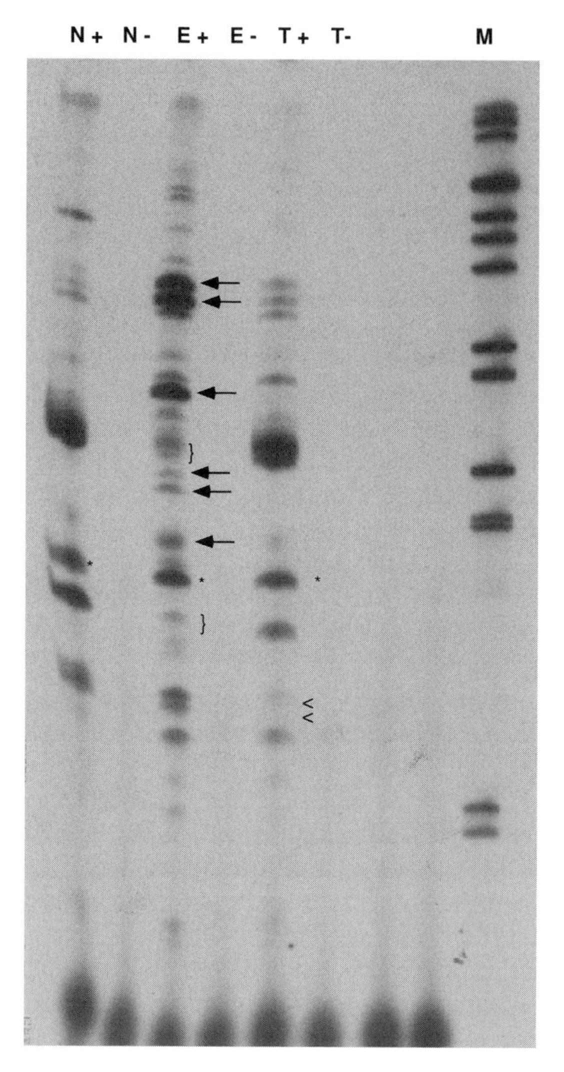

Fig. 12. Differential display regulatory fingerprints for E2 and tamoxifen in Ishikawa cells. mRNA from hormone and tamoxifen treated Ishikawa cells was reverse transcribed using the dT10**M**A primer and MMLV reverse transcriptase (+) or subjected to a mock reaction in the absence of MMLV reverse transcriptase (−). The absence of polymerase chain reaction (PCR) amplified bands in these (−) lanes indicates that amplified bands in the (+) lanes are derived from RNA and not DNA contamination. PCRs used dT10MA plus the 10′mer, dGACACCGTTC. Reactions were forty cycles of 94 (20″), 30 (1 min), and 72 (30″) with direct incorporation of $\alpha-^{32}$P-dCTP. Hormone treatment is indicated: N, no hormone; E, E2; T, tamoxifen; M, size markers. A few representative bands are identified as follows: ∗, no change with treatment; arrow, E2 induced; }, E2 repressed; <, tamoxifen repressed.

in the food chain, and through occupational exposures, may affect human health and, through developmental effects and increases in birth defects, the health of future generations.

In addition, chemical compounds from soy and other plants have been identified that also have the potential to act as mimics of the natural female steroid hormone, estradiol, and these "phytoestrogens" have received attention as being beneficial to human health.

Yet, these compounds, as they also are capable of acting as estrogens, might also cause problems by interfering with the body's own natural hormonal regulation.

These two classes of non-steroidal chemicals, both with the ability to act as estrogens, raise important questions. Mammals seem to be exposed to many potentially estrogenic signals. Why does it appear that only some of these signals disrupt endocrine signaling? Furthermore, the exposure to these compounds through soy and other foods is not a new phenomenon. If these compounds really interfered with hormonal signaling, why would not there be more obvious detrimental effects on reproduction or development—rather than apparent health benefits? This conundrum raises the possibility that novel actions of phytoestrogens and/or EDCs may be responsible for the apparently differential health effects rather than only traditional mechanisms of gene regulation. For further discussion of actions of phytoestrogens as EDCs in the brain, readers are referred to Chapter 4 by Walker and Gore.

5.2.1. Phytoestrogens

Breast cancer is known to be hormone responsive or hormone dependent (see review *109*). In addition, epidemiologic studies show Asian populations that consume large amounts of vegetables, fruits, and soy protein have a lower incidence of breast cancer *(110)*. This appears to be diet related, as the same ethnic population increases in cancer incidence to Western levels once adapting a Western diet *(111,112)*. (This fact might also suggest that dietary compounds may not promote longlasting or transgenerational epigenetic changes as proposed for EDCs *(11–14)*, a topic to be addressed below.) Furthermore, in an English study, although mortality from cardiovascular disease was reduced in vegetarians, decreased risk of breast cancer was not observed *(113)*. Even though these conflicting results may raise questions about the relationship of diet to breast cancer, these data might also suggest that a Western-style vegetarian diet differs from Asian diets, perhaps due to soy content.

Plant compounds that affect estrogenic pathways have been found, and it has been proposed that these "phytoestrogens" may be responsible for the protective epidemiologic effects *(114)*. Yet, although phytoestrogens may affect estrogenic regulation, it is not clear that direct action as estrogen agonists or antagonists is the mechanism responsible for the protective effects *(114)*. Other pathways, or indirect effects on estrogenic regulation (e.g., effects on estradiol metabolism), may be important *(114)*. It is also not clear whether all the effects of phytoestrogens are beneficial *(115)*. Animal systems indicate that these compounds, through their ability to mimic hormonal signaling, may produce adverse effects on reproduction and on development *(115)*. Further studies of phytoestrogens are increasingly important as the amount of soy and phytoestrogens in Western diets may change as soy protein is incorporated into more foods as a supplement (e.g., soy in hot dogs) and as new food trends (e.g., tofu) are adopted.

5.2.1.1. What are the Phytoestrogens and Where are They Found? Several classes of chemical compounds have been identified and partially characterized for their estrogenic effects in animals and in humans (see reviews *116,117*). As with the EDCs, each distinct compound may produce its own specific and distinct effects. In addition, as with EDCs that may occur in the environment as mixtures, the dietary sources make it likely that more than one phytoestrogen or herbal compound may be consumed together from an individual food or as part of a varied diet. Soy contains isoflavones including daidzein, equol, genistein, and glycetein. Flax, cereals, and seaweed are sources of lignans including enterodiol and enterolactone. Fungal growth on grains contains resorcylic acid lactones including zearalenone and zearalenol. Grapes and peanuts contain resveratrol, and there are a variety of compounds in plants and herbs used to make teas, including ginseng *(118)*. The activity of some of these compounds is dependent on modification of precursor molecules that are contained in the food products *(116)*. This often requires the action of gut bacteria and entero-hepatic modifications *(116)*. Red clover, although not a traditional human food, can be consumed by grazing animals and contains the coumestan, coumestrol. Clover has been associated with reproductive toxic effects *(117)*. Genistein has activity as an inhibitor of protein tyrosine kinase, an activity not associated with daidzein *(117)*. In addition, many of these estrogenic compounds are effective as anti-oxidants *(117)*. The relationship of estrogenicity, kinase inhibition, anti-oxidant activity, and effects on other pathways, to the observed in vivo effects of individual compounds may be complex.

5.2.2. Nuclear Accessory Proteins—Coactivators and Corepressors

Recently, observations regarding cell-type and ligand-specific responses have been influenced by the recognition of accessory factors and/or coregulators that interact with members of the nuclear receptor family (see review *119* and *74,75,120,121*). Many different coactivators and corepressors have been identified, and current work is classifying these factors into related groups. It is now clear that factors classified as coactivators, such as SRC-1 *(122)*, interact with many nuclear receptors to meditate transcriptional regulation through interactions with CBP/p300 *(75)*, and either intrinsic or recruited histone acetylase activity *(123,124)*. In addition, corepressors, such as silencing mediator for retinoid and thyroid-hormone receptors (SMRT) *(125)* or a related factor N-CoR *(126)*, mediate negative regulation by unoccupied receptors, such as thyroid hormone receptor, and in addition may interact with antagonist occupied progesterone and ERs *(126)*. In contrast to the histone acetylation observed with coactivators, corepressor complexes are associated with histone hypoacetylation *(123)*. These data provide a connection between gene transcription and chromatin remodeling.

In particular, these cellular factors participate with activated ERα in mediating its transcriptional effects *(74,75)*. The interaction between ERα and these factors can be ligand dependent, as successful association is observed in the presence of estradiol but not with antagonists *(74,75)*. These data extend our molecular understanding of how estrogens and the ER work. Because these interactions distinguish agonists from antagonists they can provide a novel molecular test for estrogenic ligands *(127)*.

5.2.2.1. Coactivators and Corepressors and Antagonists/Partial Agonists. This system of coregulators appears to play an important role in the activity of partial agonists. Many groups have shown that by altering the expression levels of coactivators and corepressors, gene activation by mixed agonists can be prevented or increased *(129–131)*, a result confirmed explicitly *(131)*. That is, for a fixed concentration of ligand, agonist activity, present at lower levels of corepressor, can be changed to antagonist activity as the level of expression of corepressor is increased. This has been shown for ER with tamoxifen and for GR and PR with RU486. In all cases, the activity of a true agonist (E2, Dex, and R5020) is not altered by elevated corepressor expression.

Additional information is available from the crystal structure determination of the ligand-binding domain of ERα with either estradiol or raloxifene as ligand *(133)*. The position of helix 12 is dramatically different, depending on the ligand. When E2 is bound, the helix forms a binding site for coactivator interaction, while with the antagonist raloxifene, the helix is flipped and this binding site is disrupted.

This scheme raises an additional requirement for ligand activity. Not only must a ligand successfully bind to the nuclear receptor, it must also promote the correct conformation for interactions with coregulators. This potential switch in ligand activity with alterations in corepressor levels may help explain the development of tamoxifen resistant breast cancers. Perhaps cellular levels of corepressors drop and allow tamoxifen to act as an agonist. Indeed, clinical and basic reports indicate that coactivator/corepressor levels may be predictive of the responsiveness of a breast cancer to Tamoxifen therapy *(127,134)*.

It is also possible that breast cancer cells may become more independent of estrogenic signals than normal tissues because of changes in the ratios of hormonal coregulators, a change that could occur either as a precursor to, or after, malignant transformation. This type of change is also supported by a study showing increased expression and amplification of the coactivator, AIB1, in breast cancers *(135)*. An additional study shows that a mutated ERα with constitutive activity associates with AIB1 in the absence of ligand, providing one mechanism for ligand-independent growth of breast and ovarian tumors *(136)*.

While this discrimination system has been demonstrated in the laboratory using synthetic compounds, it is not likely that this system's natural function relates to these synthetic compounds. One possibility for a more natural function or need for a secondary discrimination system would be to provide protection from adverse hormonal effects of non-steroidal estrogens that mimic the natural estrogenic hormones. Animals have been eating plants for eons, and the exposure and possibility of endocrine poisoning may have existed throughout the evolution of our steroid hormone-signaling pathways. This system of coregulators may partially restrict the activity of phytoestrogens on estrogenic pathways. Furthermore, it is possible that exposure to compounds such as phytoestrogens might actually regulate the expression of corepressors and coactivators as an adaptation to increase receptor selectivity. An altered ratio of coregulators and increased selectivity may be one aspect of the actions of natural dietary non-steroidal estrogens that limits their disruptive effects on hormonal signaling. Indeed, even the protective effects of phytoestrogens on breast cancer incidence may be related to preventing excessive estrogenic stimulation by this kind of adaptive response.

Very little is known about how the expression of coactivators and corepressors is regulated, and this represents a serious gap in existing knowledge of regulation by steroid hormones and by non-steroidal estrogens. It has been reported that levels of these factors in cultured cells and in vivo can be modified by hormonal signals including estradiol *(137)*. Specifically, the level of SRC-1 mRNA in the pituitary lactotroph cell line, GH3, was found to be decreased approximately 3-fold by treatment with 10 nM E2, an effect seen at both 6 and 24 h *(137)*. Similar in vivo experiments in male rats showed anterior pituitary SRC-1 mRNA decreased after a single E2 injection over the first 4 h, slowly returning to baseline levels over 24 h *(137)*. This effect was accompanied by an equally impressive increase in the level of SMRT mRNA, peaking at approximately 175 % of the basal level at 2 h and returning rapidly to baseline by 4–6 h *(137)*. Interestingly, the expression of p300 and of N-CoR was not changed *(137)*. The coupled decrease in SRC-1 and increase in SMRT, if also reflected in protein changes, indicates a dramatic change in the coactivator to corepressor ratios. This type of changed ratio is just the kind of change that can result in altered activity of tamoxifen and RU486 *(129–132)*.

The effects of other estrogenic compounds, including non-steroidal compounds and phytoestrogens, on coregulator expression and whether these types of changes may function as a protective cellular adaptation remain to be determined. Indeed, it has recently been shown that the aryl hydrocarbon receptor also can interact with the coactivator ERAP-140 and that gene regulatory effects of dioxin (TCDD) are modulated by the ratio of coactivator and corepressor in the MCF-7 breast cancer cell line *(138)*. This is a direct linkage between a non-steroidal chemical of environmental concern with activity modulated by coregulator ratios. Understanding how the coregulators themselves are regulated may therefore have tremendous impact on potential protective treatments for exposure and on future evaluation and risk assessment.

5.2.2.2. How Regulation of Coactivators and Corepressors Would Work to Limit Estrogenic Signaling by Non-Steroidal Compounds. It is probably worth "walking through" how the regulation of coactivators and corepressors by estrogenic compound might result in a system insensitive to estrogen mimics while retaining responsiveness to estradiol. One key fact from the data of Smith et al. is that elevated SMRT had little effect on regulation by estradiol but remarkably eliminated responsiveness to hydroxy-tamoxifen *(129)*. On the basis of our data regarding phytoestrogen activity *(139,140)* and the results with TCDD *(138)*, this modulation of activity through regulation of the coactivators and corepressors suggests that this also applies to at least some other non-steroidal estrogens and EDCs.

In this model of endocrine modulation, exposure to estrogenic compounds would have rapid effects on the expression of coactivators and corepressors, resulting in an estrogen-dependent increase in the expression of corepressor (e.g., SMRT) and a decrease in the expression of coactivator (e.g., SRC-1). Once the ratio of these regulators is changed, the activity of the non-steroidal compound would be decreased, as was seen with hydroxy-tamoxifen. E2, in contrast, will be essentially unchanged in activity. Once the non-steroidal compound is less active, the signal driving the increased expression of the corepressor will also be decreased, and the level of corepressor will begin to decrease toward baseline, and the level of expression of coactivator will begin to rise again to basal levels. As soon as the levels return to a ratio sufficient for the non-steroidal compound to have higher activity, once again the level of corepressor would increase and coactivator decrease, again terminating the signal. Indeed, chronic exposure through diet or from the environment would seem to act to achieve a new steady-state level or ratio of coregulator expression, just sufficient to keep the non-steroidal compound inactive. One might further conjecture that non-steroidal compounds that are very sensitive to coregulator ratios in the presence of pre-existing unfavorable coregulator ratios would be unlikely to have dramatic estrogenic effects in vivo and be potentially benign. Other compounds might pose a more serious problem if they were, like estradiol, relatively insensitive to coregulator ratios.

6. EPIGENETIC EFFECTS OF ESTROGENS THROUGH HISTONE ACETYLATION STATE

Epigenetic changes are an additional potential novel or non-traditional action of EDCs and other estrogens. Transgenerational effects *(11–14)* and potential changes in gene methylation will be reviewed elsewhere in this volume. The focus of this section will be potential epigenetic changes related to histone acetylation and effects on gene transcription. Readers are also referred to Chapter 7 of this book, written by Guerrero-Bosagna and Valladares.

Major health-related causes of mortality, such as cancer and heart disease, can be favorably affected by diet and lifestyle. For instance, one of the few things that has been associated with a low breast cancer risk is an Asian lifestyle including consumption of an Asian diet high in soy *(110)*. Soy diets have also been associated with lowered cholesterol and reduction in cardiac risk, and these health claims have been recognized by the Food and Drug Administration *(141)*. Similarly, consumption of red wine is associated with lowered cardiac risk, the so-called French Paradox *(142)*. One goal of research therefore has been to identify the active agents in soy and in grapes responsible for these effects, an effort that has generally focused on phytoestrogen components, including the isoflavones from soy and resveratrol from grapes, but these and other phytochemical components also have other activities including actions as antioxidants or as inhibitors of cell growth signals through action as tyrosine kinase inhibitors that may also be important *(117)*. The actions of these compounds as hormonal mimics or antagonists, through endocrine pathways or endocrine disruption, have been extensively investigated. Yet, despite the epidemiological suggestion that a lifestyle that includes consumption of the soy phytoestrogens is beneficial in preventing breast cancer, direct confirmation through dietary studies (see reviews *143,144*), or identification of a cellular or molecular mechanism through which their actions as estrogens would decrease (rather than increase) breast cancer incidence has not been forthcoming. Similar concerns also are present in regard to the multinational epidemiological evidence implicating Asian soy-containing diets and reduced prostate cancer incidence to specific effects of soy or its components (see review *145*).

Recently, advances in several seemingly unrelated areas of biology have linked diet, histone acetylation, longevity, and cancer prevention through a phenomenon called gene superinduction. Superinduction is defined as increased expression of nuclear receptor-activated genes to levels above the maximum observed with the established agonists, such as, for ERs, estradiol. These new findings may now be able to link these different observations together with the soy phytoestrogens and grape resveratrol to provide a possible non-traditional molecular mechanism to explain the observed health benefits and anticancer effects of these plant compounds. First, new insights in the field of aging and longevity include a deeper understanding of how caloric restriction and insulin/growth hormone pathways may produce its beneficial effects on life span *(146–151)*. These effects are now linked to a particular family of histone deacetylases (HDACs) including the yeast gene *Sir2* and, by inference, the human homolog, SirT1. There are also experimental data demonstrating increased life span effects of the phytoestrogen resveratrol *(149,150)*. Next, methoxyacetic acid (MAA), a metabolite of the endocrine disrupter compound, EGME, can multiply or "superinduce" the magnitude of expression of genes activated by nuclear steroid hormone receptors, including ERs and progesterone receptor *(152)*. General HDAC activity is also affected by MAA, as it is with short chain acids, such as butyrate, and trichostatin (TSA) *(152)*, an antifungal compound isolated from Streptomyces *(152)*, providing another link with histone acetylation status. Finally, the health benefits of fiber in reducing colon cancer incidence have been recognized through population studies *(154)*. These effects have been linked to butyrate, a metabolic product of fiber, again through effects on histone acetylation state *(155)*, and the resulting increased expression of p21/WAF1 *(156–158)*. Soy has been proposed to have anti-cancer effects, particularly with regard to breast and prostate, hormonally responsive cancers *(114)*. There are several lines of evidence that may suggest that the soy isoflavone phytoestrogens, genistein and daidzein, may be mechanistically linked, on the one hand, to resveratrol and lifespan effects and, on the other, to butyrate, histone acetylation status, and cancer prevention. First, both of the soy isoflavones are also "superinducers" of estrogen-signaling pathways *(158)*, producing effects that resemble those reported for MAA *(159)*. This suggests that histone acetylation status may also be affected by these compounds—or alternatively that new and distinct mechanisms may be responsible. Furthermore, similar superinduction properties are also seen with resveratrol *(160–163)*, the grape phytoestrogen that longevity studies have linked to effects on HDAC Sir2/SirT1and thus also histone acetylation status *(149,150)*. Finally, aging studies in mice have shown that a soy diet throughout life can prolong lifespan in certain genetic lines *(164)*. This suggests that soy diets, if the mechanism is like other mechanisms that affect life span in yeast, worms, and flies, may also involve HDAC/histone acetylation status. Each of these areas will be reviewed in more detail below.

6.1. Superinduction is Unusual

The superinduction phenomenon is highly unusual for estrogens. Of the many non-steroidal, plant compounds, and environmental endocrine disrupter/industrial compounds that act as estrogens, we have only observed superinduction by resveratrol, genistein, and daidzein and a recent report also noted superinduction by daidzein metabolites *(165)*. A variety of industrial compounds, endocrine disrupters, and other plant phytoestrogens, such as coumesterol, have not shown superinduction. It is even more unusual that these few compounds that exhibit superinduction are also the compounds associated with possible health benefits—resveratrol directly with longevity and indirectly through epidemiology with cardiac risk reduction (The French Paradox; *142*)—and the soy compounds through dietary studies with reductions in cholesterol and cardiac risk and by epidemiology with cross national decreased breast and prostate cancer rates.

6.2. The Consequences of a Dynamic Histone Acetylation State and Gene Expression

There is now a very clear link associating accessory factors, the coactivators and corepressors, with the changes in acetylation of histones that accompanies active gene transcription by members of the nuclear receptor family as well as other transcription factors (see Sect. 5.2.2.). The complexity of association of cyclical active coregulator complexes with ERα at an active gene the promoter has been revealed by a sophisticated chromatin immunoprecipitation (ChIP) method *(166)*. Among the cyclical changes that occur at the promoter is the recruitment of complexes, such as Brahme-related gene, BRG-1, which perform even more chromatin remodeling as an ATP-dependent process of moving core histone "beads" to new positions on the chromatin "string" that describes the DNA at the promoter *(166,167)*.

The opposing processes of histone acetylation, associated with recruitment of histone acetyltransferase HATs to an active gene promoter, and histone deacetylation, through HDACs, associated with repression or inactivity of a gene promoter, present a system that can be modified either by increasing or decreasing either acetyl transfer to histone via HATs or the removal of acetyl groups from the histones through HDAC activity. What is also true regarding this relationship is that not only is high acetylation of a promoter associated with an active promoter as a consequence of the binding of transcription factors but that as state of high acetylation of the promoter in itself, even if through the inhibition of the removal of acetyl groups, can result in the downstream events that define transcription. Sun et al. showed that the addition of butyrate to MDA MB-231 breast cancer cells, a line which is ER null, rapidly results in increases in histone H4 acetylation, without the addition of hormones or other gene inducers *(156,168)*.

HDACs are being investigated as anticancer targets as inhibition of HDACs appears to maintain high expression of p21/WAF1 *(155)*. In the analysis of colonic p21 expression stimulated by butyrate, a model has been proposed involving the butyrate-induced expression of transcription factor ZBP-89, which then replaces Sp1 and/or Sp3, and a hyperacetylated state allows activation of the p21 promoter and gene expression, a situation not supported in the basal state in the absence of butyrate *(156,169,170)*. In fact, there is evidence suggesting that the physiologic regulation of these processes of increasing acetylation state may occur more through decreases in the rate of deacetylase activity than through activation of histone acetylation *(168)*. In ER-positive human breast cancer cells MCF-7 (T5), a cell line which is responsive to E2 for growth, E2 caused a rapid increase in histone acetylation state *(168)*. What is a little unexpected is that this change in acetylation state was primarily through inhibition of HDAC activity rather than through increased HAT activity *(168)*. This suggests that sufficient HAT activity may always be present to result in sufficient acetylation of certain promoters to result in transcriptional activation and raises the possibility that repressed genes can be activated in the absence of the usually utilized metabolic signals.

One potential consequence of HDAC inhibitors might therefore be predicted to be the widespread dysregulation of cellular transcriptional regulation resulting in metabolic chaos and ultimately cellular death. Yet, while this may occur in some cases of massive cellular reprogramming and apoptosis, it is not a mandatory consequence. Also, it appears that there may be secondary effects on gene expression if the genes affected by HDAC inhibitors are key regulators, whether activators or repressors, of other genes or signaling pathways *(171)*. Often, the number of genes affected by butyrate and other HDAC inhibitors appears to be not only a very small fraction of all genes but also a set of genes that varies with the agent used as an HDAC inhibitor. For example, butyrate activates only about 2% of the genes examined *(156)*. The time course of effects by butyrate and TSA are markedly different *(171)*, and the sets of genes affected by four different HDAC inhibitors, TSA, butyrate, sulindac, and curcumin, are distinct with some overlapping gene sets, some genes that are unique and others that display regulation in opposite directions *(171)*. The recent study of MAA, an EDC shown to be an HDAC inhibitor, showed it affected only three of 19,000 genes analyzed, and, when MAA was administered together with R5020, the progesterone agonist, they altered transcription of 88 versus 61 genes by R5020 alone *(152)*. Some genes in certain cell types may be particularly affected by these agents, whereas others may not be. As an example, in neuroblastoma cells, valproic acid, tributyn, and butyrate all induced p21/WAF1, whereas p27 was induced only by valproic acid *(172)*. In the case of p21, it has been recently shown that not only is total histone acetylation state changed but the promoter of p21 itself contains hyperacetylated histones H3 and H4 after treatment with TSA or butyrate as demonstrated by a Chip assay *(157)*.

6.3. Small Molecules that Affect Lifespan—the Resveratrol Story

Caloric restriction and modulation of insulin/growth hormone pathways may produce extension of lifespan *(146–151)*. These effects were originally observed in mice through drastic diets or gene knockouts in hormonal/metabolic signaling pathways. Now, however, in yeasts *(149)*, fruit flies, and worms *(150)*, simple chemical compounds, such as the phytoestrogen resveratrol from grapes, produce the same type of effects in increasing life span with the potential for similar activities in man. Caloric restriction, resveratrol, and other compounds appear to produce these effects through a particular family of HDACs including the yeast homolog Sir2 and human gene *SirT1 (149–151,173)*. In addition, this sirtuin family of HDACs also act as deacetylases on other proteins and may produce other beneficial effects *(174)*. One key activity of the SirT1 enzyme is the deacetylation of p53 *(175,176)*, an activity that destabilizes this tumor suppressor and thus increases cell survival under stress. [This p53-related increase in cell survival might be somewhat opposed to the tumor suppressor effects of p53, including the p53-dependent increased expression of p21/WAF1 *(177)*.] Another recent role of Sir2 and SirT1 is regulation of metabolism through interactions with metabolic pathways including peroxisome proliferator activated receptor gamma coactivator-1 alpha (PGC-1 alpha) PPARs, a co-regulator of these receptors, PGC1α, and the general transcriptional integrator/regulator, p300 *(178–180)*. These also provide a more direct link between caloric restriction's metabolic effects, the metabolic knockout mice, and the other interventions that increase longevity.

It is worth noting that the particular HDAC assay used for determining SirT1 activity, for establishing a direct interaction between resveratrol and sirtuins, and for identifying additional small molecules that may enhance lifespan through activation of the SirT1 and other sirtuin enzymes *(149,150)*, has recently been questioned *(181–183)*. In particular, it appears that the increases in SirT1 enzyme activity in this assay by resveratrol are an artifact of the particular fluor-containing peptide substrates used in the HDAC assay from the original studies. Despite these data, this assay may still be useful in identifying small molecules with structures similar to resveratrol, because molecules with similar structures may behave like resveratrol in enhancing lifespan (as well as effects in the HDAC assay in question). As one example, fisetin, a compound with a similar structure to resveratrol, was both an activator in this HDAC assay and a lifespan extender in yeast and drosophila *(150)*. It is therefore possible that the HDAC assay, rather than identifying a mechanism for lifespan extension, may actually serve as a surrogate for determination of particular molecular structural features important for an as yet unidentified molecular target. Alternatively, the assay may only represent self-fulfilling expectations, in that the small molecules that were selected for testing all shared structural features with resveratrol *(150)*. Some of these structurally similar compounds may also be lifespan extenders whether or not they are active in the HDAC assay, and other compounds identified by the HDAC assay may likewise be completely inactive as longevity agents. Results showing resveratrol's effects on longevity and mutational studies indicating a dependence on functional Sir2 for these longevity effects, as has been shown for caloric restriction, are not directly invalidated by the findings that the HDAC assay for SirT1 is flawed. The studies that invalidated the assay, however, also included questions regarding the reproducibility of the yeast longevity results *(181)*. Clearly, the effects of resveratrol in yeasts and direct action on Sir2 remain unsettled at this time.

6.4. Superinduction—Non-Traditional Regulation by Resveratrol, MAA, and Soy

Since the original report that resveratrol, a phytochemical from grapes, acts as an agonist for ER-mediated reporter gene expression, its mechanism of action has been a confusing mystery *(160)*. Unlike estradiol and other ER agonists, under certain conditions and not others, the magnitude of the response produced by resveratrol was much greater than predicted and manyfold greater than that of E2, the physiologically relevant ligand *(160)*. When reviewed as recently as 2004, there still was no clear explanation for the effect itself, why it was observed in some cell types and not others, or why it appeared to be inconsistent among different laboratories studying this compound (see review *184* and references cited). Our own studies struggled with similar inconsistencies, determining that the superinduction phenomenon by resveratrol for ERα—but not ERβ—was connected to cell density and thus perhaps reflected the influence of growth factor pathways *(163)*. Furthermore, the effect appeared to be separable into a "ligand" or E2-like response typical of any weak agonist and a "multiplier effect," an effect that was present not only in the absence of EREs in the promoter but also present in Hela cells that are null for ERs *(161–162)*. This confusion was heightened by our observation that genistein and daidzein, the soy isoflavones, exhibited the same kind of superinduction effect in Hela cells for positive gene regulation by ERs *(159)*.

It was not until the publication of a particularly inspiring report by Jansen et al., evaluating the effect of MAA on nuclear receptor-mediated gene expression, that mechanisms explaining similar superinduction responses were demonstrated *(152)*. Their experiments showed two important effects. First, MAA multiplied the effects of ligands for several, if not all, nuclear receptors, including ERα and ERβ, progesterone receptor, thyroid hormone receptor β, and the androgen receptor *(152)*. Effects in the absence of ligands were reported to be minimal or absent *(152)* [although analysis of their data *(152)* shows a ligand-independent effect, very much like what we see with resveratrol *(161,162)*]. These effects were demonstrated to be independent of binding of MAA to the receptor *(152)*. There is no change in the EC50 for ligand binding in the presence of MAA, and MAA does not bind to the receptor to compete with ligand at the traditional ligand-binding domain, as determined by competition binding assays, or at other protein domains that might cause conformational changes, as demonstrated by a protease sensitivity assay *(152)*. The superinduction effect was shown to occur not only with luciferase reporter genes but also with selected progesterone-regulated endogenous genes in T47D cells, and there were effects on relatively few total (unselected) genes as determined by microarray analysis *(152)*. MAA also activated ERK 1/2 pathways as demonstrated directly by rapid increases in ERK 1/2 phosphorylation, increases in activated Ras, and activation of responses mediated by transcription factor elk-1 *(151)*. The involvement of this pathway was also shown pharmacologically by showing sensitivity of the superinduction to the mitogen-activated protein kinase (MAPK) inhibitor U0126 but not to inhibitors of JNK or p38 MAPK *(152)*. In addition to the effects on MAPK pathways, MAA was shown to have effects on histone acetylation similar to those of butyrate and TSA *(152)*. First, butyric acid, TSA, and valproic acid all superinduced ERβ responses similar to the effects seen with MAA *(152)*. In addition, direct assays of HDAC activity in nuclear extracts showed inhibition of deacetylase activity by not only TSA and valproic acid, but also by MAA, in a dose-dependent manner *(152)*. Immunoprecipitation was used to assay class I HDACs individually *(152)*. HDAC1, HDAC2, and HDAC3, were each inhibited equally by valproic acid and TSA and in these assays *(152)*. MAA also inhibited all three HDACs but was least effective on HDAC1 *(152)*. The effect of HDAC inhibition on increasing histone H4 acetylation was demonstrated by the use of a specific antibody to acetylated H4 in Western blots of histones from treated Hela cells and from spleen cells of mice treated with MAA in vivo *(152)*. Interestingly, differences in the activity of MAA versus valproic acid and butyrate were noted in the in vivo analyses, as MAA increased H4 acetylation and the other compounds did not *(152)*. All these data provide a potential explanation for superinduction of nuclear receptor-mediated gene expression and potentially for regulation of other genes as well as a consequence of MAPK pathway activation and through inhibition of class I HDAC activity and resultant increases in histone acetylation and superinduction of gene expression of a select gene population that may vary with the individual compounds used *(152)*. The mechanisms identified and responses reported in this work *(152)* are strikingly similar to our observations with resveratrol and the soy isoflavones *(159,161,162,163)*.

There are a few other reports that provide additional support to the potential novel, non-traditional activities of MAA, butyrate, resveratrol, and soy isoflavones through alterations in histone acetylation state and through inhibition of HDAC activity. Hong et al. *(185)* report that the soy isoflavones stimulate ER-mediated acetylation of core histones. These studies used isolated chromatin and cell-free preparations, apparently in the absence of HDACs, and assayed SRC-2 and p300 as the HAT activity *(185)*. It is not clear that the isoflavones were exhibiting activity distinct from E2 as a ligand in these assays, and if these results are relevant to superinduction or rather represent standard ER/ligand effects through coactivators and accessory proteins to promote gene transcription *(185)*. Farhan and Cross *(186)* report that genistein inhibits the expression of both CYP24 and CYP27B1, enzymes that metabolize vitamin D, in DU-145 prostate cancer cells. They hypothesized that inhibition of gene transcription of the two genes might reflect activation of HDACs, so, to test this, they reasoned that TSA, an HDAC inhibitor, should reverse the effects of genistein on gene expression. They treated the cells with either TSA, genistein, or both compounds together. To their surprise, TSA, the model HDAC inhibitor, also inhibited CYP24, and the combination of TSA and genistein was even more effective than either agent alone *(186)*. This suggested that the inhibition of CYP24 may be due to HDAC inhibition, not HDAC activation. In a result reminiscent of the microarray data of Mariadson et al. *(171)*, genistein and TSA had opposite effects on CYP27B1, and TSA reversed the inhibition seen with genistein alone *(186)*. These data are suggestive, but no actual acetylated histone analysis or assays of HDAC or HAT were performed *(185)*. There is also a report by Frey et al. *(187)* that genistein can upregulate expression of p21/WAF1 by 10- to 15-fold in the non-neoplastic mammary cell line, MCF-10F, a cell line that genistein causes profound but reversible cell cycle arrest. No data are reported for daidzein, and no mechanism is proposed or demonstrated for this effect *(187)*.

The link between soy isoflavones and longevity has also been demonstrated *(164)*. An aging study in mice has shown that a soy diet throughout life can prolong lifespan in certain genetic lines, but these effects of soy diets on longevity were significantly dependent on genotype and involved interesting diet X genotype interactions *(164)*. Ames dwarf mice and normal matched controls both showed 20–40% increase in lifespan on either low or high isoflavone content soy diets compared with casein controls *(164)*. Experiments in the growth hormone receptor null or knockout line (GHR-KO) and its corresponding control strain also showed 15 and 30% lifespan increases in the low isoflavone diet compared with casein control but less effectiveness of the high isoflavone soy diet *(164)*. It is not known whether these longevity effects are similar to the effects of resveratrol in yeasts, flies, or worms, but these data in mice are suggestive that similar effects may be observed in humans. Finally, a recent report identifies an aging protein hormone, dubbed "Klotho" *(188)*. Overexpression of Klotho results in longevity, and Klotho-deficient mice showed accelerated aging and shortened lifespan *(188)*. Additional experiments suggest that the function of Klotho is intimately related to suppression of insulin/IGF-1/metabolic signaling, linking this hormone to the studies and mechanisms suggested from dwarf mice and caloric restriction *(188)*. There are no reports to date that link Klotho expression to resveratrol, HDAC inhibitors including TSA butyrate and MAA, or the soy isoflavones. These remain interesting possibilities for the future.

7. CONCLUSIONS: TRADITIONAL AND NOVEL ACTIONS OF EDCs AND OTHER NON-STEROIDAL ESTROGENS

A variety of mechanisms may be relevant to assess the actions of EDCs and their effects on gene regulation in wildlife and in humans. In addition to traditional transcriptional regulation by gene activation, all aspects of hormonal signaling may be subject to disruption through actions as agonists or antagonists, through potential modulation by adaptive responses, and through novel actions.

ACKNOWLEDGMENTS

I thank the past and most recent members of my laboratory, my students, my friends, family, colleagues and teachers, and my clinical patients—all of whom served to inspire, to stimulate, and to promote consideration of unusual ideas, tempered with thoughtful critiques. Although this review has been inspired by others, its shortcomings and exclusions remain my own, and I regret any and all inadvertent omissions of significant works that should have been included. This review and the previous work of my laboratory have been supported by grants R03 CA70515 from the National Cancer Institute, NIH under the National Action Plan on Breast Cancer, grant numbers R01 ES 08301, R01 ES 08301 02S2, and RO1 ES11125 from the National Institute of

Environmental Health Sciences, NIH, the Penny Severns Breast, Cervical and Ovarian Cancer Research Fund, the United Soy Board Soy Health Research Program and generous support from the Department of Physiology, the SimmonsCooper Cancer Center, and the Office of the Associate Dean for Research and Faculty Affairs, all at the Southern Illinois University School of Medicine. Data presented represent original work performed in my laboratory both at the Washington University School of Medicine, in Saint Louis, Missouri, and at Southern Illinois University School of Medicine, Carbondale, Illinois.

With the completion of this chapter, I also have completed a chapter in my career, transitioning from an academic setting of teaching, research, and clinical activity, to my new career as a full-time clinician. I particularly thank my long time collaborators at Southern Illinois University: Dr. Cal Meyers and Dr. Yuqing Hou from the Meyers Institute for Interdisciplinary Research in Organic and Medicinal Chemistry and the Department of Chemistry and Biochemistry, and Dr. Bill Banz and Dr. Todd Winters from the Department of Physiology and the Department of Animal Science, Food and Nutrition. I thank my children, Rebecca, Sara, and Adam, who have over the years added their sense of scientific excitement, curiosity, and genuine accomplishment, from their early childhood visits to darkrooms and coldrooms, to helping in the lab, doing science fair experiments, or engaging in and then presenting their own research. Finally, I thank my wife, Molly, who has had enormous patience, first waiting for me endlessly to come home from the lab, then serving as technician in the lab with me, and finally being lab manager and commuting partner, making both lab, car, and home special by her presence.

REFERENCES

1. Colborn T, Dumanoski D, Myers JP (1996) *Our Stolen Future: Are We Threatening Our Fertility, Intelligence and Survival?* New York: Penguin Group.
2. Stone R. (1994) Environmental estrogens stir debate. *Science* 265, 308–310.
3. Sharpe RM, Skakkebaek NE. (1993) Are oestrogens involved in falling sperm counts and disorders of the male reproductive tract? *Lancet* 341, 1392–1395.
4. Fisher JS. (2004) Environmental anti-androgens and male reproductive health: focus on phthalates and testicular dysgenesis syndrome. *Reproduction* 127, 305–315.
5. Snedeker SM. (2001) Pesticides and breast cancer risk: a review of DDT, DDE, and Dieldrin. *Environ Health Perspect* 9(Suppl 1), 35–47.
6. Rubin CH, Lanier A, Kieszak S, Brock JW, Koller KR, Strosnider H, Neeedham L, Zahm S, Harpster A. (2006) Breast cancer among Alaska Native women potentially exposed to environmental organochlorine chemicals. *Int J Circumpolar Health* 65, 18–27.
7. Colborn T, vom Saal FS, Soto AM. (1993) Developmental effects of endocrine-disrupting chemicals in wildlife and humans. *Environ Health Perspect* 101, 378–384.
8. Fein GG, Jacobson JL, Jacobson JW, Schwartz PM, Fowler JK. (1984) Prenatal exposure to polychlorinated biphenyls: effects on birth size and gestational age. *J Pediatr* 105, 315–320.
9. Jacobson JL, Jacobson SW, Humphrey HEB. (1990) Effects of in utero exposure to polychlorinated biphenyls and related contaminants on cognitive functioning in young children. *J Pediatr* 116, 38–45.
10. Schmutzler C, Bacinski A, Ambrugger P, Huhne K, Grüters A, Köhrle J. (2006) Thyroid hormone biosynthesis is a sensitive target for the action of endocrine disrupting chemicals (EDC). *Exp Clin Endocrinol Diabetes* 114. http://www.thieme-connect.com/ejournals/abstract/eced/doi/10.1055/s-2006-932882.
11. Anway MD, Skinner MK. (2006) Epigenetic transgenerational actions of endocrine disruptors. *Endocrinology* 147, S43–S49.

12. Crews D, McLachlan JA. (2006) Epigenetics, evolution, endocrine disruption, health, and disease. *Endocrinology* 147, S4–S10.
13. Newbold RR, Padilla-Banks E, Jefferson WN. (2006) Adverse effects of the model environmental estrogen diethylstilbestrol are transmitted to subsequent generations. *Endocrinology* 147, S11–S17.
14. Anway MD, Cupp AS, Uzumcu M, Skinner MK. (2005) Epigenetic transgenerational actions of endocrine disrupters and male fertility. *Science* 308, 1466–1469.
15. Kelce WR, Stone CR, Laws SC, Gray LE, Kemppainen JA, Wilson EM. (1995) Persistent DDT metabolite p,p'-DDE is a potent androgen receptor antagonist. *Nature* 375, 581–585.
16. Foster PM, Mylchreest E, Gaido KW, Sar M. (2001) Effects of phthalate esters on the developing reproductive tract of male rats. *Hum Reprod Update* 7, 231–235.
17. Whitehead SA, Rice S. (2006) Endocrine-disrupting chemicals as modulators of sex steroid synthesis. *Best Pract Res Clin Endocrinol Metab* 20, 45–61.
18. Chen S. (2002) Modulation of aromatase activity and expression by environmental chemicals. *Front Biosci* 7, 1712–1719.
19. Woodhouse AJ, Cooke GM. (2004) Suppression of aromatase activity in vitro by PCBs 28 and 105 and Aroclor 1221. *Toxicol Lett* 30, 91–100.
20. Grün F, Blumberg B. (2006) Environmental obesogens: organotins and endocrine disruption via nuclear receptor signaling. *Endocrinology* 147, S50–S55.
21. Kanayama T, Kobayashi N, Mamiya S, Nakanishi T, Nishikawa J. (2005) Organotin compounds promote adipocyte differentiation as agonists of theperoxisome proliferator-activated receptor gamma/retinoid X receptor pathway. *Mol Pharmacol* 67, 766–774.
22. Giguere V, Tremblay A, Tremblay GB. (1998) Estrogen receptor beta: re-evaluation of estrogen and antiestrogen signaling. *Steroids* 63, 335–339.
23. Kuiper GGJM, Enmark E, Pelto-Huikko M, Nilsson S, Gustafsson J-Å. (1996) Cloning of a novel estrogen receptor expressed in rat prostate and ovary. *Proc Natl Acad Sci USA* 93, 5925–5930.
24. Mosselman S, Polman J, Dijkema R. (1996) ERβ: identification and characterization of a novel human estrogen receptor. *FEBS Lett* 392, 49–53.
25. Tremblay GB, Tremblay A, Copeland NG, Gilbert DJ, Jenkins NA, Labrie F, Giguère V. (1997) Cloning, chromosomal localization, and functional analysis of the murine estrogen receptor. *Mol Endocrinol* 11, 353–365.
26. Byers M, Kuiper GGJM, Gustafsson J-Å, Park-Sarge O-K. (1997) Estrogen receptor-β mRNA expression in rat ovary: down-regulation by gonadotropins. *Mol Endocrinol* 11, 172–182.
27. Kuiper GGJM, Carlsson B, Grandien K, Enmark E, Häggblad J, Nilsson S, Gustafsson J-Å. (1997) Comparison of the ligand binding specificity and transcript tissue distribution of estrogen receptors α and β. *Endocrinology* 138, 863–870.
28. Beato M. (1991) Transcriptional control by nuclear receptors. *FASEB J* 5, 2044–2051.
29. Levin ER. (2005) Integration of the extranuclear and nuclear actions of estrogen. *Mol Endocrinol* 19, 1951–1959.
30. Watson CS, Campbell CH, Gametchu B. (1999) Membrane oestrogen receptors on rat pituitary tumour cells: immuno-identification and responses to oestradiol and xenoestrogens. *Exp Physiol* 84, 1013–1022.
31. Pedram A, Razandi M, Levin ER. (2006) Nature of functional estrogen receptors at the plasma membrane. *Mol Endocrinol* 20, 1996–2009 [Epub ahead of print, April 27].
32. Carmeci C, Thompson DA, Ring HZ, Francke U, Weigel RJ. (1997) Identification of a gene (GPR30) with homology to the G-protein-coupled receptor superfamily associated with estrogen receptor expression in breast cancer. *Genomics* 45, 607–617.
33. Revankar CM, Cimino DF, Sklar LA, Arterburn JB, Prossnitz ER. (2005) A. transmembrane intracellular estrogen receptor mediates rapid cell signaling. *Science* 307, 1625–1630.
34. Filardo EJ, Thomas P. (2005) GPR30: a seven-transmembrane-spanning estrogen receptor that triggers EGF release. *Trends Endocrinol Metab* 16, 362–367.
35. Psarra AM, Solakidi S, Sekeris CE. (2006) The mitochondrion as a primary site of action of steroid and thyroid hormones: presence and action of steroid and thyroid hormone receptors in mitochondria of animal cells. *Mol Cell Endocrinol* 246, 21–33.
36. Chen JQ, Yager JD, Russo J. (2005) Regulation of mitochondrial respiratory chain structure and function by estrogens/estrogen receptors and potential physiological/pathophysiological implications. *Biochim Biophys Acta* 1746, 1–17.
37. Yager JD, Davidson NE. (2006) Estrogen carcinogenesis in breast cancer. *N Engl J Med* 354, 270–282.

38. Spink DC, Johnson JA, Connor SP, Aldous KM, Gierthy JF. (1984) Stimulation of 17β-estradiol metabolism in MCF-7 cells by bromochloro- and chloromethyl-substituted dibenzo-*p*-dioxins and dibenzofurans: correlations with antiestrogenic activity. *J Toxicol Environ Health* 41, 451–466.
39. Spink DC, Eugster H-P, Lincoln II, DW, Schuetz JD, Schuetz EG, Johnson JA, Kaminsky LS, Gierthy JF. (1992) 17β-estradiol hydroxylation catalyzed by human cytochrome P450 1A1: a comparison of the activities induced by 2,3,7,-tetrachlorodibenzo-*p*-dioxin in MCF-7 cells with those from heterologous expression of the cDNA. *Arch Biochem Biophys* 293, 342–348.
40. Beato M. (1989) Gene regulation by steroid hormones. *Cell* 56, 335–344.
41. Wilson JD, Foster DW. (eds) (1985) *Textbook of Endocrinology*. W. B. Saunders Co., Philadelphia, PA.
42. Adler S, Waterman ML, He X, Rosenfeld MG. (1988) Steroid receptor-mediated inhibition of rat prolactin gene expression does not require the receptor DNA-binding domain. *Cell* 52, 685–695.
43. Diamond MI, Miner JN, Yoshinaga SK, Yamamoto KR. (1990) Transcription factor interactions: selectors of positive or negative regulation from a single DNA element. *Science* 249, 1266–1272.
44. Tzukerman M, Zhang X-K, Pfahl M. (1991) Inhibition of estrogen receptor activity by the tumor promoter 12-O-tetradeconylphorbol-13-acetate: a molecular analysis. *Mol Endocrinol* 5, 1983–1992.
45. Jonat C, Rahmsdorf HJ, Park K-K, Cato ACB, Gebel S, Ponta H, Herrlich P. (1990) Antitumor promotion and antiinflammation: down-modulation of AP-1 (Fos/Jun) activity by glucocorticoid hormone. *Cell* 62, 1189–1204.
46. Lucibello FC, Slater EP, Jooss KU, Beato M, Müller R. (1990) Mutual transrepression of Fos and the glucocorticoid receptor: involvement of a functional domain in Fos which is absent in FosB. *EMBO J* 9, 2827–2834.
47. Yang-Yen H-F, Chambard J-C, Sun Y-L, Smeal T, Schmidt TJ, Drouin J, Karin M. (1990) Transcriptional interference between c-Jun and the glucocorticoid receptor: mutual inhibition of DNA binding due to direct protein-protein interaction. *Cell* 62, 1205–1215.
48. Schüle R, Rangarajan P, Kliewer S, Ransone LJ, Bolando J, Yang N, Verma IM, Evans RM. (1990) Functional antagonism between oncoprotein c-Jun and the glucocorticoid receptor. *Cell* 62, 1217–1226.
49. Webb P, Lopez GN, Uht RM, Kushner PJ. (1995) Tamoxifen activation of the estrogen receptor/AP-1 pathway: potential origin for the cell-specific estrogen-like effects of antiestrogens. *Mol Endocrinol* 9, 443–456.
50. Weinberg RA. (1991) Tumor suppressor genes. *Science* 254, 1138–1146.
51. Foster JS, Wimalasena J. (1996) Estrogen regulates activity of cyclin-dependent kinases and retinoblastoma protein phosphorylation in breast cancer cells. *Mol Endocrinol* 10, 488–498.
52. Greene GL, Gilna P, Waterfield M, Baker A, Hort Y, Shine J. (1986) Sequence and expression of human estrogen receptor complementary DNA. *Science* 231, 1150–1154.
53. Tora L, White J, Brou C, Tasset D, Webster N, Scheer E, Chambon P. (1989) The human estrogen receptor has two independent nonacidic transcriptional activation functions. *Cell* 59, 477–487.
54. Waterman ML, Adler S, Nelson C, Greene GL, Evans RM, Rosenfeld MG. (1988) A single domain of the estrogen receptor confers DNA binding and transcriptional activation of the rat prolactin gene. *Mol Endocrinol* 2, 14–21.
55. Tzukerman MT, Esty A, Santiso-Mere D, Danielian P, Parker MG, Stein RB, Pike JW, McDonnell DP. (1994) Human estrogen receptor transactivational capacity is determined by both cellular and promoter context and mediated by two functionally distinct intramolecular regions. *Mol Endocrinol* 8, 21–30.
56. Berry M, Metzger D, Chambon P. (1990) Role of the two activating domains of the oestrogen receptor in the cell-type and promoter context dependent agonistic activity of the anti-oestrogen 4-hydroxytamoxifen. *EMBO J* 9, 2811–2818.
57. Ramkumar T, Adler S. (1995) Differential positive and negative transcriptional regulation by tamoxifen. *Endocrinology* 136, 536–542.
58. Paech K, Webb P, Kuiper GGJM, Nilsson S, Gustafsson J-Å, Kushner PJ, Scanlan TS. (1997) Differential ligand activation of estrogen receptors ERα and ERβ at AP1 sites. *Science* 277, 1508–1510.
59. Cotton P. (1994) Environmental estrogenic agents area of concern. *JAMA* 271, 414–416.
60. Kuiper GGJM, Lemmen JG, Carlsson B, Corton JC, Safe SH, van der Saag PT, van der Burg B, Gustafsson J-Å. (1998) Interaction of estrogenic chemicals and phytoestrogens with estrogen receptor. *Endocrinology* 139, 4252–4263.

61. Barkhem T, Carlsson B, Nilsson Y, Enmark E, Gustafsson J, Nilsson S. (1998) Differential response of estrogen receptor alpha and estrogen receptor beta to partial estrogen agonists/antagonists. *Mol Pharmacol* 54, 105–112.
62. De Angelis M, Stossi F, Carlson KA, Katzenellenbogen BS, Katzenellenbogen JA. (2005) Indazole estrogens: highly selective ligands for the estrogen receptor beta. *J Med Chem* 48, 1132–1144.
63. Zaitseva M, Yue DS, Katzenellenbogen JA, Rogers PA, Gargett CE. (2004) Estrogen receptor-alpha agonists promote angiogenesis in human myometrial microvascular endothelial cells. *J Soc Gynecol Investig* 11, 529–535.
64. Compton DR, Sheng S, Carlson KE, Rebacz NA, Lee IY, Katzenellenbogen BS, Katzenellenbogen JA. (2004) Pyrazolo[1, 5-a]pyrimidines: estrogen receptor ligands possessing estrogen receptor beta antagonist activity. *J Med Chem* 47, 5872–5893. Erratum in: *J Med Chem* 48, 2724.
65. Harrington WR, Sheng S, Barnett DH, Petz LN, Katzenellenbogen JA, Katzenellenbogen BS. (2003) Activities of estrogen receptor alpha- and beta-selective ligands at diverse estrogen responsive gene sites mediating transactivation or transrepression. *Mol Cell Endocrinol* 206, 13–22.
66. Stauffer SR, Coletta CJ, Tedesco R, Nishiguchi G, Carlson K, Sun J, Katzenellenbogen BS, Katzenellenbogen JA. (2000) Pyrazole ligands: structure-affinity/activity relationships and estrogen receptor-alpha-selective agonists. *J Med Chem* 43, 4934–4947.
67. Pak TR, Chung WC, Lund TD, Hinds LR, Clay CM, Handa RJ. (2005) The androgen metabolite, 5alpha-androstane-3beta, 17beta-diol, is a potent modulator of estrogen receptor-beta1-mediated gene transcription in neuronal cells. *Endocrinology* 146, 147–155. Epub October 7, 2004.
68. Imamov O, Lopatkin NA, Gustafsson JA. (2004) Estrogen receptor beta in prostate cancer. *N Engl J Med* 351, 2773–2774.
69. Alves SE, Lopez V, McEwen BS, Weiland ng. (1998) Differential colocalization of estrogen receptor β (ERβ) with oxytocin and vasopressin in the paraventricular and supraoptic nuclei of the female rat brain: an immunocytochemical study. *Proc Natl Acad Sci USA* 95, 3281–3286.
70. Shughrue PJ, Lane MV, Merchenthaler I. (1997) Comparative distribution of estrogen receptor-α and β-mRNA in the rat central nervous system. *J Comp Neurol* 388, 507–525.
71. Dotzlaw H, Leygue E, Watson PH, Murphy LC. (1997) Expression of estrogen receptor-beta in human breast tumors. *J Clin Endocrinol Metab* 82, 2371–2374.
72. Ferguson AT, Lapidus RG, Davidson NE. (1998) The regulation of estrogen receptor expression and function in human breast cancer. *Cancer Treat Res* 94, 255–278.
73. Vladusic EA, Hornby AE, Guerra-Vladusic FK, Lupu R. (1998) Expression of estrogen receptor beta messenger RNA variant in breast cancer. *Cancer Res* 58, 210–214.
74. Halachmi S, Marden E, Martin G, MacKay H, Abbonanza C, Brown M. (1994) Estrogen receptor-associated proteins: possible mediators of hormone-induced transcription. *Science* 264, 1455–1458.
75. Kamei Y, Xu L, Heinzel T, Torchia J, Kurokawa R, Gloss B, Lin S-C, Heyman RA, Rose DW, Glass CK, Rosenfeld MG. (1996) A CBP integrator complex mediates transcriptional activation and AP-1 inhibition by nuclear receptors. *Cell* 85, 403–414.
76. Warnmark A, Almlof T, Leers J, Gustafsson JA, Treuter E. (2001) Differential recruitment: of the mammalian mediator subunit TRAP220 by estrogen receptors ERalpha and ERbeta. *J Biol Chem* 276, 23397–23404.
77. Evans RM. (1988) The steroid and thyroid hormone receptor superfamily. *Science* 240, 889–895.
78. Lam PH-Y. (1984) Tamoxifen is a calmodulin antagonist in the activation of cAMP phosphodiesterase. *Biochem Biophys Res Commun* 118, 27–32.
79. O'Brian CA, Liskamp RM, Solomon DH, Weinstein IB. (1985) Inhibition of protein kinase C by tamoxifen. *Cancer Res* 45, 2462–2465.
80. Issandou M, Faucher C, Bayard F, Darbon JM. (1990) Opposite effects of tamoxifen on in vitro protein kinase c activity and endogenous protein phosphorylation in intact MCF-7 cells. *Cancer Res* 50, 5845–5850.
81. Jordan VC, Collins MM, Rowsby L, Prestwich G (1977) A monohydroxylated metabolite of tamoxifen with potent antiestrogenic activity. *J Endocrinol* 75, 305–316.
82. Johanson G. (2000) Toxicity review of ethylene glycol monomethyl ether and its acetate ester. *Crit Rev Toxicol* 207, 149–163.
83. Cummings AM. (1997) Methoxychlor as a model for environmental estrogens. *Crit Rev Toxicol* 27, 367–379.
84. Brotons JA, Olea-Serrano MF, Villalobos M, Pedraza V, Olea N. (1995) Xenoestrogens released from lacquer coatings in food cans. *Environ Health Perspect* 103, 608–612.

85. Krishnan AV, Starhis P, Permuth SF, Tokes L, Feldman D. (1993) Bisphenol-A: an estrogenic substance is released from polycarbonate flasks during autoclaving. *Endocrinology* 132, 2279–2286.
86. Jobling S, Sumpter JP. (1993) Detergent components in sewage effluent are weakly estrogenic to fish - an in-vitro study using rainbow-trout (Oncorhynchus mykiss) hepatocytes. *Aquat Toxicol* 27, 361–372.
87. Mueller GC, Kim U-H. (1978) Displacement of estradiol from estrogen receptors by simple alkyl phenols. *Endocrinology* 102, 1429–1435.
88. Soto AM, Justicia H, Wray JW, Sonnenschein C. (1991) p-Nonyl-phenol: an estrogenic xenobiotic released from "modified" polystyrene. *Environ Health Perspect* 92, 167–173.
89. White R, Jobling S, Hoare SA, Sumpter JP, Parker MG. (1994) Environmentally persistent alkylphenolic compounds are estrogenic. *Endocrinology* 135, 175–182.
90. ENDS. (1995) Packaging industry failing to act over phthalates in food. *ENDS Rep* 245, 7–8.
91. JEH 1995. Environmental oestrogens: Consequences to human health and wildlife. Institute for Environment and Health, University of Leicester, Leicester, UK.
92. Beekman JM, Allan GF, Tsai SY, Tsai M-J, O'Malley BW. (1993) Transcriptional activation by the estrogen receptor requires a conformational change in the ligand binding domain. *Mol Endocrinol* 7, 1266–1274.
93. Metzger D, Berry M, Ali S, Chambon P. (1995) Effect of antagonists on DNA binding properties of the human estrogen receptor *in vitro* and *in vivo*. *Mol Endocrinol* 9, 579–591.
94. Brown M, Sharp PA. (1990) Human estrogen receptor forms multiple protein-DNA complexes. *J Biol Chem* 265, 11238–11243.
95. Meyers CY, Lutfi HG, Adler S. (1997) Transcriptional regulation of estrogen-responsive genes by non-steroidal estrogens: doisynolic and allenolic acids. *J Steroid Biochem Mol Biol* 62, 477–489.
96. Meyers CY, Hou Y, Winters TA, Banz WJ, Adler S. (2002) Activities of a non-classical estrogen, Z-bis-dehydrodoisynolic acid, with ERα and ERβ. *J Steroid Biochem Mol Biol* 82, 33–44.
97. Dibbs KI, Sadovsky Y, Li X-J, Koide SS, Adler S, Fuchs A-R. (1995) Estrogenic activity of RU486 (Mifepristone) in rat uterus and cultured uterine myocytes. *Am J Obstet Gynecol* 173, 134–140.
98. Adler M, Hou Y, Sandrock P, Meyers CY, Winters TA, Banz WJ, Adler S. (2006) Derivatives of Z-bisdehydrodoisynolic acid provide a new description of the binding-activity paradox and selective estrogen receptor modulator activity. *Endocrinology* 147(8), 3952–3960. Epub May 18, as doi:10.1210/en.2006-0316.
99. Berthois Y, Katzenellenbogen JA, Katzenellenbogen BS. (1986) Phenol red in tissue culture media is a weak estrogen: implications concerning the study of estrogen-responsive cells in culture. *Proc Natl Acad Sci USA* 83, 2496–2500.
100. Segaloff A. (1949) The metabolism of estrogens with particular emphasis on clinical aspects of physiology, function of ovarian hormones. In *Recent Progress in Human Research*, Vol. IV (edited by G. Pincus). Academic Press, New York, pp. 85–111.
101. Tora L, Mullick A, Metzger D, Ponglikitmongkol M, Park I, Chambon P. (1989) The cloned human oestrogen receptor contains a mutation which alters its hormone binding properties. *EMBO J* 8, 1981–1986.
102. Zhang Q-X, Borg A, Wolff DM, Oesterreich S, Fuqua SAW. (1997) An estrogen receptor mutant with strong hormone-independent activity from a metastatic breast cancer. *Cancer Res* 57, 1244–1249.
103. Krishnan V, Safe S. (1993) Polychlorinated biphenyls (PCBs), dibenzo-*p*-dioxins PCDDs), and dibenzofurans (PCDFs) as antiestrogens in MCF-7 human breast cancer cells: quantitative structure-activity relationships. *Toxicol Appl Pharmacol* 120, 55–61.
104. Jordan CV, Mittal S, Gosden B, Koch R, Lieberman ME. (1985) Structure-activity relationships of estrogens. *Environ Health Perspect* 61, 97–110.
105. Jansen HT, Cooke PS, Porcelli J, Liu T-C, Hansen LG. (1993) Estrogenic and antiestrogenic actions of PCBs in the female rat: *in vitro* and *in vivo* studies. *Reprod Toxicol* 7, 237–248.
106. Liang P, Pardee AB. (1992) Differential display of eukaryotic messenger RNA by means of the polymerase chain reaction. *Science* 257, 967–971.
107. Liang P, Averboukh L, Pardee AB. (1993) Distribution and cloning of eukaryotic mRNAs by means of differential display: refinements and optimization. *Nucleic Acids Res* 21, 3269–3275.
108. Jamil A, Croxtall JD, White JO. (1991) The effect of anti-estrogens on cell growth and progesterone receptor concentration in human endometrial cancer cells (Ishikawa). *J Mol Endocrinol* 6, 215–221.
109. Lippman ME, Dickson RB. (1989) Mechanisms of growth control in normal and malignant breast epithelium. *Recent Prog Horm Res* 45, 383–440.

110. Rose DP, Boyar AP, Wynder EL. (1986) International comparisons of mortality rates for cancer of the breast, ovary, prostate, and colon, and per capita food consumption. *Cancer* 58, 2363–2371.
111. Lee HP, Gourley L, Duffy SW, Estève J, Lee J, Day NE. (1991) Dietary effects on breast-cancer risk in Singapore. *Lancet* 337, 1197–1200.
112. Goodman MJ. (1991) Breast cancer on multi-ethnic populations: the Hawaii perspective. *Breast Cancer Res Treat* 18(Suppl 1), S5–S9.
113. Key TJ, Thorogood M, Appleby PN, Burr ML. (1996) Dietary habits and mortality in 11,000 vegetarians and health conscious people: results of a 17 year follow up. *BMJ* 313, 775–779.
114. Herman C, Adlercreutz T, Goldin BR, Gorbach SL, Höckerstedt KAV, Watanabe S, Hämäläinen EK, Markkanen MH, Mäkelä TH, Wähälä KT, Hase TA, Fotsis T. (1995) Soybean phytoestrogen intake and cancer risk. *J Nutr* 125, 757S–770S.
115. Whitten PL, Lewis C, Russell E, Naftolin F. (1995) Potential adverse effects of phytoestrogens. *J Nutr* 125, 771S–776S.
116. Knight DC, Eden JA. (1996) A review of clinical effects of phytoestrogens. *Obstet Gynecol* 87, 897–904.
117. Kurzer MS, Xu X. (1997) Dietary phytoestrogens. *Annu Rev Nutr* 17, 353–381.
118. King ML, Adler SR, Murphy LL. Extraction-dependent effects of American Ginseng (Panax quinquefolium) on human breast cancer cell proliferation and estrogen receptor activation. *Integr Cancer Ther.* Sep;5(3):236–2430.
119. Xu L, Glass CK, Rosenfeld MG. (1999) Coactivator and corepressor complexes in nuclear receptor function. *Curr Opin Genet Dev* 9, 140–147.
120. Shibata H, Spencer TE, Onate SA, Jenster G, Tsai SY, Tsai MJ, O'Malley BW. (1997) Role of co-activators and co-repressors in the mechanism of steroid/thyroid receptor action. *Recent Prog Horm Res* 52, 141–164.
121. Torchia J, Rose DW, Inostroza J, Kamei Y, Westin S, Glass CK, Rosenfeld MG. (1997) The transcriptional co-activator p/CIP binds CBP and mediates nuclear-receptor function. *Nature* 387, 677–684.
122. Oñate SA, Tsai SY, Tsai M-J, O'Malley BW. (1995) Sequence and characterization of a coactivator for the steroid hormone receptor superfamily. *Science* 270, 1354–1357.
123. Blanco JCG, Minucci S, Lu J, Yang XJ, Walker KK, Chen H, Evans RM, Nakatani Y, Ozato K. (1998) The histone acetylase PCAF is a nuclear receptor coactivator. *Genes Dev* 12, 1638–1651.
124. Chen JD, Li H. (1998) Coactivation and corepression in transcriptional regulation by steroid/nuclear hormone receptors. *Crit Rev Eukaryot Gene Expr* 8, 169–190.
125. Chen JD, Evans RM. (1995) A transcriptional co-repressor that interacts with nuclear hormone receptors. *Nature* 377, 454–457.
126. Hörlein AJ, Näär AM, Heinzel T, Torchia J, Gloss B, Kurokawa R, Ryan A, Kamei Y, Söderström M, Glass CK, Rosenfeld MG. (1995) Ligand-independent repression by the thyroid hormone receptor mediated by a nuclear receptor co-repressor. *Nature* 377, 397–404.
127. Lavinsky RM, Jepsen K, Heinzel T, Torchia J, Mullen TM, Schiff R, Delrio AL, Ricote M, Ngo S, Gemsch J, Hilsenbeck SG, Osborne CK, Glass CK, Rosenfeld MG, Rose DW. (1998) Diverse signaling pathways modulate nuclear receptor recruitment of N-CoR and SMRT complexes. *Proc Natl Acad Sci USA* 95, 2920–2925.
128. Nishikawa J, Saito K, Goto J, Dakeyama F, Matsuo M, Nishihara T. (1999) New screening methods for chemicals with hormonal activities using interaction of nuclear hormone receptor with coactivator. *Toxicol Appl Pharmacol* 154, 76–83.
129. Smith CL, Nawaz Z, O'Malley BW. (1997) Coactivator and corepressor regulation of the agonist/antagonist activity of the mixed antiestrogen, 4-hydroxytamoxifen. *Mol Endocrinol* 11, 657–666.
130. Jackson TA, Richer JK, Bain DL, Takimoto GS, Tung L, Horwitz KB. (1997) The partial agonist activity of antagonist-occupied steroid receptors is controlled by a novel hinge domain-binding coactivator L7/SPA and the corepressors N-CoR or SMRT. *Mol Endocrinol* 11, 693–705.
131. Zhang X, Jeyakumar M, Petukhov S, Bagchi MK. (1998) A nuclear receptor corepressor modulates transcriptional activity of antagonist-occupied steroid hormone receptor. *Mol Endocrinol* 12, 513–524.
132. Takimoto GS, Graham JD, Jackson TA, Tung L, Powell RL, Horwitz LD, Horwitz KB. (1999) Tamoxifen resistant breast cancer: coregulators determine the direction of transcription by antagonist-occupied steroid receptors. *J Steroid Biochem Mol Biol* 69, 45–50.

133. Brzozowski AM, Pike ACW, Dauter Z, Hubbard RE, Bonn T, Engstrom O, Ohman L, Greene GL, Gustafsson J-A, Carlquist M. (1997) Molecular basis of agonism and antagonism in the estrogen receptor. *Nature* 389, 753–758.
134. Berns EM, van Staveren IL, Klijn JG, Foekens JA. (1998) Predictive value of SRC-1 for tamoxifen response of recurrent breast cancer. *Breast Cancer Res Treat* 48, 87–92.
135. Anzick SL, Kononen J, Walker RL, Azorsa DO, Tanner MM, Guan XY, Sauter G, Kallioniemi OP, Trent JM, Meltzer PS. (1997) AIB1, a steroid receptor coactivator amplified in breast and ovarian cancer. *Science* 277, 965–968.
136. Eng FCS, Barsalou A, Akutsu N, Mercier I, Zechel C, Mader S, White JH. (1998) Different classes of coactivators recognize distinct but overlapping binding sites on the estrogen receptor ligand binding domain. *J Biol Chem* 273, 28371–28377.
137. Misiti S, Schomburg L, Yen PM, Chin WW. (1998) Expression and hormonal regulation of coactivator and corepressor genes. *Endocrinology* 139, 2493–2500.
138. Nguyen TA, Hoivik D, Lee JE, Safe S. (1999) Interactions of nuclear receptor coactivator/corepressor proteins with the aryl hydrocarbon receptor complex. *Arch Biochem Biophys* 15, 250–257.
139. Hornick JR, Zhang Z, Romans SR, Adler M, Adler S. (2004) Regulation of the SMRT Promoter by Estrogens. Endocrine Society - 86th Annual Meeting, Abstract P1-187.
140. Adler RA, Adler SR. (2002) Regulation of the Expression of Nuclear Receptor Coregulators SRC-1 and SMRT in Response to Estradiol and the Phytoestrogen Genistein. Endocrine Society - 84th Annual Meeting, Abstract P2-467.
141. [No authors listed]. (1999) Food labeling: health claims; soy protein and coronary heart disease. Food and drug administration, HHS. Final rule. *Fed Regist* 64(206), 57700–57733.
142. Wu JM, Wang ZR, Hsieh TC, Bruder JL, Zou JG, Huang YZ. (2001) Mechanism of cardioprotection by resveratrol, a phenolic antioxidant present in red wine (Review). *Int J Mol Med* 8, 3–17.
143. Peeters PH, Keinan-Boker L, van der Schouw YT, Grobbee DE. (2003) Phytoestrogens and breast cancer risk. Review of the epidemiological evidence. *Breast Cancer Res Treat* 77, 171–183.
144. Mishra SI, Dickerson V, Najm W. (2003) Phytoestrogens and breast cancer prevention: What is the Evidence? *Am J Obstet Gynecol* 188(5 Suppl), S66–S70.
145. Ganry O. (2005) Phytoestrogens and prostate cancer risk. *Prev Med* 41, 1–6.
146. Katic M, Kahn CR. (2005) The role of insulin and IGF-1 signaling in longevity. *Cell Mol Life Sci* 62, 320–43.
147. Al-Regaiey KA, Masternak MM, Bonkowski M, Sun L, Bartke A. (2005) Long-lived growth hormone receptor knockout mice: interaction of reduced insulin-like growth factor i/insulin signaling and caloric restriction. *Endocrinology* 146, 851–860.
148. Hall SS. (2003) In Vino Vitalis? compounds activate life-extending genes. *Science* 301, 1165.
149. Howitz KT, Bitterman KJ, Cohen HY, Lamming DW, Lavu S, Wood JG, Zipkin RE, Chung P, Kisielewski A, Zhang LL, Scherer B, Sinclair DA. (2003) Small molecule activators of sirtuins extend Saccharomyces cerevisiae lifespan. *Nature* 425, 191–196.
150. Wood JG, Rogina B, Lavu S, Howitz K, Helfand SL, Tatar M, Sinclair D. (2004) Sirtuin activators mimic caloric restriction and delay ageing in metazoans. *Nature* 430, 686–689.
151. Imai S, Armstrong CM, Kaeberlein M, Guarente L. (2000) Transcriptional silencing and longevity protein Sir2 is an NAD-dependent histone deacetylase. *Nature* 403, 795–800.
152. Jansen MS, Nagel SC, Miranda PJ, Lobenhofer EK, Afshari CA, McDonnell DP. (2004) Short-chain fatty acids enhance nuclear receptor activity through mitogen-activated protein kinase activation and histone deacetylase inhibition. *Proc Natl Acad Sci USA* 101, 7199–7204.
153. Yoshida M, Kijima M, Akita M, Beppu T. (1990) Potent and specific inhibition of mammalian histone deacetylase both in vivo and in vitro by trichostatin A. *J Biol Chem* 265, 17174–17179.
154. Marlett JA, McBurney MI, Slavin JL, the American Dietetic Association. (2002) Position of the American Dietetic Association: health implications of dietary fiber. *J Am Diet Assoc* 102, 993–1000.
155. Hinnebusch BF, Meng S, Wu JT, Archer SY, Hodin RA. (2002) The effects of short-chain fatty acids on human colon cancer cell phenotype are associated with histone hyperacetylation. *J Nutr* 132, 1012–1017.
156. Davie JR. (2003) Inhibition of histone deacetylase activity by butyrate. *J Nutr* 133(Suppl), 2485S–93S.
157. Chen YX, Fang JY, Zhu HY, Lu R, Cheng ZH, Qiu DK. (2004) Histone acetylation regulates p21WAF1 expression in human colon cancer cell lines. *World J Gastroenterol* 10, 2643–2646.

158. Villar-Garea A, Esteller M. (2004) Histone deacetylase inhibitors: understanding a new wave of anticancer agents. *Int J Cancer* 112, 171–178.
159. Winters TA, Banz WJ, Mallon MA, Adler S. (1999) Estrogenic gene activation and estrogenic gene repression by phytoestrogens define two functional classes. World Soybean Research Conference IV - Global Soy Forum 99, Chicago, IL. August 4–7.
160. Gehm B, McAndrews J, Chien P-Y, Jameson L. (1997) Resveratrol, a polyphenolic compound found in grapes and wine, is an agonist for the estrogen receptor. *Proc Natl Acad Sci USA* 94, 14138–14143.
161. Li W, Adler S. (2005) Resveratrol as an estrogenic superinducer: participation as ligand, signal multiplier and via the MAPK pathway. The Endocrine Society Forum on Endocrine Disrupting Chemicals, San Diego, CA. June 3.
162. Li W, Romans SR, Adler S. (2004) Superinduction of gene expression by resveratrol can occur independent of estrogen receptors. Endocrine Society - 86th Annual Meeting. Abstract P1-102.
163. Trammel AC, Adler SR. (2002) Superagonist estrogenic activity of resveratrol is related to cell density or growth effects. Endocrine Society - 84th Annual Meeting. Abstract P1-626.
164. Bartke A, Peluso MR, Moretz N, Wright C, Bonkowski M, Winters TA, Shanahan MF, Kopchick JJ, Banz WJ. (2004) Effects of soy-derived diets on plasma and liver lipids, glucose tolerance, and longevity in normal, long- lived and short lived mice. *Horm Metab Res* 36, 550–558.
165. Mueller SO, Simon S, Chae K, Metzler M, Korach KS. (2004) Phytoestrogens and their human metabolites show distinct agonistic and antagonistic properties on estrogen receptor alpha (ERalpha) and ERbeta in human cells. *Toxicol Sci* 80, 14–25. Erratum in: *Toxicol Sci* 81, 530–1.
166. Metivier R, Penot G, Hubner MR, Reid G, Brand H, Kos M, Gannon F. (2003) Estrogen receptor-alpha directs ordered, cyclical, and combinatorial recruitment of cofactors on a natural target promoter. *Cell* 115, 751–763.
167. DiRenzo J, Shang Y, Phelan M, Sif S, Myers M, Kingston R, Brown M. (2000) BRG-1 is recruited to estrogen-responsive promoters and cooperates with factors nvolved in histone acetylation. *Mol Cell Biol* 20, 7541–7549.
168. Sun JM, Chen HY, Davie JR. (2001) Effect of estradiol on histone acetylation dynamics in human breast cancer cells. *J Biol Chem* 276, 49435–49442.
169. Merchant JL, Bai L, Okada M. (2003) ZBP-89 mediates butyrate regulation of gene expression. *J Nutr* 133, 2456S–2460S.
170. Bai L, Merchant JL. (2000) Transcription factor ZBP-89 cooperates with histone acetyltransferase p300 during butyrate activation of p21waf1 transcription in human cells. *J Biol Chem* 275, 30725–30733.
171. Mariadason JM, Corner GA, Augenlicht LH. (2000) Genetic reprogramming in pathways of colonic cell maturation induced by short chain fatty acids: comparison with trichostatin A, sulindac, and curcumin and implications for chemoprevention of colon cancer. *Cancer Res* 60, 4561–4572.
172. Rocchi P, Tonelli R, Camerin C, Purgato S, Fronza R, Bianucci F, Guerra F, Pession A, Ferreri AM. (2005) p21Waf1/Cip1 is a common target induced by short-chain fatty acid HDAC inhibitors (valproic acid, tributyrin and sodium butyrate) in neuroblastoma cells. *Oncol Rep* 13, 1139–1144.
173. Vaquero A, Scher M, Lee D, Erdjument-Bromage H, Tempst P, Reinberg D. (2004) Human SirT1 interacts with histone H1 and promotes formation of facultative heterochromatin. *Mol Cell* 16, 93–105.
174. Shore D. (2000) The Sir2 protein family: a novel deacetylase for gene silencing and more. *Proc Natl Acad Sci USA* 97, 14030–14032.
175. Vaziri H, Dessain SK, ng Eaton E, Imai SI, Frye RA, Pandita TK, Guarente L, Weinberg RA. (2001) hSir2(SirT1) functions as an NAD-dependent p53 deacetylase. *Cell* 107, 149–159.
176. Luo J, Nikolaev AY, Imai S, Chen D, Su F, Shiloh A, Guarente L, Gu W. (2001) Negative control of p53 by Sir2α promotes cell survival under stress. *Cell* 107, 137–148.
177. Lee TK, Man K, Poon RT, Lo CM, Ng IO, Fan ST. (2004) Disruption of p53-p21/WAF1 cell cycle pathway contributes to progression and worse clinical outcome of hepatocellular carcinoma. *Oncol Rep* 12, 25–31.
178. Nemoto S, Fergusson MM, Finkel T. (2005) SirT1 functionally interacts with the metabolic regulator and transcriptional coactivator PGC-1α. *J Biol Chem* 280(16), 16456–16460 [Epub ahead of print, February 16].
179. Rodgers JT, Lerin C, Haas W, Gygi SP, Spiegelman BM, Puigserver P. (2005) Nutrient control of glucose homeostasis through a complex of PGC-1alpha and SirT1. *Nature* 434, 113–118.

180. Bouras T, Fu M, Sauve AA, Wang F, Quong AA, Perkins ND, Hay RT, Gu W, Pestell RG. (2005) SirT1 deacetylation and repression of p300 involves lysine residues 1020/1024 within the cell cycle regulatory domain 1. *J Biol Chem* 280, 10264–10276. Epub January 4.
181. Kaeberlein M, McDonagh T, Heltweg B, Hixon J, Westman EA, Caldwell S, Napper A, Curtis R, Distefano PS, Fields S, Bedalov A, Kennedy BK. (2005) Substrate specific activation of sirtuins by resveratrol. *J Biol Chem* 280(17), 17038–17045 [Epub ahead of print, January 31].
182. Blander G, Olejnik J, Krzymanska-Olejnik E, McDonagh T, Haigis M, Yaffe MB, Guarente L. (2005) SirT1 shows no substrate specificity in vitro. *J Biol Chem* 280, 9780–9785. Epub January 6.
183. Borra MT, Smith BC, Denu JM. (2005) Mechanism of human SirT1 activation by resveratrol. *J Biol Chem* 280(17), 17187–17195 [Epub ahead of print, March 4].
184. Gehm BD, Levenson AS, Liu H, Lee EJ, Amundsen BM, Cushman M, Jordan VC, Jameson JL. (2004) Estrogenic effects of resveratrol in breast cancer cells expressing mutant and wild-type estrogen receptors: role of AF-1 and AF-2. *J Steroid Biochem Mol Biol* 88, 223–234.
185. Hong T, Nakagawa T, Pan W, Kim MY, Kraus WL, Ikehara T, Yasui K, Aihara H, Takebe M, Muramatsu M, Ito T. (2004) Isoflavones stimulate estrogen receptor-mediated core histone acetylation. *Biochem Biophys Res Commun* 317, 259–264.
186. Farhan H, Cross HS. (2002) Transcriptional inhibition of CYP24 by genistein. *Ann N Y Acad Sci* 973, 459–462.
187. Frey RS, Li J, Singletary KW. (2001) Effects of genistein on cell proliferation and cell cycle arrest in nonneoplastic human mammary epithelial cells: involvement of Cdc2, p21(waf/cip1), p27(kip1), and Cdc25C expression. *Biochem Pharmacol* 61, 979–989.
188. Kurosu H, Yamamoto M, Clark JD, Pastor JV, Nandi A, Gurnani P, McGuinness OP, Chikuda H, Yamaguchi M, Kawaguchi H, Shimomura I, Takayama Y, Herz J, Kahn CR, Rosenblatt KP, Kuro-o M. (2005) Suppression of aging in mice by the hormone Klotho. *Science* 309, 1829–1833. Epub August 25.

7 Endocrine Disruptors, Epigenetically Induced Changes, and Transgenerational Transmission of Characters and Epigenetic States

Carlos Guerrero-Bosagna, PhD,
and Luis Valladares, PhD

Contents

1. INTRODUCTION TO EPIGENETICS

Developmental epigenetics is a broad phenomenon, which was initially described by Waddington as "the branch of biology which studies the causal interactions between genes and their products which bring phenotypes into being." Today, the study of epigenetic regulation of development has been sharpened because of recent work on molecular mechanisms of gene expression and developmental biology *(1)*, mainly focusing on how the environment produces alterations in gene expression patterns without changes in DNA sequences *(2)*. Epigenetics is now a well-accepted phenomenon by the scientific community, in large part because of recent discoveries in the field of the molecular biology, namely chromatin condensation, histone modifications, and DNA methylation, which are all well-identified processes.

From: *Endocrine-Disrupting Chemicals: From Basic Research to Clinical Practice*
Edited by: A. C. Gore © Humana Press Inc., Totowa, NJ

1.1. Mechanisms of Epigenetic Modifications

DNA methylation is by far the most widely known and most studied epigenetic modification to date. The process of DNA modification constitutes a post-replicative modification, in which a methyl group is added covalently to a DNA residue *(3)*. The reaction of DNA methylation occurs at the carbon 5 of the cytosine ring in 5′–3′-oriented CG dinucleotides (which are named as CpGs) and is catalyzed by the action of DNA methyltransferases (Dnmts) *(4)*.

Two other epigenetic mechanisms have to do with chromosome structure. Eukaryotic genomic DNA is compacted more than 10,000-fold by basic proteins named histones, which participate in the compaction of this DNA into an entity known as chromatin *(5)*. Heterochromatin is a region of the genome that is highly condensed throughout the cell cycle, in contrast to euchromatin that shows condensation exclusively during mitosis *(6)*. Because heterochromatin regions are still condensed in interphase, they are associated with repressed gene expression *(6)*. Heterochromatin is often associated with hypermethylated and hyperacetylated histones *(7,8)*. Nevertheless, there are regions (referred to as facultative chromatin) that can be transiently condensed and silenced during development *(6)*, thereby leading to variation in gene expression depending on the chromatin state (condensed or uncondensed). In turn, the chromatin state is also susceptible to modification depending on specific stimuli. For instance, there are factors responsible for the initiation of the heterochromatin formation process, such as transcriptional repressors and functional RNA, and also accessory factors that interact with many groups of proteins *(9)*. Therefore, the chromatin state may be epigenetically regulated by factors whose levels could be environmentally dependent, leading to epigenetic regulation of genes whose expression depends on chromatin state.

Histones are susceptible to a variety of post-translational modifications such as phosphorylation, acetylation, methylation, ubiquitination, sumoylation, ADP ribosylation, glycosylation biotinylation, and carbonylation *(5)*. Nevertheless, among these, methyl marks provide an epigenetic mechanism that favors the stable transfer of gene expression profiles to progeny cells *(5)*. It has been suggested that an "epigenetic conversation" exists between histones and DNA that involves cytosine methylation, histone deacetylation and methylation, all acting in synergy to generate a self-reinforcing epigenetic cycle that maintains and perpetuates a repressed chromatin state *(10)*.

Despite evidence that the aforementioned epigenetic modifications can act autonomously *(10)*, RNA factors, histone methylation, and chromatin-remodeling enzymes appear to all act together with Dnmts, resulting in the establishment and maintenance of methylation patterns and the generation of site-specific methylation and tissue-specific differences *(11)*. Among RNA factors, small RNAs (siRNA and miRNA) have recently been shown to have the ability to direct DNA methylation, a mechanism called RNA-directed DNA methylation (RdDM) that is carried out by dsRNA, which may be produced by transcription through inverted repeats *(12)*. As a whole, these data indicate that epigenetic mechanisms are not acting alone but rather are integrated such that they may be affected by stimuli, and also produce phenotypic consequences. Hence, epigenetic regulation of gene expression deals with integrating intrinsic and environmental signals *(13,14)*.

1.2. Relationship Between Epigenetics and Transgenerational Effects

In integrating the concepts of epigenetics and mechanisms of epigenetic regulation with the concept of transgenerational transmission of characters, it should be emphasized that *not every epigenetic effect can be considered as transgenerational.* This latter phenomenon depends on the susceptibility of the organism to undergo epigenetic changes and also on the capacity of the organism or environment to make those changes persist across generations. An individual's susceptibility to epigenetic changes will depend on the stages of the ontogeny when the external effect took place. Exposure to chemicals or other environmental agents may induce epigenetic changes in the genome, but only when they act during critical periods of the ontogeny *(15,16)*. In turn, persistence of the stimuli through generations will depend on the nature of the epigenetic system affected. For example, epigenetic alterations on chromatin condensation leading to the changed expression of particular genes could regulate gene expression in a variable manner during a single ontogenic event but may have a reduced capacity to transmit those changes in comparison with imprinted genes. Changing methylation in imprinted genes could have the same evolutionary value of a mutation *(15)* and may, moreover, lead to biased mutations *(17)*.

1.3. Role of Developmental Stage in Susceptibility to Epigenetic Transgenerational Effects

The timing during ontogeny when an organism is exposed to stimuli has implications on the organism's susceptibility to be affected by such stimuli. For instance, development is characterized by a high sensitivity to environmental stimuli, either external or generated by cellular productions *(13)*. Disruptions produced by a stimulus during early stages of development have more systemic consequences assessed in adulthood compared with the effect of such exposures in adulthood, which has more local and limited consequences. The reason for this is that interfering with an embryonic cell during development will produce changes in the derived cell lineage, which involves a higher number and broader types of cells than when a cell lineage is derived from an adult somatic perturbed cell. Thus, Danzo *(18)* has stated that "the greatest risks to reproductive health posed by xenobiotics would be during the embryonic, neonatal and pubertal periods, when the reproductive systems are undergoing finely tuned modulation by steroid hormones."

Any environmental effect produced in early life stages will also be of profound importance from a transgenerational perspective. Organisms in early stages of development are more susceptible to heritable structural changes that can be transmitted through the germ line, such as reprogramming of methylation patterns *(2)* or through overt mutations *(19)*, even though the effects of external agents on the early embryo may be hidden until much later in life *(20,21)*. In mammals, however, external effects in early development are strongly buffered by the uterus and placenta. Therefore, any compound interfering with mammalian early stages of development must first circumvent those barriers.

1.4. Transgenerational Effects of Endocrine-Disrupting Chemicals

Other possible transgenerational effects are those that are more related to the presence of a compound itself in the environment than with its effects on structural features of organisms. If a given environmental compound is persistently present generation after generation in a population of individuals, such a presence may lead to altered parameters consistently and transgenerationally in this population. With regard to this point, it must be emphasized that an organism is not only transmitting to the offspring the structure that permits the realization of the phenotype. The organism also transmits to the offspring the environmental conditions allowing the realization of such a phenotype *(22)*. In this sense, if environmental conditions trigger the same phenotypic response every generation, it could be considered a transgenerational transmission of characteristics. Endocrine-disrupting chemicals (EDC) achieve that condition given that it is well known that they can be persistently present in the environment and food chain *(18,23,24)*.

In this chapter, we will focus the discussion on describing situations in which EDC exposure affects epigenetic mechanisms in mammals and which possess the structural features that enable them to be passed transgenerationally to future generations. Then, the epigenetic system for which EDC effects have been more extensively studied, that is, DNA methylation, will be discussed under two conditions: first, when such changes are triggered during early stages of development and, second, during differentiation of the germ line. Both cases represent events when possibilities for the organisms to be affected by EDC are increased. Moreover, in these models, epigenetic changes have more possibilities of being transgenerationally transmitted if such changes are induced either before or during the differentiation of the germ line, which contains intrinsically transmissible structural features of organisms.

2. EDCS AND THEIR MECHANISMS OF ACTION

Nearly 14 years ago, the scientific community acknowledged the existence of chemicals capable of interfering with or mimicking endogenous hormones and other signaling molecules of the endocrine system. Moreover, these substances have the ability to cross placental and brain barriers and to interfere with development and function *(25)*. Since then, as attention to this area of investigation has grown, these compounds have been referred to by a variety of names, such as EDCs *(25)*, xenoestrogens *(26)*, environmental hormones *(18,27)*, hormonally actives agents *(28)*, and environmental agents *(29)*. These chemicals included many chemical classes and comprise an integral part of the world economy and commerce. The United States Environmental Protection Agency (USEPA) developed a screening and testing program to detect EDCs, and the Organization for Economic Cooperation and Development (OECD) has set up a task force to identify, prioritize, and validate test methods for the detection of endocrine disruptors *(30)*. Nevertheless, few of the thousands of chemicals used today have been tested systematically for endocrine-disrupting effects in organisms.

The endocrine system of vertebrates consists of an intricate web of stimulatory and inhibitory hormone signals that control basic body functions such as metabolism, growth, digestion, and cardiovascular function, as well as more specialized traits and processes such as behavior, sexual differentiation (during embryogenesis), sexual maturation (during puberty), and adult reproduction *(23)*. For example, the circulating hormone 17β-estradiol (E2) controls a variety of cellular mechanisms, including development processes and differentiation events, as well as growth in organs such as breast, ovary, and uterus. The timing and concentration of bioactive estrogen signals determine sexual maturity, ovulation, and pregnancy.

2.1. Nuclear Receptors and Endocrine Disruption

Families of nuclear receptors (NRs) are defined by both structural and functional homologies. The NR superfamily contains ligand-activated transcription factors that exert a wide variety of different cellular responses by positively and/or negatively regulating target gene expression *(31,32)*. Apart from receptors that bind steroid hormones, retinoic acid or thyroid hormone, the NR superfamily contains so-called orphan receptors for which no ligand is known *(33,34)*. Steroid/xenobiotic receptors (SXR, also know as pregnane X receptor [PXR]), which belong to the family of orphan receptors, recognize many classes of EDC and may activate responses resulting in the expression of EDC metabolizing enzymes, thereby providing a link between the internal and external environment *(35)*. Estrogen receptor-related receptors (ERRs) are a subfamily of orphan NRs that are closely related to the estrogen receptor family *(36,37)*. Research on ERRs has shown that this family shares target genes, co-regulators, and promoters with the estrogen receptors family *(38,39)*. On the contrary, ERRs seem to interfere with the classic ER-mediated estrogen-responsive signal in a variety of ways *(40,41)*. Interestingly, ERRs have been reported to be prognostic biomarkers in different types of cancer *(42,43)*. In addition to SXR and ERRs, other NRs have been shown to bind EDC. For example, chlorinated hydrocarbons such as some polychlorinated biphenyl (PCB) compounds, in theory, have the ability to bind to and activate the ligand-activated transcription factor, the aryl hydrocarbon receptor (AhR), to bind the thyroid hormone receptor *(44)*. Some studies also demonstrate binding activity of environmental agents to a thyroid hormone-binding protein similar to T4 but not to a thyroid hormone receptor *(45,46)*. A further discussion of these issues is provided in Chapter 6 (Adler).

To date, no synthetic environmental chemical has been reported to function as androgen. However, a growing number of pesticides have been recognized as androgen antagonist or anti-androgen. The anti-androgenicity of dichlorodiphenyltrichloroethane (DDT) and its metabolites and other insecticides *(47,48)* also highlights the diversity of structures underlying the hormonal antagonist activities of environmental compounds. The herbicide linuron, for example, has been shown to compete with ligand for binding with the androgen receptor in human, thereby altering androgen-dependent gene expression *(49,50)*.

In addition to those synthetic EDC previously mentioned, natural chemicals have also been shown to disrupt endocrine function. Plants produce versatile chemicals, called phytochemicals or phytoestrogens, which serve both as endogenous signals triggering color and scent production within the plant and as exogenous signals secreted for communications with other organisms, for example, to inhibit sexual reproduction of predatory herbivores *(51)*. Leguminous plants (soybeans, clover, and alfalfa) secrete phytoestrogens into the soil as a recruitment signal enabling symbiotic interactions with mycorrhizal fungi and *Rhizobium* soil-bacteria, which in turn, provide growth advantages to the host plant by increasing water/phosphate availability and fertilizing with nitrogen, respectively *(52)*. Structurally, phytoestrogens are isoflavones capable of binding to estrogen receptors alpha and beta (ERα and ERβ) and acting as a weak agonist *(53,54)*, competing with endogenous E2 for ER binding and activation of estrogen-responsive genes *(55,56)*. Despite their ability to bind these receptors, phytoestrogens exhibit only a fraction ($10^{-2}-10^{-3}$) of the estrogenic activity of estradiol *(57,58)*. In vitro binding affinities per se do not distinguish between ER agonist and antagonist, nor do they predict tissue-specific estrogen or anti-estrogenic activity. Therefore, it may be inappropriate to perform risk assessment of estrogenic compounds by estimating their potencies solely through reporter gene or binding assays. A thorough description of the mechanisms of action of phytoestrogens appears in Chapter 6 (Adler), and effects of phytoestrogens as a central nervous system EDC is provided in Chapter 4 (Walker and Gore).

Selective ER modulators (SERMs) represent another class of synthetic estrogens being developed for treatment and prevention of hormone-dependent diseases *(59)*. In human HepG2 hepatoma cells transfected with an estrogen-responsive complement C3 promoter-luciferase construct, SERMs differentially activate wild-type ERα and its variant forms expressing activation function, namely ER-AF1 and ER-AF2; these are in vitro differences that reflect SERMs' unique in vivo biologies *(60,61)*. The HepG2 cell assay has also been used to investigate the estrogenic activities of phytoestrogens and synthetic/industrial estrogenic compounds *(58,62)*. These results show that despite evidence that phyto- and synthetic estrogens have weak estrogenic activity, they induce distinct patterns of ER agonist/antagonist activities that are cell context-specific and promoter-dependent, suggesting that these compounds will induce tissue-specific in vivo ER agonist or antagonist activities. These studies suggest that other receptors such as the AhR, which also binds structurally diverse ligands, may exhibit unique responses in vivo that are not predicted in in vitro assays.

3. GENE EXPRESSION REGULATION BY EDCS THROUGH EPIGENETIC MECHANISMS

There are many ways by which EDCs could regulate gene expression *(63,64)*. Transcriptional regulation by EDC has been described, for example, in several *Hox* genes in which distinct retinoic acid-responsive elements mediate direct transcriptional regulation by retinoic acid, resulting in teratogenesis after altered transcription induction of these genes *(21)*. Nevertheless, here we will focus on available data concerning the epigenetic mechanisms for regulating gene expression by EDC. The finding that some compounds have the ability to induce alterations in DNA methylation patterns is not new [see Wachsman *(65)*]. Exposure to EDCs may result in transcriptional changes resulting from altered DNA methylation in key genes *(16)*, and this appears to be the most common mechanism for effects of EDCs. To our knowledge, the first group to report such an effect of EDC were Barrett et al. *(66)* who proposed that diethylstilbestrol (DES) could transform cells by a mechanism other than point mutations, frameshift mutations, or small deletions. By applying the current knowledge of epigenetic mechanisms, we can speculate now that such transformations reported by Barrett et al. *(66)* could have been the product of an epigenetic process. EDCs are capable of triggering impairments during the development of organs, as proposed by Li et al. *(67)*, who showed that neonatal exposure to DES produced abnormalities in the demethylation of the lactoferrin promoter. It has also been shown that the administration of the phytoestrogens coumestrol and equol to newborn mice can enhance methylation and produce inactivation of the proto-oncogene *H-ras (68)*. Later, Day et al. *(69)* demonstrated that DNA methylation patterns can be altered in 8-week-old mice that consumed high doses of genistein. All this evidence is supported by the new finding that exposure of early embryos to TCDD, DES, or PCB153 alters Dnmt activity, which has the potential to induce a change in methylation status of genes and affect further developmental processes *(70)*. Thus, the link relating EDC and DNA methylation is becoming strongly supported by scientific evidence.

With regard to EDC effects on another epigenetic system, Hong et al. *(71)* reported that genistein and equol produce effects on histone acetylation mediated by either ERα or ERβ, which takes place through stimulation of the histone acetyltransferase activity. Singleton et al. (2006) showed that treating ERα-HA breast cancer cells (which overexpress HA-tagged ERα) with bisphenol A (BPA) or estradiol leads to differential expression of a set of genes. BPA upregulated histone H2B and downregulated histone H1, on which estradiol had no effect; moreover, BPA had no effect on histone deacetylase, which also differs from the downregulating effect of estradiol in this regard *(72)*. Interestingly, from an epigenetic perspective, these histones have implications for chromatin condensation, as previously described. Histone H2B belongs to the dimer H2A/H2B that assembles with the $(H3/H4)_2$ tetramer, forming a histone octamer wrapped in the nucleosome core particle, the fundamental unit of chromatin *(73)*. Histone H1 binds to the nucleosome surface and interacts with nucleosomal DNA at the entry and exit points, determining the higher-order folding states of chromatin *(74)*.

There are not many publications examining EDC effects on chromatin condensation. Nevertheless, it has been shown that treating in vitro oocytes that have already undergone germinal vesicle breakdown with genistein (but not daidzein) produces several consequences at the chromosomal level such as retention of metaphase configuration or prevention of the spindle translocation toward the cortex *(75)*. A more recent study has shown that chronic oral treatment with the fungicide vinclozolin (30 mg/kg per day) from conception to adulthood disrupted both the sperm nuclear morphology and the chromatin texture, having a deleterious impact on chromatin condensation homogeneity *(76)*.

4. EDC EFFECTS ON DNA METHYLATION DURING DEVELOPMENT

As previously mentioned, any compound interfering with mammalian early stages of development must first circumvent the barrier represented by the uterus or placenta. Endocrine disruptors are known to act through maternal–fetal transfer, thereby having consequences on both gene expression and embryonic phenotype. With regard to the former, Nishizawa et al. *(77)* reported that mid to late embryonic exposure (organogenesis period) to BPA changes the expression levels of NRs such as AhR, RARa, and RXRa mRNAs in adult tissues such as brain (cerebrum and cerebellum) and gonads (testes and ovaries). Nielsen et al. *(78)* and Newbold *(79)* have reported that expression of ERα mRNA and protein is induced in the uterine epithelium after prenatal DES exposure. Naciff et al. *(80)* showed that early prenatal exposure to endocrine disruptors 17α-ethinyl estradiol, BPA, or genistein lead to an altered gene expression response in several genes (this latter study will be more extensively discussed below).

In section 4.2 phenotypic changes induced in embryos by EDCs have been reported by Takai et al. *(81)*, who showed that blastocysts exposed to BPA produce adult mice that are heavier at weaning than controls, despite having similar weight at birth. Other studies show the same effect by EDCs mediated by maternal transfer. Adeeko et al. *(82)* showed that pregnant maternal treatment with daily doses of tributyltin chloride (20 mg/kg) between gestational days 0 and 15, which includes the preimplantation period (until gestational day 5), leads to reduced weight gain in fetuses. In addition, maternal treatment with daily doses of 10 mg/kg produced the same effect in fetuses when treatment occurred between gestational days 8 and 19, that is, starting after implantation *(82)*. Another study showing maternal treatment with a variety of endocrine disruptors, such as genistein, resveratrol, zearalenone, BPA and DES, reported transient effects on the reproductive tract and mammary glands in offspring with maternal high doses of genistein and resveratrol; in turn, low and high doses of BPA and DES had transient effects on the reproductive tract and mammary glands, whereas high doses of zearalenone induced prolonged effects *(83)*. Markey et al. *(84)* reported a decrease in the absolute and relative weight of the vagina and also mammary gland dysgenesis due to fetal developmental exposure to BPA at doses 4,000-fold lower than those capable of inducing an uterotropic response. Another feature reported to be altered because of prenatal exposure to BPA is the number of days between vaginal opening and first vaginal estrus, which is reduced in mice *(84,85)*. We have found the same pattern for vaginal opening in the model of feeding mice mothers with a natural isoflavones concentrate containing genistein and daidzein (unpublished data, manuscript in preparation).

4.1. How do EDCs Reach the Fetus?

There are two possible ways in which EDC may undergo maternal transfer to reach the developing embryo and produce epigenetic changes; one is through oviductal and/or uterine endometrial secretions *(86)*. The hormonal environment within the uterus is of critical importance to fetal development, and thus, the way in which preexisting maternal hormones that are present in the fetus interact with added chemicals will determine how that uterine environment changes *(84)*. With regard to this, we have hypothesized that endocrine disruptors could be acting on embryos in the uterus even before implantation takes place, through altering maternal secretion of epithelial uterine steroids such as catecholestrogens, which in turn could lead to changes in the establishment of methylation patterns in the embryo *(15)*. In an in vitro study, Wu et al. *(87)* showed that exposure of preimplantation embryos to the contaminant 2,3,7,8-tetra-chlorodibenzo-*p*-dioxin can indeed alter DNA methylation in *H*19 and *IGF-2*, both imprinted genes. Thus, altering the preimplantation intrauterine environment could lead to alterations in methylation patterns not only in non-imprinted but also in imprinted genes, which are known to have relatively unchanged methylation patterns throughout generations.

The other way in which EDC acts maternally on the developing embryo is transplacentally. It has been shown that EDCs such as p,p′-Dichlorodiphenyldi chloroethylene (DDE) and α-Hexachlorobenzene (HCH) can be detected in the amniotic fluid in women between 15 and 23 weeks of gestation, leading to embryonic exposure to EDCs during organogenesis *(88)*. In a very complete study in this field, Naciff et al. *(80)* showed gene expression response to transplacental exposure to endocrine disruptors, either 17-α-ethinyl estradiol, BPA, or genistein, from gestation day 11 to 20 in rat. Of 8740 genes analyzed, they detected changes in expression in 366 genes for 17-α-ethinyl estradiol, 398 genes for BPA, and 344 genes for genistein. Moreover, among those, expression of 66 genes was consistently and significantly regulated in the same direction *(80)*. Altered expression in those genes may be related to one of the mechanisms involved in epigenetic regulation of gene expression. Newbold et al. *(89)* have shown that, after administrating DES to pregnant rats during early post-implantation development and the neonatal period, a greater susceptibility for specific tumor formation in rete testis and reproductive tract tissues occurred in F1 and reappeared in the non-exposed F2 offspring. The authors suggested that such a transgenerational phenomenon could be due to epigenetic alterations transmitted through the germ line, including changes in DNA methylation. Recent reports by Anway et al. *(90)*, Skinner and Anway *(91)*, and Anway and Skinner *(92)*, which will be discussed next, reinforce such a postulate. Moreover, this maternally mediated epigenetic effect is not limited to synthetic EDCs. We have found that feeding mice mothers with an isoflavones concentrate containing genistein and daidzein alters gene-specific methylation patterns in the offspring (unpublished data, manuscript in preparation).

5. EDC EFFECTS ON DNA METHYLATION DURING GERM LINE DIFFERENTIATION

The process of germ line segregation from somatic cells in organisms may occur through pre-formation or epigenesis; however, in metazoans, the latter is probably the main mechanism of germ cell specification *(93)*. The initiation of the functional activation of the male and female reproductive systems represents an occasion during which environmental endocrine disruptors could act to alter normal physiology *(18)*. Physiological effects due to EDC exposure have been reported to occur in germ line in both males and females during critical stages of development such as sex determination. For example, embryo exposure to methoxychlor during sex determination period affects embryonic testis cellular composition and germ cell number and survival *(94)*. Embryonic testicular cord formation is affected when embryos are exposed in vitro to vinclozolin, and transient *in utero* exposure to vinclozolin increases apoptotic germ cell numbers in the testis of pubertal and adult animals, which correlates with reduced sperm motility in the adult *(95)*. During the critical period of sexual differentiation, it is expected that exposure of a chromosomal male to antiandrogenic xenobiotics would interfere with the androgen-dependent differentiation of the Wolffian-derived structures and/or with the normal development of the male genitalia *(18)*. On the contrary, in females, it is well known that genistein has an inhibitory effect on maturation of mammalian oocytes *(96)*. Markey et al. *(84)* showed an increase in the percentage of ovarian tissue occupied by antral follicules in 3-month-old mice exposed *in utero* to 250 μg/kg BPA.

However, heritable damage can also occur in the zygote at the beginning of the embryonic development of a new individual and be transmitted to the next generation through altering features during germ line development *(19)*. Moreover, such heritable damage can be induced while germ line is developing. For example, it has been shown that either chlorambucil or melphalan is capable of inducing a high frequency of heritable deletions and other mutations in mouse germ cells *(97,98)*, thereby producing a transgenerational change because of mutations. Nevertheless, although some endogenous and exogenous agents are frequently associated with DNA mutations and transgenerational transmission, chemically induced epigenetic modifications of DNA may well have the same net effect on the phenotype of newly altered cells and on their progeny *(99)*. Regarding this, Holliday *(100)* reported that teratogens could target mechanisms that control patterns of DNA methylation in particular regions of the genome of developing embryos modifying methylation patterns of the same DNA sequence in somatic cells, leading to a developmental alteration and subsequently to changes in germ line cells. Modifications capable of being transmitted are those (i) occurring in somatic cells before germ line segregation or (ii) in germ line cells while they are in the process of differentiation. EDC effects regarding the former have been previously described in this section. With regard to the latter, Anway et al. *(90)* have shown that exposing a mother rat to either vinclozolin or methoxychlor during embryonic days 8–15 produced transgenerational defects in the spermatogenic capacity, which was transmitted throughout four generations (F1–F4). Furthermore, the authors also detected 25 different polymerase chain reaction (PCR) products that had altered methylation patterns in the F1 born to mothers subjected to the vinclozolin administration *(90)*. Therefore, exposure of a gestating mother to EDC during critical periods of sex differentiation and testis morphogenesis, which is when cord formation takes place in the embryo, triggers the germ line effect of decreased spermatogenic capacity and sperm viability; this phenotype is transgenerationally transmitted in the male and appears to be associated with altered DNA methylation of the germ line *(91)*. This interpretation is in concordance with that suggested by Newbold et al. *(89)* for the transgenerational transmission of the increased susceptibility for tumor formation because of early exposure to DES that was in section 4.1.

6. FINAL CONSIDERATIONS

There is a variable amount of evidence describing the effects of EDC on each distinct epigenetic mechanism known to date. Nevertheless, there is existing evidence of epigenetic EDC effects in histone modification, chromatin condensation and especially in DNA methylation. In any case, each epigenetic modification may or may not have transgenerational consequences, which will depend on the epigenetic system affected, on the stage during the ontogeny when this occurs and on the persistency of the stimuli throughout generations. Epigenetics integrates intrinsic and environmental factors *(14)*. Modifications of intrinsic factors are capable of being transmitted when triggered in somatic cells before germ line segregation or in germ line cells while they are in the process of differentiation. Experiments such as those performed by Newbold et al. *(89)* and Wu et al. *(87)* strongly support the possibility that EDC action occurs through mothers to the embryo, produces epigenetic changes on them, and moreover leads to the transgenerational transmission of those changes, mechanism previously described in Guerrero-Bosagna et al. *(15)*. In parallel, experiments by Anway et al. *(90)* show that EDC are also able to induce epigenetic modification during the differentiation of germ line, also with the possibility of transgenerational transmission of those changes. Moreover, such evidence suggests that EDCs are able to reach the embryo (i) during the pre-implantation period, through uterine or oviductal secretions, (ii) while implantation is taking place, through direct contact with uterine epithelia, or (iii) after implantation occurs, through the placenta.

Although, to date, there is not considerable evidence available reporting epigenetic and transgenerational effects of EDC, the consistency of the recent aforementioned findings supports the feasibility of this postulate. The implications of this may be many, ranging from public health to ecological or evolutionary issues.

ACKNOWLEDGMENTS

The authors enormously appreciate the work of Renée Hill for her linguistic revision of the manuscript.

REFERENCES

1. Jablonka E, Matzke M, Thieffry D, Van Speybroeck L. The genome in context: biologists and philosophers on epigenetics. *Bioessays* 2002; 24(4):392–4.
2. Surani MA. Reprogramming of genome function through epigenetic inheritance. *Nature* 2001; 414(6859):122–8.
3. Laird PW, Jaenisch R. The role of DNA methylation in cancer genetic and epigenetics. *Annu Rev Genet* 1996; 30:441–64.
4. Singal R, Ginder GD. DNA methylation. *Blood* 1999; 93(12):4059–70.
5. Margueron R, Trojer P, Reinberg D. The key to development: interpreting the histone code. *Curr Opin Genet Dev* 2005; 15(2):163–76.
6. Wallace JA, Orr-Weaver TL. Replication of heterochromatin: insights into mechanisms of epigenetic inheritance. *Chromosoma* 2005; 114(6):389–402.
7. Bannister AJ, Zegerman P, Partridge JF, et al. Selective recognition of methylated lysine 9 on histone H3 by the HP1 chromo domain. *Nature* 2001; 410(6824):120–4.
8. Lachner M, O'Carroll D, Rea S, Mechtler K, Jenuwein T. Methylation of histone H3 lysine 9 creates a binding site for HP1 proteins. *Nature* 2001; 410(6824):116–20.
9. Craig JM. Heterochromatin–many flavours themes. *Bioessays* 2005; 27(1):17–28.
10. Fuks F. DNA methylation and histone modifications: teaming up to silence genes. *Curr Opin Genet Dev* 2005; 15(5):490–5.
11. Chen ZX, Riggs AD. Maintenance and regulation of DNA methylation patterns in mammals. *Biochem Cell Biol* 2005; 83(4):438–8.
12. Holmes R, Soloway PD. Regulation of imprinted DNA methylation. *Cytogenet Genome Res* 2006; 113(1–4):122–9.
13. Amzallag GN. Connectance in Sorghum development: beyond the genotype-phenotype duality. *Biosystems* 2000; 56(1):1–11.
14. Jaenisch R, Bird A. Epigenetic regulation of gene expression: how the genome integrates intrinsic and environmental signals. *Nat Genet* 2003; 33(Suppl):245–54.
15. Guerrero-Bosagna C, Sabat P, Valladares L. Environmental signaling and evolutionary change: can exposure of pregnant mammals to environmental estrogens lead to epigenetically induced evolutionary changes in embryos. *Evol Dev* 2005; 7(4):341–50.
16. Li S, Hursting SD, Davis BJ, McLachlan JA, Barrett JC. Environmental exposure, DNA methylation, and gene regulation: lessons from diethylstilbestrol-induced cancers. *Ann N Y Acad Sci* 2003; 983:161–9.
17. Sved J, Bird A. The expected equilibrium of the CpG dinucleotide in vertebrate genomes under a mutation model. *Proc Natl Acad Sci USA* 1990; 87(12):4692–6.
18. Danzo BJ. The effects of environmental hormones on reproduction. *Cell Mol Life Sci* 1998; 54(11):1249–64.
19. Lewis SE. Life cycle of the mammalian germ cell: implication for spontaneous mutation frequencies. *Teratology* 1999; 59(4):205–9.
20. Markey CM, Luque EH, Munoz De Toro M, Sonnenschein C, Soto AM. In utero exposure to bisphenol A alters the development and tissue organization of the mouse mammary gland. *Biol Reprod* 2001; 65(4):1215–23.
21. Rogers MB, Glozak MA, Heller LC. Induction of altered gene expression in early embryos. *Mutat Res* 1997; 396(1–2):79–95.
22. Maturana-Romesín H, Mpodozis J. The origin of species by means of natural drift. *Rev Chil Hist Nat* 2000; 73:261–300.
23. McLachlan JA. Environmental signaling: what embryos and evolution teach us about endocrine disrupting chemicals. *Endocr Rev* 2001; 22(3):319–41.
24. Smith AG, Gangolli SD. Organochlorine chemicals in seafood: occurrence and health concerns. *Food Chem Toxicol* 2002; 40(6):767–9.

25. Colborn T, Clemen C. *Chemically Induced Alterations in Sexual and Functional Development: the Wildlife/Human Connection.* Princeton, NJ: Princeton Scientific Publishing; 1992.
26. Davis DL, Bradlow HL, Wolff M, Woodruff T, Hoel DG, Anton-Culver H. Medical hypothesis: xenoestrogens as preventable causes of breast cancer. *Environ Health Perspect* 1993; 101(5):372–7.
27. Cheek AO, McLachlan JA. Environmental hormones and the male reproductive system. *J Androl* 1998; 19(1):5–10.
28. Commission on Life sciences. *Hormonally Active Agents in the Environment.* Washington, D.C.: National Academy Press; 1999.
29. Cheek AO, Vonier PM, Oberdorster E, Burow BC, McLachlan JA. Environmental signaling: a biological context for endocrine disruption. *Environ Health Perspect* 1998; 106(Suppl 1):5–10.
30. Clode SA. Assessment of in vivo assays for endocrine disruption. *Best Pract Res Clin Endocrinol Metab* 2006; 20(1):35–43.
31. Charlier TD, Balthazart J. Modulation of hormonal signaling in the brain by steroid receptor coactivators. *Rev Neurosci* 2005; 16(4):339–57.
32. Chung AC, Cooney AJ. The varied roles of nuclear receptors during vertebrate embryonic development. *Nucl Recept Signal* 2003; 1:e007.
33. Blumberg B, Evans RM. Orphan nuclear receptors–new ligands and new possibilities. *Genes Dev* 1998; 12(20):3149–55.
34. Robinson-Rechavi M, Carpentier AS, Duffraisse M, Laudet V. How many nuclear hormone receptors are there in the human genome. *Trends Genet* 2001; 17(10):554–6.
35. Lamba J, Lamba V, Schuetz E. Genetic variants of PXR (NR1I2) and CAR (NR1I3) and their implications in drug metabolism and pharmacogenetics. *Curr Drug Metab* 2005; 6(4):369–83.
36. Giguere V, Yang N, Segui P, Evans RM. Identification of a new class of steroid hormone receptors. *Nature* 1988; 331(6151):91–4.
37. Hong H, Yang L, Stallcup MR. Hormone-independent transcriptional activation and coactivator binding by novel orphan nuclear receptor ERR3. *J Biol Chem* 1999; 274(32):22618–6.
38. Giguere V. To ERR in the estrogen pathway. *Trends Endocrinol Metab* 2002; 13(5):220–5.
39. Kraus RJ, Ariazi EA, Farrell ML, Mertz JE. Estrogen-related receptor alpha 1 actively antagonizes estrogen receptor-regulated transcription in MCF-7 mammary cells. *J Biol Chem* 2002; 277(27):24826–34.
40. Greschik H, Wurtz JM, Sanglier S, et al. Structural and functional evidence for ligand-independent transcriptional activation by the estrogen-related receptor 3. *Mol Cell* 2002; 9(2):303–13.
41. Horard B, Vanacker JM. Estrogen receptor-related receptors: orphan receptors desperately seeking a ligand. *J Mol Endocrinol* 2003; 31(3):349–57.
42. Lu D, Kiriyama Y, Lee KY, Giguere V. Transcriptional regulation of the estrogen-inducible pS2 breast cancer marker gene by the ERR family of orphan nuclear receptors. *Cancer Res* 2001; 61(18):6755–1.
43. Sun P, Wei L, Denkert C, Lichtenegger W, Sehouli J. The orphan nuclear receptors, estrogen receptor-related receptors: their role as new biomarkers in gynecological cancer. *Anticancer Res* 2006; 26(2C):1699–706.
44. Rickenbacher U, McKinney JD, Oatley SJ, Blake CC. Structurally specific binding of halogenated biphenyls to thyroxine transport protein. *J Med Chem* 1986; 29(5):641–8.
45. Arulmozhiraja S, Shiraishi F, Okumura T, et al. Structural requirements for the interaction of 91 hydroxylated polychlorinated biphenyls with estrogen and thyroid hormone receptors. *Toxicol Sci* 2005; 84(1):49–62.
46. Cheek AO, Kow K, Chen J, McLachlan JA. Potential mechanisms of thyroid disruption in humans: interaction of organochlorine compounds with thyroid receptor, transthyretin, and thyroid-binding globulin. *Environ Health Perspect* 1999; 107(4):273–8.
47. Sultan C, Balaguer P, Terouanne B, et al. Environmental xenoestrogens, antiandrogens and disorders of male sexual differentiation. *Mol Cell Endocrinol* 2001; 178(1–2):99–105.
48. Sunami O, Kunimatsu T, Yamada T, et al. Evaluation of a 5-day Hershberger assay using young mature male rats: methyltestosterone and p,p′-DDE, but not fenitrothion, exhibited androgenic or antiandrogenic activity in vivo. *J Toxicol Sci* 2000; 25(5):403–15.
49. Lambright C, Ostby J, Bobseine K, et al. Cellular and molecular mechanisms of action of linuron: an antiandrogenic herbicide that produces reproductive malformations in male rats. *Toxicol Sci* 2000; 56(2):389–99.

50. McIntyre BS, Barlow NJ, Wallace DG, Maness SC, Gaido KW, Foster PM. Effects of in utero exposure to linuron on androgen-dependent reproductive development in the male Crl:CD(SD)BR rat. *Toxicol Appl Pharmacol* 2000; 167(2):87–99.
51. Wynne-Edwards KE. Hormonal changes in mammalian fathers. *Horm Behav* 2001; 40(2):139–45.
52. Baker ME. Flavonoids as hormones. A perspective from an analysis of molecular fossils. *Adv Exp Med Biol* 1998; 439:249–67.
53. Benassayag C, Perrot-Applanat M, Ferre F. Phytoestrogens as modulators of steroid action in target cells. *J Chromatogr B Analyt Technol Biomed Life Sci* 2002; 777(1–2):233–48.
54. Kuiper GG, Carlsson B, Grandien K, et al. Comparison of the ligand binding specificity and transcript tissue distribution of estrogen receptors alpha and beta. *Endocrinology* 1997; 138(3):863–70.
55. Blair RM, Fang H, Branham WS, et al. The estrogen receptor relative binding affinities of 188 natural and xenochemicals: structural diversity of ligands. *Toxicol Sci* 2000; 54(1):138–53.
56. Ise R, Han D, Takahashi Y, et al. Expression profiling of the estrogen responsive genes in response to phytoestrogens using a customized DNA microarray. *FEBS Lett* 2005; 579(7):1732–40.
57. Barkhem T, Carlsson B, Nilsson Y, Enmark E, Gustafsson J, Nilsson S. Differential response of estrogen receptor alpha and estrogen receptor beta to partial estrogen agonists/antagonists. *Mol Pharmacol* 1998; 54(1):105–2.
58. Kuiper GG, Lemmen JG, Carlsson B, et al. Interaction of estrogenic chemicals and phytoestrogens with estrogen receptor beta. *Endocrinology* 1998; 139(10):4252–63.
59. McDonnell DP. The Molecular Pharmacology of SERMs. *Trends Endocrinol Metab* 1999; 10(8):301–11.
60. McDonnell DP, Clemm DL, Hermann T, Goldman ME, Pike JW. Analysis of estrogen receptor function in vitro reveals three distinct classes of antiestrogens. *Mol Endocrinol* 1995; 9(6):659–9.
61. Metivier R, Penot G, Flouriot G, Pakdel F. Synergism between ERalpha transactivation function 1 (AF-1) and AF-2 mediated by steroid receptor coactivator protein-1: requirement for the AF-1 alpha-helical core and for a direct interaction between the N- and C-terminal domains. *Mol Endocrinol* 2001; 15(11):1953–70.
62. Yoon K, Pellaroni L, Ramamoorthy K, Gaido K, Safe S. Ligand structure-dependent differences in activation of estrogen receptor alpha in human HepG2 liver and U2 osteogenic cancer cell lines. *Mol Cell Endocrinol* 2000; 162(1–2):211–20.
63. Lonard DM, Smith CL. Molecular perspectives on selective estrogen receptor modulators (SERMs): progress in understanding their tissue-specific agonist and antagonist actions. *Steroids* 2002; 67(1):15–24.
64. Nilsson S, Makela S, Treuter E, et al. Mechanisms of estrogen action. *Physiol Rev* 2001; 81(4):1535–65.
65. Wachsman JT. DNA methylation and the association between genetic and epigenetic changes: relation to carcinogenesis. *Mutat Res* 1997; 375(1):1–8.
66. Barrett JC, Wong A, McLachlan JA. Diethylstilbestrol induces neoplastic transformation without measurable gene mutation at two loci. *Science* 1981; 212(4501):1402–4.
67. Li S, Washburn KA, Moore R, et al. Developmental exposure to diethylstilbestrol elicits demethylation of estrogen-responsive lactoferrin gene in mouse uterus. *Cancer Res* 1997; 57(19):4356–9.
68. Lyn-Cook BD, Blann E, Payne PW, Bo J, Sheehan D, Medlock K. Methylation profile and amplification of proto-oncogenes in rat pancreas induced with phytoestrogens. *Proc Soc Exp Biol Med* 1995; 208(1):116–9.
69. Day JK, Bauer AM, DesBordes C, et al. Genistein alters methylation patterns in mice. *J Nutr* 2002; 132(8 Suppl):2419S–3S.
70. Wu Q, Zhou ZJ, Ohsako S. [Effect of environmental contaminants on DNA methyltransferase activity of mouse preimplantation embryos]. *Wei Sheng Yan Jiu* 2006; 35(1):30–2.
71. Hong T, Nakagawa T, Pan W, et al. Isoflavones stimulate estrogen receptor-mediated core histone acetylation. *Biochem Biophys Res Commun* 2004; 317(1):259–64.
72. Singleton DW, Feng Y, Yang J, Puga A, Lee AV, Khan SA. Gene expression profiling reveals novel regulation by bisphenol-A in estrogen receptor-alpha-positive human cells. *Environ Res* 2006; 100(1):86–92.
73. Luger K. Structure and dynamic behavior of nucleosomes. *Curr Opin Genet Dev* 2003; 13(2): 127–35.
74. Vignali M, Workman JL. Location and function of linker histones. *Nat Struct Biol* 1998; 5(12): 1025–8.

75. Van Cauwenberge A, Alexandre H. Effect of genistein alone and in combination with okadaic acid on the cell cycle resumption of mouse oocytes. *Int J Dev Biol* 2000; 44(4):409–20.
76. Auger J, Lesaffre C, Bazire A, Schoevaert-Brossault D, Eustache F. High-resolution image cytometry of rat sperm nuclear shape, size and chromatin status. Experimental validation with the reproductive toxicant vinclozolin. *Reprod Toxicol* 2004; 18(6):775–83.
77. Nishizawa H, Morita M, Sugimoto M, Imanishi S, Manabe N. Effects of in utero exposure to bisphenol A on mRNA expression of arylhydrocarbon and retinoid receptors in murine embryos. *J Reprod Dev* 2005; 51(3):315–24.
78. Nielsen M, Hoyer PE, Lemmen JG, van der Burg B, Byskov AG. Octylphenol does not mimic diethylstilbestrol-induced oestrogen receptor-alpha expression in the newborn mouse uterine epithelium after prenatal exposure. *J Endocrinol* 2000; 167(1):29–37.
79. Newbold R. Cellular and molecular effects of developmental exposure to diethylstilbestrol: implications for other environmental estrogens. *Environ Health Perspect* 1995; 103(Suppl 7):83–7.
80. Naciff JM, Daston GP. Toxicogenomic approach to endocrine disrupters: identification of a transcript profile characteristic of chemicals with estrogenic activity. *Toxicol Pathol* 2004; 32(Suppl 2):59–70.
81. Takai Y, Tsutsumi O, Ikezuki Y, et al. Preimplantation exposure to bisphenol A advances postnatal development. *Reprod Toxicol* 2001; 15(1):71–4.
82. Adeeko A, Li D, Forsyth DS, et al. Effects of in utero tributyltin chloride exposure in the rat on pregnancy outcome. *Toxicol Sci* 2003; 74(2):407–15.
83. Nikaido Y, Yoshizawa K, Danbara N, et al. Effects of maternal xenoestrogen exposure on development of the reproductive tract and mammary gland in female CD-1 mouse offspring. *Reprod Toxicol* 2004; 18(6):803–11.
84. Markey CM, Coombs MA, Sonnenschein C, Soto AM. Mammalian development in a changing environment: exposure to endocrine disruptors reveals the developmental plasticity of steroid-hormone target organs. *Evol Dev* 2003; 5(1):67–75.
85. Howdeshell KL, Hotchkiss AK, Thayer KA, Vandenbergh JG, vom Saal FS. Exposure to bisphenol A advances puberty. *Nature* 1999; 401(6755):763–4.
86. McEvoy TG, Robinson JJ, Ashworth CJ, Rooke JA, Sinclair KD. Feed and forage toxicants affecting embryo survival and fetal development. *Theriogenology* 2001; 55(1):113–29.
87. Wu Q, Ohsako S, Ishimura R, Suzuki JS, Tohyama C. Exposure of mouse preimplantation embryos to 2,3,7,8-tetrachlorodibenzo-p-dioxin (TCDD) alters the methylation status of imprinted genes H19 and Igf2. *Biol Reprod* 2004; 70(6):1790–7.
88. Foster W, Chan S, Platt L, Hughes C. Detection of endocrine disrupting chemicals in samples of second trimester human amniotic fluid. *J Clin Endocrinol Metab* 2000; 85(8):2954–7.
89. Newbold RR, Hanson RB, Jefferson WN, Bullock BC, Haseman J, McLachlan JA. Proliferative lesions and reproductive tract tumors in male descendants of mice exposed developmentally to diethylstilbestrol. *Carcinogenesis* 2000; 21(7):1355–63.
90. Anway MD, Cupp AS, Uzumcu M, Skinner MK. Epigenetic transgenerational actions of endocrine disruptors and male fertility. *Science* 2005; 308(5727):1466–9.
91. Skinner MK, Anway MD. Seminiferous cord formation and germ-cell programming: epigenetic transgenerational actions of endocrine disruptors. *Ann N Y Acad Sci* 2005;1061:18–32.
92. Anway MD, Skinner MK. Epigenetic transgenerational actions of endocrine disruptors. *Endocrinology* 2006; 147(6 Suppl):S43–9.
93. Extavour CG, Akam M. Mechanisms of germ cell specification across the metazoans: epigenesis and preformation. *Development* 2003; 130(24):5869–84.
94. Cupp AS, Uzumcu M, Suzuki H, Dirks K, Phillips B, Skinner MK. Effect of transient embryonic in vivo exposure to the endocrine disruptor methoxychlor on embryonic and postnatal testis development. *J Androl* 2003; 24(5):736–45.
95. Uzumcu M, Suzuki H, Skinner MK. Effect of the anti-androgenic endocrine disruptor vinclozolin on embryonic testis cord formation and postnatal testis development and function. *Reprod Toxicol* 2004; 18(6):765–4.
96. Jung T, Fulka J Jr, Lee C, Moor RM. Effects of the protein phosphorylation inhibitor genistein on maturation of pig oocytes in vitro. *J Reprod Fertil* 1993; 98(2):529–35.
97. Russell LB, Hunsicker PR, Cacheiro NL, Bangham JW, Russell WL, Shelby MD. Chlorambucil effectively induces deletion mutations in mouse germ cells. *Proc Natl Acad Sci USA* 1989; 86(10):3704–8.

98. Russell LB, Hunsicker PR, Shelby MD. Melphalan, a second chemical for which specific-locus mutation induction in the mouse is maximum in early spermatids. *Mutat Res* 1992; 282(3):151–8.
99. MacPhee DG. Epigenetics and epimutagens: some new perspectives on cancer, germ line effects and endocrine disrupters. *Mutat Res* 1998; 400(1–2):369–79.
100. Holliday R. The possibility of epigenetic transmission of defects induced by teratogens. *Mutat Res* 1998; 422(2):203–5.

II The Biology of EDCs in Humans

8 Implications of Thyroid Hormone Signaling Through the Phosphoinositide-3 Kinase for Xenobiotic Disruption of Human Health

David L. Armstrong, PhD

Contents

1. OVERVIEW AND INTRODUCTION

This chapter reviews the potential implications for human health of disrupting endocrine signaling by thyroid hormone, which has been shown recently to signal through the phosphoinositide 3 kinase (PI3K). First, the overlapping roles of thyroid hormone and PI3K in normal development and maturation are summarized, and the evidence for thyroid hormone signaling through PI3K is reviewed. Inhibition of PI3K activity is predicted to exacerbate several of the fastest growing chronic human health problems in the USA, including early learning difficulties, cardiovascular disease, obesity and diabetes, and neurodegenerative diseases of the aging brain. Many of the polyhalogenated aromatic hydrocarbons that have been produced by industrial activities are suspected of interfering with the transport, activation, and receptor binding of the thyroid hormone, L-3,5,3′ triiodothyronine. Thus, together with genetic differences and

From: *Endocrine-Disrupting Chemicals: From Basic Research to Clinical Practice*
Edited by: A. C. Gore © Humana Press Inc., Totowa, NJ

lifestyle choices, xenobiotic disruption of thyroid hormone signaling through PI3K could increase the incidence of many chronic human illnesses.

2. THYROID HORMONE IS ESSENTIAL FOR HUMAN HEALTH

Thyroid hormone is essential for normal development and metabolism of many human tissues *(1)*. In one of the earliest examples of environmental disruption of human health, when thyroid hormone synthesis is blocked completely by lack of iodine during fetal development, deafness and severe retardation of growth and human cognitive potential result. People who inherit mutations in the thyroid hormone receptor TRβετα, which reduce its affinity for thyroid hormone, experience less severe reductions in hearing and learning ability in addition to goitre and tachycardia *(2)*. In adults, hypothyroidism, as indicated by elevated serum levels of thyrotropin-stimulating hormone (TSH), is associated with cardiovascular disease and depression and is surprisingly common among the general population *(3)*.

The metabolism of thyroid hormone is summarized in Fig. 1. The prohormone thyroxine, or 3,5,3′,5′ tetraiodo L thyronine (T_4), is synthesized in the thyroid gland and transported throughout the body in the bloodstream by plasma proteins: T_4-binding globulin, transthyretin and albumin *(4)*. Thyroxine synthesis is stimulated by TSH, which is released by thyrotropes in the pituitary. In turn, TSH secretion from thyrotropes in the pituitary is stimulated by the hypothalamic neuropeptide, thyrotropin-releasing hormone (TRH). The thyroid hormone, 3,5,3′ triiodo L thyronine (T_3), is produced

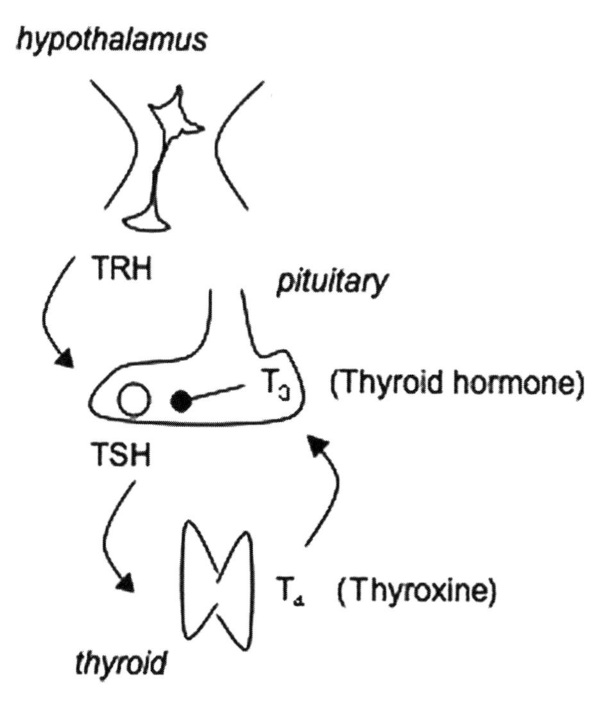

Fig. 1. Thyroxine synthesis in the thyroid is regulated by thyrotropin-stimulating hormone (TSH) production in the pituitary. Thyroxine is transported to the pituitary where it is converted by a 5′ deiodinase to thyroid hormone, which antagonizes further stimulation of TSH secretion by thyrotropin-releasing hormone (TRH).

from thyroxine by the action of a 5′ deiodinase in many tissues and inactivated by further deiodination *(5)*. Finally, thyroid hormone feeds back on thyrotropes in the pituitary to inhibit TSH secretion.

3. THYROID HORMONE SIGNALING

The two major signaling pathways for thyroid hormone signaling are summarized in Fig. 2. Most of the effects of thyroid hormone are mediated by one of the cErb-A family of nuclear zinc finger TRα and TRβ proteins encoded by alternative splicing of the *THRA* and *THRB* genes *(6)*. Unliganded dimers of these proteins have been shown to interact with specific response elements on DNA and organize complexes of transcription regulatory proteins. Thyroid hormone binding alters the composition of these complexes and consequently regulates gene expression. This model of thyroid hormone receptor signaling is consistent with the initially surprising observation that knocking out the expression of all the thyroid hormone receptors in rodents was much less deleterious than the effects of removing the hormone by blocking its synthesis *(7–9)*. Only the increase in TSH secretion and some of the sensory defects of thyroid hormone removal were reproduced in the TRβ knockout *(10)*. More recently, attention has shifted to the rapid effects of thyroid hormone on cellular physiology, which are difficult to explain by changes in gene expression *(11,12)*. Classically, these effects involved changes in oxygen consumption *(13)*, but, more recently, thyroid hormone has been reported to rapidly alter the activity of kinase cascades and transport proteins. Many of these effects appear to involve signaling through the PI3K.

We first postulated a link between the rapid effects of thyroid hormone and PI3K to explain the Rac-dependent stimulation of potassium channel activity in a pituitary cell line GH_4C_1 *(14)*. Other investigators had reported rapid effects of thyroid hormone on ion channels in cardiac myocytes *(15,16)*, but they had not studied the mechanism of thyroid hormone action. We postulated the involvement of PI3K because many Rac-activating proteins are stimulated by PIP_3 *(17)* and because other nuclear hormones had been reported to stimulate PI3K activity *(18)*. To test this hypothesis, initially we used the fungal metabolite, wortmannin, which binds to the active site of PI3K and inhibits its activity *(19)*. At concentrations selective for PI3K, wortmannin blocked potassium channel stimulation by thyroid hormone in the GH_4C_1 cells *(14)*. Subsequently, we and others published additional evidence for thyroid hormone stimulation of PI3K *(20–22)*. Although we were able to reconstitute PI3K-dependent signaling in a heterologous

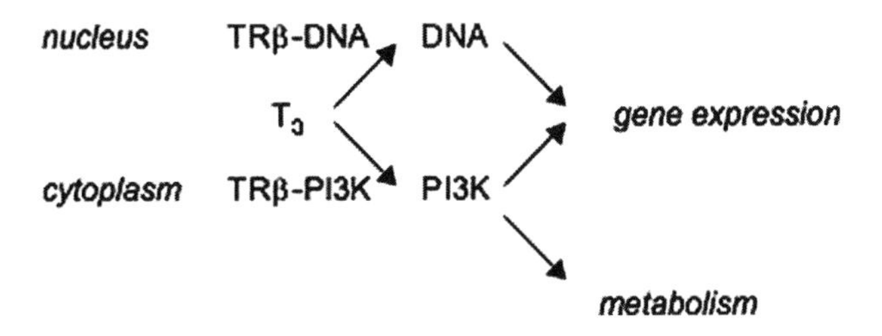

Fig. 2. Thyroid hormone signaling. The classical mechanism of thyroid hormone signaling involves binding in the nucleus to proteins of the c-erba family of zinc-finger DNA-binding transcription factors. More recently, evidence is accumulating that nuclear thyroid hormone receptors also signal in the cytoplasm and mediate phosphoinositide 3 kinase (PI3K) stimulation by thyroid hormone.

system with recombinant human TRβ receptor *(23)*, we do not yet have a precise molecular mechanism. We hypothesize that TRβ binds to the regulatory p85 subunit of PI3K through a conserved SH2 domain, but this idea remains to be tested. Nevertheless, known downstream effectors of PIP_3, which are described below, currently explain more of the classical actions of thyroid hormone than any documented effects on expression of specific genes.

4. PI3K SIGNALING

The metabolism and major effectors of 3,4,5 PIP_3 are illustrated in Fig. 3. By phosphorylating the minor membrane lipid, phosphoinositide 4,5 bisphosphate (4,5PIP_2), to produce phosphoinositide 3,4,5 trisphosphate (PIP_3), the PI3K initiates several signaling cascades involving the Rac GTPase and protein kinases, which regulate survival, growth, metabolism, motility, excitability, and gene expression *(24,25)*. PIP_3 is subsequently metabolized by the lipid phosphatases, phosphatase and tensin homologue deleted from chromosome-10 (PTEN) and Src homology domain-containing inositol phosphate (SHIP). Many microbial pathogens produce disease by inhibiting PI3K or its downstream effectors such as the Rac GTPase *(26)*. PIP_3 signals primarily by attracting proteins to the membrane with domains that recognize PIP_3 *(27)*. In addition, many of the protein kinase cascades activated by the PIP_3-dependent protein kinase (PDK) lead to changes in gene expression *(25)*. Thus, if thyroid hormone stimulates PI3K as we postulate, then one cannot a priori predict which thyroid hormone-dependent changes in gene expression are mediated by nuclear receptor signaling in the nucleus or in the cytoplasm through PI3K.

Thyroid hormone has been known for decades as the primary endocrine regulator of protein, lipid, and carbohydrate metabolism *(1)*, and selective agonists for the TRβ receptor are more effective than statins at lowering blood lipid levels and preventing obesity in rats *(28)*. Not coincidentally, PI3K plays a fundamental role in regulating metabolism at the cellular level *(29)*. For example, increasing PIP_3 levels by knocking out the lipid phosphatase, SHIP2, is just as effective at preventing obesity in rats *(30)* as the TRβ agonist, which we predicted to increase PIP_3 levels *(23)*. Metabolic disorder describes a syndrome of obesity and insulin-resistance, which contributes to the risk of cardiovascular disease and stroke. Thyroid hormone might also regulate insulin secretion from pancreatic beta cells, which express the same Kv11.1 potassium

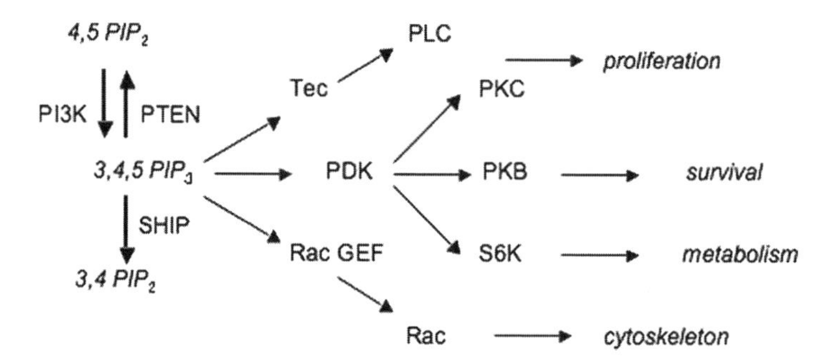

Fig. 3. Phosphoinositide 3 kinase (PI3K) initiates several signaling cascades involving the Rac GTPase and protein kinases, which regulate survival, growth, metabolism, motility, excitability, and gene expression. PIP_3 is subsequently metabolized by the lipid phosphatases, PTEN and SHIP.

channels *(31)* that thyroid hormone stimulates in pituitary cells *(14)*. Similarly in cardiac myocytes, where human polymorphisms in Kv11.1 channels are responsible for LQT syndrome *(32)*, thyroid hormone might regulate cardiac rhythmicity and hyperplasia through PI3K.

5. THE NEW RHO BIOLOGY OF THYROID HORMONE AND ITS RELATIONSHIP TO NEURAL FUNCTIONS

In the brain, PI3K promotes neural development and survival through two separate mechanisms. PIP_3-dependent stimulation of the Akt kinase (PKB) promotes survival through phosphorylation-dependent inhibition of several apoptosis-promoting proteins *(33)*. In other tissues, unrestricted signaling through PI3K often leads to cancer *(34)*. In the brain, however, where most neurons by necessity are post-mitotic to avoid dismantling and reassembling thousands of experience-dependent synapses at each cell division, neuronal proliferation is rare, but vital. One population of proliferating neuronal progenitors in particular in the dentate gyrus of the hippocampus is believed to play an essential role in clinical depression and to be the target of selective serotonin-reuptake inhibitors, which are used to treat depression *(35)*. In this context, given the long anecdotal association between hypothyroidism and clinical depression *(36)*, recent reports demonstrating that thyroid hormone stimulates neuronal proliferation in this region of the brain are highly intriguing *(37,38)*.

PIP_3 also recruits and stimulates several Rac-activating proteins or "guanine nucleotide exchange factors" (GEFs), which catalyze dissociation of guanosine 5′ diphosphate GDP *(17)*. Rac is a member of the Rho family of Ras-related monomeric GTPases in the cytoplasm that play a critical role in regulation of the neuronal cytoskeleton. In this capacity, Rac and Rho, which antagonizes Rac action, provide a central regulatory system for controlling neurite outgrowth and synaptic plasticity in the developing brain *(39)*. Consequently, human mutations that inhibit Rac signaling or potentiate Rho signaling are the most common cause of inherited mental retardation *(40)*. Furthermore, Rac and Rho are implicated in the progression of neurodegenerative disease.

The three most prevalent neurodegenerative diseases among aging humans are Alzheimer's disease, Parkinson's disease, and Gehrig's disease or amyotrophic lateral sclerosis (ALS), each of which affects a different population of neurons. In the vast majority of cases, the primary pathology that initiates neurodegeneration cannot be traced to the patient's genetic background, implicating unknown environmental factors in their etiology. However, in a small percentage of the severest cases with the earliest onset, specific disease genes have been identified. In ALS, one early onset gene, alsin, is a Rac GEF, which regulates neurite outgrowth *(41)*. This is particularly interesting for two reasons. Motor neurons, which degenerate in ALS, appear to be particularly susceptible to oxidative stress because the first susceptibility gene isolated was the superoxide dismutase (SOD) which protects cells from the accumulation of metabolically stressful molecular radicals *(42)*. Alsin also protects motor neurons from oxidative stress *(43,44)*. Cell death in response to oxidative stress is initiated in part by a protein kinase, the apoptosis signaling kinase, ASK1, which only signals when it is phosphorylated *(45)*. In the nervous system, ASK1 is maintained in an inactive state by a protein phosphatase, PP5, which dephosphorylates ASK1 *(46)*. Coincidentally, we have demonstrated that PP5 mediates the Rac-dependent effects of thyroid hormone on potassium channel activity in pituitary cells *(47)*. Thus, if thyroid hormone, and consequently PI3K and Rac, signaling were disrupted, then the motoneurons' molecular brake on ASK1 activity would be weakened, and the person would be more susceptible to developing ALS. Therefore, it now appears warranted to investigate epidemiological links between hypothyroidism and ALS.

A second potential role of thyroid hormone signaling through PI3K and Rac in neurodegenerative disease concerns the treatment of Alzheimer's disease patients with non-steroidal anti-inflammatory drugs (NSAIDs). Although it has been difficult to reproducibly detect significant effects of NSAIDs on disease progression in population studies *(48)*, it has been reported recently that the most effective NSAIDs also inhibit protein kinases stimulated by Rho *(49,50)*. In addition, Rac stimulates secretion of non-amyloidigenic soluble form of the amyloid precursor protein *(51)*. Again, it appears that Rac and Rho have opposite effects on the progression of neurodegeneration in aging as they do on neurite outgrowth during development *(40,52)*: Rac signals through a protein phosphatase and increases cognitive potential, whereas Rho signals through protein kinases and reduces cognitive potential. In this context, it is interesting that transthyretin, a plasma thyroid-binding protein, which might be predicted to increase thyroxine distribution to the brain, is a susceptibility factor for disease-like symptoms in transgenic rodent models of Alzheimer's disease *(53)*.

6. XENOBIOTICS DISRUPT THYROID HORMONE SIGNALING

Because they resemble the structure of the thyroid hormone and its precursor, thyroxine, polyhalogenated aromatic hydrocarbons, such as the polychlorinated biphenyls, dioxins, and flame retardants that are released into the environment by industrial activity, have been suspected for several years of interfering with thyroid hormone signaling *(54,55)*. In particular, attention has focused on the effects of exposure to these xenobiotics on early brain development and human cognitive potential *(56,57)*. However, thyroid hormone is equally vital for amphibian metamorphosis *(58)*, so further consideration of the role of thyroid hormone disrupting xenobiotics in the increasing incidence of frog malformations might also be warranted.

In principle, xenobiotics could interfere with thyroid hormone signaling at three different levels (Fig. 1). By inhibiting thyroxine binding to serum transport proteins, xenobiotics could reduce the amount of thyroxine available in tissues for conversion to thyroid hormone. By inhibiting the 5′ deiodinase, which produces thyroid hormone from thyroxine, xenobiotics could reduce the amount of active thyroid hormone that is produced in tissues. Finally, by inhibiting thyroid hormone binding to its receptors, xenobiotics could inhibit signaling directly. Although there have been reports of each of these examples in the literature, there are also many negative studies.

Reproducible environmentally relevant studies in humans are difficult to carry out for many reasons. Because the vast majority of thyroxine is bound to transport proteins in the serum, and because most thyroid hormone molecules are produced transiently in tissues by the local deiodinase, it is difficult to measure thyroid hormone levels from blood samples. Thyroid hormone levels may also fluctuate diurnally and seasonally, further complicating epidemiological studies. Such ambiguities also complicate laboratory studies with experimental animals. In addition, some chemicals might also function as thyroid hormone analogues, stimulating receptors directly or potentiating thyroid hormone action by blocking metabolism by other iodinases *(59)*. To overcome these difficulties, simple, quantitative molecular and physiological assays are needed urgently. While there has been some progress developing physiological assays in neuronal cell culture *(60)*, in brain slices *(61)*, and in model organisms *(62,63)*, previous molecular assays have relied exclusively on indirect, nonlinear measures of artificial expression vectors with thyroid hormone promoters. Now that thyroid hormone signaling through PI3K has been reconstituted with recombinant receptors in mammalian cell lines *(23)*, the availability of fluorescent reporters of PI3K activity *(27,64)* should allow more quantitative, real-time, high-throughput testing of large panels of xenobiotics for disruption of PI3K stimulation by thyroid hormone.

7. SUMMARY

The PI3K regulates many of the fundamental physiological processes at the cellular level, which have been attributed to thyroid hormone at the organ level, including regulation of gene expression. However, thyroid hormone is only one of many endocrine pathways regulating PI3K activity, and PI3K signaling will probably explain only a fraction of all the effects of thyroid hormone and its multiple receptors. Nevertheless, disruption of thyroid hormone signaling through PI3K is predicted to increase susceptibility to several of the most common chronic human health problems including learning disorders, neurodegenerative disease, metabolic syndrome, and cardiac arrhythmias (Fig. 4). Therefore, in the future, it will be important to clarify the molecular mechanism of PI3K stimulation by thyroid hormone and to identify the environmental chemicals that interfere with its transport, activation, and receptor binding.

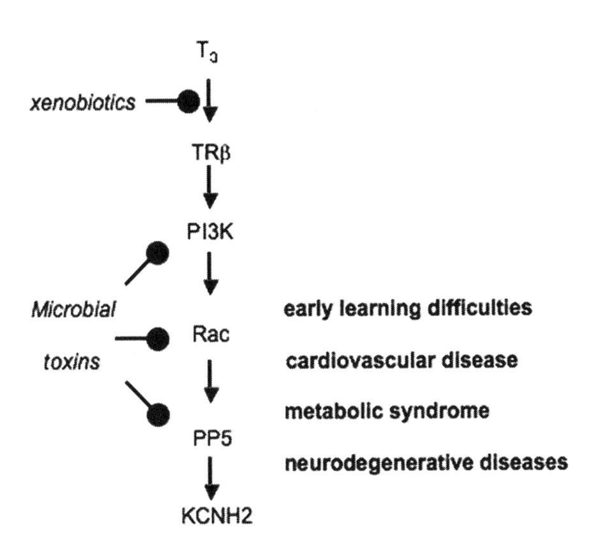

Fig. 4. Xenobiotic disruption of the transport, activation, or receptor binding of thyroid hormone could contribute to several of the most common chronic public health concerns, including early learning difficulties, cardiovascular disease, metabolic syndrome, and neurodegenerative diseases of the aging brain.

ACKNOWLEDGMENTS

I am grateful to all my current and former collaborators in the Environmental Biology Program at the National Institute of Environmental Health Sciences in Research Triangle Park, North Carolina, but I owe a special debt to Dr. Nina Margaret Storey, who had the wisdom and determination to ignore her mentor when he advised her to forget thyroid hormone because it was a nuclear hormone that did not signal through G-protein-coupled receptors.

REFERENCES

1. Yen PM. Physiological and molecular basis of thyroid hormone action. *Physiol Rev* 2001;81(3): 1097–42.
2. Yen PM. Molecular basis of resistance to thyroid hormone. *Trends Endocrinol Metab* 2003;14(7):327–3.
3. Boelaert K, Franklyn JA. Thyroid hormone in health and disease. *J Endocrinol* 2005;187(1):1–15.
4. Schreiber G. The evolutionary and integrative roles of transthyretin in thyroid hormone homeostasis. *J Endocrinol* 2002;175(1):61–73.
5. Bianco AC, Kim BW. Deiodinases: implications of the local control of thyroid hormone action. *J Clin Invest* 2006;116(10):2571–9.
6. Hollenberg AN and Jameson JL. Mechanism of thyroid hormone action. In Endocrinology, 5th ed., DeGroot LJ and Jameson JL, Ed. New York: Elsevier, 2005:1873–97.
7. O'Shea PJ, Williams GR. Insight into the physiological actions of thyroid hormone receptors from genetically modified mice. *J Endocrinol* 2002;175(3):553–70.
8. Flamant F, Samarut J. Thyroid hormone receptors: lessons from knockout and knock-in mutant mice. *Trends Endocrinol Metab* 2003;14(2):85–90.
9. Wondisford FE. Thyroid hormone action: insight from transgenic mouse models. *J Investig Med* 2003;51(4):215–20.
10. Forrest D, Reh TA, Rusch A. Neurodevelopmental control by thyroid hormone receptors. *Curr Opin Neurobiol* 2002;12(1):49–56.
11. Davis PJ, Davis FB. Nongenomic actions of thyroid hormone. *Thyroid* 1996;6(5):497–504.
12. Bassett JH, Harvey CB, Williams GR. Mechanisms of thyroid hormone receptor-specific nuclear and extra nuclear actions. *Mol Cell Endocrinol* 2003;213(1):1–11.
13. Samuels HH, Tsai JS, Cintron R. Thyroid hormone action: a cell-culture system responsive to physiological concentrations of thyroid hormones. *Science* 1973;181(106):1253–6.
14. Storey NM, O'Bryan JP, Armstrong DL. Rac and Rho mediate opposing hormonal regulation of the ether-a-go-go-related potassium channel. *Curr Biol* 2002;12(1):27–33.
15. Sakaguchi Y, Cui G, Sen L. Acute effects of thyroid hormone on inward rectifier potassium channel currents in guinea pig ventricular myocytes. *Endocrinology* 1996;137(11):4744–51.
16. Sen L, Sakaguchi Y, Cui G. G protein modulates thyroid hormone-induced Na(+) channel activation in ventricular myocytes. *Am J Physiol Heart Circ Physiol* 2002;283(5):H2119–29.
17. Welch HC, Coadwell WJ, Stephens LR, Hawkins PT. Phosphoinositide 3-kinase-dependent activation of Rac. *FEBS Lett* 2003;546(1):93–7.
18. Simoncini T, Hafezi-Moghadam A, Brazil DP, Ley K, Chin WW, Liao JK. Interaction of oestrogen receptor with the regulatory subunit of phosphatidylinositol-3-OH kinase. *Nature* 2000;407(6803):538–41.
19. Walker EH, Pacold ME, Perisic O, et al. Structural determinants of phosphoinositide 3-kinase inhibition by wortmannin, LY294002, quercetin, myricetin, and staurosporine. *Mol Cell* 2000;6(4):909–19.
20. Cao X, Kambe F, Moeller LC, Refetoff S, Seo H. Thyroid hormone induces rapid activation of Akt/protein kinase B-mammalian target of rapamycin-p70S6K cascade through phosphatidylinositol 3-kinase in human fibroblasts. *Mol Endocrinol* 2005;19(1):102–2.
21. Lei J, Mariash CN, Ingbar DH. 3,3′,5-Triiodo-L-thyronine up-regulation of Na,K-ATPase activity, cell surface expression in alveolar epithelial cells is Src kinase-, phosphoinositide 3-kinase-dependent. *J Biol Chem* 2004;279(46):47589–600.

22. Moeller LC, Dumitrescu AM, Refetoff S. Cytosolic action of thyroid hormone leads to induction of hypoxia-inducible factor-1alpha and glycolytic genes. *Mol Endocrinol* 2005;19(12):2955–63.
23. Storey NM, Gentile S, Ullah H, et al. Rapid signaling at the plasma membrane by a nuclear receptor for thyroid hormone. *Proc Natl Acad Sci USA* 2006;103(13):5197–201.
24. Di Paolo G, De Camilli P. Phosphoinositides in cell regulation and membrane dynamics. *Nature* 2006;443(7112):651–7.
25. Engelman JA, Luo J, Cantley LC. The evolution of phosphatidylinositol 3-kinases as regulators of growth and metabolism. *Nat Rev Genet* 2006;7(8):606–19.
26. Wymann MP, Marone R. Timing, location, and scaffolding. *Curr Opin Cell Biol* 2005;17(2):141–9.
27. Rusten TE, Stenmark H. Analyzing phosphoinositides and their interacting proteins. *Nat Methods* 2006;3(4):251–8.
28. Baxter JD, Webb P, Grover G, Scanlan TS. Selective activation of thyroid hormone signaling pathways by GC-1: a new approach to controlling cholesterol and body weight. *Trends Endocrinol Metab* 2004;15(4):154–7.
29. Foukas LC, Claret M, Pearce W, et al. Critical role for the p110alpha phosphoinositide-3-OH kinase in growth and metabolic regulation. *Nature* 2006;441(7091):366–70.
30. Sleeman MW, Wortley KE, Lai KM, et al. Absence of the lipid phosphatase SHIP2 confers resistance to dietary obesity. *Nat Med* 2005;11(2):199–205.
31. Rosati B, Marchetti P, Crociani O, et al. Glucose- and arginine-induced insulin secretion by human pancreatic beta-cells: the role of HERG K(+) channels in firing and release. *FASEB J* 2000;14(15):2601–10.
32. Sanguinetti MC, Tristani-Firouzi M. hERG potassium channels and cardiac arrhythmia. *Nature* 2006;440(7083):463–9.
33. Brunet A, Datta SR, Greenberg ME. Transcription-dependent and -independent control of neuronal survival by the PI3K-Akt signaling pathway. *Curr Opin Neurobiol* 2001;11(3):297–305.
34. Shaw RJ, Cantley LC. Ras, PI(3)K and mTOR signalling controls tumour cell growth. *Nature* 2006;441(7092):424–30.
35. Gage FH. Brain, repair yourself. *Sci Am* 2003;289(3):46–53.
36. Bahls SC, de Carvalho GA. [The relation between thyroid function and depression: a review]. *Rev Bras Psiquiatr* 2004;26(1):41–9.
37. Desouza LA, Ladiwala U, Daniel SM, Agashe S, Vaidya RA, Vaidya VA. Thyroid hormone regulates hippocampal neurogenesis in the adult rat brain. *Mol Cell Neurosci* 2005;29(3):414–26.
38. Montero-Pedrazuela A, Venero C, Lavado-Autric R, et al. Modulation of adult hippocampal neurogenesis by thyroid hormones: implications in depressive-like behavior. *Mol Psychiatr* 2006;11(4): 361–71.
39. Govek EE, Newey SE, Van Aelst L. The role of the Rho GTPases in neuronal development. *Genes Dev* 2005;19(1):1–49.
40. Ramakers GJ. Rho proteins, mental retardation and the cellular basis of cognition. *Trends Neurosci* 2002;25(4):191–9.
41. Tudor EL, Perkinton MS, Schmidt A, et al. ALS2/Alsin regulates Rac-PAK signaling and neurite outgrowth. *J Biol Chem* 2005;280(41):34735–40.
42. Boillee S, Vande Velde C, Cleveland DW. ALS: a disease of motor neurons and their nonneuronal neighbors. *Neuron* 2006;52(1):39–59.
43. Kanekura K, Hashimoto Y, Kita Y, et al. A Rac1/phosphatidylinositol 3-kinase/Akt3 anti-apoptotic pathway, triggered by AlsinLF, the product of the ALS2 gene, antagonizes Cu/Zn-superoxide dismutase (SOD1) mutant-induced motoneuronal cell death. *J Biol Chem* 2005;280(6):4532–3.
44. Kanekura K, Hashimoto Y, Niikura T, Aiso S, Matsuoka M, Nishimoto I. Alsin, the product of ALS2 gene, suppresses SOD1 mutant neurotoxicity through RhoGEF domain by interacting with SOD1 mutants. *J Biol Chem* 2004;279(18):19247–56.
45. Sekine Y, Takeda K, Ichijo H. The ASK1-MAP kinase signaling in ER stress and neurodegenerative diseases. *Curr Mol Med* 2006;6(1):87–97.
46. Morita K, Saitoh M, Tobiume K, et al. Negative feedback regulation of ASK1 by protein phosphatase 5 (PP5) in response to oxidative stress. *EMBO J* 2001;20(21):6028–36.
47. Gentile S, Darden T, Erxleben C, et al. Rac GTPase signaling through the PP5 protein phosphatase. *Proc Natl Acad Sci USA*. 103(13):5202–6.
48. Wyss-Coray T. Inflammation in Alzheimer disease: driving force, bystander or beneficial response. *Nat Med* 2006;12(9):1005–5.

49. Weggen S, Eriksen JL, Das P, et al. A subset of NSAIDs lower amyloidogenic Abeta42 independently of cyclooxygenase activity. *Nature* 2001;414(6860):212–6.
50. Zhou Y, Su Y, Li B, et al. Nonsteroidal anti-inflammatory drugs can lower amyloidogenic Abeta42 by inhibiting Rho. *Science* 2003;302(5648):1215–7.
51. Maillet M, Robert SJ, Cacquevel M, et al. Crosstalk between Rap1 and Rac regulates secretion of sAPPalpha. *Nat Cell Biol* 2003;5(7):633–9.
52. Van Aelst L, Cline HT. Rho GTPases and activity-dependent dendrite development. *Curr Opin Neurobiol* 2004;14(3):297–304.
53. Stein TD, Anders NJ, DeCarli C, Chan SL, Mattson MP, Johnson JA. Neutralization of transthyretin reverses the neuroprotective effects of secreted amyloid precursor protein (APP) in APPSW mice resulting in tau phosphorylation and loss of hippocampal neurons: support for the amyloid hypothesis. *J Neurosci* 2004;24(35):7707–17.
54. Brucker-Davis F. Effects of environmental synthetic chemicals on thyroid function. *Thyroid* 1998;8(9):827–56.
55. Boas M, Feldt-Rasmussen U, Skakkebaek NE, Main KM. Environmental chemicals and thyroid function. *Eur J Endocrinol* 2006;154(5):599–611.
56. Porterfield SP. Thyroidal dysfunction and environmental chemicals–potential impact on brain development. *Environ Health Perspect* 2000;108(Suppl 3):433–8.
57. Colborn T. Neurodevelopment and endocrine disruption. *Environ Health Perspect* 2004;112(9): 944–9.
58. Furlow JD, Neff ES. A developmental switch induced by thyroid hormone: Xenopus laevis metamorphosis. *Trends Endocrinol Metab* 2006;17(2):40–7.
59. Zoeller RT. Environmental chemicals as thyroid hormone analogues: new studies indicate that thyroid hormone receptors are targets of industrial chemicals. *Mol Cell Endocrinol* 2005;242(1–2):10–5.
60. Kimura-Kuroda J, Nagata I, Kuroda Y. Hydroxylated metabolites of polychlorinated biphenyls inhibit thyroid-hormone-dependent extension of cerebellar Purkinje cell dendrites. *Brain Res Dev Brain Res* 2005;154(2):259–63.
61. Gilbert ME. Alterations in synaptic transmission and plasticity in area CA1 of adult hippocampus following developmental hypothyroidism. *Brain Res Dev Brain Res* 2004;148(1):11–8.
62. Hill A, Howard CV, Strahle U, Cossins A. Neurodevelopmental defects in zebrafish (Danio rerio) at environmentally relevant dioxin (TCDD) concentrations. *Toxicol Sci* 2003;76(2):392–9.
63. Turque N, Palmier K, Le Mevel S, Alliot C, Demeneix BA. A rapid, physiologic protocol for testing transcriptional effects of thyroid-disrupting agents in premetamorphic Xenopus tadpoles. *Environ Health Perspect* 2005;113(11):1588–93.
64. Ananthanarayanan B, Ni Q, Zhang J. Signal propagation from membrane messengers to nuclear effectors revealed by reporters of phosphoinositide dynamics and Akt activity. *Proc Natl Acad Sci USA* 2005;102(42):15081–6.

9 Endocrine Disruption and Reproductive Outcomes in Women

Sarah Janssen, MD, PhD, MPH,
Victor Y. Fujimoto, MD,
and Linda C. Giudice, MD, PhD, MSc

CONTENTS

1. INTRODUCTION

1.1. Critical Windows of Exposure in Reproductive Development

Infertility and associated problems of the female reproductive tract, such as endometriosis, abnormal menstrual cycles, and premature menopause, are diagnosed in women of reproductive age. For this reason, reproductive problems are usually considered to be "adult" conditions. Whereas, some female reproductive problems are the result of contaminant exposures during adulthood, there is increasing evidence that exposures early in life can result in permanent and irreversible changes to the reproductive tract, which are not manifest until decades later.

A growing body of scientific evidence suggests that exposure to endocrine-disrupting chemicals (EDCs) early in life may alter development of the reproductive tract and hormonal responsiveness in adulthood *(1,2)*. Coupled with this evidence are a number of disturbing trends in some geographic regions, including a reduction in fertility, an increase in hormone-sensitive cancers, an earlier age of puberty in girls, and a decrease in the number of boys being born.

Evidence from animal studies indicates that these conditions are likely to originate during the prenatal period. Exposures to environmental contaminants early in life are

From: *Endocrine-Disrupting Chemicals: From Basic Research to Clinical Practice*
Edited by: A. C. Gore © Humana Press Inc., Totowa, NJ

of particular importance, because the reproductive system is undergoing an intricately orchestrated process of growth and differentiation. A fetus is vulnerable not only because of the rapid development and growth that are occurring, but also because it possesses immature and underdeveloped excretion pathways, low levels of chemical-binding proteins, and an underdeveloped blood-brain barrier that is unable to protect the nervous system from toxic exposures *(3)*.

1.2. DES as a Classic Example of Endocrine Disruption Occurring in Utero

Diethylstilbestrol (DES), a synthetic estrogen, was given to an estimated 2–10 million women in the USA and UK from the late 1940s through the early 1970s. Originally given to prevent spontaneous abortion (SAB), this drug was a failure in preventing pregnancy loss *(4)*. Furthermore, this drug has left a legacy of reproductive abnormalities in both the boys and the girls exposed *in utero*.

DES is a potent estrogen that binds to estrogen receptors in the fetal reproductive tract where it causes permanent alterations. In addition to cervico-vaginal abnormalities, women exposed to DES *in utero* have been shown in some studies to have reduced rates of fertility, vaginal and uterine abnormalities (e.g., vaginal adenosis, clear cell vaginal carcinoma, and a T-shaped uterine cavity), increased incidence of endometriosis, SAB, and poorer pregnancy outcomes *(5–8)*. Uterine fibroids were also recently associated with *in utero* DES exposure *(9)*.

Exposure to DES is now used as a model compound for studying the mechanisms and toxicities of other suspected environmental estrogens, including epigenetic alterations that are transmissible to subsequent generations *(10)*. In preliminary epidemiological studies of women whose mothers were exposed *in utero* to DES (third generation), there is evidence of irregular menstruation and possible infertility *(11)*.

2. EXPOSURE TO EDCs

Approximately 60 chemicals have been categorized as endocrine disruptors, including both synthetic and natural compounds *(12)*. Pesticides, persistent pollutants such as polychlorinated biphenyls (PCBs) and dioxins, additives used in the plastic industry, and flame retardants have all been shown to have hormonal activity *(2,13,14)*.

Exposures to EDCs happen throughout the life cycle—from preconception through adult reproductive years. Many EDCs are stored in adipose tissue and have very long half-lives, so they can persist in tissues for decades. EDCs have been identified in follicular fluid, semen, amniotic fluid, fetal cord blood, breast milk, serum, and adipose tissue. In addition to direct toxicity to oocytes and spermatozoa, endocrine disruptors may also interfere with fertilization, implantation, pregnancy, and embryonic, fetal, and pubertal development.

2.1. Preconception Exposure

EDCs may cause abnormal meiosis resulting in chromosomal aberrations in gametes or interference with fertilization or implantation. In a study of couples undergoing fertility treatments, over 50% of the women had pesticides, PCBs, DDT/DDE, and other chlorinated chemicals in their serum and follicular fluid *(15)*. In the same study, cadmium was measured in about one-third of follicular fluid samples. Bisphenol

A (BPA), an estrogenic EDC, also has been identified in follicular fluid (ng/mL range) and is reported to adversely affect embryonic development *(16)*. Cadmium and chlorinated pesticides have been measured in semen samples and may not only interfere with sperm function but also be transferred to the female where they could potentially interfere with embryonic development *(15,17)*.

2.2. Fetal Exposure

Amniotic fluid samples can provide information about exposures to EDCs during the second trimester of gestation. Similar to contaminants found in follicular fluid, chlorinated hydrocarbons, such as the DDT metabolite, DDE, PCBs, and hexachlorocyclohexane (HCH), have been measured in amniotic fluid. In one study, DDE and HCH were found to be above the limit of detection (0.1 ng/mL) in approximately one-third of amniotic fluid samples collected during routine amniocenteses *(18)*. Organophosphate and carbamate pesticides also have been measured in amniotic fluid *(19)*. In follicular and amniotic fluid, DDE is the most frequently detected pesticide, is found at the highest concentrations, and has been associated with failed in vitro fertilization (IVF) outcomes *(15,18)*. The plasticizer, BPA, also has consistently been found in amniotic fluid at low concentrations (ng/mL) in several studies from Japan *(16,20,21)*.

Cord blood samples are an easier and less invasive method of biomonitoring fetal exposures and have been used for assaying a wide array of chemicals. EDCs—including pesticides; chlorinated hydrocarbons such as PCBs, DDT, and dioxins; plasticizers such as phthalates and BPA; heavy metals such as mercury and lead; and flame retardants such as polybrominated diphenyl ethers (PBDEs)—have all been measured in cord blood samples *(16,22–24)*. In a recent study of 10 newborn cord blood samples, 100% were found to contain measurable levels of mercury [average 0.947 parts per billions (ppb), range 0.07–2.3 ppb], polyaromatic hydrocarbons (PAHs) [average 285 parts per trillion (ppt), range 217–384 ppt], perfluorinated chemicals [average 6.17 ppb, range 3.37–10.7 ppb], polychlorinated dioxins and polychlorinated furans (PCDDs and PCDFs, respectively) (average 59.4 ppt, range 37.9–102 ppt), PCBs (average 7.9 ppb, range 2.99–19.7 ppb), PBDEs (average concentration: 6.4 parts per billion, range 1.1–14.2 ppb), organochlorine pesticides (average: 18.6 ppb, range 8.7–35.40 ppb), and perchlorinated naphthalenes (617 ppt; range 295–964 ppt) *(25)*. Other cord blood studies have found similar concentrations of these chemicals; levels shown to cause reproductive abnormalities in animal models *(26–28)*.

2.3. Postnatal Exposure

After birth, newborns may continue to be exposed to EDCs in breast milk. Breast milk samples contain many of the same contaminants found in amniotic fluid and cord blood, including pesticides, PCBs and other chlorinated hydrocarbons, phthalates, BPA, and PBDEs *(29,30)*. Although these contaminants are commonly found in breast milk, it is important to emphasize that breast milk currently remains the preferred food source for newborns. Breast feeding has not only nutritional but also emotional, neurological, and immunological benefits. These benefits extend beyond the neonatal period, well into child and adult development.

Finally, there is ongoing exposure to EDCs from childhood through adult reproductive ages. Ongoing sources of exposure to EDCs occur by ingestion of food and water, inhalation of contaminants, as well as accumulation of chemicals with long half-lives. Biomonitoring of the serum, urine, and breast milk of women of reproductive age has demonstrated the presence of EDCs.

It is clear from the earliest point of development and throughout life that humans are exposed to a wide variety and mixture of chemicals, many of which have the potential to interfere with normal hormonal actions. Although it is impractical to measure body burdens in every patient and impossible to extrapolate one measurement to a health outcome, trends of EDCs exposure in humans are likely to have important public health implications.

2.4. Pubertal Development and Exposure

In the USA, there is sufficient evidence that the average age of puberty has decreased over the past 30 years. The average age of onset of menses for African-American girls is 12.1 and for Caucasian girls is 12.6 years *(31)*. Compared with statistics from 25 years ago, this is about 5 months earlier in African-American girls and 3 months earlier in Caucasian girls. The average age of onset of breast development is age 9 in blacks and age 10 in whites, compared to an average age of 11.2 years for all races 25 years earlier. Although there are less data, there is also evidence that boys are reaching puberty at earlier ages. There are a number of reasons that could account for this trend including better nutrition and higher rates of obesity in children *(31)*. There is also concern that exposure to EDCs may be responsible, in part, for this change.

In girls, earlier age at menarche was reported after exposure to PCBs, polybrominated biphenyls (PBBs), DDT, and phthalate esters *(32,33)*. However, several other studies have found no effect of these compounds on age at menarche or pubertal Tanner stages.

In the 1970s, consumption of food contaminated with the flame retardant, PBB, resulted in exposure of more than 4,000 individuals in Michigan. Breast-fed females born after the contamination had higher serum levels of PBB (>7 ppb) and underwent menarche one year earlier than those with low exposure (<1 ppb). These results were statistically significant with an OR of 3.4 (95% CI: 1.3–9.0) *(32)*. There was no change in the age of onset of breast development. Of note, the main congener contaminating the food supply, PBB-153, is estrogenic *(32)*. Long-term effects in those exposed, such as infertility, incidence of reproductive tract cancers, including breast cancer, and age at menopause, have yet to be reported.

Among women exposed postnatally and prepubertally to high levels of the potent dioxin, 2,3,7,8-tetrachlorodibenzo-*p*-dioxin (TCDD) in Seveso, Italy, there was no change in the onset of menarche *(34)*.

There are conflicting results on the association of other chlorinated hydrocarbons with the onset of puberty *(33,35)*. One study analyzed age of puberty in 327 young girls exposed *in utero* to PCBs and DDT when their mothers ate contaminated fish. An increase of 15 μg/dL DDE in maternal serum was associated with 1 year advance of menarche in daughters ($p = 0.04$) *(35)*. PCB exposure was not associated with an earlier age of menarche in this study. However, other studies have found an association of early menarche with PCB exposure. In a study of Akwesasne Mohawk Nation girls, living near a Superfund site contaminated with PCB, DDE, and the pesticides hexachlorobenzene (HCB) and mirex, it was found that only PCBs were associated with an earlier age of puberty *(33)*. In another study, a cohort of 316 girls exposed to DDE and PCBs prenatally and through breast feeding had no changes in age at menarche or other pubertal stages *(32)*.

Lead has been associated with a later age of puberty *(33)*. Consistent with this study, lead has been found in two other studies to be associated with a delay in self-assessed age at menarche and physician-determined pubic hair and breast development *(32)*.

3. MENSTRUAL-CYCLE DISRUPTION

3.1. Menstrual-Cycle Abnormalities and Other Hormonal Disturbances

Steroid and protein hormones control variation in menstrual cycle length and regularity with modification by a number of factors that can interfere with normal hormonal signaling and regulation, such as EDCs. Irregular menstrual cycles are associated with impaired fecundity, infertility, and a number of reproductive disorders such as polycystic ovarian syndrome (PCOS).

Cigarette smoking has been associated with menstrual abnormalities, and smokers have higher basal follicle-stimulating hormone (FSH) levels and shorter cycle lengths consistent with an increased prevalence of diminished ovarian reserve *(36)*. PAHs, one group of over 4000 chemicals found in tobacco smoke, are hypothesized to cause a loss of oocytes. Animal studies have confirmed that PAHs bind to the aryl hydrocarbon (Ah) receptor in the ovary, triggering apoptosis *(37)*. Similar results have been observed *in vitro* with human ovarian explants *(38)*.

Exposure to PCBs has been associated with altered menstrual cycles and hormonal changes as well. Adult women exposed to cooking oil contaminated with high levels of PCBs and PCDFs in Taiwan reported abnormal menstrual bleeding 50% more frequently than women who were not exposed *(39)*. However, these women did not report any differences in cycle regularity, fertility, or age at menopause. A smaller study of girls exposed *in utero* to the contaminated cooking oil found shorter cycles (5.5 vs. 6.5 days, $p = 0.0055$); a higher rate of irregular cycles (40 vs. 0%, $p = 0.018$), and higher serum levels of estradiol ($p = 0.016$) and FSH ($p = 0.061$) in the exposed girls compared with controls *(40)*. In contrast, a study of US women exposed to lower doses of PCBs in the 1960s found an association with longer and irregular menstrual cycles *(41)*. Another study of Southeast Asian immigrants did not find an association with PCBs and cycle length *(42)*. Differences in these studies may be due to differences in the timing of exposure, congener differences, routes of exposure, and/or racial differences.

Pesticides are known to disrupt the estrous cycle in animals, especially hormonally active pesticides such as atrazine, lindane, mancozeb, and maneb *(43)*. Among women who mixed or applied these pesticides, there was an increase in cycle length (OR = 2.7), missed periods (OR = 2.1) and intermenstrual bleeding (OR = 1.6) *(43)*. Overall, women who mixed or applied any pesticides experienced longer menstrual cycles and increased odds of missed periods compared with women who never used pesticides (OR = 1.5, 95 % CI: 1.2–1.9).

There have been conflicting results on DDT exposure and menstrual cycle characteristics. A study of women exposed in the 1960s, when DDT was commonly used, did not find any association for menstrual cycle length, but there was trend toward irregular cycles *(41)*. In a study of South East Asians, where DDT continues to be used, each doubling of the DDE level decreased cycle length by 1.1 day (95% CI: −2.4–0.23) and decreased luteal phase length by 0.6 days (95% CI: −1.1 to −0.2). In this study, a decrease in progesterone metabolite levels during the luteal phase was associated with increasing DDE concentrations *(42)*. Differences in these two studies may be due to differences in the years of exposure, routes of exposure, and/or racial differences.

High doses of the most potent dioxin, TCDD, are known to disrupt ovarian function in nonhuman primate studies *(44)*. However, chronic low-dose exposures in women have not been associated with any changes in menstrual cycle length or hormonal profiles. A study of women in Seveso, Italy, exposed to high levels of TCDD after an industrial explosion found that women exposed premenarchally had a lengthening of the menstrual cycle by 0.93 days for each 10-fold increase in serum TCDD levels *(44)*. Among women who were exposed postmenarchal, there was no change in menstrual cycle length. These findings are consistent with the hypothesis that TCDD may act as an endocrine disruptor by altering the hypothalamic–pituitary–gonadal axis resulting in an elongation of the cycle and/or a higher frequency of anovulatory cycles. After puberty has begun, the axis may be more resistant to perturbations *(44)*.

3.2. Endometriosis

The endometrium is a steroid-responsive tissue that undergoes proliferation and degradation with each menstrual cycle. Endometriosis is a condition wherein endometrial glandular and stromal components grow outside the uterus resulting in implant lesions principally within the pelvic cavity, but occasionally found elsewhere in the body. The internal bleeding associated with endometriosis can result in inflammation, pain, and the formation of pelvic adhesions. Although endometriosis sometimes occurs without symptoms, it often causes chronic pelvic pain, dysmenorrhea, and dyspareunia. Severe endometriosis has been associated with infertility *(45)*.

Endometriosis affects about 1 in 10 women in the USA, but nearly 40% of women with infertility appear to have endometriosis *(46)*. The etiology of endometriosis in not understood, although *in utero* exposures may be important. A recent prospective study found low birth weight, exposure to DES, and multiple births were associated with a greater incidence of endometriosis *(47)*.

There is an evolving body of scientific research suggesting that endometriosis may have an environmental origin. In primates, exposure to chlorinated hydrocarbons such as PCBs and dioxins can cause endometriosis and make it worse in animals that already have it *(46)*. Laboratory studies of dioxins and dioxin-like PCBs have shown that endometrial tissue responds with increased expression of estrogen (through aromatase induction), inflammation, and altered immune response leading to abnormal tissue remodeling *(46)*. The most potent dioxin, TCDD, also alters expression of an endometrial glycoprotein critical for embryonic implantation and pregnancy establishment *(48)*.

Epidemiological studies of endometriosis in humans have found inconsistent results between chlorinated hydrocarbon exposures and the incidence of endometriosis *(46,49, 50)*. Comparison of these studies is limited by small numbers of subjects, differences in study design and analytical methods, and differences in the specific organochlorines measured in each study. Many epidemiological studies of endometriosis and associated environmental exposures are further complicated by a lack of sufficient control groups. Diagnosis of endometriosis requires both laparoscopic and histological confirmation. Many women of reproductive age undergo laparoscopy only for diagnosis of gynecological conditions or for tubal ligation. Therefore, there may be selection bias in choosing a control group, because other gynecological conditions (infertility, PCOS, and dysmenorrhea) could be associated with similar environmental exposures. Women who have already had children and are undergoing laparoscopy for tubal ligation have a different reproductive history with the potential for decreases in the body burden of persistent chemicals through lactation.

Perhaps because of these limitations, there is conflicting evidence that persistent organochlorines such as PCBs and dioxins are associated with endometriosis in women *(46,49,50)*. This issue may be resolved by evaluating specific congeners rather than as a group of chemicals.

A recent case–control study of women in Belgium found an association between increased serum concentrations of dioxin-like organochlorines and endometriosis *(51)*. Serum concentrations of polychlorinated dioxins and polychlorinated furans (PCDDs and PCDFs), and dioxin-like PCBs were measured and standardized to toxic equivalent factors (TEQ) based on the toxicity of TCDD. Women with peritoneal endometriosis or deep endometriotic (adenomyotic) nodules had higher mean TEQ levels than controls. Furthermore, for each increase in total TEQ levels of 10 pg/g lipid, the risk of deep endometriotic (adenomyotic) nodules increased 3-fold (OR = 3.3, 95% CI: 1.4–7.6). In a small study comparing serum levels of PCBs with laparoscopically confirmed endometriosis, the highest tertile of anti-estrogenic PCBs was associated with a 5-fold increase in risk of endometriosis (OR = 4.78, 95% CI: 1.03–40.70) *(52)*. There was no association with the estrogenic PCBs in this study. Other studies have found that the levels of highly chlorinated PCBs (138, 153, and 180) and the pesticide, HCH, are significantly higher in women with endometriosis than those without *(53)*.

A study of women exposed to the most potent dioxin, TCDD, after an industrial accident in Seveso, Italy, found a doubled, nonsignificant, increase in endometriosis among women with serum TCDD levels of $\geq$100 ppt *(54)*. This study was limited by a lack of laparoscopically confirmed disease in all subjects.

Other environmental exposures have not been as well studied for their association with endometriosis. Phthalates, chemicals used as plasticizers in many consumer and medical products, have been associated with endometriosis in two studies. One study comparing serum phthalate ester levels in women with endometriosis to women with infertility found an association for the phthalates di-*n*-butyl phthalate (DnBP), butyl benzyl phthalate (BBP), di-*n*-octyl phthalate (DnOP), and diethylhexyl phthalate (DEH) *(55)*. DEHP also was detected in the peritoneal fluid of women with endometriosis *(56)*.

3.3. PCOS

PCOS is the most common endocrine abnormality in women and is characterized by hormonal abnormalities (primarily an excess of androgens), polycystic ovaries (PCO), and oligoovulation or anovulation. PCOS is also associated with a cluster of metabolic abnormalities including obesity, insulin resistance, and abnormal lipid profiles *(57)*. In addition to being associated with infertility and early miscarriage, PCOS is also associated with menstrual disorders, endometrial cancer, and a risk of type II diabetes later in life. There is considerable heterogeneity in the clinical presentation of the disorder with varying degrees of obesity, irregular menstrual cyclicity, hirsutism, acne, and abnormal endocrine parameters. Not all women who have PCOS are obese, but weight control can alleviate some of the symptoms of the disorder in obese women.

The etiology of PCOS is unknown, but it is generally accepted that there is a genetic predisposition. The *in utero* environment interacting with genetic predispositions is increasingly being recognized as an important determinant of this disease and may explain some of the heterogeneity in the presentations *(58–60)*.

Exposure to excess androgen in utero has been demonstrated in animal models to be associated with development of a PCOS-like picture *(59)*. It is hypothesized that the excess androgen is of fetal origin, either from the fetal ovary or adrenal gland, rather than from a maternal source. Although there is no evidence of P450 aromatase mutations in patients with PCOS *(61)*, it is possible that an aromatase disorder or fetal exposure to chemicals that interfere with aromatase activity could play a role in development of PCOS.

In both primates and sheep exposed to androgens prenatally, females developed enlarged ovaries with multiple medium-sized antral follicles *(60)*. In addition, the female reproductive tract was permanently reprogrammed to respond differently to hormonal stimulation. Animal studies suggest that *in utero* exposure to androgens causes a loss of the negative-feedback mechanism of the hypothalamic–pituitary–gonadal axis, thereby stimulating luteinizing hormone (LH) and androgen hypersecretion *(60)*.

Birth weight and gestational age have been associated with PCOS *(58)*. In normal-weight patients with normal hormonal profiles, gestational age >40 weeks and birthweight >8.5 pounds were associated with a higher incidence of PCOS *(58)*. Conversely, intrauterine growth restriction (IUGR) has been associated with insulin resistance later in life. It is possible that this could contribute to PCOS; however, no definitive relationship has been established to date *(62)*.

There is a paucity of studies on environmental exposures and development of PCOS. BPA, which is known to be metabolized differently between men and women, was found in higher concentrations in women with PCOS *(21)*. It is possible that rather than a direct association, this is due to androgen-related differences in metabolism.

3.4. Premature Menopause

The average age of menopause in the general population is 51.4, and the onset of menopause is preceded by infertility in later reproductive years. In younger women, premature ovarian failure has been associated with infertility and earlier onset of chronic diseases associated with the loss of reproductive hormones, such as osteoporosis. Damage to the follicular pool or disruption of hormonal feedback controlling follicular growth and ovulation may result in an earlier age at menopause.

There are some well-characterized causes of premature menopause including tobacco smoke exposure. Tobacco smoke contains thousands of different chemicals including heavy metals and PAHs and has been shown to be anti-estrogenic. PAHs bind to the Ah receptor and activate the Bax gene resulting in an increase in apoptosis and oocyte cell death *(37)*. Smoking has been associated with a 1–2 year earlier age at menopause, especially when smoking occurs around the time of menopause *(63,64)*.

Some EDCs bind to the Ah receptor and are anti-estrogenic. The Ah receptor–ligand complex has been shown to bind to the promoter region of estrogen-responsive genes. Dioxins bind to the Ah receptor and have been associated with ovarian dysfunction and anovulation in primates *(65)*. In a study of women exposed after an industrial explosion, serum levels of TCDD were positively associated with an earlier age at menopause *(65)*. Each 10-fold increase in serum TCDD was associated with a 6% increase in risk of early menopause. Although the association was not statistically significant, there was a linear increase in RR with doses up to 100 ppt of TCDD.

The DDT metabolite, DDE, also has been associated with premature menopause *(63)*. Although the result was not statistically significant, there was approximately a 1-year difference between the age at menopause for women with the highest DDE category and the lowest DDE category (RR = 1.4, 95% CI: 0.9–2.1) *(63)*. This study was limited by a 20-year gap between the onset of menopause and measurement of organochlorine levels for up to 25% of the subjects. Another study of Hispanic women found those with the highest serum levels of DDE reached menopause 1.7 years earlier than women in the lowest DDE category *(66)*. However, no consistent dose–response effect was evident across the exposure categories. DDE is anti-androgenic and may interfere with follicular development, leading to increased apoptosis or atresia *(63)*.

In contrast, a study of women who mixed or used "hormonally active" pesticides, such as atrazine, DDT, lindane, or mancozeb/maneb, found a 5-month delay for menopause *(67)*. The differences in these studies may be due to small sample sizes or different sampling or exposure-assessment techniques.

Several studies investigating exposure to PCBs have not found an association with age of onset of menopause *(63)*. This includes high levels of exposure associated with widespread contamination of cooking oil in Taiwan (Yucheng) and contamination of food by the flame retardant, PBB *(39,68)*.

4. FEMALE REPRODUCTIVE EXPOSURES AND OUTCOMES

4.1. Fecundity Rates (Time to Pregnancy)

Compared with 1970, in 2002, fertility rates were lower in both developing and industrialized countries *(69)*. Lower fertility rates could be due to sociocultural differences in decisions not to have children and/or increased rates of infertility adjusted for female age. Infertility rates are harder to estimate, because there is no national reporting system in most countries. In the USA, the National Survey of Family Growth has found a steadily increasing rate of women who report difficulty conceiving (impaired fecundity) *(70)*. In 2002, about 7.3 million US women self-reported impaired fecundity compared with 6.1 million women in 1995 and 4.9 million in 1988. Although the trend of couples to delay childbirth until later ages may account for some of the increases, a previous survey found an increasing number of younger women reporting difficulty conceiving. Between 1982 and 1995 in women under age 25, there was a 42% increase as compared with a 12% increase in women aged 25–34 and 6% in women aged 35 and older *(71)*.

Fecundity is the rate of conception in the absence of contraception. Time to pregnancy (TTP) is often used as a measure of fecundity and is estimated by assessing the time that elapses between stopping use of contraceptive and clinical recognition of pregnancy. TTP takes into account both male and female reproductive capacity as well as early survival of the fetus. An increased TTP is also associated with pregnancy loss and extra-uterine pregnancies *(72)*.

The causes of increased TTP are varied and can include both male and female factors. Although there are some well-recognized causes such as endometriosis, PCOS, and tubal defects, a significant proportion remains unexplained. Exposure to endocrine disruptors such as chlorinated hydrocarbons (PCBs, dioxins and furans, DDT, and other organochlorine pesticides), pesticides, heavy metals, and cigarette smoke have been hypothesized to interfere with female reproduction *(73)*.

Cigarette smoke consistently has been associated with impaired fecundity. A meta-analysis of active smoking in women of reproductive age found the relative risk of infertility in smokers versus nonsmokers is 1.6 (95% CI: 1.34–1.91) *(73)*. Smoking is also with associated lower numbers of oocytes retrieved and fertilization failure in infertility treatments *(73)*. Cigarette smoke is known to be anti-estrogenic and may interfere with feedback at the hypothalamic–pituitary–gonadal axis. Lower FSH levels have been measured in women smokers, which may result in impaired oocyte development and ovulation. Moreover, as noted previously, animal studies have found that fetal exposures to PAHs in tobacco smoke are associated with a higher rate of apoptosis and fewer oocytes at birth *(38)*.

There is no clear association between organochlorine exposures and impaired fertility. One study found regional differences in TTP and serum levels of the PCB congener-153 and the DDT metabolite, DDE *(74)*. A positive association was found for these contaminants in serum and TTP in an Inuit population, but no association was found in women from Poland or Sweden. Differences in diet, contraceptive use, and contraceptive failure may account for these discrepancies. Another study of women from the 1960s, a time when PCB and DDT levels were historically high, found positive but nonsignificant associations between PCBs, DDE, and TTP *(75)*.

Some studies have found that elevated DDT levels were associated with increased TTP *(53,76)*, but other studies have found no association between DDE levels and increased TTP or failure of IVF *(15,76)*. These inconsistencies require further study.

Studies of consumption of fish contaminated with PCBs have found conflicting results on TTP with some finding an association and others finding no association *(74)*. The New York Angler Cohort Study reported an association between increased TTP and maternal consumption of fish caught from Lake Ontario associated with a high PCB index *(77)*. Overall, epidemiological studies on associations between the chlorinated hydrocarbons, DDT and PCBs, and increased TTP are weak and inconclusive.

However, animal studies have found an association between chlorinated hydrocarbons and reduced female fertility. PCBs may adversely affect folliculogenesis by decreasing circulating thyroid hormone, which is important for steroidogenesis in the ovary and steroid metabolism *(78)*. Organochlorine pesticides, such as methoxychlor and kepone, were shown in animal studies to cause infertility, anovulation, and abnormalities in steroidogenesis *(79)*. However, there have not been any studies investigating these pesticides in the TTP in women. TCDD has been shown to decrease fertility in rodents and nonhuman primates, but to date, there are no studies demonstrating an association of increased TTP and dioxin exposures in women or men *(73,79)*.

Work in an agricultural environment, which involves exposures to many factors in addition to pesticides, has been associated with an ovulatory or tubal cause of infertility and an increased TTP *(80,81)*. There are a number of small studies that have found associations between occupational exposures to pesticides and increased TTP, although individual pesticides were not identified in these studies *(82,83)*. Studies of TTP in women exposed to pesticides at lower levels of exposure are lacking as are studies of other types of pesticides such as organophosphates (parathion, malathion, and diazinon) and herbicides (atrazine and dimethyldithiocarbamate), which have been found to cause infertility in animal models *(79)*.

Exposure to the heavy metals, lead and mercury, has been associated with impaired fertility. In a case–control study of infertile couples in Hong Kong, women with unexplained infertility had average blood mercury concentrations 2-fold higher than their fertile counterparts (37 mmol/L vs. 17.5 mmol/L, $p < 0.001$) *(84)*. The elevated mercury levels were associated with seafood consumption.

4.2. Pregnancy Loss

Spontaneous abortion (SAB) is defined as loss of a pregnancy before 20 weeks of gestation. Most SABs occur in the first trimester and may occur before a woman even realizes she is pregnant. Approximately 15% of recognized pregnancies and up to 50% of all pregnancies end with SAB. Stillbirth is defined as loss of a pregnancy and fetal death after the 20th week of gestation and occurs in about one in 200 pregnancies. Although the causes of SAB and stillbirth are not entirely known, chromosomal abnormalities, anatomic uterine abnormalities, autoimmunity, infections, and low progesterone levels are all considered important causes.

Aneuploidy is an important cause of pregnancy loss accounting for 35% of SAB. It is also a significant cause of birth defects. In a small study of Japanese women, BPA, an estrogenic compound widely used in the production of polycarbonate plastics and epoxy resins, was associated with recurrent miscarriage *(85)*. BPA levels were more than three times higher in women with a history of recurrent miscarriage (2.59 ng/mL vs. 0.77 ng/mL, $p = 0.024$). Of the 13 aborted fetuses available for analysis, four were found to be aneuploid. Although this was a small preliminary study, the results are consistent with findings from animal studies. BPA has been shown in mice studies to cause an increased rate of aneuploidy in oocytes *(86)*. The doses the mice were exposed to in these studies were quite low and comparable to those measured in humans, providing biological plausibility for the ability of BPA to cause aneuploidy and miscarriages.

In nonhuman primates, exposure to PCBs and TCDD is associated with pregnancy loss *(87,88)*. However, in humans, the association between exposure to the chlorinated hydrocarbons, PCBs, dioxins and furans, and early pregnancy loss is weak and inconclusive *(88)*.

Women exposed to high levels of PCBs and PCDFs from contaminated cooking oil in Taiwan reported a 2.5-fold increased rate of stillbirth (4.2 vs. 1.7%, $p = 0.068$) but not SABs *(39)*. Likewise, exposure to PCBs through consumption of contaminated fish has not been associated with SAB *(89)*. Exposure to TCDD in Seveso, Italy, was not associated with an increased risk of SAB, although many of the most highly exposed women had not reached reproductive age at the time of the study *(90)*. Other retrospective studies of organochlorine exposures have found positive associations with SAB *(89)*. Many of these studies are limited by small numbers of subjects, incomplete exposure assessment, and are subject to confounding.

Studies of exposure to DDT and other organochlorine pesticides also have found conflicting results *(35,89,91)*. Historically high levels of DDE have been associated with pregnancy loss but lower concentrations are not consistently associated *(92)*. A preliminary study in Chinese women found that each ng/g serum increase in DDE was associated with an OR 1.13 (95% CI: 1.02–1.26) for SAB, but in this study,

DDE levels were measured 2 years after exposure *(93)*. A follow-up prospective study in Chinese women found a linear trend with preconception DDT levels and increasing odds of early pregnancy loss *(91)*. The relative odds of early pregnancy loss associated with a 10 ng/g lipid increase in serum total DDT was 1.17 (95% CI: 1.05–1.29). Another organochlorine pesticide, HCH, was found to be significantly higher in women with history of miscarriage than in women with no history *(53)*. Women exposed as children to grain contaminated with the fungicide, HCB, were found to have an increased rate of SAB *(94)*. HCB has been shown in nonhuman primate models to alter steroidogenesis *(95)*.

A large study of Ontario farm families found that SAB was associated with pesticide exposure. This study stratified results by preconception and postconception exposure with early ($<$12 weeks gestation) and late loss of pregnancy ($>$12 weeks) *(96)*. Preconception exposures (3 months before and up to month of conception) to the phenoxy-acetic and triazine herbicides were associated with early pregnancy loss. Late pregnancy loss was associated with a different type of herbicide, glyphosate, and another class of pesticides, thiocarbamates. Postconception exposures were generally associated with late SABs. The study interviewed both male and female partners and hypothesized that early pregnancy loss associated with preconception exposures could be male-mediated due to sperm chromosomal damage. Other studies have implicated pregnancy loss with paternal exposure to pesticides, such as Dibromochloro-propane DBCP, dithiocarbamates, and carbaryl *(97,98)*. Organophosphates have been associated with aneuploidy in sperm *(99)*.

Most studies of pesticides have small numbers of subjects and do not identify or quantify pesticide exposure but rather infer exposure from the living or working environment. A retrospective cohort study of farming households in the Philippines found an increased risk of SAB with pesticide users compared with organic farmers (RR = 6.2) *(100)*. A moderate increase in SAB was observed in both female workers and the wives of male workers in the floriculture industry in Colombia *(101)*. However, other studies have found no association [reviewed in *(102)*].

There is biological plausibility for the ability of many common pesticides to cause embryo loss. A murine model of preimplantation embryos exposed to common mixtures of pesticides at concentrations simulating exposures encountered by handling pesticides, inhaling drift, or ingesting contaminated groundwater found an increase in embryonic loss, reduced blastocyst development, and increased apoptosis *(103)*.

Heavy metals also have been associated with pregnancy loss. Both high and moderate (10–25 μg/dL) exposures to lead have been associated with SAB. In a prospective study in Mexico City, a dose–response relation between blood lead levels and risk of SAB was found *(104)*. The OR for SAB was 1.8 (95% CI: 1.1–3.1) for every 5 μg/dL increase in blood lead. In studies of rodents, lead suppresses FSH, affects gonadotropin–receptor binding in the ovary, and alters steroid metabolism *(79)*. Exposure to both inorganic and organic forms of mercury has been associated with SAB *(79)*. Exposure to high concentrations of another heavy metal, arsenic, in drinking water ($\geq$200 μg/L) during pregnancy increases the risk of stillbirth but not SAB *(105)*. Another study with 100 μg/L arsenic levels in drinking water found an association with both SAB and stillbirth *(106)*.

5. PERINATAL OUTCOMES

5.1. Preterm Delivery

Preterm delivery is one of the leading causes of perinatal mortality. Over the past two decades, preterm births have increased by 3.6% among blacks (from 15.5% in 1975 to 16.0% in 1995) and by 22.3% among whites (from 6.9% to 8.4%) *(107)*. Although extremes of maternal age and prenatal nutrition are known to be important factors, the causes of preterm delivery are largely unknown. Multiple gestations also commonly result in preterm delivery. Women who were born preterm are more likely to have preterm labor suggesting a genetic predisposition *(107)*. Environmental exposures to chemicals such as endocrine disruptors that interfere with hormonal signals may also be associated with preterm delivery.

Cigarette smoke, from both active smoking and secondhand smoke, is associated with preterm delivery and low birth weight *(108)*. It is hypothesized that these birth outcomes are due to exposures to the tobacco smoke constituents nicotine and carbon monoxide, resulting in vasoconstriction and reduced oxygen to the placenta, respectively.

Exposure to chlorinated hydrocarbons has not been consistently associated with preterm delivery. Exposures to high levels of PCBs and PCDFs in contaminated cooking oil in Taiwan was associated with high proportion of premature (<37 weeks) births (24.6 vs. 8.1%) *(89)*. However, in other studies of relatively high exposure to PCBs, there was no association with preterm delivery *(109)*. In studies of high exposures to TCDD, maternal but not paternal exposure was associated with preterm delivery *(90,110)*. Exposure in women to high levels of TCDD in Seveso, Italy, was associated with a 1.0–1.3 day nonsignificant decrease in gestational age and a 20–50% nonsignificant increase in the odds of preterm delivery *(90)*.

There is good evidence that exposure to historically high levels of DDT and the metabolite, DDE, is associated with preterm delivery *(35,111)*. In a study of pregnancy outcomes between 1959 and 1966, there was a linear trend for increased odds of delivery prior to 37 weeks with increasing maternal concentrations of DDE above 10 μg/L *(111)*. For serum concentrations above 60 μg/L, levels currently seen in countries using DDT, the adjusted OR was 3.1 (95% CI: 1.8–5.4). At levels currently seen in the USA where the average serum concentration of DDE is now less than 15 μg/L, the OR for preterm delivery was 1.5 (95% CI: 1.0–2.3).

There are limited studies on effects of other pesticide exposures on preterm delivery. Decreased length of gestation by several days has been found for organophosphates and the organochlorine pesticide, HCB *(112,113)*. However, these results were not clinically relevant because the study population had a very low rate of preterm delivery. The Ontario Farm Family Health Study found a weak association for paternal exposure to mixing or applying herbicides and preterm delivery (OR = 2.1, 95% CI: 1.0–4.4) *(98)*. However, other studies have not found an association between parental pesticide exposure and preterm labor *(100)*.

Heavy metal exposure has been associated with preterm delivery. Exposures to high levels of arsenic in drinking water (>100 μg/L) and lead were associated with preterm delivery *(79,106)*.

Phthalates are chemicals used as plasticizers in many consumer and medical products. Di-(2-ethylhexyl)phthalate (DEHP) is one of the more commonly used phthalates, with Mono(z-ethylhexyl)phthalate MEHP as its primary metabolite. In a preliminary study of gestational DEHP exposure, cord blood levels of MEHP were associated with an approximate 1 week shorter gestational age compared with MEHP-negative infants ($p =$ 0.033) *(24)*. Cord blood levels of MEHP were not associated with preterm delivery however, (<37 weeks) in this study. In addition to being anti-androgenic, DEHP is known to be pro-inflammatory and potentially could cause uterine inflammation, a known cause of preterm delivery *(24)*.

5.2. *Low Birth Weight/Small for Gestational Age (IUGR)*

Low birth weight is associated with greater perinatal morbidity and mortality as well as a number of adult conditions such as obesity, diabetes, hypertension, and cardiovascular disease. Although there are good data on the links between IUGR and adult disease, there are little data on the reproductive health of IUGR infants. IUGR has been associated with a smaller uterus and ovaries in adolescent girls *(62)*.

There are conflicting studies on the associations between chlorinated hydrocarbons and impaired fetal growth. Exposures to high levels of PCBs and PCDFs in contaminated cooking oil and other industrial exposures have been associated with low birth weight, but lower levels of environmental exposure have not shown consistent results *(89)*. Furthermore, in a study of pregnant women from the 1960s, maternal exposure to relatively high levels of PCBs was not associated with low birth weight *(109)*. Infants, especially boys, with high cord blood PCB levels tend to have lower birth weight and smaller head circumference *(89,114)*. Animal studies have also found low birth weight to be associated with PCBs and the oral reference dose for the commercial PCB mixture Aroclor 1016 is based on birth weight effects in primates *(109)*.

Low birth weight has been associated with consumption of PCB-contaminated fish in a number of studies although other studies have found no association [reviewed in *(89)*]. The differences in these fish consumption studies may be due to the omega-3 fatty acid content of the fish being consumed, which has been demonstrated to have a positive affect on birth weight *(89)*.

Exposure to relatively high levels of DDE (60 μg/L) in maternal blood was associated with a decrease in birth weight (OR = 2.6, 95% CI: 1.3–5.2) *(111)*. The present levels of DDE in the population are considerably lower and have not been associated with low birth weight *(89)*.

There are conflicting studies of the pregnancy outcomes of pesticide exposure and IUGR. The National Natality Survey and National Fetal Mortality Survey in the USA, found that the risk of SGA infant was somewhat elevated with maternal pesticide exposure at home, OR = 1.5 (95% CI: 1.1–2.0) [reviewed in *(115)*]. A recent study measuring maternal urinary and cord blood organophosphates and cholinesterase levels found no association with IUGR *(113)*. A preliminary study found that IUGR was associated with elevated levels of atrazine and other herbicides in drinking water in rural Iowa *(116)*. A small but significant decrease in birth weight was associated with maternal pyrethroid use in the first or second trimester *(115)*. Other studies have found no association with agricultural occupations and low birth weight *(102)*.

6. CONCLUSIONS

A number of synthetic chemicals have been identified as EDCs, including a variety of pesticides, chlorinated hydrocarbons (dioxins/furans and PCBs), plasticizers (BPA and phthalates), and heavy metals. Exposure to EDCs occurs throughout the human life cycle from preconception to postnatal life, during childhood and pubertal development, and into the adult reproductive years.

Although controversial, EDCs have been associated with increased abnormalities in menstrual cycles, endometriosis, hormonal changes, altered onset of puberty, premature menopause, reduced fecundity, preterm labor, and IUGR. Emerging evidence indicates that human body burdens of many EDCs are approaching or are similar to levels shown to cause effects in animal studies. Exposures during adulthood have been associated with adverse outcomes; however, the developing reproductive tract may be exquisitely vulnerable to EDCs. Associations between chemical exposures during development and during adulthood are summarized in Table 1.

Table 1
Chemical Exposures and Female Reproductive Outcomes

Exposure (sources)	*Developmental exposure*	*Adulthood exposure*
Bisphenol A (BPA): monomer used to make polycarbonate plastic and resins	Altered puberty onset (A) Abnormal reproductive organ development (A)	Oocyte chromosomal abnormalities (A) Recurrent miscarriage (H)
Chlorinated hydrocarbons: dioxins/furans, PCBs, some pesticides (DDT/DDE and HCH), and wood preservative (pentachlorophenol)	Reproductive tract malformations[a] (A) Menstrual irregularities[b] (H, A) Reduced fertility[c] (A) Hormonal changes (H, A) Altered puberty onset x(H, A) Low birth weight (H, A)	Menstrual irregularities[b] (H, A) Hormonal changes (H, A) Premature menopause (H) Reduced fertility[c] (H, A) Endometriosis (H, A) *(50,54,79)* Fetal loss[d] (H, A) Preterm delivery (H) *(35,91,117)*
Polybrominated biphenyls	Altered puberty onset (H)	
Polybrominated diphenyl ethers (PBDEs)	Altered puberty onset (A)	
Heavy metals: lead, mercury, manganese, and cadmium	Hormonal changes (A) Altered puberty onset (H)	Fetal loss[d] (H, A) *(79,117–119)* Reduced fertility[c] (H) Hormonal changes (A) *(79)* Preterm delivery (H)
Pesticides: broad category that includes many classes of insecticides, fungicides, herbicides, rodenticides, and fumigants	Altered puberty onset (A) Delayed time to pregnancy (H, A) Hormonal changes (A) Estrous cycle, (ovulatory) irregularities irregularities[b] (A)	Menstrual irregularities[b] (H) *(43)* Reduced fertility[c] (H, A) Fetal loss[d] (H, A) Reduced gestational age (H)

(*Continued*)

Table 1
(*Continued*)

Exposure (sources)	*Developmental exposure*	*Adulthood exposure*
Phthalates: plasticizers added to soften plastics like PVC, also found in cosmetics, toys, pharmaceuticals, and medical devices	Altered puberty onset (A)	Fetal loss[d] (A) Estrous cycle (ovulatory) irregularities[b] (A) Reduced fertility[c] (A) Reduced oocyte number (A) *(120)* Endometriosis (H) Preterm delivery (H)
DES	Reproductive tract malformations[a] (H,A) Altered hormone response (A) Menstrual irregularities[b] (H, A) Reduced fertility[c] (H, A) Uterine fibroids (H, A) Fetal loss (H, A)	
Cigarette smoke includes active and/or secondhand smoke	Reduced oocyte numbers (A)	Reduced fertility[c] (H) Miscarriage (H) Premature menopause (H) Hormonal changes (H) Menstrual irregularities[b] (H) Preterm delivery (H) Intrauterine growth restriction (H) Reduced oocyte number (H, A)

Adapted from Carlson (2005). *(121)*

A summary of epidemiological studies of endocrine-disrupting chemicals (EDCs) and relevant animal studies. A, evidence from animal studies; H, evidence from human studies; H, A, evidence from human and animal studies.

[a]Malformations of the reproductive tract: In males could include shortened anogenital distance in animals or hypospadias (humans), undescended testicles (cryptorchidism), small testicles (hypoplasia), and structural abnormalities of the epididymis. In females could include small ovaries, reduced number of follicles (eggs), and structural abnormalities of the oviducts, uterus, cervix, and/or vagina.

[b]Menstrual irregularities could include short or long menstrual cycles, missed periods, abnormal bleeding, and anovulation.

[c]Reduced fertility could include infertility and increased time to pregnancy (reduced fecundity).

[d]Fetal loss (typical in animal studies) is used as shorthand also for early pregnancy loss, miscarriage, or stillbirth (human).

Conflicting data are present for nearly all of these outcomes. Many studies are limited by small sample size, poor exposure assessment, differing methods of outcome assessment, recruitment, or other forms of bias. Additionally, the timing of exposure, exposures to mixtures or different congener profiles, and the genetic context of the exposure all influence the outcome of a study. Animal studies have provided support for many of these associations and have elucidated mechanisms of toxicity, adding some biological plausibility. However, further well-designed research studies are necessary to clarify these associations.

Although it is often impossible to apply results from epidemiological or animal studies to individual patient outcomes, clinicians can provide guidance on limiting exposures to many common EDCs and avoiding unnecessary risks. Chapter 13 of this book (Solomon and Janssen) specifically addresses how to communicate with patients and the general public about EDCs. In addition, Chapter 14 (Brenner and Galvez) discusses a community intervention to reduce exposures to EDCs.

REFERENCES

1. Gore A.C., J.J. Heindel, and R.T. Zoeller. Endocrine disruption for endocrinologists (and others). *Endocrinology* 2006, **147**(6 Suppl): S1–3.
2. McLachlan J.A., E. Simpson, and M. Martin. Endocrine disrupters and female reproductive health. *Best Pract Res Clin Endocrinol Metab* 2006, **20**(1): 63–75.
3. National Research Council Committee on Pesticides in the Diets of Infants and Children. *Pesticides in the Diets of Infants and Children.* 1993, Washington, D.C: National Academy Press.
4. Brackbill Y. and H.W. Berendes. Dangers of diethylstilboestrol: review of a 1953 paper. *Lancet* 1978, **8088**(2): 520.
5. Goldberg J.M. and T. Falcone. Effect of diethylstilbestrol on reproductive function. *Fertil Steril* 1999, **72**(1): 1–7.
6. Palmer J.R., et al. Infertility among women exposed prenatally to diethylstilbestrol. *Am J Epidemiol* 2001, **154**(4): 316–21.
7. Kaufman R.H., et al. Continued follow-up of pregnancy outcomes in diethylstilbestrol-exposed offspring. *Obstet Gynecol* 2000, **96**(4): 483–9.
8. Missmer S.A., et al. In utero exposures and the incidence of endometriosis. *Fertil Steril* 2004, **82**(6): 1501–8.
9. Baird D.D. and R. Newbold. Prenatal diethylstilbestrol (DES) exposure is associated with uterine leiomyoma development. *Reprod Toxicol* 2005, **20**(1): 81–4.
10. Newbold R.R., E. Padilla-Banks, and W.N. Jefferson. Adverse effects of the model environmental estrogen diethylstilbestrol (DES) are transmitted to subsequent generations. *Endocrinology* 2006, **147** (6 Suppl): S11–17.
11. Titus-Ernstoff L., et al. Menstrual and reproductive characteristics of women whose mothers were exposed in utero to diethylstilbestrol (DES). *Int J Epidemiol* 2006, **35**(4): 862–8.
12. Brevini T.A., S.B. Zanetto, and F. Cillo. Effects of endocrine disruptors on developmental and reproductive functions. *Curr Drug Targets Immune Endocr Metabol Disord* 2005, **5**(1): 1–10.
13. Whitehead S.A. and S. Rice. Endocrine-disrupting chemicals as modulators of sex steroid synthesis. *Best Pract Res Clin Endocrinol Metab* 2006, **20**(1): 45–61.
14. Zoeller R.T. Environmental chemicals as thyroid hormone analogues: new studies indicate that thyroid hormone receptors are targets of industrial chemicals. *Mol Cell Endocrinol* 2005, **242**(1–2): 10–5.
15. Younglai E.V., et al. Levels of environmental contaminants in human follicular fluid, serum, and seminal plasma of couples undergoing in vitro fertilization. *Arch Environ Contam Toxicol* 2002, **43**: 121–6.
16. Ikezuki Y., et al. Determination of bisphenol A concentrations in human biological fluids reveals significant early prenatal exposure. *Hum Reprod* 2002, **17**(11): 2839–41.
17. Arbuckle T.E., et al. 2,4-Dichlorophenoxyacetic acid residues in semen of Ontario farmers. *Reprod Toxicol* 1999, **13**(6): 421–9.
18. Foster W., et al. Detection of endocrine disrupting chemicals in samples of second trimester human amniotic fluid. *J Clin Endocrinol Metab* 2000, **85**(8): 2954–7.
19. Bradman A., et al. Measurement of pesticides and other toxicants in amniotic fluid as a potential biomarker of prenatal exposure: a validation study. *Environ Health Perspect* 2003, **111**: 1779–82.
20. Yamada H., et al. Maternal serum and amniotic fluid bisphenol A concentrations in the early second trimester. *Reprod Toxicol* 2002, **16**(6): 735–9.
21. Tsutsumi O. Assessment of human contamination of estrogenic endocrine-disrupting chemicals and their risk for human reproduction. *J Steroid Biochem Mol Biol* 2005, **93**(2–5): 325–30.

22. Mazdai A., et al. Polybrominated diphenyl ethers in maternal and fetal blood samples. *Environ Health Perspect* 2003, **111**: 1249–1252.
23. Dallaire F., et al. Time trends of persistent organic pollutants and heavy metals in umbilical cord blood of Inuit infants born in Nunavik (Quebec, Canada) between 1994 and 2001. *Environ Health Perspect* 2003, **111**(13): 1660–4.
24. Latini G., et al. In utero exposure to di-(2-ethylhexyl)phthalate and duration of human pregnancy. *Environ Health Perspect* 2003, **111**(14): 1783–5.
25. Houlihan J., et al. *BodyBurden: The Pollution in Newborns*. 2005, Washington D.C.: Environmental Working Group.
26. Schecter A., I. Kassis, and O. Papke. Partitioning of dioxins, dibenzofurans, and coplanar PCBS in blood, milk, adipose tissue, placenta and cord blood from five American women. *Chemosphere* 1998, **37**(9–12): 1817–23.
27. Butler Walker J., et al. Organochlorine levels in maternal and umbilical cord blood plasma in Arctic Canada. *Sci Total Environ* 2003, **302**(1–3): 27–52.
28. Rhainds M., et al. Lead, mercury, and organochlorine compound levels in cord blood in Quebec, Canada. *Arch Environ Health* 1999, **54**(1): 40–7.
29. Main K.M., et al. Human breast milk contamination with phthalates and alterations of endogenous reproductive hormones in infants three months of age. *Environ Health Perspect* 2006, **114**(2): 270–6.
30. Solomon G.M. and P.M. Weiss. Chemical contaminants in breast milk: time trends and regional variability. *Environ Health Perspect* 2002, **110**: A339–47.
31. Herman-Giddens M.E. Recent data on pubertal milestones in United States children: the secular trend toward earlier development. *Int J Androl* 2006, **29**(1): 241–6; discussion 286–90.
32. Den Hond E. and G. Schoeters. Endocrine disrupters and human puberty. *Int J Androl* 2006, **29**(1): 264–71.
33. Denham M., et al. Relationship of lead, mercury, mirex, dichlorodiphenyldichloroethylene, hexachlorobenzene, and polychlorinated biphenyls to timing of menarche among Akwesasne Mohawk girls. *Pediatrics* 2005, **115**(2): 127–34.
34. Warner M., et al. Serum dioxin concentrations and age at menarche. *Environ Health Perspect* 2004, **112**(13): 1289–92.
35. Rogan W.J. and A. Chen. Health risks and benefits of bis(4-chlorophenyl)-1,1,1-trichloroethane (DDT). *Lancet* 2005, **366**(9487): 763–73.
36. Windham G.C., et al. Cigarette smoking and effects on hormone function in premenopausal women. *Environ Health Perspect* 2005, **113**(10): 1285–90.
37. Matikainen T.M., et al. Ligand activation of the aromatic hydrocarbon receptor transcription factor drives Bax-dependent apoptosis in developing fetal ovarian germ cells. *Endocrinology* 2002, **143**(2): 615–20.
38. Matikainen T., et al. Aromatic hydrocarbon receptor-driven Bax gene expression is required for premature ovarian failure caused by biohazardous environmental chemicals. *Nat Genet* 2001, **28**(4): 355–60.
39. Yu M.L., et al. Menstruation and reproduction in women with polychlorinated biphenyl (PCB) poisoning: long-term follow-up interviews of the women from the Taiwan Yucheng cohort. *Int J Epidemiol* 2000, **29**(4): 672–7.
40. Yang C.Y., et al. The endocrine and reproductive function of the female Yucheng adolescents prenatally exposed to PCBs/PCDFs. *Chemosphere* 2005, **61**(3): 355–60.
41. Cooper G.S., et al. Polychlorinated biphenyls and menstrual cycle characteristics. *Epidemiology* 2005, **16**(2): 191–200.
42. Windham G.C., et al. Exposure to organochlorine compounds and effects on ovarian function. *Epidemiology* 2005, **16**(2): 182–90.
43. Farr S., et al. Pesticide use and menstrual cycle characteristics among premenopausal women in the Agricultural Health Study. *Am J Epidemiol* 2004, **160**(12): 1194–204.
44. Eskenazi B., et al. Serum dioxin concentrations and menstrual cycle characteristics. *Am J Epidemiol* 2002, **156**(4): 383–92.
45. Giudice L.C., Kao and L.C. Endometriosis. *Lancet* 2004, **364**(9447): 1789–99.
46. Rier S. and W.G. Foster. Environmental dioxins and endometriosis. *Semin Reprod Med* 2003, **21**(2): 145–54.
47. Missmer S.A., et al. Reproductive history and endometriosis among premenopausal women. *Obstet Gynecol* 2004, **104**(5 Pt 1): 965–74.

48. Mueller M.D., et al. 2,3,7,8-Tetrachlorodibenzo-p-dioxin increases glycodelin gene and protein expression in human endometrium. *J Clin Endocrinol Metab* 2005, **90**(8): 4809–15.
49. Tsukino H., et al. Associations between serum levels of selected organochlorine compounds and endometriosis in infertile Japanese women. *Environ Res* 2005, **99**(1): 118–25.
50. Louis G., et al. Environmental polychlorinated biphenyl exposure and risk of endometriosis. *Obstet Gynecol Surv* 2005, **60**(4): 243–4.
51. Heilier J.F., et al. Increased dioxin-like compounds in the serum of women with peritoneal endometriosis and deep endometriotic (adenomyotic) nodules. *Fertil Steril* 2005, **84**(2): 305–12.
52. Louis G.M., et al. Environmental PCB exposure and risk of endometriosis. *Hum Reprod* 2005, **20**(1): 279–85.
53. Gerhard I., et al. Chlorinated hydrocarbons in infertile women. *Environ Res* 1999, **80**(4): 299–310.
54. Eskenazi B., et al. Serum dioxin concentrations and endometriosis: a cohort study in Seveso, Italy. *Environ Health Perspect* 2002, **110**(7): 629–34.
55. Reddy B.S., et al. Association of phthalate esters with endometriosis in Indian women. *BJOG* 2006, **113**(5): 515–20.
56. Cobellis L., et al. High plasma concentrations of di-(2-ethylhexyl)-phthalate in women with endometriosis. *Hum Reprod* 2003, **18**(7): 1512–5.
57. Legro R.S., R. Azziz and L. Giudice. A twenty-first century research agenda for polycystic ovary syndrome. *Best Pract Res Clin Endocrinol Metab* 2006, **20**(2): 331–6.
58. Davies M.J. and R.J. Norman. Programming and reproductive functioning. *Trends Endocrinol Metab* 2002, **13**(9): 386–92.
59. Franks S., M.I. McCarthy, and K. Hardy. Development of polycystic ovary syndrome: involvement of genetic and environmental factors. *Int J Androl* 2006, **29**(1): 278–85.
60. Abbott D.H., D.A. Dumesic, and S. Franks. Developmental origin of polycystic ovary syndrome - a hypothesis. *J Endocrinol* 2002, **174**(1): 1–5.
61. Soderlund D., et al. No evidence of mutations in the P450 aromatase gene in patients with polycystic ovary syndrome. *Hum Reprod* 2005, **20**(4): 965–9.
62. Sharpe R.M. and S. Franks. Environment, lifestyle and infertility–an inter-generational issue. *Nat Cell Biol* 2002, **4**(Suppl): s33–40.
63. Cooper G.S., et al. Organochlorine exposure and age at natural menopause. *Epidemiology* 2002, **13**(6): 729–33.
64. Asselt K.M., et al. Current smoking at menopause rather than duration determines the onset of natural menopause. *Fertil Steril* 2004, **15**(5): 634–9.
65. Eskenazi B., et al. Serum dioxin concentrations and age at menopause. *Environ Health Perspect* 2005, **113**(7): 858–62.
66. Akkina J., et al. Age at natural menopause and exposure to organochlorine pesticides in Hispanic women. *J Toxicol Environ Health A* 2004, **67**(18): 1407–22.
67. Farr S., et al. Pesticide exposure and timing of menopause: the Agricultural Health Study. *Am J Epidemiol* 2006, **163**(8): 731–42.
68. Blanck H.M., et al. Age at menarche and tanner stage in girls exposed in utero and postnatally to polybrominated biphenyl. *Epidemiology* 2000, **11**(6): 641–47.
69. Skakkebaek N.E., et al. Is human fecundity declining? *Int J Androl* 2006, **29**(1): 2–11.
70. Chandra A., et al. Fertility, family planning, and reproductive health of U.S. women: data from the 2002 National Survey of Family Growth. *Vital Health Stat* 2005, **23**(25): 1–160.
71. Chandra A. and E.H. Stephen. Impaired fecundity in the United States: 1982–1995. *Fam Plann Perspect* 1998, **30**(1): 34–42.
72. Axmon A. and L. Hagmar. Time to pregnancy and pregnancy outcome. *Fertil Steril* 2005, **84**(4): 966–74.
73. Hruska K.S., et al. Environmental factors in infertility. *Clin Obstet Gynecol* 2000, **43**(4): 821–9.
74. Axmon A., et al. Time to pregnancy as a function of male and female serum concentrations of 2,2′4,4′5,5′-hexachlorobiphenyl (CB-153) and 1,1-dichloro-2,2-bis (p-chlorophenyl)-ethylene (p,p′-DDE). *Hum Reprod* 2006, **21**(3): 657–65.
75. Law D.C., et al. Maternal serum levels of polychlorinated biphenyls and 1,1-dichloro-2,2-bis(p-chlorophenyl)ethylene (DDE) and time to pregnancy. *Am J Epidemiol* 2005, **162**(6): 523–32.
76. Cohn B.A., et al. DDT and DDE exposure in mothers and time to pregnancy in daughters. *Lancet* 2003, 361: 2205–6.
77. Buck G.M., et al. Parental consumption of contaminated sport fish from Lake Ontario and predicted fecundability. *Epidemiology* 2000, **11**(4): 388–93.

78. Baldridge M.G., et al. Modulation of ovarian follicle maturation in Long-Evans rats exposed to polychlorinated biphenyls (PCBs) in-utero and lactationally. *Reprod Toxicol* 2003, **17**(5): 567–73.
79. Sharara F.I., D.B. Seifer, and J.A. Flaws. Environmental toxicants and female reproduction. *Fertil Steril* 1998, **70**(4): 613–22.
80. Fuortes L., et al. Association between female infertility and agricultural work history. *Am J Ind Med* 1997, **31**(4): 445–51.
81. Idrovo A.J., et al. Time to first pregnancy among women working in agricultural production. *Int Arch Occup Environ Health* 2005, **78**(6): 493–500.
82. Abell A., S. Juul, and J.P. Bonde. Time to pregnancy among female greenhouse workers. *Scand J Work Environ Health* 2000, **26**(2): 131–6.
83. Greenlee A.R., T.E. Arbuckle, and P.H. Chyou. Risk factors for female infertility in an agricultural region. *Epidemiology* 2003, **14**(4): 429–36.
84. Choy C.M.Y., et al. Infertility, blood mercury concentrations and dietary seafood consumption: a case–control study. *BJOG* 2002, **109**(10): 1121–5.
85. Sugiura-Ogasawara M., et al. Exposure to bisphenol A is associated with recurrent miscarriage. *Hum Reprod* 2005, **20**(8): 2325–9.
86. Hunt P.A., et al. Bisphenol A exposure causes meiotic aneuploidy in the female mouse. *Curr Biol* 2003, **13**(7): 546–53.
87. McNulty W.P. Toxicity and fetotoxicity of TCDD, TCDF and PCB isomers in rhesus macaques (Macaca mulatta). *Environ Health Perspect* 1985, 60: 77–88.
88. Toft G., et al. Fertility in four regions spanning large contrasts in serum levels of widespread persistent organochlorines: a cross-sectional study. *Environ Health* 2005, 4: 26.
89. Toft G., et al. Epidemiological evidence on reproductive effects of persistent organochlorines in humans. *Reprod Toxicol* 2004, **19**(1): 5–26.
90. Eskenazi B., et al. Maternal serum dioxin levels and birth outcomes in women of Seveso, Italy. *Environ Health Perspect* 2003, **111**(7): 947–53.
91. Venners S., et al. Preconception serum DDT and pregnancy loss: a prospective study using a biomarker of pregnancy. *Am J Epidemiol* 2005, **162**(8): 1–8.
92. Longnecker M.P., et al. Maternal serum level of the DDT metabolite DDE in relation to fetal loss in previous pregnancies. *Environ Res* 2005, **97**(2): 127–33.
93. Korrick S.A., et al. Association of DDT with spontaneous abortion: a case-control study. *Ann Epidemiol* 2001, **11**(7): 491–6.
94. Jarrell J., et al. Evaluation of reproductive outcomes in women inadvertently exposed to hexachlorobenzene in southeastern Turkey in the 1950s. *Reprod Toxicol* 1998, **12**(4): 469–76.
95. Foster W.G., et al. Alterations in circulating ovarian steroids in hexachlorobenzene-exposed monkeys. *Reprod Toxicol* 1995, **9**(6): 541–8.
96. Arbuckle T., Z. Lin, and L.S. Mery. An exploratory analysis of the effect of pesticide exposure on the risk of spontaneous abortion in an Ontario farm population. *Environ Health Perspect* 2001, **109**(8): 851–7.
97. Goldsmith J.R. Dibromochloro-propane: epidemiological findings and current questions. *Ann N Y Acad Sci* 1997, **837**: 300–6.
98. Savitz D.A., et al. Male pesticide exposure and pregnancy outcome. *Am J Epidemiol* 1997, **146**(12): 1025–36.
99. Recio R., et al. Organophosphorous pesticide exposure increases the frequency of sperm sex null aneuploidy. *Environ Health Perspect* 2001, **109**(12): 1237–40.
100. Crisostomo L. and V.V. Molina. Pregnancy outcomes among farming households of Nueva Ecija with conventional pesticide use versus integrated pest management. *Int J Occup Environ Health* 2002, **8**(3): 232–42.
101. Restrepo M., et al. Prevalence of adverse reproductive outcomes in a population occupationally exposed to pesticides in Colombia. *Scand J Work Environ Health* 1990, **16**(4): 232–8.
102. Nurminen T. Maternal pesticide exposure and pregnancy outcome. *J Occup Environ Med* 1995, **37**(8): 935–40.
103. Greenlee A.R., T.M. Ellis, and R.L. Berg. Low-dose agrochemicals and lawn-care pesticides induce developmental toxicity in murine preimplantation embryos. *Environ Health Perspect* 2004, **112**(6): 703–9.
104. Hertz-Picciotto I. The evidence that lead increases the risk for spontaneous abortion. *Am J Ind Med* 2000, **38**(3): 300–9.

105. von Ehrenstein O.S., et al. Pregnancy outcomes, infant mortality, and arsenic in drinking water in West Bengal, India. *Am J Epidemiol* 2006, **163**(7): 662–9.
106. Ahmad S.A., et al. Arsenic in drinking water and pregnancy outcomes. *Environ Health Perspect* 2001, **109**(6): 629–31.
107. Ananth C.V., et al. Rates of preterm delivery among Black women and White women in the United States over two decades: an age-period-cohort analysis. *Am J Epidemiol* 2001, **154**(7): 657–65.
108. Kharrazi M., et al. Environmental tobacco smoke and pregnancy outcome. *Epidemiology* 2004, **15**(6): 660–70.
109. Longnecker M.P., et al. Maternal levels of polychlorinated biphenyls in relation to preterm and small-for-gestational-age birth. *Epidemiology* 2005, **16**(5): 641–7.
110. Lawson C.C., et al. Paternal occupational exposure to 2,3,7,8-tetrachlorodibenzo-p-dioxin and birth outcomes of offspring: birth weight, preterm delivery, and birth defects. *Environ Health Perspect* 2004, **112**(14): 1403–8.
111. Longnecker M.P., et al. Association between maternal serum concentration of the DDT metabolite DDE and preterm and small-for-gestational-age babies at birth. *Lancet* 2001, **358**(9276): 110–4.
112. Fenster L., et al. Association of *in utero* organochlorine pesticide exposure and fetal growth and length of gestation in an agricultural population. *Environ Health Perspect* 2006, **114**(4): 597–602.
113. Eskenazi B., et al. Association of in Utero organophosphate pesticide exposure and fetal growth and length of gestaion in an agricultural population. *Environ Health Perspect* 2004, **112**: 1116–24.
114. Hertz-Picciotto I., et al. In utero polychlorinated biphenyl exposures in relation to fetal and early childhood growth. *Epidemiology* 2005, **16**(5): 648–56.
115. Hanke W., et al. The use of pesticides in a Polish rural population and its effect on birth weight. *Int Arch Occup Environ Health* 2003, **76**(8): 614–20.
116. Munger R., et al. Intrauterine growth retardation in Iowa communities with herbicide-contaminated drinking water supplies. *Environ Health Perspect* 1997, **105**(3): 308–14.
117. Miller K.P., et al. In utero effects of chemicals on reproductive tissues in females. *Toxicol Appl Pharmacol* 2004, **198**(2): 111–31.
118. Thomas MU and Thomas DA. *Casarett and Doull's Toxicology: The Basic Science of Poisons*. 6 ed, ed. C.D. Klaassen, 2001, New York: McGraw Hill.
119. Winker R. and H.W. Rudiger. Reproductive toxicology in occupational settings: an update. *Int Arch Occup Environ Health* 2006, **79**(1): 1–10.
120. Lovekamp-Swan T. and B.J. Davis. Mechanisms of phthalate ester toxicity in the female reproductive system. *Environ Health Perspect* 2003, **111**(2): 139–45.
121. Carlson, A., et al. Challenged concepts: Environmental chemicals and Fertility. Vallombrosa Consensus Statement on Enviromental Contaminants and Human Fertility Compromise. October 2005 available on-line at: www.health and environment.org.

10 Epidemiologic Evidence on the Relationship Between Environmental Endocrine Disruptors and Male Reproductive and Developmental Health

Russ Hauser, MD, ScD, MPH,
Julia S. Barthold, MD,
and John D. Meeker, MS, ScD

CONTENTS

1. INTRODUCTION

There is scientific, governmental, and public concern over the potential adverse human health risks of exposure to environmental endocrine-disrupting chemicals (eEDCs). Human health endpoints of concern include, among others, (i) disrupted reproductive function, manifest as infertility or early pregnancy loss, (ii) altered fetal development, manifest as urogenital tract abnormalities in male newborns, including hypospadias and cryptorchidism, (iii) altered thyroid function, and (iv) increased risk of reproductive cancers including breast in women and testicular and prostate in men. Other endpoints that have also been studied include increased risk of endometriosis and altered birth sex ratios.

Given the wide range of potential health endpoints, it would not be possible to synthesize the existing epidemiologic evidence in a single chapter. Therefore, we made the decision to focus on a subset of literature, specifically male reproductive and developmental health endpoints, for which there is some human epidemiologic

From: *Endocrine-Disrupting Chemicals: From Basic Research to Clinical Practice*
Edited by: A. C. Gore © Humana Press Inc., Totowa, NJ

evidence. In addition, because it has been hypothesized that several of the male reproductive and developmental endpoints may be linked through a common casual pathway, testicular dysgenesis syndrome (TDS) *(1)*, which may include exposure to environmental endocrine-disrupting chemicals (eEDCs), we welcomed the opportunity to explore their relationship with eEDCs in depth.

Our understanding of the mechanisms through which environmental chemicals alter the endocrine system is derived largely from experimental studies in laboratory animals and in vitro systems. In observational epidemiologic studies, it is generally not possible to explore potential mechanisms of the eEDCs. Nevertheless, epidemiologic studies are essential to our understanding of the potential risks, or lack thereof, of eEDCs on human reproductive function and development. Because this chapter is a synthesis of the epidemiologic literature on eEDCs and male reproductive function and development, experimental data on mechanisms of eEDCs are only briefly discussed. We refer readers to Chapter 3 of this book by Cowin, Foster, and Risbridger for further discussion of the basic biology of effects of eEDCs on male reproductive health. Environmental eEDCs can alter male reproductive function and development through multiple mechanisms. Two relevant mechanisms include anti-androgenic and estrogenic effects. Anti-androgenic chemicals can act directly as androgen receptor (AR) antagonists. This includes *p,p′*-DDE, the primary metabolite of DDT *(2)*. Chemicals can also act as anti-androgens without directly interacting with the AR. For example, some phthalates can inhibit fetal testosterone synthesis by the fetal rat Leydig cells *(3)*.

Chemicals that are estrogenic can, at high doses, alter male development by down-regulating ARs *(4)*. This may lower androgen action and elevate estrogen action simultaneously, leading to an altered androgen–estrogen balance that may then alter reproductive tract development *(5)*. Evidence for this has been seen in experimental studies with diethylstilbestrol (DES), a potent synthetic estrogen. However, weaker estrogens, such as bisphenol A (BPA), at environmentally relevant levels have not been shown to induce similar effects following neonatal exposure *(5)*. These weaker estrogens at lower doses had positive effects on the rat testis and accelerated normal onset of pubertal spermatogenesis *(6)*.

Polychlorinated biphenyls (PCBs) can alter endocrine function through estrogenic agonist or antagonist activities that may be at the level of the estrogen receptor or may influence biological effects of estradiol by altering its metabolism. For instance, PCB metabolites (OH-PCBs) may alter metabolism of endogenous estrogens by inhibiting sulfotransferases (SULT) *(7)*. Furthermore, dioxin-like PCBs, as well as polychlorinated dibenzo-*p*-dioxins (PCDDs) and polychlorinated dibenzofurans (PCDFs), may induce activation of AhR, which induces the expression of CYP450 genes that are involved in estrogen metabolism *(8)*. Studies have shown that 2,3,7,8-tetrachlorodibenzo-*p*-dioxin (TCDD) exposure in rats reduces plasma testosterone and dihydrotestosterone (DHT) levels by inhibiting steroidogenesis and the mobilization of cholesterol to CYP450scc (a mitochondrial enzyme that converts cholesterol into pregnenolone) *(9)*. TCDD has also been shown to have anti-estrogenic effects by down-regulating ER *(10)*.

Because androgens and estrogens play key roles in regulating reproduction and development, it is biologically plausible that environmental EDCs that possess anti-androgenic activity or estrogenic activity (either directly at the level of the receptor or indirectly through steroid synthesis or metabolism) may influence the human reproductive system. However, despite the experimental data on the endocrine-disrupting mechanisms of eEDCs, human evidence of altered male reproductive and developmental health in relation to eEDCs is limited.

Furthermore, although well appreciated by the scientific community, the potential effect on human health from exposure to mixtures of eEDCs is not well studied. As has been shown in the recent Third Report by CDC *(11)*, humans are exposed to at least hundreds of environmental chemicals, of which dozens are known EDCs. A major limitation of epidemiologic studies is that they generally only measure human exposure to a single EDC or at best to a set of isomers or congeners within a family of EDCs. A fuller understanding of potential human health risks will require studying the complex mixtures to which we are exposed. This limitation should be kept at the forefront as we discuss the current epidemiologic evidence on health risks from eEDCs.

2. MALE REPRODUCTIVE FUNCTION AND DEVELOPMENT

Some epidemiologic studies suggest that human semen quality has declined during the previous 50 years *(12–14)*. Further evidence for this comes from a recent study among healthy young men that found an unexpectedly high proportion of poor semen quality *(15)*. However, there are other epidemiologic studies that have not reported a decline in semen quality *(16–18)*. Epidemiologic studies also suggest that there are temporal upward trends in the rates of testicular germ cell cancers *(19–21)* and male urogenital tract abnormalities *(22)*, such as cryptorchidism and hypospadias.

Some scientists hypothesize that environmental exposures to low levels of EDCs may be partially responsible for these trends. Skakkebaek et al. *(1)* proposed that the trends in semen quality, testicular germ cell cancer, and male urogenital tract anomalies may share a common causal pathway and defined this triad as the TDS. The hypothesis invokes a common pathway through which EDCs, and other environmental chemicals and genetic factors, may lead to abnormal development of the fetal testis, producing testicular dysgenesis. This in turn can manifest as an increased risk of urogenital abnormalities in newborn males, as well as altered semen quality and testicular germ cell cancer in young men.

Despite the concern raised by the trends in male reproductive and developmental health endpoints, the epidemiologic data on their potential relationships with exposure to eEDCs remain limited and generally inconclusive. There are epidemiologic studies on the putative relationship between semen quality and adult exposure to three classes of eEDCs: PCBs, pesticides (persistent and non-persistent), and phthalates. In addition, there are also epidemiologic studies of the relationship between pesticides and phthalates and the prevalence of hypospadias and/or cryptorchidism in newborn males. Finally, a provocative recent study on the relationship between reduced anogenital distance among infant boys and exposure to phthalates will be discussed as a potential link between human exposure and alterations in urogenital tract development.

3. SEMEN QUALITY AND eEDCs

Although semen quality is measured in the adult male, it may be affected by EDC exposure during earlier life stages, such as during gestation or peri-pubertal development. In addition, as recently shown by Anway et al. *(23)*, there are transgenerational effects of anti-androgenic chemicals, whereby exposure of the maternal or paternal (or even the grandparents) gametes to chemicals may confer an increased risk of altered semen quality in the offspring. Although evidence from studies in laboratory animals and human studies on prenatal exposure of men to DES supports early life exposure effects on spermatogenesis, the epidemiologic evidence on the relationship between semen quality and exposure to eEDCs is limited to the assessment of adult exposure to eEDCs. The explanation for this is simply that studies on adult exposures are more straightforward to design and implement.

We specifically focus on epidemiologic studies that explored the relationship of semen quantity and quality with PCBs, DDT, non-persistent pesticides, and phthalates. The majority of the epidemiologic studies described are cross-sectional designs in which exposure and semen parameters were assessed at the same time, in the adult. One of the limitations of a cross-sectional design is that the onset and timing of exposure in relation to outcome is not always clear.

In the majority of studies, subject participation rates were generally very low, raising concern with selection bias *(24)*. Recently, we showed that selection bias in the setting of an infertility clinic was unlikely to be important *(25)*. Finally, most studies collected a single semen sample. Because of the known within individual variability in semen parameters, there may be misclassification of semen quality, which would generally tend to bias associations, if present, to the null hypothesis.

3.1. Phthalates and Semen Quality

The diesters of 1,2-benzenedicarboxylic acid (phthalic acid), commonly known as phthalates, are a group of man-made chemicals widely used in industrial applications. High-molecular weight phthalates [e.g.,di(2-ethylhexyl) phthalate (DEHP)], are primarily used as plasticizers in the manufacture of flexible vinyl plastic which, in turn, is used in consumer products, flooring and wall coverings, food contact applications, and medical devices *(26–28)*. Manufacturers use low-molecular weight phthalates [e.g., diethyl phthalate (DEP) and dibutyl phthalate (DBP)] in personal-care products (e.g., perfumes, lotions, cosmetics), as solvents and plasticizers for cellulose acetate, and in making lacquers, varnishes, and coatings, including those used to provide timed releases in some pharmaceuticals *(26,29,30)*.

As a result of the ubiquitous use of phthalates in personal-care and consumer products, human exposure is widespread. Exposure through ingestion, inhalation, and dermal contact are considered important routes of exposure for the general population *(27–32)*. Parenteral exposure from medical devices and products containing phthalates are important sources of high exposure to phthalates, primarily DEHP *(28,33)*. Upon exposure, phthalates are rapidly metabolized and excreted in urine and feces *(27–30)*. The most common biomonitoring approach for investigating human exposure to phthalates is the measurement of urinary concentrations of phthalate metabolites.

As compared with the laboratory animal data on the reproductive toxicity of phthalates, the human data are limited (Table 1). In an early study on phthalates and semen quality, Murature and coworkers *(34)* measured DBP concentrations in the cellular fractions of ejaculates from 21 university students and found an inverse relationship with sperm concentration. The study was small and did not adjust for potential confounders. In another small study, conducted in India, Rozati and coworkers *(35)* studied 21 infertile men with poor semen quality and 32 "control" men with normal semen parameters. Phthalate esters were measured in seminal plasma, and the results

Table 1
Summary Table of Epidemiologic Studies (in Chronological Order) on the Relationship Between Phthalates and Semen Quality

Author (country)	*Study population*	*Exposure*	*Results*	*Comments*
Murature et al., 1987 *(34)* (USA)	21 young men	DBP in cellular fractions of ejaculates	In men with "low ability to metabolize DBP," inverse relationship between sperm concentration and DBP ($r = -0.4$; slope of regression was −0.7). In the "men with a greater ability to metabolize DBP," there was also an inverse correlation of −0.4 (slope of regression −0.6) between DBP and sperm concentration	Small sample size, no adjustment for confounders, and measured diester
Rozati et al., 2002 *(35)* (India)	53 men (21 infertile and 32 controls)	Seminal plasma levels of phthalates (DBP, BBzP, DEHP, and DnOP)	Sum of phthalates was inversely correlated with sperm morphology ($r = -0.77$, $p < 0.001$) and positively correlated with the percentage of single-stranded DNA in sperm ($r = 0.86$, $p < 0.001$) assessed with the sperm nuclear chromatin condensation test. The concentration of phthalates was not correlated with ejaculate volume, sperm concentration, or motility	Measured total phthalate diesters concern with contamination
Duty et al., 2003 *(36)* (USA)	168 men from an infertility clinic (semen parameters)	Urinary levels of phthalate metabolites (MBP, MBzP, MEP, MEHP, and MMP)	Dose–response relationships (after adjusting for age, abstinence time, and smoking status) between MBP and sperm motility (OR per tertile = 1.0, 1.8, and 3.0; p for trend = 0.02) and sperm concentration (OR per tertile = 1.0, 1.4, and 5.5; p for trend = 0.07). Dose–response relationship between MBzP and sperm concentration (OR per tertile = 1.0, 1.4, and 5.5; p for trend = 0.02)	Confounders considered: age, BMI, abstinence time, smoking status, and race
Jonsson et al., 2005 *(37)* (Sweden)	234 young men	Urinary levels of MEP, MEHP, MBzP, MBP, and phthalic acid	No relationships of MBP, MBzP, or MEHP with any of the semen parameters. The highest quartile for MEP had fewer motile sperm (mean difference was 8.8%, 95% CI = 0.8, 17) and more immotile sperm (8.9%, 95% CI = 0.3–18). Phthalic acid was associated with improved function as measured by more motile sperm and fewer immotile sperm	Confounders considered: abstinence time and smoking status

BBzP, butylbenzyl phthalate; DBP, dibutyl phthalate; DEHP, di(2-ethylhexyl) phthalate; DEP, diethyl phthalate; DiNP, di-isononyl phthalate; DnOP, di-*n*-octyl phthalate; MBP, monobutyl phthalate; MBzP, monobenzyl phthalate; MEHP, monoethylhexyl phthalate; and MEP, monoethyl phthalate.

were reported as the sum of a mixture of dimethyl phthalate, DEP, DBP, butylbenzyl phthalate (BBzP), DEHP, and di-*n*-octyl phthalate. The concentration of phthalates was inversely correlated with sperm morphology but not correlated with ejaculate volume, sperm concentration, or motility. In this study, as in the Murature study, the measurement of phthalate diesters raises concern with sample contamination from the ubiquitous presence of the diester in the environment.

More recently, a larger study using urinary levels of phthalate metabolites was conducted by Duty and colleagues *(36)*. Study subjects consisted of male partners of subfertile couples who presented to an infertility clinic in Massachusetts, USA. At the time of the clinic visit, one sample of semen and of urine were collected from 168 men. There were dose–response relationships (after adjusting for age, abstinence time, and smoking status) between monobutyl phthalate (MBP, the hydrolytic metabolite of DBP) and below World Health Organization (WHO) reference value sperm motility and sperm concentration *(36)*. There was also a dose–response relationship between monobenzyl phthalate (MBzP, the primary hydrolytic metabolite of BBzP) and below WHO reference value sperm concentration.

In a recently published study from Sweden, Jonsson and colleagues *(37)* recruited 234 young Swedish men at the time of their medical conscript examination. Each man provided a single urine sample used to measure concentrations of MEP, MEHP, MBzP, MBP and phthalic acid.

In contrast to the US study, in the Swedish study there were no relationships of MBP or MBzP with any of the semen parameters. MEHP was also not associated with any of the semen parameters. Men in the highest quartile for MEP had fewer motile sperm and more immotile sperm than men in the lowest MEP quartile. Contrary to their hypothesis, phthalic acid was associated with improved function as measured by more motile sperm and fewer immotile sperm. Phthalic acid is a non-specific marker of phthalate exposure, formed as the result of the hydrolysis of any of the phthalates measured. Interactions between urinary phthalate levels and PCB 153 (measured previously in serum samples from these men) were assessed by including an interaction term in the models. There was no evidence of multiplicative interactions between PCB 153 and any of the phthalates with the reproductive markers (data was not shown). This is in contrast to a previous study by Hauser et al. *(38)*, who reported interactions of MBP and MBzP with congener PCB 153 in relation to sperm motility.

Although the Swedish study had some similarities to the US study, as they were both cross-sectional studies in which a single urine and semen sample were collected, there were many important differences. One of the primary differences was in the age of the study population and the method of recruitment. The Swedish study population consisted of young men (median age 18 years, range 18–21 years) undergoing a medical examination before military service. As approximately 95% of young men in Sweden undergo the conscript examination, these young men reflected the general population of young Swedish males. In contrast, in the US study, the median age of the men recruited from an infertility clinic was 35.5 years and ranged from 22 to 54 years. None of the men from the infertility clinic were 21 years of age or younger. The differences across studies in the ages and source of the men may account for some of the differences in the results between studies. For instance, it is unclear whether men presenting to an infertility clinic are more "susceptible" to reproductive toxicants, including phthalates, than men from the general population. Furthermore, it is also unclear whether middle-aged men, as compared with young men, are more "susceptible" to reproductive toxicants because of an age-related response to the toxicant.

Other differences across studies include major differences in participation rates (14% in the Swedish study and 65% in the US study). This is unlikely to introduce selection bias, because the Swedish young men would have to participate differentially in relation to both reproductive function and phthalate levels, which is unlikely as they did not know their phthalate exposure or semen quality status. Another difference across studies is differences in the analytical methods used to measure urinary phthalate metabolites. The higher limits of detection and lower analytical precision in the Swedish study may have contributed to measurement error of urinary phthalate levels and may result in bias to the null hypothesis. However, by categorizing the phthalate levels into quartiles for the statistical analysis, this would minimize concern with measurement error resulting from the analytical imprecision and low detection limits.

In summary, the epidemiologic data on the relationship between semen quality and phthalate exposure remains limited and inconsistent. Although the two recent studies by Duty et al. *(36)* and Jonsson et al. *(37)* had some similarities, important differences existed. Additional studies are critically needed to help elucidate possible explanations for differences across studies and most importantly to address whether phthalate exposure alters semen quality.

3.1.1. Semen Quality and Environmental PCB Exposure

PCBs are a class of synthetic, persistent, lipophilic, halogenated aromatic compounds that were widely used in industrial and consumer products for decades before their production was banned in the late 1970s. PCBs were used in cutting oils, lubricants, and as electrical insulators. As a result of their extensive use and persistence, PCBs remain ubiquitous environmental contaminants. They are distributed worldwide and have been measured in air, water, aquatic, and marine sediments, fish, and wildlife *(39)*. Furthermore, they are biologically concentrated and stored in human adipose tissue. The general population is exposed primarily through ingestion of contaminated foods (e.g., fish, meat, and dairy products), as PCBs can bioaccumulate up the food chain. However, exposure may also occur through dermal contact (soil and house dust) and inhalation (indoor air in residential buildings and workplaces, as well as outdoor air). For example, in the 1960s and 1970s, PCBs were used in sealants for commercial building construction, and high levels of PCBs (up to 36, 000 ppm) have been found to remain in the caulking of some public buildings, which may lead to contamination of indoor air and dust *(40)*. As a result of their persistence and ubiquity, measurable levels of serum PCBs are found in the majority of the US general population *(41)*. Serum levels of PCBs are an integrated measure of internal dose, reflecting exposure from all sources over the previous years; depending on the congener, the half-life of PCBs in the blood ranges from 1 to 10 or more years *(42,43)*.

In one of the early studies on environmental exposure to PCBs and semen quality, Bush and coworkers *(44)* collected semen samples from fertile men ($n = 33$), men with oligozoospermia ($n = 50$) or azoospermia ($n = 50$) and men status postvasectomy ($n = 25$) (Table 2). The seminal plasma levels of the PCB congeners 153, 138, and 118 were inversely related to sperm motility but only among semen samples with a sperm count <20 million/ml. Because an association was only found among a subset of the subjects, caution should be used to interpret these results.

Table 2
Summary Table of Epidemiologic Studies (in Chronological Order) on the Relationship of Polychlorinated Biphenyls and *p,p*-DDE with Semen Quality

Author	*Study population*	*Exposure*	*Results*	*Comments*
Bush et al., 1986 *(44)*	33 fertile, 50 subfertile, 50 infertile, and 25 post-vasectomy men	Seminal plasma levels of PCBs and p,p'-DDE	PCB 153, 138, and 118 were inversely related to sperm motility only among samples with a sperm count <20 million/ml. No associations of semen parameters with p,p'-DDE.	Association found only among a subset of men
Rozati et al., 2002 *(35)*	53 men from India (21 infertile and 32 controls)	Seminal plasma levels of PCBs	PCBs detected in the seminal plasma of infertile men but not controls. Negative correlation between PCBs and total progressive motility ($r = -0.5$) and positive correlation with percentage of single-stranded DNA in sperm ($r = 0.6$). No correlations with sperm count, rapid progressive motility, or normal morphology	Data on individual PCB congeners not presented. No statistical adjustment for potential confounders
Dallinga et al., 2002 *(45)*	65 Dutch men from an infertility clinic	Serum and semen levels of PCB 118, 138, 153, 180, and their metabolites	Seminal plasma PCB levels among men with good semen quality were higher than among men with poor semen quality (0.071 ng/ml and 0.022 ng/ml, respectively, $p = 0.06$). In men with good semen quality, there were inverse associations between serum levels of sum of PCB metabolites and sperm count ($p = 0.04$) and progressive motile sperm concentration ($p = 0.02$). There were also negative non-significant corresponding associations in men with poor semen quality	Confounders considered: age and smoking status. Measured PCB metabolites
Richthoff et al., 2003 *(46)*	305 Swedish young men	Serum levels of PCB 153	Inverse association between PCB 153 and percent motile sperm (10 ng/g lipid increase in PCB 153 associated with a 1.0% decline in percent CASA motile sperm (95% CI = −2.0 to −0.13)]. No association of PCB 153 with sperm concentration	Confounders considered: BMI, abstinence period, and smoking status
Hauser et al., 2003 *(48)*	212 US men from an infertility clinic	Serum levels of PCBs and p,p'-DDE	Dose–response relationships (odds ratio per tertile adjusted for age, abstinence time, and smoking status) between PCB 138 and below reference sperm motility (1.00, 1.68, and 2.35, respectively; p-value for trend is 0.04) and sperm morphology (1.00, 1.36, and 2.53; $p = 0.04$). DDE had a non-significant association with sperm motility	Confounders considered: BMI, age, abstinence period, and smoking status

Rignell-Hydbom et al., 2004 *(47)*	195 Swedish fishermen	Serum levels of PCB 153 and p,p'-DDE	The highest PCB 153 quintile had decreased sperm motility as compared with men in the lowest quintile. The age adjusted mean difference was 9.9% (95% CI = −1.0 to − 21%, $p = 0.08$). No significant associations of p,p'-DDE with semen parameters	Confounders considered: age, smoking status, abstinence time, BMI, and reproductive hormones
High-exposure studies				
Guo et al., 2000 *(50)*	35 young men from Taiwan (12 prenatally exposed to contaminated rice oil and 23 unexposed men)	Maternal ingestion (yes/no) of rice oil contaminated with PCBs and PCDFs	Increased abnormal morphology in exposed men (37.5%) as compared with unexposed men (25.9%). Exposed men had decreased percentage of motile sperm (35.1% compared with 57.1% in unexposed men) and rapidly motile sperm (25.5% compared with 42.4% in unexposed men). Reduced hamster oocyte penetration in exposed men	Age and percent smokers in exposed and unexposed groups were similar. No statistical adjustment for confounders
Hsu et al., 2003 *(51)*	68 men from Taiwan (40 exposed to contaminated rice oil and 28 unexposed)	Ingestion (yes/no) of rice oil contaminated with PCBs and PCDFs	Exposed men had higher percentage of sperm with abnormal morphology (27.5%) compared with unexposed men (23.3%) and a higher oligospermia rate (9% compared with 1%, respectively). Ability of sperm to penetrate the hamster oocyte was reduced in exposed men	Age and percent smokers in exposed and unexposed groups were similar. No statistical adjustment for confounders

BMI, body mass index; PCBs, polychlorinated biphenyls; and PCDFs, polychlorinated dibenzofurans.

In a more recent study in the Netherlands, Dallinga and colleagues *(45)* studied men who were partners in couples visiting an infertility treatment center. Based on progressive motile sperm concentration, they identified a group of men with good semen quality ($n = 31$) and a group of men with very poor semen quality ($n = 34$). Blood and semen were analyzed for PCB 118, 138, 153, and 180 and their hydroxylated metabolites. Contrary to expectations, the sum of PCBs in seminal plasma of men with good semen quality were higher than among men with poor semen quality ($p = 0.06$). However, within the group of men with good semen quality, there were inverse associations between serum levels of sum of PCB metabolites and sperm count ($p = 0.04$) and progressive motile sperm concentration ($p = 0.02$). There were also inverse non-significant corresponding associations in the men with poor semen quality. Because associations with semen quality were found for PCB metabolites and not the parent PCBs, these results suggested that the PCB metabolites were the biologically active compounds.

In Sweden, Richthoff and coworkers *(46)* studied 305 young men 18–21 years of age undergoing a conscript examination for military service. There were significant inverse associations between PCB 153, a good biomarker of exposure to total PCBs, and percent motile sperm. There were no associations between PCB 153 and sperm concentration or total sperm count. Although the participation rate was very low, only 13.5% of eligible subjects agreed to participate, it is unlikely that this would introduce bias as young men are likely to be unaware of their fertility or exposure levels.

In India, Rozati and coworkers *(35)* studied infertile men and controls (study details are provided Section 3.1.). They reported a negative correlation between seminal plasma PCB levels and total progressive motility and a positive correlation with percentage of single-stranded DNA in sperm. No correlations were found between PCBs and sperm count, rapid progressive motility, or normal morphology. The authors reported results for total PCBs and not for individual congeners. Potential confounders were considered in the methods section, but no adjustments were made in the analysis.

Rignell-Hydbom et al. *(47)* studied 195 Swedish fishermen (median age 50.6 years, ranged from 24–65 years) from the east and west coasts. The highest PCB 153 quintile had decreased sperm motility compared with men in the lowest quintile. There were no consistent associations of PCB 153 with sperm concentration. Although p,p'-DDE was inversely associated with sperm motility, when age was included in the models, the association became weaker and non-significant.

In the US, Hauser and colleagues *(48)* studied 212 male (mean age was 36.0 years) partners of subfertile couples visiting an infertility clinic. Fifty-seven PCB congeners including PCB 118, 138, 153, and 180 were measured. There were significant dose–response relationships (increasing odds ratios with increasing PCB tertile adjusted for age, abstinence time, and smoking status) between PCB 138 and below WHO reference sperm motility and sperm morphology. Associations between semen parameters and PCB 153 were not consistent. p,p'-DDE showed a weak non-significant relationship with sperm motility.

Ayotte and coworkers *(49)* reported on the association between p,p'-DDE, a major biologically persistent metabolite of DDT, and semen quality in 24 young men from Chiapas, Mexico. The men, 16–28 years of age, were non-occupationally exposed to DDT. The mean concentration of p,p'-DDE was several hundred fold higher than levels in men from other countries, such as the US and Canada, where DDT was not recently used. p,p'-DDE was inversely correlated with both semen volume and sperm count. Although the study was small and did not control for potential confounders, the results are intriguing and worthy of replication in other cohorts.

3.1.2. Semen Quality and High PCB Exposure

Guo and colleagues *(50)* studied the relationship between semen quality and prenatal exposure to PCBs and PCDFs after the poisoning episode in Taiwan in 1979 in which PCB-contaminated rice oil was ingested. Twelve men with prenatal exposure to contaminated rice oil and 23 healthy unexposed subjects of comparable age provided a semen sample. The unexposed men had no unusual chemical exposure and were recruited from a local high school. The mean (SD) age of the exposed men was 17.3 (1.2) years and 17.6 (1.0) for the unexposed men. The proportion of sperm with abnormal morphology was increased in the exposed men. In the exposed men, the percentage of motile sperm and rapidly motile sperm were reduced. In addition, sperm from exposed men had reduced hamster oocyte penetration as compared with unexposed men. This small study provided the opportunity to explore high prenatal exposure to PCBs and PCDFs.

In a second study on men from the Taiwan PCB poisoning episode, Hsu and coworkers *(51)* studied the relationship between semen quality and levels of PCBs among men that consumed contaminated rice oil some 20 years earlier. They identified 40 exposed men and 28 unexposed men. Mean age of exposed (37.9 years) and unexposed (40.4) were similar. Exposed men had a higher percentage of sperm with abnormal morphology and a higher oligozoospermia rate. The ability of sperm to penetrate the hamster oocyte was reduced in exposed men. The results of this study provide evidence of adverse effects of exposure to PCBs and PCDFs among men exposed 20 years earlier to the contaminated rice oil.

The data on the relationship between PCBs and semen quality support an inverse association of PCBs with reduced semen quality, specifically reduced sperm motility. The associations found were generally consistent across studies performed in different countries (India, Netherlands, Taiwan, Sweden, and USA) that used different methods to measure semen quality and PCBs. Furthermore, associations were consistently found despite a range of PCB levels, that is there did not appear to be a threshold. The PCB levels in these studies ranged from low-background levels *(45,46,48)* to high-background levels due to consumption of contaminated fish *(47)*, and to even higher exposure levels due to ingestion of contaminated rice oil *(50,51)*. Although the data across studies generally support a relationship between PCBs and poor semen quality, there are possible alternative explanations. One potential alternative explanation is that PCBs are a surrogate for exposure to other environmental factors that may predict semen quality. Although this is possible, there is currently no evidence identifying potential alternative exposures. Another explanation is that there may be confounding of the associations by some currently unrecognized or unmeasured confounders. However, the more recent studies considered important potential confounders, and the results were consistent across studies. In conclusion, although PCBs are no longer used, these data, along with ongoing human exposure, albeit at lower levels than several decades ago, raise concerns regarding altered human fertility due to adverse affects on semen quality.

3.2. Semen Quality and Non-persistent Pesticides

Non-persistent pesticides (also referred to as "contemporary-use pesticides") are chemical mixtures that are currently available for application to control insects (insecticides), weeds (herbicides), fungi (fungicides), or other pests (e.g., rodenticides), as opposed to pesticides that have been banned from use in most countries (e.g., many of the formerly popular organochlorine pesticides such as DDT). Three common classes of non-persistent pesticides in use today include organophosphates, carbamates, and pyrethroids. Though environmentally non-persistent, owing to the extensive use of pest control in these various settings, a majority of the general population is exposed to some of the more widely used pesticides at low levels. Exposure among the general population occurs primarily through the ingestion of foods that contain low levels of pesticide residue or through inhalation and/or dermal exposure in or around the home and in other indoor environments. Non-persistent pesticides that are applied indoors or tracked in from outdoors may persist for extended periods while protected from sunlight, rain, temperature extremes, and most microbial action *(52)*. For example, chlorpyrifos, a semivolatile organophosphate insecticide the use of which was recently restricted in the USA by the US Environmental Protection Agency, was measured in indoor air 4 years after pest control application in a home *(53)*.

There are several epidemiologic studies on men exposed to non-persistent pesticides during agricultural work (Table 3). A cross-sectional study on testicular function measured sperm concentration, motility, and morphology in 122 greenhouse workers defined as low, medium, or highly exposed to more than a dozen pesticides *(54)*. Adjusting for abstinence time and other potential confounders, a higher proportion of abnormal sperm was found in the high exposure group compared with the group with low exposure. Lower median sperm concentration was also observed in workers with more than 10 years of work in the greenhouse compared with men with less than 5 years of experience. In a cross-sectional study on traditional and organic farmers, Juhler and coworkers *(55)* investigated the relationship between dietary exposure to pesticides and semen quality. Estimating exposure through food frequency questionnaires and data from pesticide monitoring programs, the authors found that men with a lower intake of organic food had a lower proportion of normal-shaped sperm according to the strict criteria (2.5 vs. 3.7%, $p = 0.003$). However, organic food intake was not associated with the other 14 semen parameters measured in the study. Results in the study were adjusted for age, urogenital disease, spillage, abstinence time, smoking, and alcohol intake. Oliva and coworkers *(56)* recently investigated the impact of environmental factors on infertility among 177 men in Argentina. Adjusting for age, body mass index (BMI), abstinence time, income, health center, and smoking, a dose-related response was observed in (primary) infertile men occupationally exposed to pesticides. Significantly elevated odds ratios were reported for sperm concentration ($< 1 \times 10^6$/ml, OR = 3.4, 95% CI = 1.2 – 7.4), motility (<50% motile, OR = 3.6, 95% CI = 1.1 – 11.4), and morphology (<30% normal, OR = 4.1, 95% CI = 1.4 – 12.0) for men exposed to pesticides compared with occupationally non-exposed men. Conversely, in models adjusting for many of the same variables, Larsen and coworkers *(57)* found only marginal differences among 15 semen quality parameters from Danish farmers who sprayed pesticides compared with farmers who did not spray pesticides. These studies show a possible association between pesticide exposure and human semen

Table 3
Summary Table of Epidemiologic Studies (in Chronological Order) on the Relationship Between Non-Persistent Pesticides and Semen Quality

Author	*Study population*	*Exposure*	*Results*	*Comments*
Whorton et al., 1979 *(61)*	47 US carbaryl production workers plus 90 unexposed controls	Subjective exposure classification based on job tasks	Greater proportion of oligozoospermic men among the carbaryl workers (15%) as compared with the chemical workers (5.5%, $p = 0.07$)	No adjustment for potential confounders. Sperm motility not measured
Wyrobek et al., 1981 *(62)*	50 US carbaryl production workers plus 34 unexposed controls	Exposure ranks/groups based on job type held for previous year	Elevated percent of abnormal sperm in carbaryl workers (52%) as compared with comparison subjects (42%, $p < 0.005$). The proportion of men defined as teratospermics (>60% abnormal sperm) was higher among the carbaryl workers (28.6%) than in the comparison group (11.8%, $p = 0.06$)	Confounders considered: smoking, medical history, and previous exposure to hazardous agents
Padungtod et al., 2000 *(58)*	43 pesticide factory workers in China; 20 high exposed and 23 with no or very low exposure	Occupational exposure to ethyl parathion and methami-dophos	Exposure associated with reduction in sperm concentration and motility but not sperm morphology. Adjusted means for exposed and non-exposed workers were 28.5 and 49.4 million sperm/ml ($p = 0.01$), respectively, for sperm concentration and 64 and 74% ($p = 0.03$), respectively, for percentage of motile sperm	Confounders considered: age, abstinence period, and current smoking status
Swan et al., 2003 *(64)*	86 male partners from fertile couples attending US prenatal clinics	Urinary levels of pesticides or metabolites (IMPY, 1N, TCPY, and others)	Increased odds ratios (95% CI) for below reference semen parameters associated with high exposure group for alachlor mercapturate 30.0 (4.3–210), IMPY 16.7 (2.8–98), atrazine mercapturate 11.3 (1.3–99), 1-naphthol 2.7 (0.2–34), and TCPY 6.4 (0.5–86)	Small study size limited statistical power; odds ratios were unadjusted for potential confounders

(*Continued*)

Table 3
(***Continued***)

Author	*Study population*	*Exposure*	*Results*	*Comments*
Meeker et al., 2004 *(65)*	272 male partners from couples attending US infertility clinic	Urinary levels of insecticide metabolites (1N and TCPY)	Inverse association between urinary carbaryl metabolite (1N) and sperm concentration and motility. IQR increase in 1N associated with 16% decline in sperm concentration and 3.8% decline in motile sperm. Suggestive inverse association between chlorpyrifos metabolite (TCPY) and sperm motility	Confounders considered: age, BMI, abstinence time, smoking status, race, and season
Tan et al., 2005, 2006 *(60,63)*	31 workers in a Chinese pesticide factory exposed to carbaryl and fenvalerate; 46 internal and 22 external control subjects	Men defined as exposed or unexposed based on job tasks, air, and dermal monitoring	In exposed workers, seminal volume and sperm motility were lower than in the control groups ($p < 0.05$) as were the following sperm motion parameters: linearity, straightness, straight-line velocity, and beat cross frequency ($p < 0.05$)	Reported risk factors between groups were similar (health, age, smoking, and alcohol) but not included in the models

1N, 1-naphthol; BMI, body mass index; IMPY, 2-isopropoxy-4-methyl-pyrimidinol; and TCPY, 3,5,6-trichloro-2-pyridinol.

quality. However, the broad nature of the exposure assessments makes it impossible to determine which pesticides, if any, were responsible for the observed effects.

Few studies have been conducted that provide information on specific chemicals or classes of contemporary-use insecticides and altered testicular function. Padungtod and coworkers *(58)* studied the relationship between occupational exposure to organophosphates (parathion and methamidophos) and testicular function among Chinese pesticide factory workers. They found a significant reduction in adjusted mean sperm concentration (28.5 vs. 49.4 million sperm/ml, $p = 0.01$), and percentage of motile sperm (64 vs. 74%, $p = 0.03$) in the 20 exposed workers as compared with the 23 unexposed workers. In a recent Japanese study, pesticide sprayers exposed primarily to organophosphates and pyrethroids showed spraying season-dependant reductions in motile sperm velocity measures compared with unexposed controls *(59)*. Another recent study of 32 men occupationally exposed to the synthetic pyrethroid fenvalerate in a Chinese pesticide factory found that the exposed workers had decreased sperm counts, as well as declined sperm movement and progression, compared with men in both internal and external comparison groups *(60)*.

Two publications reported the results from a study on a small cohort of men exposed to the insecticide carbaryl (1-naphthyl methyl carbamate, commonly known as Sevin®) during the production and packaging of the compound *(61,62)*. Although analyses using sperm counts as a continuous measure failed to find significant differences based on carbaryl exposure, the authors found a greater proportion of oligozoospermic men among the carbaryl workers as compared with the chemical workers *(61)*. In a subsequent publication on the same cohort of carbaryl production workers, Wyrobek and coworkers *(62)* studied the relationship between sperm shape abnormalities and carbaryl. Morphological analyses showed an elevated percent of abnormal sperm in carbaryl workers as compared with comparison subjects, which remained after stratifying on potential confounders such as smoking, medical history, or previous exposure to hazardous agents. The proportion of men defined as teratospermics (>60% abnormal sperm) was higher among the carbaryl workers than in the comparison group (28.6 and 11.8%, respectively). More recently, a Chinese study of 31 carbaryl-exposed workers in a pesticide factory found significantly lower sperm motility and sperm motion parameters among the exposed men compared with men in internal and external comparison groups *(63)*.

Researchers are now utilizing urinary and serum biomarkers of pesticide exposure to explore associations with reduced semen quality. In a US study on the male partners of pregnant women, Swan and coworkers *(64)* compared urinary levels of pesticide biomarkers in 34 men with sperm concentration, motility, and morphology below the median (defined as cases) with 52 men with above median semen parameters (defined as controls). They found elevated odds ratios [OR (95% confidence interval)] for alachlor mercapturate [30.0 (4.3–210)], 2-isopropoxy-4-methyl-pyrimidinol (IMPY, diazinon metabolite) [16.7 (2.8–98)], atrazine mercapturate [11.3 (1.3–99)], 1-naphthol (carbaryl and naphthalene metabolite) [2.7 (0.2–34)] and 3,5,6-trichloro-2-pyridinol (TCPY, chlorpyrifos metabolite) [6.4 (0.5–86)]. However, a small study size led to the wide confidence intervals that restrict interpretation of the study results.

Also using urinary biomarker data representative of low environmental levels of pesticides commonly encountered among the general population, Meeker et al. *(65)* studied 272 men who were partners of an infertile couple. They found inverse associations between urinary levels of 1-naphthol, a metabolite of both carbaryl and naphthalene, and sperm concentration and motility. They also found a suggestive inverse relationship between the urinary metabolite of chlorpyrifos and sperm motility. When insecticide metabolite levels were categorized into tertiles, odds ratios (95% CI) for medium and high tertiles of 1-napthol were 4.2 (1.4–13.0) and 4.2 (1.4–12.6) compared with the lowest tertile for below reference concentration (<20 million sperm/ml), and 2.5 (1.3–4.7) and 2.4 (1.2–4.5) for below reference motility (<50% motile sperm). In multiple linear regression analyses, an interquartile range increase in 1N (1.8–5.0 μg/l) was associated with a 3.9% (95% CI = −7.3 to −0.5%) decline in motile sperm and a 16% (−29 to +1.0%) decline in sperm concentration. An interquartile range increase in chlorpyrifos metabolite (TCPY, also 1.8–5.0 μg/l) was suggestively associated with a 2.2% (−5.1 to +0.7%) decline in motile sperm. In two follow-up studies conducted among an overlapping group of the same men, both TCPY and 1N were associated with declined serum testosterone levels *(66)*, whereas TCPY was also associated with decreased free thyroxine (T_4) and increased thyroid-stimulating hormone (TSH) *(67)*.

In summary, there are human data supporting an association between non-persistent pesticide exposure and altered semen quality though it is somewhat limited. Although the human data are suggestive, they are mostly derived from occupational studies involving simultaneous exposure to several pesticides. The relationship between reproductive health and exposure to specific non-persistent pesticides, and/or low-level mixed exposures through diet and residential use among the general male population, is not well understood. Additional research using biomarkers of exposure to specific pesticides is needed to further our understanding of the potential reproductive health risks associated with non-persistent pesticides.

4. CRYPTORCHIDISM, HYPOSPADIAS, AND eEDCs

4.1. Rationale for Linkage of Genital Anomalies to eEDC Exposure

The published literature addressing the possible role of eEDCs in the etiology of two common male genital anomalies, cryptorchidism, and hypospadias has grown rapidly in recent years. Several factors contributing to increasing levels of concern about the etiologic basis of these anomalies include (i) their induction by certain classes of eEDCs in rodents after prenatal exposure; (ii) their clinical correlation with risk of subfertility and testicular malignancy, other conditions that have been hypothesized to be linked to eEDCs; and (iii) reports of their increasing frequency in defined populations. The significance of each of these observations will be discussed in turn, followed by specific epidemiologic data regarding the association of specific eEDCs with cryptorchidism or hypospadias in human populations (Table 4).

Several classes of endocrine-disrupting chemicals including phthalates, pesticides and polychlorinated aromatic compounds are reproductive toxicants that can inhibit genital development and testicular descent in male rat offspring after maternal exposure *(68,69)* (see also Chapter 3 of this book by Cowin, Foster, and Risbridger). The mechanisms of action of these agents include but may not be limited to inhibition of AR activation, androgen synthesis inhibition, estrogenic effects, and/or inhibition of insl3 expression, effects that may be strain specific. Testosterone and its metabolite,

Table 4

Summary Table of Epidemiologic Studies (in Chronological Order) of the Relationship Between Pesticides and Cryptorchidism and/or Hypospadias

Author (country)	*Study population*	*Exposure*	*Results*	*Comments*
Longnecker et al., 2002 *(94)* (USA)	Birth cohort study 1959–1966; 22,347 boys, 219 with cryptorchidism, 199 with hypospadias, and 552 controls	Maternal serum levels of DDE during pregnancy	Adjusted OR of highest level of serum DDE with risk for: cryptorchidism, 1.2 (95% CI = 0.6–2.4) and hypospadias, 1.3 (0.7–2.4)	No consistent association found
Bhatia et al, 2005 *(95)* (San Francisco)	Birth cohort study 1959–1967; 9345 males followed until age 5, 101 with cryptorchidism, 73 with hypospadias, and 6 with both	Maternal serum levels of DDT and DDE during pregnancy	Adjusted OR of highest serum level of DDT with risk for cryptorchidism, 1.01 (0.44–2.28) and hypospadias, 0.79 (0.33–1.89). Adjusted OR of highest serum level of DDT with risk for cryptorchidism, 1.34 (0.51–3.48) and hypospadias, 1.18 (0.46–3.02)	No association of DDT or DDE with either anomaly
Garry et al., 1996 *(96)* (USA)	4935 births, 1989–1992, 20 cases of urogenital anomalies	34,772 parental pesticide appliers	Age adjusted OR for urogenital anomalies 1.89 (95% CI = 1.06–2.64)	Birth registry data. Authors did not explore associations separately for cryptorchidism and hypospadias
Garcia-Rodriquez et al., 1996 *(97)* (Spain)	270 cases of orchidopexy for cryptorchidism, 1980–1991	Predicted pesticide exposure level based on a four-point agricultural density scale	Logistic regression analysis of orchidopexy rates per 10,000 males aged 1–16: Rates increased with pesticide use (maximum 5.74 in highest pesticide use region) but highest rate (7.13) in urban low pesticide use region	Highest rate in urban area not consistent with relationship to pesticide exposure

(*Continued*)

Table 4
(***Continued***)

Author (country)	*Study population*	*Exposure*	*Results*	*Comments*
Kristensen et al., 1997 *(98)* (Norway)	192,417 births, 77 cases of cryptorchidism, and 70 cases of hypospadias	Parental occupation as farmer 1967–1991	Overall adjusted OR for cryptorchidism, 0.77 (0.58–1.03) and hypospadias, 1.00 (0.75–1.34); OR for pesticide purchase/cryptorchidism, 1.70 (1.16–2.5) and tractor spraying/hypospadias, 1.38 (0.95–1.99)	Confounders considered: year of birth, maternal age, region, and parental consanguinity
Weidner et al., 1998 *(99)* (Denmark)	Surgical records, 4226 cases of cryptorchidism, and 1345 cases of hypospadias	Maternal occupation as farmer or gardener 1983–1992	Risk of cryptorchidism: OR = 1.36 (95% CI = 1.1–1.73) for maternal farming or gardening and OR = 1.67 (95% CI = 1.14–2.47) for gardening only	No increased risk for paternal farming or gardening for hypospadias
Pierik et al, 2005 *(100)* (Netherlands)	8,698 births; 78 cases of cryptorchidism and 56 cases of hypospadias	Parental questionnaire data: self-reported and occupational exposures	Probable paternal exposure to pesticides, OR = 3.8 (95% CI = 1.1–13.4) for cryptorchidism	No association of pesticide exposure with hypospadias
Carbone et al, 2006 *(101)* (Sicily)	8,199 births; 59 cases of cryptorchidism and 54 cases of hypospadias	Geographical "pesticide impact" score based on use of spraying equipment and farm activity	Progressive increase in prevalence of hypospadias ($p = 0.003$, test for trend) or either anomaly ($p = 0.001$) but not cryptorchidism alone ($p = 0.096$) in relation to "pesticide" impact of geographical location	Registry data and review of all pediatric records, 1998–2002

DHT, are essential (particularly DHT) for masculinization of the genitalia but participate to a lesser degree in the process of testicular descent. Insl3 is critical for development of the gubernaculum, a fibromuscular fetal structure that guides the testis to a scrotal position. In rats, inhibition of nipple development and shortening of anogenital distance are the most sensitive markers of anti-androgenic effect *(70)*. By contrast, phthalates, particularly DBP, are potent in inhibiting testicular descent, an effect that may be related to specific inhibition of insl3 expression in addition to inhibition of fetal testosterone production *(71)*. However, the effective exposure levels are markedly higher than those estimated for the general human population *(72)*. Although of concern, the applicability of these animal studies to human disease remains unclear, not only because experimental exposures are extremely high, but also because the genetic susceptibility of rats to cryptorchidism and hypospadias appears to be extremely low, unlike that of larger mammals and man.

In human males, the risk of non-syndromic cryptorchidism in full-term newborn males is 2–4% in most series *(73)* compared with a prevalence of hypospadias of 0.3–0.4%; 4–13% of males with hypospadias also have cryptorchidism *(74)*. Cryptorchid males are at approximately 10 times increased risk for testicular germ cell cancer compared with the general population although the overall risk remains well below 1% *(75)*. Reported paternity rates in previously unilaterally and bilaterally cryptorchid men are 90 and 65%, respectively *(76)*. In view of the associations amongst these conditions, Skakkebæk et al. *(1)* hypothesized that a primary testicular abnormality, defined as "testicular dysgenesis syndrome" and possibly induced by eEDCs, is the common link between testicular germ cell tumors, reduced semen quality, and other reproductive tract abnormalities. They propose that TDS is due to prenatal Leydig and Sertoli cell dysfunction with secondary androgen insufficiency and impaired germ cell development. This entity should not be confused with the well-defined clinical entity known as dysgenetic testes, which is associated with genital ambiguity and a high risk of testicular malignancy *(77)*. Reductions in the normal postnatal surge in androgen and LH production *(78)* and in germ cell maturation *(79)* in cryptorchidism provide support for a TDS-like global testicular dysfunction; however, other studies suggest that these findings are absent or less common than initially reported *(75,80–82)*. Similarly, hypospadias alone has not been defined as a clear separate risk factor for infertility *(74)* or testicular cancer. The existence of TDS as a distinct clinical entity and a possible association with eEDCs, if present, have yet to be defined.

Recent trends over relatively short time periods showing increases in the prevalence of cryptorchidism and hypospadias are cited as evidence supporting an environmental cause of these diseases *(1)*. For cryptorchidism, the prevalence appears to be highly variable and geographically specific *(83)*, with temporal upward trends noted in some locations but not others *(22,84,85)*. Boisen et al. *(83)* correlated geographical differences in Denmark and Finland with correspondingly similar trends in testicular cancer and fertility rates. However, in Lithuania, the prevalence of cryptorchidism was higher than expected based on relatively low cancer and high fertility rates *(86)*. Prevalence data for cryptorchidism are hard to interpret, because of the limitations of registry-based data and how they are obtained, changes in clinical practice that emphasize earlier diagnosis and treatment, confounding factors such as birth weight and prematurity, and inaccurate diagnosis related to changes in testicular position (spontaneous descent or secondary "ascent") over time *(73)*. Similarly, data for hypospadias prevalence are conflicting. Although rates increased in some locations over time, trends were not consistent, and most recent reports identified no changes over time *22,87–89*. Ascertainment bias may also easily exist for this anomaly, particularly for milder forms, as both false-negative and false-positive diagnoses may be made in newborns based on circumcision status.

In summary, animal studies suggest that many eEDCs are capable of altering genital development at high exposure levels. Feminization of the genitalia (shortened anogenital distance with or without hypospadias) is a more common outcome in these studies than prevention of testicular descent (cryptorchidism). By contrast, in human males, cryptorchidism is about 10 times more common than hypospadias, and there have been no consistent trends in the frequency of these common anomalies over time. Nevertheless, susceptibility to the adverse reproductive effects of eEDCs may vary with species and with race/ethnicity differences in human populations. Such differences are compatible with the possibility that eEDCs contribute to the risk of cryptorchidism and hypospadias in human populations.

4.2. Epidemiological Evidence for eEDC Exposure in Males with Cryptorchidism or Hypospadias

Data that directly correlate specific eEDC exposure during pregnancy with fetal reproductive tract anomalies are limited to women treated with DES during pregnancy. Although DES exposure is no longer clinically relevant, study of its effects is useful, as it is a synthetic estrogen agonist that does not bind to maternal plasma proteins and has both anti-androgenic and anti-insl3 effects in animal models. In a large cohort of boys born to women who received high dose DES in a prospective study over 50 years ago, the most common anomalies were epididymal cysts and testicular atrophy, with a risk of cryptorchidism of only 5% *(90)*. Long-term follow-up of this group identified an increased risk of pathological semen but not of failed paternity, other measures of infertility, or of testicular cancer *(91,92)*. Similarly, in a recent review, Storgaard et al. *(93)* found evidence for an association between indices of estrogen exposure and testicular cancer but not cryptorchidism or hypospadias.

Evidence for specific eEDC exposure in males with cryptorchidism or hypospadias is limited. Several studies have failed to identify consistently increased levels of DDT or DDE in serum of mothers bearing offspring with cryptorchidism or hypospadias (Table 4). Longnecker et al. *(94)* interpreted their data as showing a modest association (OR = 1.3, 95% CI = 0.7–2.4) between maternal levels of DDE and cryptorchidism in a large case–control study. In a smaller birth cohort study, maternal serum levels of DDT and DDE during pregnancy did not correlate with the presence of cryptorchidism or hypospadias in male offspring *(95)*. These studies show only small increases in odds ratios for the highest levels of exposure. However, the ability to detect associations may be limited by population heterogeneity and size.

There are very limited data from epidemiologic studies linking exposure to organochlorine compounds with cryptorchidism or hypospadias. Mol et al. *(102)* performed a cohort study of children born in the Faroe Islands, a location of confirmed increased PCB exposure due to dietary intake. Umbilical cord PCB levels were available for the majority of a cohort of 196 boys (75% of those born during a 21-month period in 1986–1987) evaluated for timing of puberty. Nineteen boys (10%) had a history of cryptorchidism, and half were bilateral. Although this is of high prevalence compared with historical controls, no comparable low exposure group was studied, and PCB levels did not differ between the groups with and without undescended testes. In a small study conducted in Germany *(103)*, fat samples were obtained at surgery from boys undergoing orchidopexy for cryptorchidism ($n = 18$) and controls without genital anomalies ($n = 30$). Of a panel of organochlorine compounds measured in these samples including PCBs and DDT and its metabolites, only levels of heptachlorepoxide and hexachlorobenzene, compounds not previously associated with genital anomalies, were significantly elevated in the case group.

The relationship of parental or general community pesticide exposure with hypospadias or cryptorchidism has been explored in several epidemiologic studies (Table 4). Exposures were estimated in these studies based on geographical location (proximity to agricultural activity), parental occupation in agriculture, and/or self-report of exposures. Ascertainment of cases was also varied and based either on newborn exams or medical record documentation of surgical treatment for cryptorchidism. These populations are different, as a proportion of boys undergoing surgery did not have documented cryptorchidism at birth, whereas others identified at birth have spontaneous resolution and do not require surgery *(73)*. As such, these studies provide only indirect evidence for exposure to a variety of eEDCs with varying degrees of reproductive toxicity. Even in cases of direct known exposure (e.g., parental occupation as a gardener or farmer), the timing and degree of exposure cannot be retrospectively quantified. Also, as noted by Vidaeff and Sever *(104)*, caution must be exercised in interpretation of these studies because of sample size limitations, weak or inconsistent associations, the possibility of recall bias, and/or misclassification of exposure. However, the reported data suggest a possible association between pesticides and the occurrence of one or both genital anomalies. In another series, maternal vegetarianism was significantly correlated with hypospadias in a cohort of 51 affected males (OR = 3.53, 95% CI = 1.56–7.98) and attributed to phytoestrogen exposure, although increased risk related to concomitant pesticide exposure could not be ruled out *(105)*. Additional work that directly addresses the association of maternal and/or paternal pesticide exposure with urogenital anomalies by measurement of exposure levels of specific compounds is indicated.

Recent studies indicate that certain phthalate diesters and their metabolites are measurable in breast milk, cord blood, and at higher than expected levels in certain pregnant women *(31,106,107)*, and indirect evidence suggests that certain phthalates may affect Leydig cell function in the perinatal period. In a study of phthalate concentrations in breast milk and hormone levels in 3-month-old male offspring, Main et al. *(106)* identified a significant negative correlation between MBP, the active metabolite of DBP, and free testosterone levels (determined by total testosterone: sex hormone binding globulin ratio). However, phthalate levels did not correlate with the occurrence of cryptorchidism that was present in approximately half of this cohort of boys. Interestingly, levels of MBP were significantly higher in breast milk from Finnish as compared with Danish mothers despite a significantly lower prevalence of cryptorchidism and hypospadias *(86)* in Finland. In another study of phthalate exposure and male genital development, Swan and colleagues *(108)* determined "anogenital index" (anogenital distance/body weight) and testicular position in young boys (mean age 16 months) and corresponding maternal levels of urinary phthalate metabolites at three separate clinical sites. In this study, the authors found significant inverse relationships between the highest maternal levels of MBP, MBzP, MEP, MiBP, and anogenital index [OR for MBP: 10.2 (95% CI = 2.5–42.2)], although MEP has not been linked to reproductive anomalies in rodent studies. The prevalence of "incomplete testicular descent" ranged from 6 to 20% with decreasing anogenital index, levels that are unusually high and may reflect the high incidence of non-pathological testicular retractility in normal boys of this age *(73)*. Moreover, in boys, stretched penile length is the standard measure used to assess male genital masculinization; the validity of anogenital index in this regard is well defined in rats but not confirmed in human males. Scrotal anatomy, another endpoint studied that may in part determine anogenital distance, is highly variable and does not necessarily correlate with penile development or testicular descent.

In summary, the strongest epidemiological data that link eEDC exposure to cryptorchidism and/or hypospadias are those suggesting an association between residency in agricultural areas and/or measures of direct parental exposure to non-organochlorine pesticides, without providing insight into specific potentially causative agents. However, these data are not necessarily consistent for both anomalies or congruent with observations made in animal experiments. Further studies will be needed to provide a clearer understanding of the role(s), if any, of specific eEDCs in the etiology of genital anomalies in man.

5. CONCLUSIONS AND CHALLENGES

Throughout this chapter, we have tried to provide insights into the current state of the epidemiologic evidence on the relationship between eEDC and male reproductive and developmental health. The chapter was not meant to be an exhaustive review of the evidence but rather a synthesis of the current knowledge in an ever-changing field of inquiry and discovery. The classes of chemicals included in this chapter represent a number of contemporary and widely used compounds such as pesticides and phthalates. In addition, although PCBs were banned several decades ago, they were also discussed because recently many well-designed epidemiologic studies have been published.

Although there is current scientific, public, and governmental interest in the potential health risks of exposure to eEDCs, the human evidence on associations of eEDCs with altered male reproductive health endpoints remains limited. In addition, the quality and quantity of studies varies widely across exposures and endpoints. For example, although there are hundreds of different pesticides currently in use worldwide, limited human data exists on risks to male reproductive endpoints for only a select few. Also, for some of these chemicals, such as phthalates, the data across studies were not entirely consistent. For instance, one study found associations of semen quality with MBP and MBzP while another large epidemiologic study did not. The limited human data, and in certain instances inconsistent data across studies, highlight the need for further epidemiological research on these classes of eEDCs.

In addition to the classes of eEDCs discussed in this chapter, there are other classes of chemicals that require further study as to their relation with human reproductive and developmental health. These chemicals include, among others, alkylphenols, such as 4-nonylphenol, BPA, and fluorinated organic compounds such as perfluorooctane octanoate (PFOA) and perfluorooctane sulfonate (PFOS). Alkylphenols are used as surface active agents in cleaning/washing agents, paints, and cosmetics, whereas BPA is used in the manufacture of polycarbonate plastics and epoxy resins. The perfluorinated compounds are used to make fabrics stain resistant/water repellent and in coatings on cookware and other products. Although human exposure to these chemicals has been demonstrated, the epidemiologic evidence on potential health effects remains very limited.

A future challenge in understanding the relationship between eEDC and male reproductive and developmental health includes the changes in exposure levels among populations over time because of the ever-changing patterns of production and use of these compounds. Another challenge is to understand how simultaneous coexposures to these chemicals may affect male reproductive health and development. It is well known that most men are exposed to all of these classes of eEDC simultaneously, as well as to many other chemicals. However, there is limited data on the interactions between chemicals within a class or across classes of chemicals. Chemicals may interact additively or multiplicatively, or antagonistically. An understanding of the potential human health risks of exposure to mixtures of eEDC is important but remains very understudied. Despite these challenges, evolving and innovative technologies designed to improve the assessment of human exposure and male reproductive health endpoints should provide enhanced opportunities for improving our understanding of the relationship between eEDC and health. Innovations include improved biomarkers of exposure and more sophisticated statistical methods that deal with multiple exposures simultaneously.

REFERENCES

1. Skakkebaek NE, Rajpert-De Meyts E, Main KM. Testicular dysgenesis syndrome: an increasingly common developmental disorder with environmental aspects. *Hum Reprod* 2001;16:972–8.
2. Kelce WR, Stone CR, Laws SC, Gray LE, Kemppainen JA, Wilson EM. Persistent DDT metabolite p,p'-DDE is a potent androgen receptor antagonist. *Nature* 1995;375:581–5.
3. Parks LG, Ostby JS, Lambright CR, et al. The plasticizer diethylhexyl phthalate induces malformations by decreasing fetal testosterone synthesis during sexual differentiation in the male rat. *Toxicol Sci* 2000;58:339–49.
4. McKinnell C, Atanassova N, Williams K, et al. Suppression of androgen action and the induction of gross abnormalities of the reproductive tract in male rats treated neonatally with diethylstilbestrol. *J Androl* 2001;22:323–38.
5. Williams K, Fisher JS, Turner KJ, McKinnell C, Saunders PT, Sharpe RM. Relationship between expression of sex steroid receptors and structure of the seminal vesicles after neonatal treatment of rats with potent or weak estrogens. *Environ Health Perspect* 2001;109:1227–35.
6. Atanassova N, McKinnell C, Turner KJ, et al. Comparative effects of neonatal exposure of male rats to potent and weak (environmental) estrogens on spermatogenesis at puberty and the relationship to adult testis size and fertility: evidence for stimulatory effects of low estrogen levels. *Endocrinology* 2000;141:3898–907.
7. Kester MH, Bulduk S, Tibboel D, et al. Potent inhibition of estrogen sulfotransferase by hydroxylated PCB metabolites: a novel pathway explaining the estrogenic activity of PCBs. *Endocrinology* 2000;141:1897–900.
8. Rowlands JC, Gustafsson JA. Aryl hydrocarbon receptor-mediated signal transduction. *Crit Rev Toxicol* 1997;27:109–34.
9. Moore RW, Jefcoate CR, Peterson RE. 2,3,7,8-Tetrachlorodibenzo-p-dioxin inhibits steroidogenesis in the rat testis by inhibiting the mobilization of cholesterol to cytochrome P450scc. *Toxicol Appl Pharmacol* 1991;109:85–97.

10. Tian Y, Ke S, Thomas T, Meeker RJ, Gallo MA. Transcriptional suppression of estrogen receptor gene expression by 2,3,7,8-tetrachlorodibenzo-p-dioxin (TCDD). *J Steroid Biochem Mol Biol* 1998;67:17–24.
11. CDC. *Third National Report on Human Exposure to Environmental Chemicals*. Washington, DC: Centers for Disease Control and Prevention, 2005.
12. Carlsen E, Giwercman A, Keiding N, Skakkebaek NE. Evidence for decreasing quality of semen during past 50 years. *BMJ* 1992;305:609–13.
13. Auger J, Kunstmann JM, Czyglik F, Jouannet P. Decline in semen quality among fertile men in Paris during the past 20 years. *N Engl J Med* 1995;332:281–5.
14. Swan SH, Elkin EP, Fenster L. Have sperm densities declined? A reanalysis of global trend data. *Environ Health Perspect* 1997;105:1228–32.
15. Andersen AG, Jensen TK, Carlsen E, et al. High frequency of sub-optimal semen quality in an unselected population of young men. *Hum Reprod* 2000;15:366–72.
16. Bujan L, Mansat A, Pontonnier F, Mieusset R. Time series analysis of sperm concentration in fertile men in Toulouse, France between 1977 and 1992. *BMJ* 1996;312:471–2.
17. Fisch H, Goluboff ET, Olson JH, Feldshuh J, Broder SJ, Barad DH. Semen analyses in 1, 283 men from the United States over a 25-year period: no decline in quality. *Fertil Steril* 1996;65:1009–14.
18. Paulsen CA, Berman ng, Wang C. Data from men in greater Seattle area reveals no downward trend in semen quality: further evidence that deterioration of semen quality is not geographically uniform. *Fertil Steril* 1996;65:1015–20.
19. Adami HO, Bergstrom R, Mohner M, et al. Testicular cancer in nine northern European countries. *Int J Cancer* 1994;59:33–8.
20. Bergstrom R, Adami HO, Mohner M, et al. Increase in testicular cancer incidence in six European countries: a birth cohort phenomenon. *J Natl Cancer Inst* 1996;88:727–33.
21. Huyghe E, Matsuda T, Thonneau P. Increasing incidence of testicular cancer worldwide: a review. *J Urol* 2003;170:5–11.
22. Paulozzi LJ. International trends in rates of hypospadias and cryptorchidism. *Environ Health Perspect* 1999;107:297–302.
23. Anway MD, Cupp AS, Uzumcu M, Skinner MK. Epigenetic transgenerational actions of endocrine disruptors and male fertility. *Science* 2005;308:1466–9.
24. Cohn BA, Overstreet JW, Fogel RJ, Brazil CK, Baird DD, Cirillo PM. Epidemiologic studies of human semen quality: considerations for study design. *Am J Epidemiol* 2002;155:664–71.
25. Hauser R, Godfrey-Bailey L, Chen Z. Does the potential for selection bias in semen quality studies depend on study design? Experience from a study conducted within an infertility clinic. *Hum Reprod* 2005;20:2579–83.
26. David RM, McKee RH, Butala JH, Barter RA, Kayser M. Esters of aromatic mono-, di-, and tricarboxylic acids, aromatic diacids, and di-, tri-, or polyalcohols. In: Bingham E, Cohrssen B, Powell CH, eds. *Patty's Toxicology*. New York: John Wiley and Sons, 2001:635–932.
27. ATSDR. *Toxicological Profile for Di-n-Octyl Phthalate DNOP)*. Atlanta, GA: Agency for Toxic Substances and Disease Registry, 1997.
28. ATSDR. *Toxicological Profile for Di(2-Ethylhexyl)Phthalate (DEHP)*. Atlanta, GA: Agency for Toxic Substances and Disease Registry, 2002.
29. ATSDR. *Toxicological Profile for Diethyl Phthalate (DEP)*. Atlanta, GA: Agency for Toxic Substances and Disease Registry, 1995.
30. ATSDR. *Toxicological Profile for Di-n-Butyl Phthalate (DBP)*. Atlanta, GA: Agency for Toxic Substances and Disease Registry, 2001.
31. Adibi JJ, Perera FP, Jedrychowski W, et al. Prenatal exposures to phthalates among women in New York City and Krakow, Poland. *Environ Health Perspect* 2003;111:1719–22.
32. Rudel RA, Camann DE, Spengler JD, Korn LR, Brody JG. Phthalates, alkylphenols, pesticides, polybrominated diphenyl ethers, and other endocrine-disrupting compounds in indoor air and dust. *Environ Sci Technol* 2003;37:4543–53.
33. Green R, Hauser R, Calafat AM, et al. Use of di(2-ethylhexyl) phthalate-containing medical products and urinary levels of mono(2-ethylhexyl) phthalate in neonatal intensive care unit infants. *Environ Health Perspect* 2005;113:1222–5.
34. Murature DA, Tang SY, Steinhardt G, Dougherty RC. Phthalate esters and semen quality parameters. *Biomed Environ Mass Spectrom* 1987;14:473–7.
35. Rozati R, Reddy PP, Reddanna P, Mujtaba R. Role of environmental estrogens in the deterioration of male factor fertility. *Fertil Steril* 2002;78:1187–94.

36. Duty SM, Silva MJ, Barr DB, et al. Phthalate exposure and human semen parameters. *Epidemiology* 2003;14:269–77.
37. Jonsson BA, Richthoff J, Rylander L, Giwercman A, Hagmar L. Urinary phthalate metabolites and biomarkers of reproductive function in young men. *Epidemiology* 2005;16:487–93.
38. Hauser R, Williams P, Altshul L, Calafat AM. Evidence of interaction between polychlorinated biphenyls and phthalates in relation to human sperm motility. *Environ Health Perspect* 2005;113:425–30.
39. De Voogt P, Brinkman UAT. Production properties and usage of polychlorinated biphenyls. In: Kimbrough RD, Jensen AA, eds. *Halogenated Biphenyls, Terphenyls, Naphthalenes, Dibenzodioxins and Related Products*, 2nd ed. Amsterdam: Elsevier-North Holland; 1989;325–344.
40. Herrick RF, McClean MD, Meeker JD, Baxter LK, Weymouth GA. An unrecognized source of PCB contamination in schools and other buildings. *Environ Health Perspect* 2004;112:1051–3.
41. Longnecker MP, Rogan WJ, Lucier G. The human health effects of DDT (dichlorodiphenyltrichloroethane) and PCBS (polychlorinated biphenyls) and an overview of organochlorines in public health. *Annu Rev Public Health* 1997;18:211–44.
42. Brown JF. Determination of Pcb metabolic, excretion, and accumulation rates for use as indicators of biological response and relative risk. *Environ Sci Technol* 1994;28:2295–305.
43. Phillips DL, Smith AB, Burse VW, Steele GK, Needham LL, Hannon WH. Half-life of polychlorinated biphenyls in occupationally exposed workers. *Arch Environ Health* 1989;44:351–4.
44. Bush B, Bennett AH, Snow JT. Polychlorobiphenyl congeners, p,p′-DDE and sperm function in humans. *Arch Environ Contam Toxicol* 1986;15:333–41.
45. Dallinga JW, Moonen EJ, Dumoulin JC, Evers JL, Geraedts JP, Kleinjans JC. Decreased human semen quality and organochlorine compounds in blood. *Hum Reprod* 2002;17:1973–9.
46. Richthoff J, Rylander L, Jonsson BA, et al. Serum levels of 2,2′,4,4′,5,5′-hexachlorobiphenyl (CB-153) in relation to markers of reproductive function in young males from the general Swedish population. *Environ Health Perspect* 2003;111:409–13.
47. Rignell-Hydbom A, Rylander L, Giwercman A, Jonsson BA, Nilsson-Ehle P, Hagmar L. Exposure to CB-153 and p,p′-DDE and male reproductive function. *Hum Reprod* 2004;19:2066–75.
48. Hauser R, Chen Z, Pothier L, Ryan L, Altshul L. The relationship between human semen parameters and environmental exposure to polychlorinated biphenyls and p,p′-DDE. *Environ Health Perspect* 2003;111:1505–11.
49. Ayotte P, Giroux S, Dewailly E, et al. DDT spraying for malaria control and reproductive function in Mexican men. *Epidemiology* 2001;12:366–7.
50. Guo YL, Hsu PC, Hsu CC, Lambert GH. Semen quality after prenatal exposure to polychlorinated biphenyls and dibenzofurans. *Lancet* 2000;356:1240–1.
51. Hsu PC, Huang W, Yao WJ, Wu MH, Guo YL, Lambert GH. Sperm changes in men exposed to polychlorinated biphenyls and dibenzofurans. *JAMA* 2003;289:2943–4.
52. Lewis RG. Pesticides. In: Samet J, Spenger JD, eds. *Indoor Air Quality Handbook*. New York: McGraw-Hill, 2000:35.1–35.21.
53. Schenk G, Rothweiler H, Schlatter C. Human exposure to airborne pesticides in homes treated with wood preservatives. *Indoor Air* 1997;7:135–42.
54. Abell A, Ernst E, Bonde JP. Semen quality and sexual hormones in greenhouse workers. *Scand J Work Environ Health* 2000;26:492–500.
55. Juhler RK, Larsen SB, Meyer O, et al. Human semen quality in relation to dietary pesticide exposure and organic diet. *Arch Environ Contam Toxicol* 1999;37:415–23.
56. Oliva A, Spira A, Multigner L. Contribution of environmental factors to the risk of male infertility. *Hum Reprod* 2001;16:1768–76.
57. Larsen SB, Giwercman A, Spano M, Bonde JP. A longitudinal study of semen quality in pesticide spraying Danish farmers. The ASCLEPIOS Study Group. *Reprod Toxicol* 1998;12:581–9.
58. Padungtod C, Savitz DA, Overstreet JW, Christiani DC, Ryan LM, Xu X. Occupational pesticide exposure and semen quality among Chinese workers. *J Occup Environ Med* 2000;42:982–92.
59. Kamijima M, Hibi H, Gotoh M, et al. A survey of semen indices in insecticide sprayers. *J Occup Health* 2004;46:109–18.
60. Lifeng T, Shoulin W, Junmin J, et al. Effects of fenvalerate exposure on semen quality among occupational workers. *Contraception* 2006;73:92–6.
61. Whorton MD, Milby TH, Stubbs HA, Avashia BH, Hull EQ. Testicular function among carbaryl-exposed employees. *J Toxicol Environ Health* 1979;5:929–41.

62. Wyrobek AJ, Watchmaker G, Gordon L, Wong K, Moore D, II, Whorton D. Sperm shape abnormalities in carbaryl-exposed employees. *Environ Health Perspect* 1981;40:255–65.
63. Tan LF, Sun XZ, Li YN, et al. Effects of carbaryl production exposure on the sperm and semen quality of occupational male workers [abstract]. *Zhonghua Lao Dong Wei Sheng Zhi Ye Bing Za Zhi* 2005;23:87–90.
64. Swan SH, Kruse RL, Liu F, et al. Semen quality in relation to biomarkers of pesticide exposure. *Environ Health Perspect* 2003;111:1478–84.
65. Meeker JD, Ryan L, Barr DB, et al. The relationship of urinary metabolites of carbaryl/naphthalene and chlorpyrifos with human semen quality. *Environ Health Perspect* 2004;112:1665–70.
66. Meeker JD, Ryan L, Barr DB, Hauser R. Exposure to nonpersistent insecticides and male reproductive hormones. *Epidemiology* 2006;17:61–8.
67. Meeker JD, Barr DB, Hauser R. Thyroid hormones in relation to urinary metabolites of non-persistent insecticides in men of reproductive age. *Reprod Toxicol* 2006;22:437–442.
68. Gray LE Jr, Wolf C, Lambright C, et al. Administration of potentially antiandrogenic pesticides (procymidone, linuron, iprodione, chlozolinate, p,p'-DDE, and ketoconazole) and toxic substances (dibutyl- and diethylhexyl phthalate, PCB169, and ethane dimethane sulphonate) during sexual differentiation produces diverse profiles of reproductive malformations in the male rat. *Toxicol Ind Health* 1999;15:94–118.
69. Gray LE Jr, Ostby J, Furr J, et al. Effects of environmental antiandrogens on reproductive development in experimental animals. *Hum Reprod Update* 2001;7:248–64.
70. Imperato-McGinley J, Sanchez RS, Spencer JR, Yee B, Vaughan ED. Comparison of the effects of the 5 alpha-reductase inhibitor finasteride and the antiandrogen flutamide on prostate and genital differentiation: dose-response studies. *Endocrinology* 1992;131:1149–56.
71. Lehmann KP, Phillips S, Sar M, Foster PM, Gaido KW. Dose-dependent alterations in gene expression and testosterone synthesis in the fetal testes of male rats exposed to di (*n*-butyl) phthalate. *Toxicol Sci* 2004;81:60–8.
72. Mylchreest E, Wallace DG, Cattley RC, Foster PM. Dose-dependent alterations in androgen-regulated male reproductive development in rats exposed to Di(n-butyl) phthalate during late gestation. *Toxicol Sci* 2001;55:143–51.
73. Barthold JS, González R. The epidemiology of congenital cryptorchidism, testicular ascent and orchidopexy. *J Urol* 2003;170:2396–401.
74. Mieusset R, Soulié M. Hypospadias. Psychosocial, sexual and reproductive consequences in adult life. *J Androl* 2005;26:163–8.
75. Cortes D, Thorup JM, Visfeldt J. Cryptorchidism: aspects of fertility and neoplasms. A study including data of 1, 334 consecutive boys who underwent testicular biopsy simultaneously with surgery for cryptorchidism. *Horm Res* 2001;55:21–7.
76. Lee PA. Fertility after cryptorchidism: epidemiology and other outcome studies. *Urology* 2005;66:427–31.
77. Slowikowska-Hilczer J, Szarras-Czapnik M, Kula K. Testicular pathology in 46, XY dysgenetic male pseudohermaphroditism: an approach to pathogenesis of testis cancer. *J Androl* 2001;22:781–92.
78. Gendrel D, Roger M, Job JC. Plasma gonadotropin and testosterone values in infants with cryptorchidism. *J Pediatr* 1980;97:217–20.
79. Hadziselimovic F, Herzog B. The importance of both an early orchidopexy and germ cell maturation for fertility. *Lancet* 2001;358:1156–7.
80. McAleer IM, Packer MG, Kaplan GW, Scherz HC, Krous HF, Billman GF. Fertility index analysis in cryptorchidism. *J Urol* 1995;153:1255–8.
81. Barthold JS, Manson J, Regan V, et al. Reproductive hormone levels in infants with cryptorchidism during postnatal activation of the pituitary-testicular axis. *J Urol* 2004;172:1736–41.
82. Suomi AM, Main KM, Kaleva M, et al. Hormonal changes in 3-month-old cryptorchid boys. *J Clin Endocrinol Metab* 2006;91:953–8.
83. Boisen KA, Kaleva M, Main KM, et al. Differences in prevalence of congenital cryptorchidism in infants between two Nordic countries. *Lancet* 2004;363:1264–9.
84. Toppari J, Kaleva M, Virtanen HE. Trends in the incidence of cryptorchidism and hypospadias, and methodlogical limitations of registry-based data. *Hum Reprod Update* 2001;7:282–6.
85. Thonneau PF, Candia P, Mieusset R. Cryptorchidism: incidence, risk factors and potential role of environment; an update. *J Androl* 2003;24:155–62.

86. Preikša RT, Žilaitienk B, Matulevičius V, et al. Higher than expected prevalence of congenital cryptorchidism in Lithuania: a study of 1204 boys at birth and 1 year follow-up. *Hum Reprod* 2002;20:1928–32.
87. Dolk H, Vrijheid M, Scott JES, et al. Toward the effective surveillance of hypospadias. *Environ Health Perspect* 2004;112:398–402.
88. Aho M, Koivisto A-M, Tammela TLJ, Auvinen A. Is the incidence of hypospadias increasing? Analysis of Finnish hospital discharge data 1970-1994. *Environ Health Perspect* 2000;108:463–5.
89. Martínez-Frias ML, Prieto D, Prieto L, Bermejo E, Rodriguez-Pinilla E, Cuevas L. Secular decreasing trend of the frequency of hypospadias among newborn male infants in Spain. *Birth Defects Res (Part A)* 2004;70:75–81.
90. Gill WB, Schumacher GF, Bibbo M, Straus 2nd FH, Schoenberg HW. Association of diethylstilbestrol exposure in utero with cryptorchidism, testicular hypoplasia and semen abnormalities. *J Urol* 1979;122:36–9.
91. Wilcox AJ, Baird DD, Weinberg CR, Hornsby PP, Herbst AL. Fertility in men exposed prenatally to diethylstilbestrol. *N Engl J Med* 1995;332:1411–16.
92. Strohsnitter WC, Noller KL, Hoover RN, et al. Cancer risk in men exposed *in utero* to diethylstilbestrol. *J Natl Cancer Inst* 2001;93:545–51.
93. Storgaard L, Bonde JP, Olsen J. Male reproductive disorders in humans and prenatal indicators of estrogen exposure: a review of published epidemiological studies. *Reprod Toxicol* 2006;21:4–15.
94. Longnecker MP, Klebanoff MA, Brock JW, et al. Maternal serum level of 1, 1-dichloro-2-2-bis(*p*-chlorophenyl)ethylene and risk of cryptorchidism, hypospadias and polythelia among male offspring. *Am J Epidemiol* 2002;155:313–22.
95. Bhatia R, Shiau R, Petreas M, Weintraub JM, Farhang L, Eskenazi B. Organochlorine pesticides and male genital anomalies in the child health and development studies. *Environ Health Perspect* 2005;113:220–4.
96. Garry VR, Schreinemachers D, Harkins ME, Griffith J. Pesticide appliers, biocides and birth defects in rural Minnesota. *Environ Health Perspect* 1996;104:394–9.
97. Garcia-Rodriquez J, Garcia-Martin M, Nogueras-Ocaña M, et al. Exposure to pesticides and cryptorchidism: geographical evidence of a possible association. *Environ Health Perspect* 1996;104:1090–5.
98. Kristensen P, Irgens LM, Andersen A, Snellingen Bye A, Sundheim L. Birth defects among offspring of Norwegian farmers, 1967-1991. *Epidemiology* 1997;8:537–44.
99. Weidner IS, Moller H, Jensen TK, Skakkebæk NE. Cryptorchidism and hypospadias in sons of gardeners and farmers. *Environ Health Perspect* 1998;106:793–6.
100. Pierik FH, Burdork A, Deddens JA, Juttman RE, Weber RFA. Maternal and paternal risk factors for cryptorchidism and hypospadias: a case-control study in newborn boys. *Environ Health Perspect* 2004;112:1570–6.
101. Carbone P, Giordano F, Nori F, et al. Cryptorchidism and hypospadias in the Sicilian district of Ragusa and the use of pesticides. *Reprod Toxicol* 2006;22:8–12.
102. Mol NM, Sorensen N, Weihe P, et al. Spermaturia and serum hormone concentrations at the age of puberty in boys prenatally exposed to polychlorinated biphenyls. *Eur J Endocrinol* 2002;146:357–63.
103. Hosie S, Loff S, Witt K, Niessen K, Waag KL. Is there a correlation between organochlorine compounds and undescended testes. *Eur J Pediatr Surg* 2000;10:304–9.
104. Vidaeff AC, Sever LE. In utero exposure to environmental estrogens and male reproductive health: a systematic review of biological and epidemiologic evidence. *Reprod Toxicol* 2005;20:5–20.
105. North K, Golding J, ALSPAC Study Team. A maternal vegetarian diet in pregnancy is associated with hypospadias. *BJU Int* 2000;85:107–13.
106. Main KM, Mortensen GK, Kaleva MM, et al. Human breast milk contamination with phthalates and alterations of endogenous reproductive hormones in infants three months of age. *Environ Health Perspect* 2006;114:270–6.
107. Latini G, de Felice C, Presta G, et al. *In utero* exposure to di-(2-ethylhexyl)phthalate and duration of human pregnancy. *Environ Health Perspect* 2003;111:1783–85.
108. Swan SH, Main KM, Liu F, et al. Decrease in anogenital distance among male infants with prenatal phthalate exposure. *Environ Health Perspect* 2005;113:1056–61.

11 Human Exposures and Body Burdens of Endocrine-Disrupting Chemicals

Antonia M. Calafat, PhD,
and Larry L. Needham, PhD

Contents

1. INTRODUCTION AND OVERVIEW

In today's industrial societies, humans may be exposed to a wide variety of environmental chemicals, including those with potential endocrine-disruptive properties. Although for many endocrine-disrupting chemicals (EDCs) the health significance of this exposure in humans is unknown, studies to investigate the prevalence of exposure are warranted. As often indicated in animal studies, exposure to EDCs may have potentially harmful health effects. To assess the effects of exposure, researchers have used three tools: exposure history/questionnaire information, environmental monitoring, and biomonitoring (i.e., measuring concentrations of the chemicals, their metabolites, or adducts in human specimens). Assessing exposure to most environmental chemicals of public health concern through a combination of biomonitoring data and indirect exposure measures is usually adequate. In this chapter, we present an overview on the use of biomonitoring in exposure assessment to EDCs using as examples phthalates, bisphenol A (BPA), and phytoestrogens, among others. We discuss some factors relevant to interpreting and understanding biomonitoring data, including selection of

From: *Endocrine-Disrupting Chemicals: From Basic Research to Clinical Practice*
Edited by: A. C. Gore © Humana Press Inc., Totowa, NJ

biomarkers of exposure and human matrices, and selection of toxicokinetic information. To link biomarker measurements to exposure, internal dose, or health outcome, additional information (e.g., toxicokinetics and inter- and intra-individual differences) is needed. We will not discuss assessing potential health risks from biomonitoring exposure data (human risk assessment) or risk management. For more information on this latter subject, see Chapter 12 in this book by Woodruff.

2. USING BIOMONITORING IN ASSESSMENT OF HUMAN EXPOSURE TO EDCS

Given their high production volumes and wide use, the probability of nonoccupational human exposure to chemicals used in consumer products—including potential EDCs—is high. For the most part, little information exists about the extent of human exposure to many such chemicals, and their potential toxic health effects in humans are largely unknown. Because information on risk to human health from exposure to EDCs is limited, studies to investigate the prevalence of these exposures are warranted. Exposure assessment in epidemiologic studies is, however, complex; under controlled conditions of dose–response (exposure or health effect) evaluations associated with animal studies, human exposures do not occur.

Historically, exposure assessments for EDCs and other environmental chemicals have relied on

- surveys of product use and food consumption,
- measurement of EDCs in food and various environmental media (e.g., air, water, and dust),
- estimates of human contact, and
- pharmacokinetic assumptions based on animal data.

In other words, to assess human exposure to EDCs, researchers have generally used indirect measures of exposure that combine environmental monitoring with exposure history and questionnaire data. Advances in analytical chemistry have made measuring trace levels of multiple environmental chemicals in biological tissues (i.e., biological monitoring or biomonitoring) possible and have contributed to increased use of biomonitoring in exposure assessment *(1,2)*. Biomonitoring data may in fact permit more accurate assessments of human exposure—concentrations of EDCs or their metabolites in biological matrices represent an integrative measure of exposure to these chemicals from multiple sources and routes. Therefore, using biomarkers for EDC exposure in combination with indirect measures of exposure are the most appropriate tools for exposure assessment.

Biomonitoring data on EDCs (i.e., internal dose measurements) can also be used in risk assessment and risk management. For risk assessment, information on EDC concentrations or their metabolites concentrations in biological media—as well as on the EDC pharmacokinetics—are required to obtain exposure estimates. These estimates are compared with toxicological parameters (e.g., NOAELs—no-observed-adverse-effect limits), obtained generally from studies in animal (most frequently rodent) models. Considerations relating to these hazard data include, among others, choice of species, identification of critical endpoints, and relevance to humans. Biomonitoring in risk management is also useful for ascertaining the effectiveness of risk management practices; although, to develop effective risk management strategies, exposure pathway information is needed. However, discussing in detail the issues pertaining to biomonitoring in risk assessment and management is beyond the scope of this chapter. Rather, we focus on biomonitoring for exposure assessment—we will not address biomonitoring in risk assessment or risk management.

3. ANALYTICAL CONSIDERATIONS FOR BIOMONITORING

Biomonitoring for assessing exposure to EDCs and other environmental chemicals generally requires measuring the analytes of interest at relatively low concentrations (e.g., at or below parts per billion). Therefore, biomonitoring must employ state-of-the-art analytical methods. These often include stable isotope-labeled internal standards and mass spectrometry *(3)*.

Biological matrices are complex; they can also be difficult to obtain and may be available only in small amounts. Moreover, although environmental chemicals are normally present in the matrix at trace levels, other matrix components occur at higher concentrations. Thus, highly sensitive, specific, and selective multianalyte methods for the extraction, separation, and quantification of these chemicals must be developed *(3)*.

Furthermore, when developing sampling, storage, and analysis protocols for biomonitoring studies, researchers must consider (i) the chemical properties of the compounds of interest; (ii) the composition of the matrix; and (iii) the matrix's potential effects on concentrations of selected analytes. Phthalates, a group of EDCs widely used as plasticizers in consumer goods and in personal care products *(4)*, illustrate these considerations. Because some phthalates are environmentally ubiquitous, their direct measurement in biological specimens is subject to error; during sample collection, storage, and throughout the analytical measurement process, contamination can occur. Consequently, because environmental exposure levels are normally much lower than those resulting from sampling contamination, human studies using phthalate diesters as biomarkers of exposure were limited to highly exposed populations *(5–9)*. Phthalates are nonpersistent compounds that are rapidly hydrolyzed, metabolized, and excreted *(10–13)*. To minimize contamination, the preferred biomonitoring approach is to measure urinary levels of phthalate metabolites *(5,14)*. Phthalates can, however, be hydrolyzed to their monoesters by esterases present in milk *(15)*, serum *(16)*, and saliva *(17,18)*. Other matrices such as amniotic fluid and meconium may also contain esterases. Concentration of phthalate monoesters in matrices other than urine can be used to estimate exposure, but only with safeguards in place to minimize contamination with phthalates during sampling, storage, and analysis. Otherwise, measured concentrations may include an unknown contribution from hydrolysis of contaminant phthalates by endogenous esterases. This phthalate example demonstrates that when developing protocols for biomonitoring sampling, storage, and analysis in addition to the chemical properties of the compounds of interest, the composition of the matrix and its potential effects on the concentrations of the selected analytes must be considered.

4. SELECTION OF BIOLOGICAL MATRIX

After exposure, environmental chemicals may enter the body. Once they reach the blood systemic circulation, they can distribute into various body compartments, where they can be in equilibrium with blood concentrations, secretion concentrations (e.g., milk), or both. To compare concentrations in blood and other matrices, information is needed on partitioning these chemicals from blood into tissues *(19)*.

Blood (or its components) and urine are the most common matrices for biomonitoring. In general, the concentrations of persistent organic chemicals are measured in blood or blood products, whereas those of nonpersistent chemicals are most frequently measured in urine *(20)*. Furthermore, traditional persistent lipophilic compounds [e.g., organochlorine pesticides, polychlorinated biphenyls (PCBs), and polychlorinated dibenzo-*p*-dioxins (PCDDs)] can partition from blood into adipose tissue and into milk for lactating women *(21–24)*. Nonpersistent chemicals can also be found in milk. For example, phthalates and BPA, used to manufacture polycarbonate plastics and epoxy resins, have been found in human milk (*15,25–28*).

In the assessment of exposure to EDCs and other environmental contaminants, the use of unconventional matrices is becoming an increasingly important area of research *(29,30)*. Urine is generally the matrix of choice for nonpersistent chemicals—urinary concentrations of metabolites are higher than blood concentrations, urine is a relatively abundant matrix, and its collection is, in general, simple and noninvasive *(20)*. In turn, blood is the preferred matrix for persistent compounds. Amounts of blood available for analysis are, however, normally limited, and blood collection is complicated and invasive *(20)*. Depending upon the time period of concern for a particular exposure or health effect, alternative matrices to urine or blood may be preferable. For example, meconium and amniotic fluid may be used for assessing prenatal exposure to EDCs and other environmental chemicals *(31–36)*. Although alternative matrices may not be practical for large screening programs (vide infra), they may be very useful in specific situations. Some matrices are, for example, particularly useful in assessing exposure during fetal and early childhood life, when humans are most susceptible to potential adverse health effects of EDCs and other environmental chemicals. Amniotic fluid, cord blood, and meconium are promising matrices for monitoring prenatal exposures *(31,32,36)*. Breast milk can be used to monitor neonatal exposures, and its analysis can also provide an estimate of fetal exposures to some chemicals (*15,21–24,37–43*).

5. SELECTION OF BIOMARKERS OF EXPOSURE

The extent of metabolism for nonpersistent chemicals can vary substantially. Therefore, the choice of exposure biomarkers for nonpersistent chemicals is wider than for persistent chemicals, for which the biomarker of exposure is generally the parent compound. Nonpersistent EDCs can be partially or fully metabolized (e.g., phase I biotransformations) to increase their hydrophilic character. Both parent compound and phase I metabolites can be excreted unchanged or can undergo phase II biotransformations and may be used as potential biomarkers of exposure.

Low-molecular-weight phthalates mostly metabolize to their hydrolytic monoesters *(10,12,44)*. High-molecular-weight phthalates, such as those with eight or more carbons in the alkyl chain (e.g., di[2-ethylhexyl] phthalate, DEHP), metabolize to their hydrolytic monoesters, which can be further transformed to oxidative products (*11,13,44–48*). Oxidative metabolites are more water-soluble than are the corresponding hydrolytic monoesters. This explains, for a given phthalate, why the urinary oxidative metabolite concentrations are higher than hydrolytic monoester concentrations in human populations *(49–54)*. Using hydrolytic monoesters of some high-molecular-weight phthalates as biomarkers of exposure may permit exposure comparison to the parent phthalate among studies. That said, however, using these metabolites as sole biomarkers to compare relative exposures to various phthalates can be misleading, especially when comparing monoester concentrations of high- and low-molecular-weight phthalates. Metabolism of the former results in more metabolites (*11,13,44–48*), thus decreasing the relative amounts of their hydrolytic monoester metabolites. To date, research on the oxidative metabolism of phthalates has been largely limited to phthalates with a defined chemical composition (e.g., DEHP). Most high-molecular-weight phthalates are complex mixtures of isomers (e.g., di-isononyl phthalate, DiNP). Their composition varies depending on the nature of the mixture of alcohols used for their synthesis, which, in turn, may vary with manufacturers. Metabolism of isomeric high-molecular-weight phthalates will result in multiple hydrolytic and oxidative monoesters. In rats, hydrolytic monoesters represent a very small percentage of the phthalate dose *(47,55)*. Although metabolic differences among species are possible, oxidative metabolites are the most abundant urinary metabolites of isomeric high-molecular-weight phthalates in humans *(56)*. To assess properly the prevalence of exposure to these phthalates, research needs to focus on identifying and characterizing suitable oxidative metabolites. Until then, researchers will likely underestimate exposure to isomeric high-molecular-weight phthalates.

Oxidative metabolism is a common metabolic pathway in mammals. In a demographically diverse sample of US adults, the frequency of detection and urinary concentration ranges of nonyl phenol (NP) were lower than those of BPA *(57)*. Oxidative metabolism may have contributed, at least in part, to the relatively low urinary concentrations and frequency of detection of NP *(57)*. Although ingested BPA is completely recovered in urine as BPA-glucuronide within approximately 24 h after exposure in humans (Volkel et al., 2002), only 10% of the ingested NP is excreted in the urine as NP or conjugated NP within 8 h—the rest are unidentified NP metabolites *(58)*. If oxidative metabolism of NP prevails in humans as it does in animals, oxidative metabolites of NP may be the preferred biomarkers to assess exposure to NP rather than NP itself. Moreover, 4-n-NP, the measured NP isomer, represents a small percentage of the NP used in commercial mixtures *(59)*. The point is that unless the most appropriate urinary biomarker(s) is used, exposure to NP may be underestimated.

Conjugation with β-D-glucuronide and sulfate (phase II biotransformations) may reduce the bioactivity of nonpersistent EDCs and other chemicals while facilitating their urinary excretion. Phenols with ED properties, such as the phytoestrogen genistein, BPA, and parabens, are mostly excreted as conjugates *(60–62)*. In these cases, the parent compound or its conjugated metabolites can potentially be used as biomarkers of exposure *(62–64)*. Identifying and measuring the conjugated species of EDCs may, however, be challenging. Conjugated standards are not always readily available, and sensitive and accurate analytical methods are required to measure the concentrations of these species at trace levels. An alternate approach is to measure the total concentration of the compound (i.e., free plus conjugated species) after an enzymatic hydrolysis of the conjugates. This approach has been used in the measurement of EDCs such as phthalates, phytoestrogens, and BPA (*14,56,62,64–72*).

Individual variability to EDC exposure can result from changes in diet, daily lifestyle activities, and physical condition. Although biomarkers of nonpersistent chemicals in urine can accurately assess a person's exposure at a single point in time, determining exposure over an extended time period may require multiple measurements. Therefore, optimizing the design of exposure assessment in epidemiologic studies requires information on the temporal variability of urinary levels of biomarkers of nonpersistent EDCs. To date, two published studies have addressed this issue. The first study documented relatively good reproducibility of phthalate monoester concentrations in two first-morning urine specimens collected for two consecutive days from 46 African-American women. Day-to-day intra-class correlation coefficients ranged from 0.5 to 0.8 *(73)*. In the second report, the temporal variability in phthalate metabolite concentrations was evaluated among 11 men who provided up to nine urine samples each during a 3-month period *(74)*. Although substantial day-to-day and month-to-month variability in each individual's urinary phthalate metabolite concentrations existed, a single urine sample was moderately predictive of each subject's exposure over 3 months, with sensitivities ranging from 0.56 to 0.74. Between- and within-subject variances as well as the predictive ability of a single urine sample differed among phthalate metabolites. This suggests that the most efficient exposure assessment strategy for a particular epidemiologic study may depend on the chemicals of interest. Furthermore, a cross-sectional study of more than 2540 people in the USA showed variations in the population distributions of several phthalate metabolites depending on the time of day the urine samples were collected *(75)*. These observations, along with the nonpersistent nature of phthalates, may reflect differences in the timing of exposure to phthalates during the day.

Because collecting 24-h urine samples is not practical for epidemiological studies, consideration should be given to standardizing the sample collection time. When the goal is to compare relative exposures to various chemicals across populations or among different studies, collecting first-morning urine samples is preferred. Nevertheless, studies designed to explore potential health risks of EDCs should not restrict sample collection to first-morning urine samples. The above data suggest that relevant exposure opportunities in the course of a day may be missed and exposure misclassified. At minimum, always record urine-collection timing.

6. BIOMONITORING PROGRAMS

Biomonitoring programs are useful for investigating human exposure to EDCs and other environmental chemicals. In the USA, the Centers for Disease Control and Prevention (CDC) annually conducts one of these programs. The National Health and Nutrition Examination Survey (NHANES) is designed to collect data on the health and nutritional status of the noninstitutionalized civilian US population *(76)*. The survey includes a physical examination and collection of detailed medical history and biological specimens from participants. Although biological specimens are used mostly for clinical and nutritional testing, some can be used to assess exposure to environmental chemicals, including EDCs such as phthalates, phytoestrogens, PCBs, and dioxins.

Beginning with NHANES 1999, concentrations of selected chemicals in urine and blood of NHANES participants, presented by age group, sex, and race/ethnicity, have been reported in the National Reports on Human Exposure to Environmental Chemicals *(77,78)*. These reports provide the most comprehensive biomonitoring assessment of the US population's exposure to environmental chemicals. They may also help prioritize and foster research on the human health risks that result from exposure, which for many of the chemicals included in the reports are today largely unknown.

Data estimates from NHANES are probability-based and hence are representative of the US population. NHANES data can therefore be used to establish reference ranges for selected chemicals, provide exposure data for risk assessment (e.g., set intervention and research priorities, evaluate effectiveness of public health measures), and monitor exposure trends. Reference ranges can be used to assist epidemiologic investigations to correlate the levels to other NHANES parameters/measurements (including potential health effects) and to identify (i) populations with the highest exposures, (ii) potential sources/routes of exposure, and (iii) chemicals with highest prevalence/frequency *(1)*. Still, even a comprehensive program such as NHANES has limitations: persons under 1 year of age and older than 60 years of age are not included, and no data are collected on fetal exposures. Therefore, a pressing need exists for assessing exposure during critical periods of development—a period of increased susceptibility to the potential adverse effects of EDCs. Furthermore, NHANES by design provides only cross-sectional data; it intentionally excludes population groups that might be highly exposed to various point sources but which could be examined to evaluate possible associations between high exposures and adverse health effects.

In Germany, German Environmental Surveys (GerESs) have since the mid-1980s conducted large-scale representative population studies for assessing general population exposure to environmental chemicals (e.g., lead, mercury, pentachlorophenol, polycyclic aromatic hydrocarbons (PAHs), cotinine). GerESs has used various tools, including questionnaires, human biomonitoring, and both indoor and outdoor environmental samplings (http://www.umweltbundesamt.de/survey-e/index.htm). Although previous surveys focused mostly on adult exposures, the ongoing GerES IV will also provide biomonitoring data on children between 3 and 5 years of age. For example, di(2-ethylhexyl) phthalate metabolites were measured in a subset of samples collected for the GerES IV pilot study *(79)*.

7. BIOMONITORING OF EDCS (PHTHALATES, PHYTOESTROGENS, BPA, AND HALOGENATED CHEMICALS)

Since the late 1990s, both in Europe and in the USA, urinary concentrations of environmental chemicals, including EDCs or their metabolites, have been used as biomarkers to calculate human exposure among the general population to EDCs such as phytoestrogens, phthalates, and BPA *(51,57,75,78,80,81)*.

In the USA, CDC reported the levels of seven urinary phthalate monoester metabolites in three subsets of NHANES participants *(75,78,80)*. The first study included the analysis of a nonrepresentative call-back cohort of 289 urine samples, collected from adults during 1988–1994 for NHANES III *(80)*. NHANES 1999–2000 and 2001–2002 *(75,78)* provided nationally representative, population-based, phthalate metabolite concentrations in urine for selected demographic groups in the USA. Although the NHANES III and NHANES 1999–2002 data sets are not directly comparable, the frequencies of detection of the phthalate monoesters were similar. The high-molecular-weight phthalate monoesters (e.g., mono-2-ethylhexylphthalale (MEHP), mono-*n*-octyl phthalate, mono-isononyl phthalate) were detected less frequently and at lower levels than the low-molecular-weight phthalates [e.g., mono-ethylphthalate (MEP), mono-butyl phthalate (MBP), and mono-benzyl phthalate (MBzP)]. Still, the mean and median concentrations of MEP, MBP, and MBzP were lower in the NHANES 1999–2002 than in the NHANES III subset population while the MEHP concentrations remained essentially unchanged. These findings may be related to the small size and nonrepresentative nature of the sampling for the NHANES III call-back cohort or to reduced exposure to some phthalates for the NHANES 1999–2002 population. The relatively low detection frequency of the high-molecular-weight phthalate monoesters may have been, at least in part, because, unlike the oxidative phthalate monoester metabolites, these hydrolytic metabolites may not be the most sensitive urinary biomarkers (vide infra). Nonetheless, these investigations, which confirmed that human exposure to selected phthalates is widespread, are also supported by studies assessing exposure to phthalates in specific groups of individuals *(52–54,56,82–92)*.

Similarly, urinary concentrations of four isoflavones (genistein, daidzein, equol, and *O*-desmethylangolensin) and two lignans (enterolactone and enterodiol) were measured in three subsets of NHANES participants *(78,93,94)*. In the first report, urinary concentrations of these six phytoestrogens were measured in 199 adult participants of a call-back cohort for NHANES III *(93)*. NHANES 1999–2000 and 2001–2002 each included over 2500 persons 6 years of age and older and provided nationally representative, population-based, urinary phytoestrogen concentrations for selected demographic groups in the USA. In these three studies, all phytoestrogens were frequently detected, and enterolactone was detected in the highest concentrations, with daidzein being the isoflavone found at the highest median concentrations *(78,94)*. This research suggests that human exposure to phytoestrogens is widespread.

Another EDC of interest is BPA, used in the manufacture of polycarbonate plastic and epoxy resins, which in turn are used in commercial products such as baby bottles, as protective coatings on food containers, and for composites and sealants in dentistry. BPA was measured in 394 archived urine samples, from a nonrepresentative call-back cohort of US adult residents, collected during 1988–1994 for NHANES III *(57)*. BPA was detected in 95% of the samples examined at concentrations at or above 0.1 μg/L of urine. The BPA concentration ranges were similar to those observed in other human populations outside the USA. Despite the relatively small size of the sample population and its nonrepresentative character, this study provided the first reference range of human internal dose BPA levels in a demographically diverse human population. The study also confirmed widespread exposure to this compound in residents of the USA.

In addition to these examples of nonpersistent chemicals as potential EDCs, certain persistent chemicals, such as PCBs, dibenzo-*p*-dioxins, and furans have been linked with endocrine activity *(95–97)*. The manufacturing of PCBs has been banned in most countries since the 1970s. Nevertheless, because of these chemicals' continued use and environmental persistence, people continue to be exposed—although in general to much lower amounts than a generation ago *(98,99)*. For toxicological purposes, the PCBs have been divided into three main classes: those containing no *ortho* chlorine atoms (coplanar); those with one chlorine in the *ortho* position (mono-*ortho*); and those containing chlorine substitution in two or more *ortho* positions. The first two classes mentioned above are deemed to have dioxin-like activity and are discussed more with the dioxin-related chemicals. Although PCBs generally enter the environment as commercial mixtures of congeners, the relative proportion of the various congeners changes because of environmental degradation and animal metabolism. In humans, generally the highest PCB concentrations are those with two or more *ortho* positions with chlorine substitution and to a lesser extent mono-*ortho* PCBs. In addition to the PCBs themselves, PCB-hydroxylated metabolites (OH-PCBs) have been reported to have endocrine-disrupting properties. The levels of the OH-PCBs are generally in the range of 10–20% of the PCB concentrations in blood lipids.

Exposure to PCBs is through the food chain, including high-fat foods and some fish and wildlife. Therefore, the highest human PCB exposures are generally found among high consumers of contaminated fish, wildlife, and dairy products *(100–103)*. People consuming these foods in the 1960–1970 decades have much higher internal concentrations of PCBs today than do present-day consumers *(98)*. The reasons for this include much higher PCB environmental concentrations during the last century and also the bioaccumulation and long half-lives of many of the PCB congeners. Today, most of the human toxicological concerns center on newborns, infants, and young adults. Because PCBs are lipophilic and because the within-person lipid blood concentration can vary widely (e.g., because of recently eating a high-fat meal), PCB blood concentrations are generally reported on a lipid-adjusted basis as well as on the more traditional whole-weight basis.

PCDDs, dibenzofurans, and PCBs have been measured in NHANES since 1999. The NHANES 1999–2002 concentrations for these compounds were comparable with those reported in other countries *(104,105)* and were much lower than those reported in occupational settings. Most importantly, the NHANES 1999–2002 data demonstrate that human serum concentrations of these EDCs have decreased by more than 80% since the 1980s *(78,98,99)*. For example, in NHANES 2001–2002 *(78)*, the 95th percentile concentrations of 2,3,7,8-tetrachlorodibenzo-*p*-dioxin (TCDD)—one of the most potent of the dioxin-like chemicals—were 6.4 and 7.4 pg/g of lipid for women and non-Hispanic blacks, respectively. The concentrations in the remainder of the US population are most likely even lower than those reported for these two population groups.

Biomonitoring may also be valuable when a point source for exposure to low levels of EDCs is expected. For examples, exposure to BPA may occur following sealant placement. Leaching of BPA from the dental sealant into patients may be related to the quality of the sealant used *(106)*. One important biomonitoring use is to assess exposures not specifically addressed in large-survey situations that could otherwise lead to biomarker concentrations well above the mean values found in the general population. For example, some phthalates [e.g., diethyl phthalate (DEP) and dibutyl phthalate (DBP)] can be used in enteric coatings for pharmaceutical agents, whereas DEHP is extensively used in medical tubing and devices. Therefore, use of certain medications *(107,108)* and medical interventions such as platelet donations *(109,110)* or intensive care therapeutic treatments *(111–113)* can result in exposures much higher than those experienced by the general population. Similarly, industrial accidents or other catastrophes may result in very high exposures to EDCs, among other chemicals. One example is the exposure to TCDD after an explosion at a chemical plant near Seveso, Italy, in 1976, that released a mixture of chemicals, including TCDD and 2,4,5-trichlorophenol *(114–116)*. Another is the accidental poisoning by PCB-contaminated cooking oil in Yusho, Japan, and in Yucheng, Taiwan *(117–124)*. Some of the concentrations found in such cases are among the highest ever reported. Thus, these populations can serve as benchmarks for comparison of human exposure and potential adverse EDC health effects *(125–137)*.

8. SUMMARY

Biomonitoring provides a reliable estimate of internal dose. Comprehensive biomonitoring programs, such as NHANES in the USA, must continue. However, understanding of toxicokinetics of the environmental chemicals (e.g., distribution among body compartments and metabolism) and of their bioactivity at environmental exposure levels is required to properly interpret biomonitoring measurements. Age; diet; route, frequency, and magnitude of exposure; potential synergistic or antagonistic interactions among chemicals; and genetic factors, among others, are critical in determining health outcomes associated with exposure to environmental chemicals.

Because biomonitoring provides an integrated measure of exposure from all sources and routes, adequate sampling and storage protocols, and validated analytical methods that take into account both the nature of the matrix and the biomarkers must be used. For relatively lipophilic nonpersistent compounds (e.g., phthalates and alkyl phenols), formation of oxidative metabolites may be critical for facilitating their urinary excretion, and these oxidative metabolites may be the most appropriate biomarkers of exposure.

To maximize the impact of biomonitoring in public health, future research should focus on (i) identifying the compounds best suited for use as biomarkers, in particular those that may provide the greatest analytical sensitivity (e.g., oxidative metabolites), (ii) characterizing their potential bioactivity in humans, (iii) improving the understanding of their toxicokinetics in different populations and with different doses with emphasis on fetal and neonatal exposures, when susceptibility to potential adverse health effects of environmental chemicals may be highest, and (iv) studying targeted populations with known source(s) of exposure to facilitate relating internal exposure to potential health effects.

REFERENCES

1. Pirkle JL, Needham LL, Sexton K. Improving exposure assessment by monitoring human tissues for toxic chemicals. *J Expo Anal Environ Epidemiol* 1995; 5(3):405–424.
2. Pirkle JL, Sampson EJ, Needham LL, Patterson DG, Ashley DL. Using biological monitoring to assess human exposure to priority toxicants. *Environ Health Perspect* 1995; 103(Suppl 3):45–48.
3. Needham LL, Patterson DG, Barr DB, Grainger J, Calafat AM. Uses of speciation techniques in biomonitoring for assessing human exposure to organic environmental chemicals. *Anal Bioanal Chem* 2005; 381(2):397–404.
4. David RM, McKee RH, Butala JH, Barter RA, Kayser M. Esters of aromatic mono-, di-, and tricarboxylic acids, aromatic diacids, and di-, tri-, or polyalcohols. In: Bingham E, Cohrssen B, Powell CH, eds. *Patty's Toxicology*. New York: John Wiley and Sons; 2001:635–932.
5. Dirven HAAM, VandenBroek PHH, Jongeneelen FJ. Determination of 4 metabolites of the plasticizer di(2-ethylhexyl)phthalate in human urine samples. *Int Arch Occup Environ Health* 1993; 64(8):555–560.
6. Ching NPH, Jham GN, Subbarayan C, Bowen DV, Smit ALC, Grossi CE, et al. Gas chromatographic-mass spectrometric detection of circulating plasticizers in surgical patients. *J Chromatogr* 1981; 222(2):171–177.

7. Faouzi MA, Dine T, Gressier B, Kambia K, Luyckx M, Pagniez D, et al. Exposure of hemodialysis patients to di-2-ethylhexyl phthalate. *Int J Pharm* 1999; 180(1):113–121.
8. Mettang T, Thomas S, Kiefer T, Fischer FP, Kuhlmann U, Wodarz R, et al. Uraemic pruritus and exposure to di(2-ethylhexyl)phthalate (DEHP) in haemodialysis patients. *Nephrol Dial Transplant* 1996; 11(12):2439–2443.
9. Pollack GM, Buchanan JF, Slaughter RL, Kohli RK, Shen DD. Circulating concentrations of di(2-ethylhexyl) phthalate and its de-esterified phthalic-acid products following plasticizer exposure in patients receiving hemodialysis. *Toxicol Appl Pharmacol* 1985; 79(2):257–267.
10. ATSDR. *Toxicological Profile for Diethyl Phthalate (DEP)*. Atlanta, GA: Agency for Toxic Substances and Disease Registry, 1995. Available at http://www.atsdr.cdc.gov/toxprofiles/tp73.html
11. ATSDR. *Toxicological Profile for Di-n-Octyl Phthalate (DNOP)*. Atlanta, GA: Agency for Toxic Substances and Disease Registry, 1997. Available at http://www.atsdr.cdc.gov/toxprofiles/tp95.html
12. ATSDR. *Toxicological Profile for Di-n-Butyl Phthalate (DBP)*. Atlanta, GA: Agency for Toxic Substances and Disease Registry, 2001. Available at http://www.atsdr.cdc.gov/toxprofiles/tp135.html
13. ATSDR. *Toxicological Profile for Di(2-Ethylhexyl)Phthalate (DEHP)*. Atlanta, GA: Agency for Toxic Substances and Disease Registry, 2002. Available at http://www.atsdr.cdc.gov/toxprofiles/tp9.html
14. Blount BC, Milgram KE, Silva MJ, Malek NA, Reidy JA, Needham LL, et al. Quantitative detection of eight phthalate metabolites in human urine using HPLC-APCI-MS/MS. *Anal Chem* 2000; 72(17):4127–4134.
15. Calafat AM, Slakman AR, Silva MJ, Herbert AR, Needham LL. Automated solid phase extraction and quantitative analysis of human milk for 13 phthalate metabolites. *J Chromatogr B* 2004; 805(1):49–56.
16. Kato K, Silva MJ, Brock JW, Reidy JA, Malek NA, Hodge CC, et al. Quantitative detection of nine phthalate metabolites in human serum using reversed-phase high-performance liquid chromatography-electrospray ionization-tandem mass Spectrometry. *J Anal Toxicol* 2003; 27(5):284–289.
17. Silva MJ, Reidy JA, Samandar E, Herbert AR, Needham LL, Calafat AM. Detection of phthalate metabolites in human saliva. *Arch Toxicol* 2005; 79(11):647–652.
18. Niino T, Ishibashi T, Ishiwata H, Takeda K, Onodera S. Characterization of human salivary esterase in enzymatic hydrolysis of phthalate esters. *J Health Sci* 2003; 49(1):76–81.
19. Needham LL, Barr DB, Calafat AM. Characterizing children's exposures: beyond NHANES. *Neurotoxicology* 2005; 26(4):547–553.
20. Needham LL, Sexton K. Assessing children's exposure to hazardous environmental chemicals: an overview of selected research challenges and complexities. *J Expo Anal Environ Epidemiol* 2000; 10(6 Pt 2):611–629.
21. Solomon GM, Weiss PM. Chemical contaminants in breast milk: Time trends and regional variability. *Environ Health Perspect* 2002; 110(6):A339–A347.
22. Landrigan PJ, Sonawane B, Mattison D, McCally M, Garg A. Chemical contaminants in breast milk and their impacts on children's health: An overview. *Environ Health Perspect* 2002; 110(6):A313–A315.
23. Campoy C, Jimenez M, Olea-Serrano MF, Moreno-Frias M, Canabate F, Olea N, et al. Analysis of organochlorine pesticides in human milk: preliminary results. *Early Hum Dev* 2001; 65:S183–S190.
24. Hooper K, Chuvakova T, Kazbekova G, Hayward D, Tulenova A, Petreas MX, et al. Analysis of breast milk to assess exposure to chlorinated contaminants in Kazakhstan: Sources of 2,3,7,8-tetrachlorodibenzo-p-dioxin (TCDD) exposures in an agricultural region of southern Kazakhstan. *Environ Health Perspect* 1999; 107(6):447–457.
25. Mortensen GK, Main KM, Andersson AM, Leffers H, Skakkebwk NE. Determination of phthalate monoesters in human milk, consumer milk, and infant formula by tandem mass spectrometry (LC-MS-MS). *Anal Bioanal Chem* 2005; 382(4):1084–1092.
26. Otaka H, Yasuhara A, Morita M. Determination of bisphenol A and 4-nonylphenol in human milk using alkaline digestion and cleanup by solid-phase extraction. *Anal Sci* 2003; 19(12):1663–1666.
27. Sun Y, Irie M, Kishikawa N, Wada M, Kuroda N, Nakashima K. Determination of bisphenol A in human breast milk by HPLC with column-switching and fluorescence detection. *Biomed Chromatogr* 2004; 18(8):501–507.
28. Ye X, Kuklenyik Z, Needham LL, Calafat AM. Measuring environmental phenols and chlorinated organic chemicals in breast milk using automated on-line column-switching-high performance

liquid chromatography-isotope dilution tandem mass spectrometry. *J Chromatogr B* 2006; 831(1–2):110–115.

29. Needham LL, Ozkaynak H, Whyatt RM, Barr DB, Wang RY, Naeher L, et al. Exposure assessment in the National Children's Study: Introduction. *Environ Health Perspect* 2005; 113(8):1076–1082.
30. Barr DB, Wang RY, Needham LL. Biologic monitoring of exposure to environmental chemicals throughout the life stages: Requirements and issues for consideration for the National Children's Study. *Environ Health Perspect* 2005; 113(8):1083–1091.
31. Burse VW, Najam AR, Williams CC, Korver MP, Smith BF Jr, Sam PM, et al. Utilization of umbilical cords to assess in utero exposure to persistent pesticides and polychlorinated biphenyls. *J Expos Anal Environ Epidemiol* 2000; 10(6 Pt 2):776–788.
32. Foster W, Chan S, Platt L, Hughes C. Detection of endocrine disrupting chemicals in samples of second trimester human amniotic fluid. *J Clin Endocrinol Metab* 2000; 85(8):2954–2957.
33. Pichini S, Pacifici R, Pellegrini M, Marchei E, Perez-Alarcon E, Puig C, et al. Development and validation of a liquid chromatography mass spectrometry assay for the determination of opiates and cocaine in meconium. *J Chromatogr B* 2003; 794(2):281–292.
34. Pichini S, Pacifici R, Pellegrini M, Marchei E, Lozano J, Murillo J, et al. Development and validation of a high-performance liquid chromatography - Mass spectrometry assay for determination of amphetamine, methamphetamine, and methylenedioxy derivatives in meconium. *Anal Chem* 2004; 76(7):2124–2132.
35. Silva MJ, Reidy JA, Herbert AR, Preau JL, Needham LL, Calafat AM. Detection of phthalate metabolites in human amniotic fluid. *Bull Environ Contam Toxicol* 2004; 72(6):1226–1231.
36. Whyatt RM, Barr DB. Measurement of organophosphate metabolites in postpartum meconium as a potential biomarker of prenatal exposure: a validation study. *Environ Health Perspect* 2001; 109(4):417–420.
37. Kuklenyik Z, Reich JA, Tully JS, Needham LL, Calafat AM. Automated solid-phase extraction and measurement of perfluorinated organic acids and amides in human serum and milk. *Environ Sci Technol* 2004; 38(13):3698–3704.
38. LaKind JS, Wilkins AA, Berlin CM Jr. Environmental chemicals in human milk: a review of levels, infant exposures and health, and guidance for future research. *Toxicol Appl Pharmacol* 2004; 198(2):184–208.
39. Needham LL, Wang RY. Analytic considerations for measuring environmental chemicals in breast milk. [Review] [55 refs]. *Environ Health Perspect* 2002; 110(6):A317–A324.
40. Needham LL, Ryan JJ, Furst P. Guidelines for analysis of human milk for environmental chemicals. *J Toxicol Environ Health Part A* 2002; 65(22):1893–1908.
41. Rohrig L, Meisch HU. Application of solid phase micro extraction for the rapid analysis of chlorinated organics in breast milk. *Fresenius J Anal Chem* 2000; 366(1):106–111.
42. Thomsen C, Leknes H, Lundanes E, Becher G. A new method for determination of halogenated flame retardants in human milk using solid-phase extraction. *J Anal Toxicol* 2002; 26(3):129–137.
43. Pronczuk J, Akre J, Moy G, Vallenas C. Global perspectives in breast milk contamination: Infectious and toxic hazards. *Environ Health Perspect* 2002; 110(6):A349–A351.
44. Albro PW, Moore B. Identification of the metabolites of simple phthalate diesters in rat urine. *J Chromatogr* 1974; 94:209–218.
45. Albro PW, Thomas RO. Enzymatic hydrolysis of di-(2-ethylhexyl) phthalate by lipases. *Biochim Biophys Acta* 1973; 306(3):380–390.
46. Albro PW, Lavenhar SR. Metabolism of di(2-ethylhexyl)phthalate. *Drug Metabol Rev* 1989; 21(1):13–34.
47. McKee RH, El Hawari M, Stoltz M, Pallas F, Lington AW. Absorption, disposition and metabolism of di-isononyl phthalate (DINP) in F-344 rats. *J Appl Toxicol* 2002; 22(5):293–302.
48. Koch HM, Bolt HM, Angerer J. Di(2-ethylhexyl)phthalate (DEHP) metabolites in human urine and serum after a single oral dose of deuterium-labelled DEHP. *Arch Toxicol* 2004; 78(3):123–130.
49. Barr DB, Silva MJ, Kato K, Reidy JA, Malek NA, Hurtz D, et al. Assessing human exposure to phthalates using monoesters and their oxidized metabolites as biomarkers. *Environ Health Perspect* 2003; 111(9):1148–1151.
50. Kato K, Silva MJ, Reidy JA, Hurtz D, Malek NA, Needham LL, et al. Mono(2-ethyl-5-hydroxyhexyl) phthalate and mono-(2-ethyl-5-oxohexyl) phthalate as biomarkers for human exposure assessment to di-(2-ethylhexyl) phthalate. *Environ Health Perspect* 2004; 112(3):327–330.

51. Koch HM, Rossbach B, Drexler H, Angerer J. Internal exposure of the general population to DEHP and other phthalates - determination of secondary and primary phthalate monoester metabolites in urine. *Environ Res* 2003; 93(2):177–185.
52. Koch HM, Drexler H, Angerer J. Internal exposure of nursery-school children and their parents and teachers to di(2-ethylhexyl)phthalate (DEHP). *Int J Hyg Environ Health* 2004; 207(1):15–22.
53. Silva MJ, Reidy A, Preau JL, Samandar E, Needham LL, Calafat AM. Measurement of eight urinary metabolites of di(2-ethylhexyl) phthalate as biomarkers for human exposure assessment. *Biomarkers* 2006; 11(1):1–13.
54. Silva MJ, Samandar E, Preau JL, Needham LL, Calafat AM. Urinary oxidative metabolites of di(2-ethylhexyl) phthalate in humans. *Toxicology* 2006; 219(1–3):22–32.
55. Silva MJ, Kato K, Wolf C, Samandar E, Silva SS, Gray LE, et al. Urinary biomarkers of di-isononyl phthalate in rats. *Toxicology* 2006; 223(1–2):101–112.
56. Silva MJ, Reidy JA, Preau JLJ, Needham LL, Calafat AM. Oxidative metabolites of diisononyl phthalate as biomarkers for human exposure assessment. *Environ Health Perspect* 2006; 114:1158–1161.
57. Calafat AM, Kuklenyik Z, Reidy JA, Caudill SP, Ekong J, Needham LL. Urinary concentrations of bisphenol A and 4-nonylphenol in a human reference population. *Environ Health Perspect* 2005; 113(4):391–395.
58. Muller S, Schmid P, Schlatter C. Pharmacokinetic behavior of 4-nonylphenol in humans. *Environ Toxicol Pharmacol* 1998; 5(4):257–265.
59. EC. Risk Assessment Report and Summary Vol. 10. 4-nonylphenol (branched) and nonylphenol. [http://ecb.jrc.it/DOCUMENTS/Existing-Chemicals/RISK_ASSESSMENT/REPORT/4-nonylphenol_nonylphenolreport017.pdf]. 2002. Luxembourg, European Commission. Risk-Assessment Report.
60. Volkel W, Colnot T, Csanady GA, Filser JG, Dekant W. Metabolism and kinetics of bisphenol A in humans at low doses following oral administration. *Chem Res Toxicol* 2002; 15(10):1281–1287.
61. CERHR. NTP-CERHR Expert Panel Report on the Reproductive and Developmental Toxicity of Genistein. Center for the Evaluation of Risks to Human Reproduction, editor. [http://cerhr.niehs.nih.gov/chemicals/genistein-soy/genistein/ Genistein_Report_FR.pdf]. 2006. Research Triangle Park, NC, National Toxicology Program, U.S. Department of Health and Human Services.
62. Ye X, Kuklenyik Z, Needham LL, Calafat AM. Quantification of urinary conjugates of bisphenol A, 2,5-dichlorophenol, and 2-hydroxy-4-methoxybenzophenone in humans by online solid phase extraction-high performance liquid chromatography-tandem mass spectrometry. *Anal Bioanal Chem* 2005; 383(4):638–644.
63. Volkel W, Bittner N, Dekant W. Detection of bisphenol A in human urine by LC-MS/MS. *Toxicol App Pharm* 2004; 197(3):190.
64. Volkel W, Bittner N, Dekant W. Quantitation of bisphenol A and bisphenol A glucuronide in biological samples by high performance liquid chromatography-tandem mass spectrometry. *Drug Metabol Dispos* 2005; 33(11):1748–1757.
65. Kuklenyik Z, Ekong J, Cutchins CD, Needham LL, Calafat AM. Simultaneous measurement of urinary bisphenol A and alkylphenols by automated solid-phase extractive derivatization gas chromatography/mass spectrometry. *Anal Chem* 2003; 75(24):6820–6825.
66. Kuklenyik Z, Ye X, Reich JA, Needham LL, Calafat AM. Automated on-line and off-line solid phase extraction methods for measuring isoflavones and lignans in urine. *J Chromatogr Sci* 2004; 42(9):495–500.
67. Ye X, Kuklenyik Z, Needham LL, Calafat AM. Automated on-line column-switching HPLC-MS/MS method with peak focusing for the determination of nine environmental phenols in urine. *Anal Chem* 2005; 77(16):5407–5413.
68. Kato K, Silva MJ, Needham LL, Calafat AM. Determination of 16 phthalate metabolites in urine using automated sample preparation and on-line preconcentration/high-performance liquid chromatography/tandem mass spectrometry. *Anal Chem* 2005; 77(9):2985–2991.
69. Silva MJ, Malek NA, Hodge CC, Reidy JA, Kato K, Barr DB, et al. Improved quantitative detection of 11 urinary phthalate metabolites in humans using liquid chromatography-atmospheric pressure chemical ionization tandem mass spectrometry. *J Chromatogr B* 2003; 789(2):393–404.
70. Silva MJ, Slakman AR, Reidy JA, Preau JL, Herbert AR, Samandar E, et al. Analysis of human urine for fifteen phthalate metabolites using automated solid-phase extraction. *J Chromatogr B* 2004; 805(1):161–167.

71. Setchell KDR, Brown NM, Zimmer-Nechemias L, Brashear WT, Wolfe BE, Kirschner AS, et al. Evidence for lack of absorption of soy isoflavone glycosides in humans, supporting the crucial role of intestinal metabolism for bioavailability. *Am J Clin Nutr* 2002; 76(2):447–453.
72. Adlercreutz H, Vanderwildt J, Kinzel J, Attalla H, Wahala K, Makela T, et al. Lignan and isoflavonoid conjugates in human urine. *J Steroid Biochem Mol Biol* 1995; 52(1):97–103.
73. Hoppin JA, Brock JW, Davis BJ, Baird DD. Reproducibility of urinary phthalate metabolites in first morning urine samples. *Environ Health Perspect* 2002; 110(5):515–518.
74. Hauser R, Meeker JD, Park S, Silva MJ, Calafat AM. Temporal variability of urinary phthalate metabolite levels in men of reproductive age. *Environ Health Perspect* 2004; 112(17):1734–1740.
75. Silva MJ, Barr DB, Reidy JA, Malek NA, Hodge CC, Caudill SP, et al. Urinary levels of seven phthalate metabolites in the US population from the National Health and Nutrition Examination Survey (NHANES) 1999–2000. *Environ Health Perspect* 2004; 112(3):331–338.
76. CDC. National Health and Nutrition Examination Survey. National Center for Health Statistics. [http://www.cdc.gov/nchs/nhanes.htm]. 2003.
77. CDC. Second National Report on Human Exposure to Environmental Chemicals. [http://www.cdc.gov/exposurereport]. 2003. Atlanta, GA, Centers for Disease Control and Prevention; National Center for Environmental Health; Division of Laboratory Sciences.
78. CDC. Third National Report on Human Exposure to Environmental Chemicals. [http://www.cdc.gov/exposurereport/3rd/pdf/thirdreport.pdf]. 2005. Atlanta, GA, Centers for Disease Control and Prevention; National Center for Environmental Health; Division of Laboratory Sciences.
79. Becker K, Seiwert M, Angerer J, Heger W, Koch HM, Nagorka R, et al. DEHP metabolites in urine of children and DEHP in house dust. *Int J Hyg Environ Health* 2004; 207(5):409–417.
80. Blount BC, Silva MJ, Caudill SP, Needham LL, Pirkle JL, Sampson EJ, et al. Levels of seven urinary phthalate metabolites in a human reference population. *Environ Health Perspect* 2000; 108(10):979–982.
81. Koch HM, Drexler H, Angerer J. An estimation of the daily intake of di(2-ethylhexyl)phthalate (DEHP) and other phthalates in the general population. *Int J Hyg Environ Health* 2003; 206:1–7.
82. Adibi JJ, Perera FP, Jedrychowski W, Camann DE, Barr D, Jacek R, et al. Prenatal exposures to phthalates among women in New York City and Krakow, Poland. *Environ Health Perspect* 2003; 111(14):1719–1722.
83. Brock JW, Caudill SP, Silva MJ, Needham LL, Hilborn ED. Phthalate monoesters levels in the urine of young children. *Bull Environ Contam Toxicol* 2002; 68(3):309–314.
84. Duty SM, Singh NP, Silva MJ, Barr DB, Brock JW, Ryan L, et al. The relationship between environmental exposures to phthalates and DNA damage in human sperm using the neutral comet assay. *Environ Health Perspect* 2003; 111(9):1164–1169.
85. Duty SM, Silva MJ, Barr DB, Brock JW, Ryan L, Chen ZY, et al. Phthalate exposure and human semen parameters. *Epidemiology* 2003; 14(3):269–277.
86. Duty SM, Calafat AM, Silva MJ, Brock JW, Ryan L, Chen ZY, et al. The relationship between environmental exposure to phthalates and computer-aided sperm analysis motion parameters. *J Androl* 2004; 25(2):293–302.
87. Duty SM, Calafat AM, Silva MJ, Ryan L, Hauser R. Phthalate exposure and reproductive hormones in adult men. *Human Reprod* 2005; 20(3):604–610.
88. Duty SM, Ackerman RM, Calafat AM, Hauser R. Personal care product use predicts urinary concentrations of some phthalate monoesters. *Environ Health Perspect* 2005; 113(11):1530–1535.
89. Jonsson BAG, Richthoff J, Rylander L, Giwercman A, Hagmar L. Urinary phthalate metabolites and biomarkers of reproductive function in young men. *Epidemiology* 2005; 16(4):487–493.
90. Koch HM, Preuss R, Drexler H, Angerer J. Exposure of nursery school children and their parents and teachers to di-n-butylphthalate and butylbenzylphthalate. *Int Arch Occup Environ Health* 2005; 78(3):223–229.
91. Swan SH, Main KM, Liu F, Stewart SL, Kruse RL, Calafat AM, et al. Decrease in anogenital distance among male infants with prenatal phthalate exposure. *Environ Health Perspect* 2005; 113(8):1056–1061.
92. Main KM, Mortensen GK, Kaleva MM, Boisen KA, Damgaard IN, Chellakooty M, et al. Human breast milk contamination with phthalates and alterations of endogenous reproductive hormones in infants three months of age. *Environ Health Perspect* 2006; 114(2):270–276.

93. Valentin-Blasini L, Blount BC, Caudill SP, Needham LL. Urinary and serum concentrations of seven phytoestrogens in a human reference population subset. *J Expo Anal Environ Epidemiol* 2003; 13(4):276–282.
94. Valentin-Blasini L, Sadowski MA, Walden D, Caltabiano L, Needham LL, Barr DB. Urinary phytoestrogen concentrations in the U. S. population (1999-2000). *J Expo Anal Environ Epidemiol* 2005; 15(6):509–523.
95. Fierens S, Mairesse H, Heilier JF, De Burbure C, Focant JF, Eppe G, et al. Dioxin/polychlorinated biphenyl body burden, diabetes and endometriosis: findings in a population-based study in Belgium. *Biomarkers* 2003; 8(6):529–534.
96. Johnson ES, Shorter C, Bestervelt LL, Patterson DG, Needham LL, Piper WN, et al. Serum hormone levels in humans with low serum concentrations of 2,3,7,8-TCDD. *Toxicol Ind Health* 2001; 17(4):105–112.
97. Persky V, Turyk M, Anderson HA, Hanrahan LP, Falk C, Steenport DN, et al. The effects of PCB exposure and fish consumption on endogenous hormones. *Environ Health Perspect* 2001; 109(12):1275–1283.
98. Aylward LL, Hays SM. Temporal trends in human TCDD body burden: Decreases over three decades and implications for exposure levels. *J Expo Anal Environ Epidemiol* 2002; 12(5):319–328.
99. Lorber M. A pharmacokinetic model for estimating exposure of Americans to dioxin-like compounds in the past, present, and future. *Sci Total Environ* 2002; 288(1–2):81–95.
100. Anderson HA, Falk C, Hanrahan L, Olson J, Burse VW, Needham L, et al. Profiles of great lakes critical pollutants: A sentinel analysis of human blood and urine. *Environ Health Perspect* 1998; 106(5):279–289.
101. Goldman LR, Harnly M, Flattery J, Patterson DG, Needham LL. Serum polychlorinated dibenzo-p-dioxins and polychlorinated dibenzofurans among people eating contaminated home-produced eggs and beef. *Environ Health Perspect* 2000; 108(1):13–19.
102. Hanrahan LP, Falk C, Anderson HA, Draheim L, Kanarek MS, Olson J. Serum PCB and DDE levels of frequent Great Lakes sport fish consumers - A first look. *Environ Res* 1999; 80(2):S26–S37.
103. Shadel BN, Evans RG, Roberts D, Clardy S, Jordan-Izaguirre D, Patterson DG, et al. Background levels of non-ortho-substituted (coplanar) polychlorinated biphenyls in human serum of Missouri residents. *Chemosphere* 2001; 43(4–7):967–976.
104. Papke O. PCDD/PCDF: Human background data for Germany, a 10-year experience. *Environ Health Perspect* 1998; 106:723–731.
105. Bates MN, Buckland SJ, Garrett N, Ellis H, Needham LL, Patterson DG, et al. Persistent organochlorines in the serum of the non-occupationally exposed New Zealand population. *Chemosphere* 2004; 54(10):1431–1443.
106. Joskow R, Barr DB, Barr JR, Calafat AM, Needham LL, Rubin C. Exposure to bisphenol a from bis-glycidyl dimethacrylate-based dental sealants. *J Am Dent Assoc* 2006; 137:353–362.
107. Hauser R, Duty S, Godfrey-Bailey L, Calafat AM. Medications as a source of human exposure to phthalates. *Environ Health Perspect* 2004; 112(6):751–753.
108. Koch HM, Muller J, Drexler H, Angerer J. Dibutylphthalate (DBP) in medications: are pregnant women and infants at risk. *Umweltmed Forsch Prax* 2005; 10(2):144–146.
109. Koch HM, Angerer J, Drexler H, Eckstein R, Weisbach V. Di(2-ethylhexyl)phthalate (DEHP) exposure of voluntary plasma and platelet donors. *Int J Hyg Environ Health* 2005; 208(6):489–498.
110. Koch HM, Bolt HM, Preuss R, Eckstein R, Weisbach V, Angerer J. Intravenous exposure to di(2-ethylhexyl)phthalate (DEHP): metabolites of DEHP in urine after a voluntary platelet donation. *Arch Toxicol* 2005; 79(12):689–693.
111. Calafat AM, Needham LL, Silva MJ, Lambert G. Exposure to di-(2-ethylhexyl) phthalate among premature neonates in a neonatal intensive care unit. *Pediatrics* 2004; 113(5):e429–e434.
112. Green R, Hauser R, Calafat AM, Weuve J, Schettler T, Ringer S, et al. Use of di(2-ethylhexyl) phthalate-containing medical products and urinary levels of mono(2-ethylhexyl) phthalate in neonatal intensive care unit infants. *Environ Health Perspect* 2005; 113(9):1222–1225.
113. Weuve J, Sanchez BR, Calafat AM, Schettler T, Green RA, Hu H, et al. Exposure to phthalates in neonatal intensive care unit infants: urinary concentrations of monoesters and oxidative metabolites. *Environ Health Perspect* 2006; 114:1424–1431.
114. Needham LL, Gerthoux PM, Patterson DG, Brambilla P, Smith SJ, Sampson EJ, et al. Exposure assessment: Serum levels of TCDD in Seveso, Italy. *Environ Res* 1999; 80(2):S200–S206.
115. Needham LL, Gerthoux PM, Patterson DG, Brambilla P, Turner WE, Beretta C, et al. Serum dioxin levels in Seveso, Italy, population in 1976. *Teratog Carcinog Mutagen* 1997; 17(4–5):225–240.

116. Mocarelli P, Needham LL, Marocchi A, Patterson DG, Brambilla P, Gerthoux PM, et al. Serum concentrations of 2,3,7,8-tetrachlorodibenzo-para-dioxin and test-results from selected residents of Seveso, Italy. *J Toxicol Environ Health* 1991; 32(4):357–366.
117. Ryan JJ, Levesque D, Panopio LG, Sun WF, Masuda Y, Kuroki H. Elimination of Polychlorinated Dibenzofurans (Pcdfs) and Polychlorinated-Biphenyls (Pcbs) from Human Blood in the Yusho and Yu-Cheng Rice Oil Poisonings. *Arch Environ Contam Toxicol* 1993; 24(4):504–512.
118. Ryan JJ, Hsu CC, Boyle MJ, Guo YLL. Blood-serum levels of Pcdfs and Pcbs in Yu-Cheng children peri-natally exposed to a toxic rice oil. *Chemosphere* 1994; 29(6):1263–1278.
119. Soong DK, Ling YC. Reassessment of PCDD/DFs and Co-PCBs toxicity in contaminated rice-bran oil responsible for the disease "Yu-Cheng". *Chemosphere* 1997; 34(5–7):1579–1586.
120. Masuda Y, Schecter A, Papke O. Concentrations of PCBs, PCDFs and PCDDs in the blood of Yusho patients and their toxic equivalent contribution. *Chemosphere* 1998; 37(9–12):1773–1780.
121. Iida T, Hirakawa H, Matsueda T, Takenaka S, Yu ML, Guo YLL. Recent trend of polychlorinated dibenzo-p-dioxins and their related compounds in the blood and Sebum of Yusho and Yu-Cheng patients. *Chemosphere* 1999; 38(5):981–993.
122. Aoki Y. Polychlorinated biphenyls, polychloronated dibenzo-p-dioxins, and polychlorinated dibenzofurans as endocrine disrupters - What we have learned from Yusho disease. *Environ Res* 2001; 86(1):2–11.
123. Masuda Y. Fate of PCDF/PCB congeners and change of clinical symptoms in patients with Yusho PCB poisoning for 30 years. *Chemosphere* 2001; 43(4–7):925–930.
124. Furue M, Uenotsuchi T, Urabe K, Ishikawa T, Kuwabara M. Overview of Yusho. *J Derm Sci* 2005; S3–S10.
125. Eskenazi B, Warner M, Marks AR, Samuels S, Gerthoux PM, Vercellini P, et al. Serum dioxin concentrations and age at menopause. *Environ Health Perspect* 2005; 113(7):858–862.
126. Warner M, Samuels S, Mocarelli P, Gerthoux PM, Needham L, Patterson DG, et al. Serum dioxin concentrations and age at menarche. *Environ Health Perspect* 2004; 112(13):1289–1292.
127. Eskenazi B, Mocarelli P, Warner M, Needham L, Patterson DG, Samuels S, et al. Relationship of serum TCDD concentrations and age at exposure of female residents of Seveso, Italy. *Environ Health Perspect* 2004; 112(1):22–27.
128. Eskenazi B, Mocarelli P, Warner M, Chee WY, Gerthoux PM, Samuels S, et al. Maternal serum dioxin levels and birth outcomes in Women of Seveso, Italy. *Environ Health Perspect* 2003; 111(7):947–953.
129. Warner M, Eskenazi B, Mocarelli P, Gerthoux PM, Samuels S, Needham L, et al. Serum dioxin concentrations and breast cancer risk in the Seveso Women's Health Study. *Environ Health Perspect* 2002; 110(7):625–628.
130. Eskenazi B, Mocarelli P, Warner M, Samuels S, Vercellini P, Olive D, et al. Serum dioxin concentrations and endometriosis: a cohort study in Seveso, Italy. *Environ Health Perspect* 2002; 110(7):629–634.
131. Eskenazi B, Warner M, Mocarelli P, Samuels S, Needham LL, Patterson DG, et al. Serum dioxin concentrations and menstrual cycle characteristics. *Am J Epidemiol* 2002; 156(4):383–392.
132. Eskenazi B, Mocarelli P, Warner M, Samuels S, Vercellini P, Olive D, et al. Seveso Women's Health Study: a study of the effects of 2,3,7,8-tetrachlorodibenzo-p-dioxin on reproductive health. *Chemosphere* 2000; 40(9–11):1247–1253.
133. Mocarelli P, Gerthoux PM, Ferrari E, Patterson DG, Kieszak SM, Brambilla P, et al. Paternal concentrations of dioxin and sex ratio of offspring. *Lancet* 2000; 355(9218):1858–1863.
134. Guo YL, Lin CJ, Yao WJ, Ryan JJ, Hsu CC. Musculoskeletal changes in children prenatally exposed to polychlorinated-biphenyls and related-compounds (Yu-Cheng children). *J Toxicol Environ Health* 1994; 41(1):83–93.
135. Guo YLL, Chen YC, Yu ML, Hsu CC. Early development of Yu-Cheng Children Born 7 to 12 years after the Taiwan PCB outbreak. *Chemosphere* 1994; 29(9–11):2395–2404.
136. Hsieh SF, Yen YY, Lan SJ, Hsieh CC, Lee CH, Ko YC. A cohort study on mortality and exposure to polychlorinated biphenyls. *Arch Environ Health* 1996; 51(6):417–424.
137. Yoshimura T, Kaneko S, Hayabuchi H. Sex ratio in offspring of those affected by dioxin and dioxin-like compounds: the Yusho, Seveso, and Yucheng incidents. *Occup Environ Med* 2001; 58(8):540–541.

III Implications and Mitigations of EDCs

12 Policy Implications of Endocrine-Disrupting Chemicals in Humans

*Tracey J. Woodruff,** PHD, MPH

CONTENTS

1. INTRODUCTION

The significance of endocrine-disrupting chemical (EDC) effects on human health is becoming more apparent. EDCs are known or suspected to play a role in many illnesses, such as breast cancer *(1,2)*, which has been increasing, and other markers of population health, such as age at onset for puberty, which has been going down *(3)*. Ideally, the role of public policies is to reduce and prevent these adverse health effects in the human population. An environmental health framework (Fig. 1) shows how chemicals, such as EDCs, are released from sources, such as industry or transportation. They can then enter into environmental media, such as air, food, water, and consumer products, where there is opportunity for human exposures. Levels of some contaminants can be measured directly in people and eventually could manifest into illness. Reducing and preventing illness requires a policy focus on the upstream events: sources and releases into the environment. Monitoring for EDC-mediated health effects in people is important for measuring policy success and assessing needs for further interventions.

This chapter will discuss current policies and their limitations in identifying EDCs. The process of collecting and synthesizing the available science for use in developing policies and regulations to protect human health and alternative policy options at the

* Written while on sabbatical from the US Environmental Protection Agency. The views in this chapter are those of the author and not of the US Environmental Protection Agency.

From: *Endocrine-Disrupting Chemicals: From Basic Research to Clinical Practice*
Edited by: A. C. Gore © Humana Press Inc., Totowa, NJ

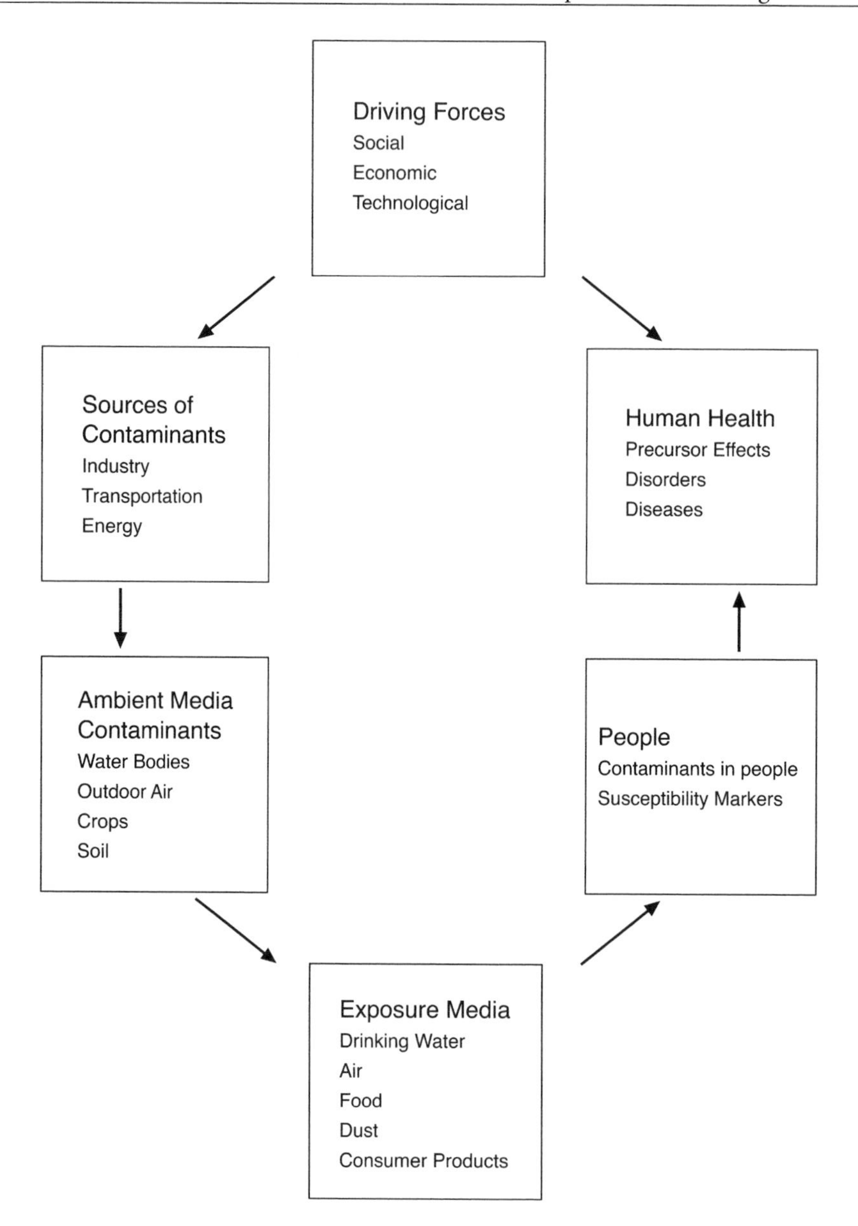

Fig. 1. Conceptual framework to represent relationships between environmental factors and health [adapted from kyle et al. *(44)*].

national level. A complementary discussion of biomonitoring for exposure assessment is provided in Chapter 11 (Calafat and Needham).

2. WHO REGULATES EDCs?

Most laws addressing environmental contaminants passed by the US Congress are administered by the US Environmental Protection Agency (USEPA). These laws cover environmental contaminants and their sources primarily for air, drinking water, pesticides in food and other exposures, contaminated waste, and land. USEPA develops regulations and policies intended to implement these laws. There are also important sources of exposure to chemicals through consumer products, cosmetics, pharmaceuticals, and chemical additives to food. The laws passed to address these routes of exposure are primarily administered by the Food and Drug Administration (FDA). Many of the laws governing environmental contaminants require assessing the potential for harm to human health as support for the development of policies and regulations designed to limit chemical exposures. It should be noted, however, that there are other tangible factors, such as technology and costs and benefits, and intangible factors, such as public comment and political pressures, that can also play a role, but will not be covered in this chapter *(4)*.

Assessing the potential for harm to human health requires assembling and synthesizing the available science. The available scientific information comes primarily from toxicological studies using animals as test subjects and epidemiology studies that evaluate relationships between exposures and effects in humans. Toxicological studies are the primary source of information for the vast majority of chemicals. A routine set of toxicological information includes tests that evaluate toxicity from acute, subchronic, and chronic exposures and evaluates health effects for reproductive and developmental effects, neurodevelopment effects, effects on major organ systems, and cancer only from chronic exposure *(5)*. Although basic toxicity information may include testing for some endpoints that could be endocrine-mediated, such as those related to reproductive and developmental toxicity, conventional toxicological protocols for reproductive and developmental effects are not generally designed to detect whether observed effects are endocrine related *(6)*. In addition, the tests do not adequately address exposures and effects that are important for EDCs. For example, the routine tests do not include evaluating the toxic effects that may occur from prenatal or early postnatal exposures *(5)*.

As of 2006, the USEPA estimated that there are approximately 75,000 chemical substances in commerce *(7)*. Although many of these chemicals are used in small quantities or within closed manufacturing systems, about 2,800 are used or imported in high volumes (over 1 million pounds) *(8)*. In addition, there are also pesticides (about 900 active ingredients and 2,500 inert ingredients), nuclear material, and chemicals in food (about 3,000), and chemicals in drugs and cosmetics (about 5,000) *(9)* This brings the total number of chemicals to about 87,000, although it is difficult to ascertain the complete number of chemicals, because no single entity records the universe of chemicals used in the US.

Of the 87,000 chemicals, there is limited basic toxicity information for many chemicals. For example, as of 1998, there were 2800 high production volume chemicals in commerce and just 7% had the full set of basic toxicological data and 43% had no testing data *(10)*. Because testing data for most chemicals are incomplete, existing reproductive and developmental tests provide limited help to identify EDCs. A voluntary industry initiative was started in 1998 to collect basic screening data and some information has been added *(8)*.

2.1. Identifying EDCs

Concern for EDC-mediated health effects in humans emerged in the early 1990s, along with growing awareness that the existing laws and regulations were insufficient for EDCs. This prompted a focus on improving the science base for EDCs as one means of improving policies and regulations related to EDCs. Congress passed the Food

Quality Protection Act in 1996, in large part because of concerns about the inadequacy of the existing pesticide law to protect infants and children from pesticide exposures. In the Food Quality Protection Act, Congress also recognized the emerging concern for endocrine-related health effects and inadequate testing for potential endocrine disruption. The Food Quality Protection Act requires USEPA to "develop a screening program, using appropriate validated test systems and other scientifically relevant information, to determine whether certain substances may have an effect in humans that is similar to an effect produced by a naturally occurring estrogen, or such other endocrine effect" *(11)*. It further requires that the screening "(i) shall provide for the testing of all pesticide chemicals and (ii) may provide for the testing of any other substance that may have an effect that is cumulative to an effect of a pesticide chemical if the Administrator determines that a substantial population may be exposed to such substance" *(11)*. Further language in the amendments to the Safe Drinking Water Act also authorized USEPA to screen "contaminants in drinking water to which substantial number of people would be exposed" *(12)*.

USEPA established the Endocrine Disruptor Screening and Testing Advisory Committee (EDSTAC), made up of a broad range of constituencies, to advise the agency on developing a practical and scientifically defensible endocrine disruptor-screening strategy *(6)*. In 1998, EDSTAC proposed 71 consensus recommendations for the agency *(9)*, from which USEPA has developed an Endocrine Disruptor Screening Program (EDSP) *(13)*. USEPA adopted EDCSTAC's recommendation to screen for estrogen, androgen, and thyroid effects and to screen for human and ecological effects. In addition, EDSP will include a broad universe of chemicals in the screening program beyond pesticides and drinking water contaminants *(9)*. As USEPA does not have regulatory authority over cosmetics, food additives, and nutritional supplements, the focus of the screening program will be on pesticides, industrial chemicals, and environmental contaminants *(6)*, although USEPA states it is committed to collaborate with other appropriate federal agencies to facilitate examining the other chemicals *(6)*.

USEPA proposed in 1998 a two-tiered approach to toxicity testing based on the EDSTAC recommendations *(9,13)*. Tier 1 is a screening battery of assays designed to test for potential interactions with the endocrine system and includes in vitro and in vitro screens (Table 1) *(6,9)*. Those chemicals identified in Tier 1 as interacting with the endocrine system will be tested in Tier 2 to identify adverse health effects from endocrine disruption and to identify dose–response relationships (mammal-testing requirements summarized in Table 1). This includes testing of the most sensitive developmental life stages and multi-generational effects for five major taxonomic groups (mammals, birds, fish, amphibians, and invertebrates) *(6,9)*. The information from the Tier 2 tests would then be used in a hazard assessment, which can then lead to regulation *(9)*.

In September of 2005, USEPA published the approach it will take for selecting an initial group of chemicals to screen under the EDSP *(14)*. The approach proposed to select 50–100 chemicals for screening, with the list being named in a future proposal. These chemicals will be selected from (i) pesticide active ingredients and (ii) chemicals that are pesticide inerts that have relatively large overall volume considering both pesticide and nonpesticide use *(14)*. There were two types of chemicals that are excluded from consideration. The first are chemicals that USEPA is using as "positive controls" to validate screening assays. The second are chemicals which are anticipated

Table 1
US Environmental Protection Agency-Proposed Testing under the Endocrine Disruptor-Screening Program based on the Recommendations of the Endocrine Disruptor Screening and Testing Advisory Committee

Tier 1 screening batter
In vitro screens
Estrogen Receptor (ER) binding/reporter gene assay
Androgen Receptor (AR) binding/reporter gene assay
Steroidogenesis assay
In vivo screens
Rodent 3-day uterotrophic assay (s.c.)
Rodent 20-day pubertal female assay with thyroid
Rodent 5–7 day Hershberger assay
Frog metamorphosis assay
Fish reproduction-screening assay
Alternative screening assays
In vitro placental aromatase
In vivo rodent 20-day pubertal male assay with thyroid
In vivo adult male assay
Tier 2 mammalian toxicity testing
Two-Generation Mammalian Reproductive Toxicity Study (currently required for pesticides with widespread outdoor exposures expected to affect reproduction). Alternative tests (somewhat less comprehensive, but can include in utero and/or lactational exposures)
Alternative mammalian reproduction test (AMRT)
1-Generation reproductive test

to have low potential to cause endocrine disruption: polymers with numerical average molecular weight greater than 1000 Daltons (these are considered unlikely to pass through biological membranes) and strong acids or strong mineral bases (these are considered highly reactive and would cause damage at the point of entry but would not be expected to be associated with systemic toxicity) *(14)*.

USEPA initially proposed in 1998 to develop the candidate list for the initial endocrine-screening program based primarily on exposure and hazard-related data *(13)*. To supplement the lack of toxicity data and facilitate future endocrine-screening priority setting activities, USEPA also proposed to investigate using a high throughput prescreening for chemicals which are produced in excess of 10,000 pounds per year (about 15,000 chemicals) *(9,14,15)*. The prescreening method is an automated cell-based test system that can detect estrogen and androgen receptor interactions on thousands of chemicals *(15)*. However, as of 2006, USEPA indicates that the high throughput methods are "not yet sufficiently developed for regulatory purposes" *(15)*, thus limiting the toxicological data available for prioritizing chemicals for an endocrine-screening program.

The final proposal in 2005 based the primary consideration for prioritizing the initial group of chemicals on the potential for human and ecological exposures. The types of exposure data considered for selecting pesticide active ingredients include presence in food, presence in drinking water, residential use, and occupational contact *(14)*.

The initial high production/inert ingredients will be prioritized based on monitoring data showing presence in human tissue, fish tissue, drinking water, or indoor air *(14)*. Although it is reasonable to primarily consider the potential for exposure, as many of the chemicals will lack sufficient toxicological data for screening, the approach is limited by the lack of comprehensive monitoring data in the four areas. For example, chemical presence in human tissue is based primarily on biomonitoring data from the National Health and Nutrition Examination Survey (NHANES), which includes about 150 chemicals *(14)*. There are more chemicals monitored in the fish tissue and drinking water databases, but the number is typically in the range of 100–300 chemicals, whereas the potential universe of chemicals is in the thousands *(14)*.

There has been great effort developing a screen to identify EDCs, and the program could eventually produce a large amount of data on the endocrine potential for many chemicals. However, the time frame for testing has been long and difficult to predict. USEPA proposed in 1998 to develop a method for priority setting for Tier 1 phase 1 chemicals, which was to be completed by 2000 *(13)*. This was pushed back to 2005, when USEPA published the final priority-setting method *(14)*. This initial effort will just identify the first 50–100 chemicals for Tier 1 screening within the universe of 87,000. USEPA proposes to complete the Tier 1 screening for these approximately 100 chemicals by 2006 *(14)* compared with the original proposal of 2001 *(13)*. Even if 100 chemicals could eventually be screened through Tier 1 every year, it would still take 28 years to screen just the high production volume chemicals using current technologies. And this does not include the Tier 2 screens, which take more time. The Tier 2 screens are required before making a hazard identification, which comes before any regulatory activity.

There are many important issues to be deliberated in this process, and there are many different interested stakeholders, including industry and environmental groups, who are weighing in on the steps in the process, all of which add considerably to the length of time for producing relevant scientific information. In addition, this only considers the development of the screening methodology. Once the data start to come in, evaluating the information adds another layer of deliberation and discussion. Although it is important to have an open process, it also increases the length of time before any action takes place, which has consequences for health.

The length of time to decision is compounded by USEPA's reluctance to use existing information to identify "potential endocrine disruptor" chemicals *(14)*. In the initial 1998 proposal, USEPA proposed to include two other categories of chemicals (in addition to those that are candidates and those that are excluded). These categories comprise (i) chemicals that have sufficient data to bypass the Tier 1 screening, but need further testing and (ii) chemicals for which there was adequate data to be referred for hazard assessment *(13)*. It was estimated that there would be some chemicals in these categories (hundreds), which is still small compared with the thousands considered for screening *(6)*.

In the final 2005 proposal, USEPA decided to eliminate the categories where some data existed that would allow a chemical to go either into the Tier 2 screen or directly to hazard assessment *(14)*. USEPA noted it "generally believes that it lacks sufficient information and experience to determine whether a chemical should be designated as a 'potential endocrine disruptor.' " *(14)*. In addition, it also states for pesticides:

EPA has decided for this initial list that it would be impractical to establish criteria for judging whether a chemical should be designated as a "potential endocrine disruptor" and removed from the initial group for screening. Although a relatively broad range of toxicity data are available for pesticide active ingredients regulated under FIFRA, in most cases EPA has not yet established how the available data might be confidently used to predict the endocrine disruption potentials of these chemicals. This may be due to the non-specific nature of an effect or effects observed, questions related to whether the mode of action in producing a given effect or effects is or are endocrine system-mediated in whole or in part, or the lack of relevant data to make a judgment altogether (14)

While the text primarily refers to pesticides, it has implications for other chemicals as well. There appears to be a shift on how existing information is considered for identifying chemicals as potential EDCs. In 1998, USEPA noted that the Tier 2 testing for reproductive effects is the same as that currently used for pesticides with widespread outdoor exposures that are expected to affect reproduction *(13)*. Now, it appears that even these pesticides are part of the chemicals required to be screened both in Tier 1 and in Tier 2 testing. This establishes a relatively high level of information before deciding and regulating based on endocrine-related effects. Finally USEPA indicates how it will evaluate the data, adding a further layer of complexity and time.

Although identifying chemicals that are potential EDCs requires some scientific knowledge, for many chemicals there appears to be sufficient data to determine whether they are potential endocrine disruptors as outlined in 1998. No official government body has a list of chemicals that are potential endocrine disruptors, although scientific reviews have identified many chemicals as potential endocrine disruptors, for example, dioxins, PCBs, atrazine, and phthalates *(16–18)*. How available scientific information for potential EDCs is evaluated illustrates several important features that should be considered in the policy context, which include differences between use of science in the laboratory and policy setting and weighing the scientific evidence, and identifying adverse effects.

2.2. Assessing Science Information and the Policy Process

Deciding whether a chemical is a "potential endocrine disruptor" is based on whether there is available scientific information and on the synthesis of that information. The synthesis is typically done through a weight of evidence evaluation, which considers the scientific information as a whole and provides judgment on the potential for harm from the chemical. This requires a judgment about the strength of the scientific evidence, or level of proof required, to make a decision. Determining the level of proof within the policy arena considers different factors than within the laboratory setting.

Weighing the evidence in a laboratory setting typically requires a higher level of proof, which emphasizes reducing uncertainties and minimizing false positives. These are appropriate emphases in the laboratory setting, as the goal is to assure that observed relationships are true. However, meeting these goals can require more time and effort to gather sufficient evidence, and this can have consequences in the public policy setting.

The policy process considers other factors in deciding on the level of proof when weighing the evidence, such as risks from chemical exposure to the health of the public, benefits of the proposed chemical, alternatives to the chemical, and public policy goals *(19)*. Waiting to address all uncertainties in the science of EDCs before developing regulatory policy can be time-consuming and without an identified end, which allows for exposures to continue, potentially putting the public health at continued risk. In addition, minimizing false positives can lead to increasing the number of false negatives, also with potential adverse impacts on public health.

An example of this would be Diethylstibestrol (DES), a synthetic estrogen, which was prescribed for women from the late 1940s through the 1970s *(20)*. When it was prescribed starting in 1947, it had undergone limited toxicological investigation, and it was assumed to be safe, even beneficial, during pregnancy *(20)*. However, it was later discovered in 1971 that DES lead to vaginal clear-cell adenocarcinoma in the prenatally exposed daughters *(21)*, a clear case of a false negative. In addition, in the years after discovery of the prenatal effects of DES, there have been over 20,000 publications; yet, there are still uncertainties about the exact nature of DES and subsequent health effects *(19)*.

As has been described by D. Gee, there are many features of the methods used in the scientific study of the relationship between environmental contaminants and health, and a higher proportion of these features increase the chances of finding a false negative (Table 2). Although minimizing false positives can lead to more comprehensive science, it can also have untoward consequences for public health. In a public health context, an emphasis on reducing false negatives may be more important if the goal is to reduce harm.

In the context of identifying EDCs, USEPA states that that it lacks sufficient information and experience to identify potential EDCs, which suggest a greater emphasis is placed on a higher level of proof or scientific evidence before decision-making. This has implications for policies addressing EDCs and other chemicals.

Part of the tension in deciding the nature of the chemical, such as whether it is a potential endocrine disruptor, is the dichotomous decision, either yes or no, when the evidence may be more graded. An example of another process that acknowledges different levels of information is for assessment of hazard of carcinogenesis. This is discussed in further detail below (section 2.2.1).

2.2.1. An Example of Weighing the Evidence Under Different Levels of Available Scientific Information

There are other existing processes for using a gradation when assessing the weight of evidence. For example, for cancer, the process used by the USEPA [which is similar to that of the International Agency for Research on Cancer (IARC)] recognizes that there are different levels of information linking chemicals and cancer, and USEPA had adopted different descriptors that "represent points along a continuum of evidence" *(22)*. The highest level of evidence is for those carcinogens that are well-established "Carcinogenic to Humans," with lesser evidence for "Likely to Be Carcinogenic to Humans," "Suggestive Evidence of Carcinogenic Potential," and "Inadequate Information to Assess Carcinogenic Potential" (Table 3). An exception is for "Not Likely to be Carcinogenic to Humans," similar to "Carcinogenic to Humans," requires a high level of evidence.

Table 2
Methodological Features in Environmental and Health Sciences and Their Directions of Error [a]

Scientific study type	Some methodological features	Main[b] directions of error increase chances of detecting	
		False positive	False negative
Experimental studies (animal laboratory)	High doses	X	
	Short (in biological terms) range of doses		X
	Low genetic variability		X
	Few exposures to mixtures		X
	Few fetal-lifetime exposures (high fertility strains)		X
			X (developmental/reproductive end points
Observational studies (wildlife and humans)	Confounders	X	
	Inappropriate controls	X	X
	Nondifferential exposure misclassification		X
	Inadequate follow-up		X
	Lost cases		X
	Simple models that do not reflect complexity		X
Both experimental and observational studies	Publication bias toward positives	X	
	Scientific/cultural pressure to avoid false positives e		X
	Low statistical power (e.g., from small studies)		X
	Use of 5 % probability level to minimize chances of false positives		X

[a]Adapted from D. Gee *(19)*.
[b]Some features can go either way (e.g., inappropriate controls), but most of the features err mainly in the direction shown.

Table 3
US Environmental Protection Agency Cancer Descriptors Based on Weight of Evidence

Carcinogenic to Humans

This descriptor indicates strong evidence of human carcinogenicity. It covers different combinations of evidence.

Convincing epidemiologic evidence of a causal association between human exposure and cancer.

Exceptionally, with a lesser weight of epidemiologic evidence strengthened by the following conditions:

(1) there is strong evidence of an association between human exposure and either cancer or the key precursor events of the agent's mode of action but not enough for a causal association;

(2) there is extensive evidence of carcinogenicity in animals;

(3) the mode(s) of carcinogenic action and associated key precursor events have been identified in animals; and

(4) there is strong evidence that the key precursor events that precede the cancer response in animals are anticipated to occur in humans and progress to tumors, based on available biological information.

Likely to Be Carcinogenic to Humans

This descriptor is appropriate when the weight of the evidence is adequate to demonstrate carcinogenic potential to humans but does not reach the weight of evidence for the descriptor "Carcinogenic to Humans." Adequate evidence consistent with this descriptor covers a broad spectrum.

(1) An agent demonstrating a plausible (but not definitively causal) association between human exposure and cancer, often with some supporting biological, experimental evidence, although not necessarily carcinogenicity data;

(2) An agent that has tested positive in animal experiments in more than one species, sex, strain, site, or exposure route, with or without evidence of carcinogenicity in humans;

(3) A positive study that raises additional biological concerns beyond that of a statistically significant result, for example, a high degree of malignancy or an early age at onset;

(4) A rare animal tumor response in a single experiment that is assumed to be relevant to humans; or

(5) A positive study that is strengthened by other lines of evidence, for example, either plausible (but not definitively causal) association between human exposure and cancer or evidence that the agent or an important metabolite causes events generally known to be associated with tumor formation (such as DNA reactivity or effects on cell growth control) likely to be related to the tumor response in this case.

Suggestive Evidence of Carcinogenic Potential

This descriptor of the database is appropriate when the weight of evidence is suggestive of carcinogenicity; a concern for potential carcinogenic effects in humans is raised, but the data are judged not sufficient for a stronger conclusion. This descriptor covers a spectrum of evidence associated with varying levels of concern for carcinogenicity, ranging from a positive result in the only study on an agent to a single positive result in an extensive database that includes negative studies in other species. Some examples include

(1) A small increase in tumors observed in a single animal or human study that does not reach the weight of evidence for the descriptor "Likely to Be Carcinogenic to Humans."

(2) A small increase in a tumor with a high background rate in that sex and strain, when there is some but insufficient evidence that the observed tumors may be due to intrinsic factors that cause background tumors and not due to the agent being assessed;

(*continued*)

Table 3
(*Continued*)

(3) Evidence of a positive response in a study whose power, design, or conduct limits the ability to draw a confident conclusion, but where the carcinogenic potential is strengthened by other lines of evidence (such as structure-activity relationships); or

(4) A statistically significant increase at one dose only, but no significant response at the other doses and no overall trend.

Inadequate Information to Assess Carcinogenic Potential
This descriptor of the database is appropriate when available data are judged inadequate for applying one of the other descriptors.

(1) Little or no pertinent information;

(2) Conflicting evidence, that is, some studies provide evidence of carcinogenicity but other studies of equal quality in the same sex and strain are negative. Differing results, that is, positive results in some studies and negative results in one or more different experimental systems, do not constitute conflicting evidence, as the term is used here. Depending on the overall weight of evidence, differing results can be considered either suggestive evidence or likely evidence; or

(3) Negative results that are not sufficiently robust for the descriptor, "Not Likely to Be Carcinogenic to Humans."

Not Likely to Be Carcinogenic to Humans
This descriptor is appropriate when the available data are considered robust for deciding that there is no basis for human hazard concern. The judgment may be based on data such as

(1) Animal evidence that demonstrates lack of carcinogenic effect in both sexes in well-designed and well-conducted studies in at least two appropriate animal species (in the absence of other animal or human data suggesting a potential for cancer effects);

(2) Convincing and extensive experimental evidence showing that the only carcinogenic effects observed in animals are not relevant to humans;

(3) Convincing evidence that carcinogenic effects are not likely by a particular exposure route (see Section 2.3.); or

(4) Convincing evidence that carcinogenic effects are not likely below a defined dose range.

Although the cancer descriptors acknowledge different levels of certainty about the nature of the relationship between exposure and cancer, they also allow for action in situations where the science has not reached a definitive conclusion but is sufficient to ascertain that a hazard is likely to exist. For example, regulatory decisions are routinely made for carcinogens that are described as "Likely to be Carcinogenic to Humans."

A similar but slightly different hazard identification scheme is used for identifying chemicals "known . . . to cause reproductive toxicity" under California's Safe Drinking Water and Toxic Enforcement Act of 1986, informally known as Proposition 65, the ballot initiative number under which it was passed. In this hazard analysis, the finalclassification is "known ... to cause," but the types of information that can be used for classifying a chemical as "known ... to cause" include the same types of scientific information as for "Carcinogenic to Humans" and "Likely to be Carcinogenic to Humans." For example, a single good animal or epidemiologic study could be used to classify a chemical as a carcinogen under USEPA's cancer classifications or a reproductive toxicant under California's Proposition 65. It is useful to evaluate how evidence is judged for other types of health effects when considering hazard assessments for EDCs. The cancer classification scheme acknowledges differing levels of knowledge while also acknowledging the needs of the policy process. The policy process need not depend on complete proof, and action can proceed with reasonable assurance of harm.

It is important to note that the analysis and synthesis can only be fulfilled if information exists. For example, one cancer classification is "Inadequate Information to Assess Carcinogenic Potential" which includes those chemicals that have little or no pertinent evidence (Table 3). In the policy context, this can have the unfortunate side effect of being interpreted as no harm, as there is no classification and no risk number. How do we address these chemicals where we have ongoing exposures yet no information to ascertain their potential effects on humans? This quandary is highly salient to EDCs.

2.3. Implications of Small Individual Effects on Large Populations and Findings of Adverse Health Outcomes

The current program to identify EDCs emphasizes identifying adverse effects before deciding which chemicals are potential EDCs (implemented through the two Tier testing system described above section 2.1). USEPA proposed in 1998 "The purpose of Tier 2 is to determine whether the substance causes adverse effects, identify the adverse effects caused by the substance, and establish a quantitative relationship between the dose and the adverse effect. At this stage of the science, only after completion of Tier 2 tests will EPA be able to determine whether a particular substance may have an effect in humans that is similar to an effect produced by a naturally occurring EAT, that is, that the substance is an endocrine disruptor" *(13)*. Some EDCs can contribute to health outcomes that are clearly adverse, such as prenatal exposures to DES resulting in vaginal cancers *(21)*. However, for many of the EDCs, current ambient exposures are relatively low, and the observed effects can be subtle or are not discernable as an overt effect. For example, a study by Swan et al. *(23)* of pregnant women found that certain phthalates were associated with reduced anogenital distance in the male infants. The actual clinical or health meaning of reduced anogenital distance was not defined, although reduced anogenital distance was significantly correlated with degree of testicular descent, as well as penile volume and scrotal size in the Swan et al. study *(23)*.

Evaluating these types of outcomes, which are upstream from more overteffects (such as undescribed testiclies) and are typically measured on a continuous scale, represent challenges to assessing adversity not always obvious when we have reached an adverse outcome. Or outcomes measured on a continous scale certainly, at some low (or high) enough value on the distribution of many measurements, most health scientists would agree that the measured value is outside the norm and would be considered adverse, even if there are no overt clinically observed effects. For example, small for gestational age in newborns is defined as lowest 10th percentile of birth weight for gestational age, and although this is not an adverse effect in itself, it is a risk factor for diseases later in child and adulthood *(24–26)*.

In addition, these subtle outcomes can be an indicator of a future risk or an event along the pathway to overt disease. Consider the prenatal phthalate exposure and the reduced anogenital distance. Reduced anogenital distance is thought to be a marker for reduction of androgen action, which is linked to several other observed health disorders in males, such as cryptorchidism and hypospadia, testicular germ cell cancer, and low sperm count *(27)*. Preventry reduced AGD also prevents more downstream counts from occuring.

Another aspect in assessing, adversity is that subtle changes at the individual level may not be of significant consequence, but they can have important implications over an entire population. Within a population, there is a distribution of clinical or other health measurements. Exposures that are ubiquitous, as is the case for many EDCs, can shift the distribution of response toward the adverse end of the distribution, and small changes can have large consequences for those at the end of the distribution (Fig. 2). An example is IQ. Lower IQ has been found to be linked to many adverse outcomes, such as increased need of remedial education services and reduced life-time earning potential *(28)*, and individuals who have an IQ less than 70 are considered mentally retarded. An increase from 10 to 20 ug/dL in blood lead is associated with a 1–3 point reduction in IQ score *(29–31)*. While a 1–3 point reduction in IQ does not seem much

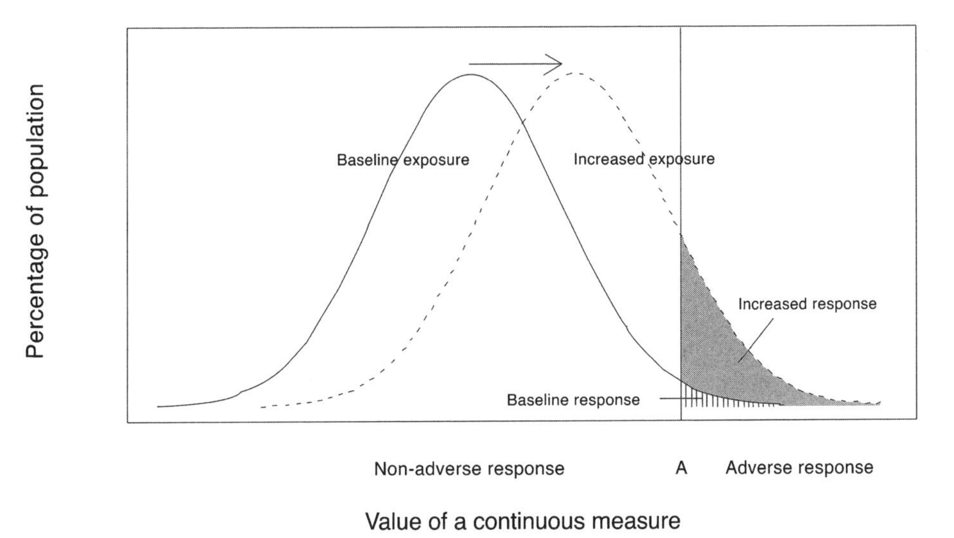

Fig. 2. Example of the response distribution among a baseline or unexposed (solid line) population and an exposed population. The arrow depicts the change in mean response between the baseline and the exposed populations. Shaded areas represent the proportion of the population with a response past a level ("A") that is considered abnormal. For EDCs, as well as other chemicals, making policy, regulatory, hazard and risk decision that reduce or prevent the incidence of upstream endpoints will in turn mitigate more downstream overt effects.

for the individual, it has the effect of shifting the population IQ down and has been observed to triple the number of children with IQs below 80 *(32)*. A similar argument might be made the AGD.

Another feature of small changes at the individual level is that interventions producing a small improvement in biological measurement has a small benefit to the individual but great benefits at the population level, which support widespread policy interventions *(33)*. This is salient to EDCs where exposures can produce subtle effects at the individual level and exposures are ubiquitous.

3. OTHER REGULATORY AND POLICY IMPLICATIONS

The current process of identifying and then regulating EDCs at the national level is constrained by the current testing system and is likely to take many years to implement. This does not mean that EDCs are not being addressed in the regulatory process. There are ongoing regulatory and policy activities directed at chemicals in air, food, drinking water, and hazardous waste. These regulatory programs often regulate on the basis of health, and promulgate action to reduce risks of adverse health consequences, often independent of how the adverse response is mediated. In addition, there is some overlap in the required toxicity testing between the standard battery of toxicity tests and those for EDCs (e.g., there are overlapping reproductive endpoints for the standard reproductive test for pesticides and the proposed Tier 2 tests for EDCs). PCBs are an example of an EDC that can cause adverse effects that are captured through other testing methods. PCB exposure is linked to endocrine-related effects such as feminization *(18)*. PCBs were also banned for production in the USA in 1979 largely because of the concerns about cancer under the Toxic Substances Control Act *(34–36)*. There are some regulatory programs that are not health based, such as those based on reducing emissions in the most technologically feasible manner and would reduce the emissions of any EDCs along with other chemicals. But these programs cover EDCs somewhat precemeal and there some remaining concern systematic that EDCs can uniquely affect human health. For example, exposures that occur during critical windows of prenatal and postnatal development are an area of potentially unique concern *(37)* and would require uniform testing.

In addition there are potentially important exposures to EDCs that occur outside the regulatory and policy process. For examples, There are many potentially important exposures to EDCs that can occur through consumer products, such as cosmetics. Cosmetics are regulated by FDA and cosmetic products, and chemical ingredients are not subject to FDA approval before putting on the market (except for color additives). FDA can pursue enforcement actions against products after they are on the market *(38)*. Cosmetic firms are responsible for verifying the safety of their products and cosmetic ingredients before releasing them onto the market. In addition, reporting the chemical ingredients of cosmetics is voluntary *(38)*. The lack of independent review and oversight of chemicals in cosmetics before release suggests cosmetics are not stringently regulated and there is very little control of exposures *(37)*.

A further limitation for addressing potential EDCs is that not all sources of EDCs are well described. For example, for phthalates, we know from measurements in humans across the USA that there are ubiquitous exposures to many phthalates including di (2-ethylhexyl) phthalate (DEHP) and dibutyl phthalates (DBP) *(39)*, both of which have numerous animal and some human studies finding effects on the endocrine system

(40–49). The sources of exposure for these chemicals are not well known, although it is suspected that exposure occurs largely through consumer products such as plastic packaging, personal care products, and building products *(40,41)*.

4. FUTURE CONSIDERATIONS

The issue of EDCs raises a larger issue of how chemical policy is conducted in the USA. Chemicals in the environment are addressed through numerous existing statutes, and this has resulted in important progress in many of the areas for which it was intended. For example, air and water pollution for important contaminants have been reduced; many substances have been banned because of their long persistence in the environment (e.g., PCBs) or because they can inflict permanent damage on the environment (e.g., chlorofluorocarbons). However, new emerging science, the rapidly increasing ability to manufacture more chemicals and device new substances previously unthought-of (e.g., nanotechnology), and our growing ability to both measure and observe previously unknown effects of environmental contaminants requires that policy and regulation also should evolve.

Some of the challenges that remain for EDCs are lack of adequate testing data and the limitations of current technologies to achieve adequate data in a timely manner, pressure to raise the level of proof before regulatory and policy decisions are made, and a need for more comprehensive monitoring and source information related to EDCs.

There are emerging new initiatives in Europe and in some of the states that could have implications for US chemical policy. The European Union has proposed the Registration, Evaluation, and Authorization of Chemicals (REACH), which is the European effort to increase the number of chemicals with basic toxicity information. It will require chemical producers and importers of chemicals in volumes of 1 ton or more per year to register and supply basic health and environmental information to the European Union *(42)*. REACH also requires that "high risk" chemicals, those that cause cancer or problems with reproduction or accumulate in the environment, can only continue to be used if the companies show that the risks are controlled or that the social and economic benefits outweigh the risks *(42)*. REACH essentially puts the onus on the chemical manufacturers to produce basic health and hazard information for their chemicals and encourages them to find substitutes for those that are more risky. Similarly, there have been recent efforts in California to address the gap in chemical information, where a recent white paper commissioned by members of the legislature recommends a similar strategy as REACH in California *(43)*.

While these initiatives can result in great progress in closing the chemical data gap, considerations remain about the level of detail of the information, and the process and considerations in weighing the scientific information. In addition, it is unclear if the basic toxicity information would be sufficiently sensitive to cover the same type of concerns that prompted the USEPA EDSP.

Consideration should be given to a multipronged strategy that includes (i) new methods for high throughput screening for EDCs, possibly using new omics technologies; (ii) using existing information with a more flexible weight of evidence system for identifying EDCs; (iii) continue research to continue to assess the relationships between EDCs and effects in humans; (iv) increased emphasis on ambient monitoring and source identification for EDCs; and (v) advance models for comprehensive chemicals evaluation, such as those discussed above, at the local, state, and national levels and develop effective; and (vi) improve chemicals regulations policy.

REFERENCES

1. Edwards BK, Brown ML, Wingo PA, et al. Annual report to the nation on the status of cancer, 1975–2002, featuring population-based trends in cancer treatment. *J Natl Cancer Inst* 2005;97(19):1407–27.
2. King M-C, Marks JH, Mandell JB. Breast and Ovarian Cancer Risks Due to Inherited Mutations in BRCA1 and BRCA2. *Science* 2003;302(5645):643–6.
3. Herman-Giddens ME. Recent data on pubertal milestones in United States children: the secular trend toward earlier development. *Int J Androl* 2006;29(1):241–6.
4. Oliver TR. The politics of public health policy. *Annu Rev Public Health* 2006;27:195–233.
5. US Environmental Protection Agency. A Review of the Reference Dose and Reference Concentration Process Washington, DC; 2002. Report No.: EPA/630/P-02/002F.
6. Maciorowski A, Timm G. Endocrine Disruptor Research, Regulation in the United States. In: Hester RE and Harrison RM, eds. *Issues in Environmental Science and Technology: Endocrine Disrupting Chemicals.* Cambridge, UK: The Royal Society of Chemistry; 1999;135–146.
7. What is the TSCA Chemical Substance Inventory. 2006. (Accessed June 30, 2006, at http://www.epa.gov/opptintr/newchems/pubs/invntory.htm)
8. US Environmental Protection Agency. Testing of certain high production volume chemicals, final rule. *Fed Regist* 2006;71(51):13708–35.
9. US Environmental Protection Agency. Endocrine Disruptor Screening, Testing Advisory Committee (EDSTAC) Final Report. Washington, DC, 1998.
10. US Environmental Protection Agency. Chemical Hazard Data Availability Study. In 1998.
11. The Food Quality Protection Act. In: USC; 1996.
12. Amendments to the Safe Drinking Water Act - Public Law In: USC; 1996.
13. US Environmental Protection Agency. Endocrine disruptor screening program: statement of policy; notice. *Fed Regist* 1998;63:71542–68.
14. US Environmental Protection Agency. Endocrine disruptor screening program; chemical selection approach for initial round screening. *Fed Regist* 2005;70(186):56449–65.
15. High Throughput Pre-Screening (HTPS) and Quantitative Structure Activity Relationships (QSAR) 2006. (Accessed July 3, 2006, at http://www.epa.gov/scipoly/oscpendo/prioritysetting/models.htm)
16. Brody JG, Rudel RA. Environmental pollutants and breast cancer. *Environ Health Perspect* 2003;111(8):1007–19.
17. Fenton SE. Endocrine-Disrupting Compounds and Mammary Gland Development: Early Exposure and Later Life Consequences. *Endocrinology* 2006;147(6):s18–24.
18. McLachlan JA, Simpson E, Martin M. Endocrine disrupters and female reproductive health. *Best Pract Res Clin Endocrinol Metab* 2006;20(1):63–75.
19. Gee D. Late lessons from early warnings: toward realism and precaution with endocrine-disrupting substances. *Environ Health Perspect* 2006;114(Suppl 1):152–60.
20. Ibarreta D, Swan SH. The DES Story: Long-term Consequences of Prenatal Exposure. In: P Harremoes DG, M MacGarvin, A Stirling, J Keys, B Wynne, S Guedes Vaz, eds. *Late Lessons from Early Warnings: the Precautionary Principle 1896-2000: European Environment Agency* 2001.
21. Herbst AL, Ulfelder H, Poskanzer DC. Adenocarcinoma of the vagina. Association of maternal stilbestrol therapy with tumor appearance in young women. *N Engl J Med* 1971;284(15):878–81.
22. US Environmental Protection Agency. Guidelines for Carcinogen Risk Assessment. Washington DC: U.S. Environmental Protection Agency, National Center for Environmental Assessment; 2005. Report No.: NCEA-F-0644A.
23. Swan SH, Main KM, Liu F, et al. Decrease in anogenital distance among male infants with prenatal phthalate exposure. *Environ Health Perspect* 2005;113:1056–61.
24. Barker DJP, Eriksson JG, Forsen T, Osmond C. Fetal origins of adult disease: strength of effects and biological basis. *Int J Epidemiol* 2002;31(6):1235–9.
25. Kiely J, Brett K, Yu S, Rowley D. Low Birthweight, Intrauterine Growth Retardation. In: Wilcox L, Marks J, eds. *From Data to Action: CDC's Public Health Surveillance for Women, Infants, and Children.* Atlanta, GA: Centers for Disease Control and Prevention; 1994.

26. Tambyraja RL, Ratnam SS. The small fetus: growth-retarded and preterm. *Clin Obstet Gynaecol* 1982;9(3):517–37.
27. Skakkebak NE, Rajpert-De Meyts E, Main KM. Testicular dysgenesis syndrome: an increasingly common developmental disorder with environmental aspects: opinion. *Hum Reprod* 2001;16(5):972–8.
28. Bellinger DCDC. What is an adverse effect? A possible resolution of clinical, epidemiological perspectives on neurobehavioral toxicity. *Environ Res* 2004;95(3):394–405.
29. International Programme on Chemical Safety. Environmental Health Criteria 165: Inorganic Lead. Geneva: World Health Organization; 1995.
30. Pocock SJ, Smith M, Baghurst P. Environmental lead, children's intelligence: a systematic review of the epidemiological evidence. *BMJ* 1994;309(6963):1189–97.
31. Schwartz J. Low-level lead exposure, children's IQ: a metaanalysis and search for a threshold. *Environ Res* 1994;65(1):42–55.
32. Needleman HL, Leviton A, Bellinger D. Lead-associated intellectual deficit. *N Engl J Med* 1982;306(6):367.
33. Rose G. Sick individuals, sick populations. *Int J Epidemiol* 1985;14(1):32–8.
34. EPA Bans PCB Manufacture, Phases Out Uses. 2006. (Accessed July 7, 2006, at http://www.epa.gov/history/topics/pcbs/01.htm)
35. International Programme on Chemical Safety. Overall Evaluations of Carcinogenicity. Lyon, France; 1987.
36. Integrated Risk Information System: Polychlorinated biphenyls (PCBs) (CASRN 1336-36-3). 2006. (Accessed July 7, 2006, at http://cfpub.epa.gov/iris/quickview.cfm?substance_nmbr=0294)
37. Harvey PW, Everett DJ. Regulation of endocrine-disrupting chemicals: critical overview and deficiencies in toxicology and risk assessment for human health. *Best Pract Res Clin Endocrinol Metab* 2006;20(1):145–65.
38. FDA Authority Over Cosmetics. 2006. (Accessed July 1, 2006, at http://www.cfsan.fda.gov/~dms/cos-206.html)
39. National Center for Environmental Health. Third National Report on Human Exposure to Environmental Chemicals. Atlanta: Centers for Disease Control and Prevention; 2005. Report No.: NCEH Pub. No. 05-0570.
40. ATSDR. Toxicological Profile for Di(2-Ethylhexyl) Phthalate 2003.
41. ATSDR. Toxicological Profile for Di-n-butyl Phthalate 2001.
42. Council of the European Union. Registration, Evaluation, Authorisation, Restriction of Chemicals (REACH). Brussels 2006. Report No.: Interinstitutional File: 2003/0256 (COD).
43. Wilson M, Chia D, Ehlers B. *Green Chemistry in California: a Framework for Leadership in Chemicals Policy and Innovation*. Berkeley, CA: California Policy Research Center, 2006.
44. Kyle AD, Woodruff TJ, Axelrad DA. Integrated assessment of environment, health: America's children and the environment. *Environ Health Perspect* 2006;114(3):447–52.

13 Talking With Patients and the Public About Endocrine-Disrupting Chemicals

Gina Solomon, MD, MPH,
and Sarah Janssen, MD, PhD, MPH

CONTENTS

1. COMMUNICATION OF ENVIRONMENTAL HEALTH RISKS

Health care providers often discuss issues of risk and uncertainty with patients in the context of surgery, diagnostic tests, immunizations, or treatment. Communication of potential risks and benefits before making a voluntary decision is the foundation of informed consent *(1)*.

Communication about environmental health risk differs from medical informed consent in that the hazard usually involves involuntary exposure, may be unfamiliar, provides no benefit to those exposed, and there often is much less known about the potential risks. Health care providers may not know where to find information to answer questions related to environmental health generally, or endocrine disruption specifically. Even when scientific information is available about the health effects of chemical exposure, it is almost never enough to answer questions pertaining to individual risk (as opposed to population risk) and rarely addresses issues related to the long-term effects of poorly quantified exposure during vulnerable periods of development of the fetus, young child, or adolescent *(2)*.

Despite the paucity of relevant information, health care providers have an important role to play in communication of risks associated with endocrine disruptors and

From: *Endocrine-Disrupting Chemicals: From Basic Research to Clinical Practice*
Edited by: A. C. Gore © Humana Press Inc., Totowa, NJ

other environmental toxicants. There is widespread concern among the general public about environmental health risks, especially risks to infants and children *(3)*. Patients frequently come to their health care provider's office with questions about environmental hazards. A survey of pediatric patients found that exposure to "environmental poison" leads the list of issues that parents worry about but that pediatricians rarely gave advice on this topic *(4)*.

There are three major settings in which health care providers may need to communicate with people about environmental health risks. The most obvious setting is the clinic or office, where discussions occur about personal risk to specific individuals. Second, health care providers are sometimes called upon to provide information in workplaces, schools, or community settings, where there is a potential environmental risk to a group of people, or a perceived cluster of disease. Third, health care providers may occasionally be called upon to address the public at large and to comment on the public health significance of scientific findings. The latter situations may involve conversations with the media or policymakers.

This chapter will present a framework for conversations in all three of the general settings outlined above. The key components for successfully communicating about environmental health issues include (i) anticipating the general categories of questions that may arise, (ii) preparing approaches to common specific questions, (iii) understanding the principles of risk communication in the environmental health context, and (iv) rapidly gathering information from reliable sources to help answer questions that arise.

2. COMMUNICATING WITH THE INDIVIDUAL PATIENT

In the patient care or community setting, environmental health concerns tend to focus on questions about individual risk. People bring worries about specific exposures or illnesses to their personal physician or to a meeting at their workplace or school or in their community. Because the science on endocrine disruptors and other environmental health hazards does not pertain to individual risk, the challenge to the health care professional is substantial. Even assuming that the physician is familiar with the scientific data relevant to the issue in question, there remains the challenge of translating a combination of results from sources such as *in vitro* receptor-binding assays, laboratory rodent studies, ecological epidemiology, and occasionally limited human epidemiological research into something relevant to a patient's individual situation. This problem is further complicated by difficulties in exposure assessment, the fact that most people are exposed to mixtures, and uncertainties about the different effects of chemicals at different times during the lifespan. The resulting conversation must therefore move away from a focus on trying to "answer the question" toward a more open discussion of scientific uncertainty, risk, and prevention.

2.1. Categories of Questions

There are three major categories of questions that health care providers may encounter in the office or community setting. These categories include

(1) Anticipatory guidance: Patients often come to their physician with questions or concerns about potential environmental hazards in the absence of any specific medical complaints. These concerns may be sparked by an article in the newspaper, something

on the Internet, a conversation with a friend, or an observation they have made in their home or community. Often, physicians do not know the answers to these questions and may feel frustrated by their inability to respond. In addition to responding to questions patients bring, physicians have the opportunity to provide anticipatory guidance on environmental health issues such as endocrine disruption, yet rarely do so.

(2) Future risk: Patients may present after a known exposure in the workplace or community. They are interested in understanding the future health risks faced either by themselves or by their family related to this exposure. The specific concerns may relate to whether they will develop cancer, they will have reproductive problems, or their child will be healthy. In addition, patients are often interested in ways they can take action to lower their risk. In the context of body burden monitoring for endocrine-disrupting chemicals, patients may want to know how to interpret personal biomonitoring results and how they can eliminate the chemicals from their bodies.

(3) Causation: Individuals who have experienced a disease or other health problem may wish to know whether the disorder was likely to have been caused by an environmental or occupational exposure. In some cases, there is an interest in determining causation within a legal framework, and in other cases the question may be more general. Nonetheless, the challenge remains in determining the likelihood that any individual case of an illness may have been caused (entirely or in part) by a historical exposure.

It is impossible to prepare for all questions that may occur related to endocrine disruptors within these three categories. In general, the questions that arise under anticipatory guidance are easier to predict and should fall within the routine purview of the primary care practitioner. Questions about future risk after an exposure or questions of causation may require referral or consultation with a specialist, or additional focused research to formulate a response.

3. RESPONDING TO COMMON QUESTIONS ABOUT ENVIRONMENTAL EXPOSURES

It is difficult to predict what environmental health concerns or questions a health provider may encounter. Some questions can be anticipated because of their ongoing prevalence in the news media, popular press, or on the Internet. Others may be based on specific community concerns. The questions that an individual patient asks of their health provider may be very similar to those asked in a public meeting. Likewise, whether speaking to an individual or group, the clinician's response to these questions is often very similar. The overall approach involves having some knowledge of the toxicity of the contaminant of concern, assessing the route and likelihood of exposure, and being able to communicate a science-based approach to reducing unnecessary exposures. Furthermore, health providers also can offer advice to groups or individuals about reducing exposure to other contaminants they were not previously aware of.

Patients often have concerns regarding exposures in their home or work environment without a specific associated complaint or illness. Health professionals can easily and efficiently answer common questions focusing on healthy food and drinking water during a routine office visit. It is useful for health care providers to prepare science-based responses for common questions, either in written form available as a handout to patients in the office or for discussion when such questions arise. Approaches to some illustrative situations are addressed in this section.

3.1. Common Office Questions

A young mother comes with her toddler for a routine check-up. At her weekly playgroup meeting, she was told that and it was the local tap water was polluted and suggested she have a water filtration system installed in her home or drink only bottled water. She lives on a limited budget and wonders if she should be concerned.

Before investing in a water filter, check the local water utility company's annual water quality report. (For help interpreting water quality reports, this web site can help: http://**www.safe-drinking-water.org/rtk.html**) In most cities, healthy adults can drink tap water without concern. Pregnant women and young children may be more vulnerable to some contaminants in water, such as lead or trihalomethanes. People who have private wells should get their water tested for common contaminants. If any pollutants are identified in drinking water, a filter that is appropriate for removal of the specific contaminants can be chosen. Different types of filters take out different contaminants, so there is no "one size fits all" solution (see Box 1 for information about water filtration systems).

Bottled water is not necessarily a better alternative to filtered water. About one-quarter of bottled water is ordinary tap water that has been filtered and packaged. Bottled water quality is actually subject to less stringent regulatory standards than tap water *(5)*. Bottled water also can contain residues of the plastic it is bottled in. Some bottled water was found to contain bisphenol A, a known endocrine disruptor in animal studies *(6)*.

Box 1: Types of water filtration

In general, there are two types of water filters, point of entry and point of use. All filtration systems require regular maintenance for proper functioning.

(1) *Point of entry units* are more expensive, are installed in the pipes outside the home, and treat all the water before it enters the house;
(2) *Point of use filters* such as countertop filters (e.g., filter pitchers), faucet filters, and under-the-sink units generally use activated charcoal to remove bad tastes and odors and chemical contaminants. Charcoal water filters are simple to install, relatively economical, and effectively remove many toxins found in the environment and comprise the majority of filters in use;
(3) For many people, an activated carbon filter bearing NSF Standard 53 certification will filter out most pollutants of concern, including endocrine disruptors such as heavy metals and pesticides. However, some contaminants that are suspected endocrine disruptors, such as arsenic or perchlorate, may not be removed by charcoal filters;
(4) In reverse osmosis filtration, water is forced through a membrane and then filtered through charcoal; a method that removes most contaminants, including arsenic and perchlorate. However, this filtration system wastes a lot of water and is much more expensive.

A Vietnamese child comes to your office for a well-child check. His father asks you if it is OK for the family to eat fish caught in the local bay where he fishes everyday. He saw a warning sign the last time he was there but had problems reading it. The family eats the fish at least 4 times a week.

Fish contains many beneficial nutrients, including omega-3 fatty acids, which are important for brain development in the fetus *(7)*. However, certain types of fish are known to contain high levels of endocrine disruptors such as mercury, polychlorinated biphenyls (PCBs), and dioxins. The Environmental Protection Agency (EPA) and Food and Drug Administration (FDA) have advised pregnant women, women of reproductive age, and young children to avoid eating commercial fish high in mercury including tilefish, king mackerel, shark, and swordfish. Tuna (especially ahi, bigeye, and canned albacore) also contains high levels of mercury and consumption should be limited before and during pregnancy. Several organizations have issued lists of fish that are high or low in mercury *(8,9)*.

Freshwater fish caught in waters contaminated with PCBs or mercury should not be consumed at all. Contaminated fresh water fish are identified by state fish advisories (http://**www.epa.gov/ost/fish/**) fish that may have high levels of PCBs, include bluefish, striped bass (wild), farmed or Atlantic salmon, and croaker *(9)*. When cooking fish, the fatty portions should be removed to reduce exposure to contaminants that accumulate in fat, such as PCBs.

A middle-aged woman comes in for her yearly check-up. She will be babysitting her grandchildren for much of the summer and her daughter-in-law insists that they eat only organic food. She wonders if this is really necessary.

Many foods contain pesticide residues. Some pesticides are known or suspected endocrine disruptors and although the levels are often low, consumption should be limited as a precautionary measure. Peeling or washing can reduce some surface residues, although peeling can also remove some nutrients and fiber. The outer leaves of vegetables such as lettuce and cabbage should be discarded. Organic foods or pesticide-free foods can be more expensive and are not always readily available. Reviews of government residue-testing data suggest that certain foods—such as apples, bell peppers, celery, imported grapes, cherries, peaches, potatoes, pears, raspberries, spinach, and strawberries—tend to be high in pesticide residues and should be priorities for purchasing organically, whereas others—such as asparagus, avocado, bananas, broccoli, sweet corn, onions, and peas—rarely contain residues even if grown conventionally. More information on pesticide residues is available at http://**www.foodnews.org/walletguide.php**. Patients should be encouraged to eat a variety of fruits and vegetables to provide a variety of nutrients and to limit exposure to a single pesticide.

A pregnant woman comes to your office concerned about the plastic toys she has received as gifts from well-meaning friends and family. She has tried to have only natural products in her home and has heard plastic toys are toxic. She feels guilty throwing them away but does not want to give them to someone else if they are toxic. She also was given many plastic baby bottles and wonders if they are safe to use.

Many soft plastic toys are made from polyvinylchloride (PVC), which often contains phthalates to soften the plastic and make it more pliable. Testing has shown these chemicals are not "bound" to the plastic but leach out of it over time *(10)*. Chewing and sucking on toys during play may increase the rate at which these toxic chemicals are released and increase children's exposure to them. There are many different types of phthalates, but those found in most consumer products are either known or suspected endocrine disruptors. In the past several years, US toy manufacturers have voluntarily agreed to remove phthalates from "mouthing toys" such as teethers and rattlers but not from other plastic toys that might end up in a child's mouth. Some soft plastic toys are sold as "phthalate-free". As a precautionary measure, use of soft toys, especially those that are heavily mouthed, should be avoided in small children under the age of 3 years.

Many baby bottles are made from polycarbonate—a resin made with bisphenol A. Bisphenol A is an endocrine-disrupting compound, and there are concerns that human exposures are occurring at levels known to cause effects in animal studies *(11)*. Bisphenol A is known to leach from polycarbonate plastic bottles, including drinking water bottles and baby bottles, especially as the plastic ages or if it is washed in harsh detergents or bleach *(11)*. Although they are not always marked, the type of plastic used to manufacture a bottle or children's cup can be identified by the number inside the recycling triangle, often found at the bottom of a container. Polycarbonate (recycling symbol #7) bottles are generally clear and rigid. When a baby bottle must be used, it is preferable to choose an alternative, such as glass or the plastics polyethylene or polypropylene (recycling symbols #1, #2, or #5).

3.2. Special Concerns and Issues Around Breastfeeding and Infant Formula

A pregnant woman who is near her due date is in for a check up. She has been preparing for her new baby and has many questions about breast feeding. She is concerned because she recently read that many chemicals have been found in breast milk. She worries about the effects of passing these contaminants onto her new baby and wonders if it would be better to use formula.

Breast milk has been found to contain many contaminants including endocrine disruptors such as PCBs, dioxins and furans, pesticide residues, flame retardants polybrominated biphenyls (PBBs) and polybrominated diphenyl ether (PBDEs) and the plasticizers phthalates and bisphenol A *(12–14)*. Providers should reassure their patients that despite this issue, the benefits of breast feeding outweigh the risks of contamination. Breast-feeding may even protect a baby against the adverse effects of exposures that occurred *in utero (15)*.

Because of the benefits to baby and mother, the American Academy of Pediatrics recommends breast feeding for at least the first 6 months of life. Breast milk provides vital nutrients and antibodies that are passed from the mother to infant. These help prevent infections and promote growth of the brain and nervous system. Some studies have also shown that breast-fed babies are less likely to develop chronic illnesses as adults such as asthma and cancer. *(16)* Breast-feeding also is beneficial to the mother as it promotes bone strength, weight loss, and reduces the chances of pre-menopausal breast and ovarian cancer *(17)*.

Baby formula is not an equivalent substitute for breast milk. Formula is lacking in many of the vital trace nutrients and antibodies found in breast milk. Studies have demonstrated that formula-fed babies get sick more often than breast-fed babies *(18)*. While formula may not contain many of the contaminants found in breast milk, such as PCBs and dioxins, infant formula may contain other toxins such as manganese, lead, or cadmium *(19–21)*. In addition, exposure to toxins can occur if infant powder formula is diluted with water contaminated with pesticides, heavy metals, or microorganisms. Soy formulas are a particular concern because of very high levels of plant-derived estrogens (phytoestrogens) in soy products. The amounts of phytoestrogens are 2200–4500 times greater in soy milk than in breast milk and the long-term health effects are not very well studied *(22)*.

3.3. Questions About Future Risk that may Arise After an Exposure

A young woman comes to her annual physical with questions about her mercury level. She recently participated in a study that measured hair mercury levels and sent the results to the participants. She is hoping to become pregnant in the next 6 months and is worried that her mercury level is too high and will harm the baby. She wonders what she can do to lower her mercury body burden.

Patients may present with specific concerns about a known exposure in their workplace or home or may have testing results for contaminants in their body. They may be worried that an exposure will harm them or their children and may specifically wonder if they are at an increased risk for adverse pregnancy outcomes or developing a particular disease such as cancer. For the health care provider, addressing these concerns often requires obtaining more history from the patient about the exposure—including the specific chemical(s) of concern, the dose, timing, route of exposure, number of exposures, and whether there were any associated symptoms that might herald a substantial exposure. An example of an environmental exposure history for a typical pregnant or breast-feeding woman is summarized in Table 1.

It is often impossible to quantify or predict how much greater risk a person faces from an environmental exposure. In most situations, the exposure happens only once or a few times, is at a low concentration, and it is not likely to substantially increase the risk of adverse effects above that seen in the general population. In addition, there is often little or nothing that can be done in retrospect about the exposure incident. Providers can use this opportunity to offer reassurance and to educate the patient on how to reduce future exposures.

On the contrary, some exposures occur at higher doses or on an ongoing basis—for example, in an industrial or agricultural environment where higher concentrations of chemicals or pesticides may be used. In these situations, the provider will need to obtain a more thorough exposure history and more information about the toxicity of the chemical to address the specific patient concerns. In the case of a reproductive toxicant, any indication that an exposure may be significant or recurrent may warrant closer monitoring, additional testing, or precautionary action to remove the patient from the exposure.

Some people may ask about undergoing special treatments to reduce the level of contaminants in their body. Some common treatments include chelation therapy, special diets, or medicines. In general, most of these treatments are not effective, can have serious side effects, and should be regarded with caution. Chelation therapy can be used appropriately for treating acute, severe metal poisoning. However, it is not generally accepted for use to decrease body burdens because of past exposures, especially in adults with no symptoms of toxicity. All chelating agents have side effects; most

Table 1
Environmental Exposure History [a]

Work/hobbies
  What is your occupation? What are your hobbies?
  Are you exposed to any of these substances at work, home, or school? Fumes, vapors, dusts, pesticides, painting materials, strong odors, lead, mercury, or other metals?
  Have you ever felt sick after contact with a chemical?
  Have you ever been in the military or worked on a farm?
  Do you wear personal protective equipment at work or while doing hobbies?
  Do your symptoms get better away from work/hobbies?
Residence
  Was your home built before 1978? If so, has it been tested for lead paint? If your home has lead paint, is it flaking? Have you done any recent remodeling?
  Where does your drinking water come from?
  Have you had your water tested for lead?
  If you have a private well, has the water been tested?
  What is the occupation of your spouse or others at home? (toxicants can be brought home on clothing)
  Do you know of any major industrial emissions in your neighborhood (hazardous waste sites, dry cleaners, auto repair)?
  Do you live in an agricultural area?
  What kinds of chemicals are in your home/garage?
  Do you use pesticides? In your home? Garden? On pets?
  Do you have a mercury thermometer in your home?
  Do you use any traditional healing or cultural practices?[b]
  Do you ever smell chemical odors while you are at home?
  Do your symptoms get better away from home?
Diet
  What kind of fish do you eat? How often do you eat fish?
  Do you or anyone in your home fish in local waters?
  Do you eat foods high in animal fat (fast food, ice cream, cheese, whole milk, fatty meats)?
  Do you grow your own vegetables? (possibility of contaminated soil)
  Do you take any dietary supplements?[b]

[a]An example of questions that might be asked of a pregnant or nursing woman to identify current or past exposures, reduce or eliminate current exposures, and reduce health impacts.
[b]May involve exposure to heavy metals such as mercury or lead.

commonly, these include abdominal discomfort, nausea, liver damage, neutropenia, decreased blood pressure, and allergic reactions. In addition, chelators may bind other divalent mineral cations essential for normal physiologic function.

In the case of mercury, the half-life of the metal in the human body is about 60 days. Appropriate management of an asymptomatic patient with a confirmed elevated mercury level would include advice about avoiding mercury from fish and a review of other possible sources of mercury exposure. Repeat testing to assure that the level is declining is also useful. A woman of reproductive age may be counseled to delay pregnancy for 6–8 months as a precautionary measure.

When there is concern about an exposure that may result in adverse effects, consultation with a specialist in occupational/environmental medicine or toxicology may be warranted. In all situations, patient education on how to prevent and reduce future exposures is essential.

3.4. Questions of Causality After an Adverse Health Event

A migrant farmworker was recently diagnosed with testicular cancer. His job involves mixing and applying pesticides, and he does not always wear protective equipment. He wonders if the pesticides could have caused his cancer.

Patients may approach their health provider with concerns that a past exposure is related to a specific condition, such as infertility or cancer loss. In some cases, the patient is simply struggling to understand a bad health outcome; in other situations, a legal case may be pending. This is often a challenging case for a physician as most health outcomes have multi-factorial causes, including, in some cases, chemical exposures.

Although health providers may not be able to give an immediate answer, they can take a thorough history to determine whether the exposure was substantial (Table 1), give education about the health condition and the associated etiologies, discuss the uncertainty and challenges in determining individual risk, and provide guidance on how to avoid future exposure. Referral to an appropriate consulting specialist may be necessary for complex exposures and determination of causality.

3.5. Precautionary Anticipatory Guidance that Clinicians can Offer

People are routinely exposed to a wide variety of chemicals, some of which are endocrine disruptors. Endocrine disruptors may be in the food people eat, the water they drink, the air they breathe, and in consumer products.

Patients may not be aware of their ongoing exposures, and a visit to their health care provider can be an opportunity for patient education on topics they were not previously aware of. Instead of focusing the conversation on past exposures and unpredictable health outcomes, health care providers can use this opportunity to educate patients about prevention. The provider does not need to be an expert in environmental health to discuss environmental exposures. Simple, common sense guidance for reducing exposure to known or suspected endocrine disruptors can prevent unnecessary risks.

Provided here are a few topics that health providers can discuss with their patients during the course of a routine office visit. Not all topics can be covered in one visit, nor is there one best way to avoid all exposures. By starting the conversation, providers empower a patient with information and a course of action.

Dietary Advice. Eat more fruits, vegetables, grains, and reduce consumption of fatty animal products (beef, pork, and dairy). Eat a variety of fish and limit consumption of those known to be high in contaminants. Many endocrine disruptors, such as PCBs and dioxins, are lipophilic and accumulate in fat. Eating fatty foods can increase the body burden and fat accumulation of these chemicals in humans over years and even decades. When a woman breast-feeds, these contaminants are mobilized from fat and end up in breast milk. Therefore, animal fat intake should be reduced beginning in childhood.

Reducing Pesticide Exposure. Many pesticides are known or suspected endocrine disruptors. Pesticides are commonly used around the home and garden and on pets. Pesticides are also found in head lice treatments. To reduce pesticide exposure, choose non-chemical alternatives for home, garden, and pet use. If the use of a pesticide is necessary, use the least toxic alternative. For examples of some alternatives, see http://**www.pesticide.org/factsheets.html**. For insect and rodent control, baits and traps are the best approach in conjunction with sealing cracks and cleanliness. Least preferred are insecticide sprays or "bombs." Pregnant women and children should not mix or apply pesticides and should not be present in a home when treatments are applied. Head lice can be treated with nit combs or other non-toxic alternatives. Further discussion of reduction of pesticide exposure in the community is provided in Chapter 14 by Brenner and Galvez.

Reduce Exposure to Heavy Metals—Such as Lead and Mercury. Lead is still commonly found in the paint and pipes of older housing and can be found in the consumer marketplace in pottery with lead glaze, some costume jewelry, and personal care products such as hair dyes. People can reduce their exposure to lead by having their drinking water tested for lead, by having a professional remove any lead paint in their home, and avoiding use of products containing lead. Lead may be present in garden soil and can end up in vegetables or be tracked into the home on shoes.

In addition to dietary exposure, mercury exposure can occur through occupations, hobbies, magic-religious practices, and consumer products. The most common occupational exposure is in dentistry. Hobbies that may involve mercury include gold panning. Some Caribbean religious practices involve sprinkling metallic mercury inside the home or car as a purification ceremony. The resulting exposures to mercury vapor can be very high. Imported skin-lightening creams and acne remedies may contain inorganic mercury. Finally, thermometers, fluorescent lamps, and some types of batteries may all contain mercury. People should recycle lamps and batteries, and exchange their mercury thermometer for a digital one. Many communities sponsor collections of these products. Further evidence for heavy metals as endocrine-disrupting chemicals is found in Chapter 5 by Dyer.

Smoking. Both active and secondhand tobacco smoke exposure have been associated with many adverse reproductive outcomes, including infertility, low sperm counts, spontaneous abortion, low birth weight, and preterm labor. All couples should be advised to stop smoking and to avoid secondhand smoke exposure when attempting to become pregnant, while pregnant, or when there is an infant or child in the home.

4. COMMUNICATING WITH GROUPS

Although health care providers tend to focus on communication about environmental health issues with individual patients, health professionals are also often called upon to communicate with groups of people in a variety of settings, such as workplaces, schools, and communities. These types of communications are somewhat different than the discussions that occur in the medical office.

4.1. The Workplace Setting

An electronics company contracts with a local hospital for employee health services. In less than 1 year, three women working in one area of the manufacturing process have spontaneous abortions, and the company requests that a physician from the hospital come speak to the employees to answer their questions about whether the miscarriages were related to chemical exposures at work.

Workplaces may be settings where people are exposed to significantly higher concentrations of industrial chemicals or pesticides than are usually encountered in the general environment. In addition, some chemicals used in workplaces are not found in the community, and the toxicity may be distinct. OSHA standards generally focus on preventing undue acute toxicity but rarely are set on the basis of chronic effects. In fact, for only four chemicals are the OSHA standards designed to protect against adverse reproductive outcomes *(23)*. No chemicals are regulated by OSHA specifically on the basis of endocrine effects. The working population in the USA now includes a large percentage of women, and it is not unusual for women to work during pregnancy or lactation.

Communicating with workers about workplace exposures requires a careful job history, some understanding of the workplace setting, collection of material safety data sheets (MSDS) for the products or chemicals handled, and an effort to gather as much information as possible about exposure pathways, duration, and magnitude. It is important to be aware that OSHA compliance does not necessarily imply that a workplace is safe, especially for pregnant women, and that MSDS information is frequently incomplete, especially for reproductive and developmental toxicity and endocrine disruption. In fact, one review of MSDS for lead and ethylene glycol ethers (both known reproductive toxicants) found that 60 % of the 700 MSDS surveyed failed to even mention reproductive effects *(24)*. Before communicating with a group of workers, it is helpful to tour the workplace and review the scientific literature on the chemicals used.

Although it can be very difficult to answer questions about causation, especially for multifactorial health outcomes such as spontaneous abortion, it is possible to use communication opportunities to offer precautionary guidance to the employer and workers about reducing exposures within the workplace. In this context, the hierarchy of controls for management of risk in the workplace specifies elimination of the hazard, such as by substitution of less toxic chemicals or processes whenever possible *(25)*. When elimination of the hazard is not possible, engineering controls that prevent worker exposure are far preferable to administrative changes (e.g., rotating workers through the most dangerous jobs) or to personal protective equipment (PPE) such as respirators (which can fail or may not be worn properly). If workers do need to wear PPE, it is important that they are properly trained in the reasons for its use and in proper use procedures. Because the workplace is a more controlled environment than the community, if an employer is willing to make the effort, it is often possible to substantially reduce or eliminate exposures. If exposure is truly prevented, then there is little need to worry about health risk.

4.2. The School and Community Setting

A worried mother calls about her child's preschool. The building is in a low-income predominantly African-American community and is situated next to a closed semiconductor wafer fabrication facility, which is now a designated Superfund site. During the cleanup process, workers accidentally ruptured an underground storage tank containing trichloroethylene (TCE). Soil gas testing and air testing at the preschool showed elevated levels of this chlorinated solvent. The parents are looking for a physician who can talk to them at a community meeting.

Conversations about environmental health risks in a community setting require that the health care provider consider and address the exposures or illnesses that have already occurred, the future risks to people in the local community, and the need for public health action to protect the health of the people in the community. Physicians are one of the most trusted and credible sources of information about environmental health risks *(26)*. It is important that health care providers listen to community concerns and respond with honesty. Blanket reassurances are rarely appropriate and rarely believed.

Understanding different perceptions of risk is important to help understand how to communicate about risk in a community setting. If the physician who is attempting to explain a risk does not realize that the community may perceive risks differently, the discussion is less likely to be productive and effective.

One important issue related to discussing risks in the community setting is the history of environmental injustice in the USA. Low-income communities of color have become increasingly concerned about a disproportionate and unfair burden of environmental risk in their communities. Even a relatively small risk may be seen in the context of a history of racial and socioeconomic discrimination in the distribution of environmental risks and is perceived as adding to an already unacceptable background of risk.

The most significant predictor for the location of hazardous waste facilities nationwide is the race of the local community *(27)*. Regulation of facilities in communities of color is also deficient, with fines imposed for pollution in non-white communities averaging 54 % lower, and timelines for listing sites on the National Priority List for cleanup 20 % longer *(28)*. Similar disparities have been reported for exposures to toxic air contaminants *(29)*. Low income communities and communities of color are also more likely to contain multiple environmental pollution sources. Reviews of research have shown that children of color suffer disproportionate burdens of disease with potential environmental aspects, including asthma, neurodevelopmental disorders, and childhood cancer *(30)*. In part because of these disparities, people living in these communities may see a single incident such as a ruptured chemical tank near a day care center as part of a bigger picture of environmental hazards and a history of environmental injustice. If the health care provider ignores the history and context behind an individual incident, the conversation can feel frustrating and confusing to all parties involved.

In many cases, the actual risk to people in the community is essentially unquantifiable. As a result, it may not be possible to assign a risk of an adverse outcome, even at a community level. Therefore, support and precautionary guidance may be the most useful information a health care provider can offer to a community. Whenever possible, vulnerable populations such as pregnant women and children should be removed from situations of potential exposure to endocrine disruptors or environmental toxicants. Likewise, environmental pollution should be cleaned up and minimized whenever possible.

5. COMMUNICATING WITH A BROADER AUDIENCE

Endocrine disruption is an environmental health issue that generally has effects at a population level rather than to identifiable individuals. For example, environmental agents that slightly lower average sperm concentration in the population, slightly increase the risk of cancer of hormone-sensitive tissues, or that cause children to reach sexual maturity slightly earlier, are not likely to result in health effects that will be readily discerned at the level of the individual patient. This characteristic of endocrine disruptors creates particular difficulty for health care providers trying to communicate about this issue to their patients or even to small groups of people.

The population-level implications of endocrine disruptors can instead make it far easier for health care providers to communicate to a larger audience. In an expanded role, physicians and other health care workers can communicate about the broader public health implications of large-scale, subtle shifts in hormone-sensitive endpoints. Contexts in which such discussions might occur include comments to the media, policy discussions at medical and nursing societies, and health policy discussions at the local, state, and national level.

Although many health care providers are reluctant to make public statements about population risk, there are at least three reasons why they should consider doing so. First, the scientific basis for extrapolating the results of current endocrine disruptor research is far stronger when the extrapolation is at the population level rather than at the level of the individual patient; second, health care providers are a trusted and important voice that is rarely heard in public discussions about environmental health policy; and third, the foundation of medicine is prevention, and the most useful prevention activities around endocrine disruptors can occur at the population, rather than individual level.

In October of 2005, the governor of California signed legislation to require manufacturers of cosmetics to disclose any product ingredient that is on state or federal lists of chemicals that cause cancer or birth defects *(31)*. The legislation also allowed the state Department of Health Services to demand that manufacturers supply health-related information about other cosmetic ingredients and authorized a program to regulate these products to protect beauty salon workers. The legislation was sponsored by several breast cancer groups and was introduced because of studies that showed phthalates in cosmetics and (high) levels of certain phthalates in the urine of women of reproductive age *(32,33)*. The discovery that cosmetics may contain chemicals that are considered to be known endocrine disruptors and developmental toxicants raised significant public concern, especially when it became clear that these chemicals are generally not listed on the product label.

Health care providers played a role in the passage of this legislation. Physicians and nurses testified before committees in the California Senate and Assembly, and health organizations and individual health care providers sent letters of support. Some health care providers stated that it was easier for them to support broad public health protection and consumer information regarding phthalate exposure than to attempt to address this issue with each individual patient. Communications in the press emphasized that exposure of millions of women to phthalates on a daily basis could have subtle, long-term effects on the health of these women or their babies. However, it was also clear that the risk to any individual woman and child was unquantifiable and probably very small.

Communicating with the media or with policymakers requires a different set of considerations than communicating with individual patients or small groups of people. It is important to realize that the health care provider does not necessarily need to be an expert in toxicology or endocrine disruption to be a spokesperson on these issues, but there is a need to understand the basic state of the science, the level of scientific uncertainty, and the potential public health implications of the scientific data. Before speaking with a reporter or a policymaker, it is important to also review the policy proposal at issue and to determine whether it seems like a reasonable, precautionary, science-based step toward protecting public health.

If the science and the policy seem reasonable, the next step is to either write out a statement of support or develop key talking points. Communications experts suggest identifying three or four major “messages” that summarize the main points that need to be conveyed. For the legislative scenario discussed above, examples of such messages could include

(1) Several phthalates have been linked to subtle abnormalities in fetal reproductive system development, so widespread exposure to women of reproductive age should be minimized whenever possible;
(2) The Centers for Disease Control and Prevention reports that women of reproductive age have high concentrations of certain phthalates in their urine, and these chemicals have been detected in cosmetics;
(3) Consumers should have a right to know about potential endocrine disruptors or developmental toxicants in products so they can make informed choices, especially during pregnancy;
(4) The Department of Health Services needs more information about chemicals in products to carry out its responsibility to protect the public health.

The talking points must be tailored both to the issue and to the perspective of the person who will be speaking publicly. Although it is useful to know the science supporting the points, and even to have anecdotes or stories to support the points, the overall messages should remain broad and clear to the non-scientific listener. Health care providers should take care to speak only within the talking points that they have developed to avoid making mistakes or going astray into issues that are either irrelevant or outside the speaker’s area of expertise. It is generally possible to answer almost any question by restating a talking point, even if it means saying something such as: “I don’t know the answer to your question, but the real issue here is....” By pre-identifying a set of talking points within a scientific and policy comfort zone and staying within those points, the health care provider can assure both that the major issues will come through clearly in any news story or policy hearing and also that his or her credibility will remain intact.

6. PRINCIPLES OF RISK COMMUNICATION

Risk communication is the exchange of information about the nature, magnitude, significance, and control of a risk. Different groups of people view risks differently. Government agencies and industry scientists often engage in quantitative risk assessment that involves a series of steps (hazard identification, exposure assessment, dose–response assessment, and risk characterization) to generate a number that can be used for regulatory purposes *(34)*. Officials tend to use the resulting numerical regulatory limits to either dismiss a given chemical exposure as insignificant or take it seriously.

Health care providers are more likely to engage in qualitative risk assessment. A physician’s exposure assessment, for example, is less likely to involve actual measurements or mathematical modeling of exposure and is more likely to involve questions about the frequency and duration of the exposure, coupled with a rough assessment of the magnitude of the exposure based on the medical history. Physicians however often defer to governmental regulatory limits without necessarily scrutinizing the basis for the number.

The general public often distrusts quantitative risk assessment. This sentiment is not without foundation, as the process is limited by failure to account for multiple chemical exposures, failure to consider highly vulnerable groups (particularly fetuses and infants), the use of highly uncertain assumptions about species differences in extrapolating from high-dose toxicity studies in animals, and an absence of data on important health outcomes such as endocrine disruption or developmental toxicity. In addition, risk assessors often do not agree on which studies to include in their calculations, and different studies often find different levels of concern.

Risk is not an objective issue but rather has both scientific and social components, which are subject to interpretation. Historical and social context, as well as ethical issues, may lead to greater concern about some risks relative to others. Therefore a scientist's perception of risk is not necessarily "correct," and a lay person's perception is not necessarily "incorrect" *(35)*. It is important to approach questions of risk humbly with an understanding of the limitations of the science and the importance of the social context. It is also important to understand factors that contribute to different perceptions of risk to anticipate ways patients or communities may react to a hazard. Although each person probably perceives risk differently, extensive research has identified some common characteristics that influence risk perception: the nature of the hazard, the characteristics of the person perceiving the risk, and the social context in which the risk occurs *(36)*.

Hazards that are seen as potentially catastrophic, although unlikely, are generally perceived as posing a greater risk than hazards that are more likely but would result in less serious or reversible outcomes *(37)*. For example, the risk from a nuclear power plant may be seen as greater than the risk from coal power plants, although the likelihood of emissions that are hazardous to health is higher from coal plants. Similarly, the risk of a dreaded outcome (such as cancer, birth defects, or brain damage) is often seen as worse than the risk of a disease that is less universally dreaded (such as liver, lung, or kidney disease) *(38)*. Unfamiliar hazards are generally seen as riskier than familiar hazards, and manmade hazards may be perceived as riskier than those that occur naturally. The population affected by the hazard is also important. For example, a hazard to infants is often judged worse than a similar hazard to adults *(39)*. Finally, hazards that are involuntary are almost always judged more serious than hazards that are faced by choice *(40)*. Thus, comparison of the risks associated with skiing or drinking alcohol with risks from a hazardous waste incinerator will not be seen as equivalent because the former are voluntary and under the control of the individual, whereas the latter is imposed from outside.

The social context of risk communication is extremely important to perceptions of risk. If the individual or organization imposing the risk is trusted by the community (i.e., a local company that has provided jobs in the community for many years and is well known to the community), the risk is often perceived as less than if the risk is imposed by an outsider. Similarly, the level of trust in government regulatory officials is important in the perception of risk *(41)*. Risks seen as unfair are often seen as larger than risks seen as fairly distributed *(42)*.

Communication about risk yields optimal results only when there is a back-and-forth dialogue. The affected people need to feel that their concerns have been heard and addressed. In the most progressive view, those involved not only develop a good understanding of the risks but also search together for solutions that can mitigate their concerns *(43)*.

7. RESOURCES

Physicians who are faced with questions of environmental or occupational exposures in their patients need quick, reliable sources of information. Provided here are scientifically based resources that physicians can use to help in the care of their patients with exposures to contaminants, including endocrine disruptors.

7.1. Clinical Referrals

7.1.1. The Association of Occupational and Environmental Clinics (http://www.aoec.org)

Network of greater than 60 clinics and 250 individuals committed to improving the practice of occupational and environmental medicine. Has clinical directory for finding specialists in Occupational and Environmental Medicine: www.aoec.org/directory.

7.1.2. Pediatric Environmental Health Specialty Units (PEHSU) (http://www.aoec.org/pehsu.htm)

Provides education and consultation for health professionals, public health professionals, and others about the topic of children's environmental health.

7.2. Governmental Organizations

7.2.1. Centers for Disease Control and Prevention (CDC) (www.cdc.gov)

7.2.1.1. Agency for Toxic Substances Disease Registry (ATSDR) (http://www.atsdr.cdc.gov). The principal federal public health agency charged with responsibility for evaluating the human health effects of exposure to hazardous substances. Produces "toxicological profiles" for hazardous substances found at National Priorities List (NPL) sites: www.atsdr.cdc.gov/toxpro2.html

7.2.1.2. The National Institute for Occupational Safety and Health (NIOSH) (http://www.cdc.gov/niosh/homepage.html). Information on chemical safety, workplace health hazard evaluations, and reproductive health and occupational exposures.

7.2.2. The National Library of Medicine (http://www.nlm.nih.gov/)

Has links to databases including

PubMed (www.pubmed.gov)—references abstracts from thousands of biomedical journals.
ToxNet (toxnet.nlm.nih.gov)—network of databases on toxicology, hazardous chemicals, and environmental health.
Household Hazardous Substance Database (householdproducts.nlm.nih.gov/ products.htm)—links over 6000 consumer brands to health effects from MSDS and allows scientists and consumers to research products based on chemical ingredients.

7.2.3. The US Environmental Protection Agency (EPA)

Integrated Risk Information System (IRIS) (www.epa.gov/iris/) A database of human health effects that may result from exposure to various substances found in the environment.

7.3. Non-Governmental Organizations

7.3.1. The Collaborative on Health and the Environment (CHE, healthandenvironment.org) (http://health and environment.org)

Tracks emerging scientific evidence on links between diseases, disorders and disabilities, and possible environmental causes. Has produced many peer-reviewed overview papers on environmental causes of disease and a large database showing the associations between contaminants and human disease: database.healthandenvironment.org

7.3.2. Pesticide Action Network, North America (PANNA) (http://www.panna.org/)

Maintains a pesticides database: www.pesticideinfo.org/Index.html

7.4. Other Useful Websites

7.4.1. Our Stolen Future (http://www.ourstolenfuture.org)

Provides regular updates about the cutting edge of science related to endocrine disruption and information about ongoing policy debates, as well as new suggestions about what people can do to minimize risks related to hormonally disruptive contaminants.

7.4.2. The National Pesticide Information Center (http://npic.orst.edu/)

A cooperative effort of Oregon State University and the USEPA—provides on-line information about pesticide safety and toxicity. The organization also runs a toll-free hotline for pesticide questions (1 800 858 7378).

7.4.3. E.Hormone (http://e.hormone.tulane.edu)

Hosted and run by the Center for Bioenvironmental Research at Tulane/Xavier Universities—provides background and up-to-date information about endocrine disruption and other environmental signaling.

7.4.4. Em-Com (http://www.emcom.ca)

Information resource about endocrine-disrupting substances directed by a group of faculty at six Canadian universities.

REFERENCES

1. Evans G, Bostrom A, Johnston RB. *MA Risk Communication and Vaccination Institute of Medicine*. Washington, D.C.: National Academy Press, 1997.
2. Bearer C. How are children different from adults. *Environ Health Perspect* 1995;103(Suppl 6):7–12.
3. Princeton Survey Research Associates. National survey of public perceptions of environmental health risks. April 2000. http://healthyamericans.org/docs/index.php?DocID=18 [Accessed 30 May, 2006].

4. Stickler GB, Simmons PS. Pediatricians' preferences for anticipatory guidance topics compared with parental anxieties. *Clin Pediatr* 1995;34(7):384–387.
5. Natural Resources Defense Council. Bottled Water: Pure Drink or Pure Hype? Washington DC, March 1999. http://www.nrdc.org/water/drinking/bw/exesum.asp [Accessed 18 June 2006].
6. Consumer Reports, Drinking water safety: What's in that bottle? January 2003 http://www.consumerreports.org/cro/food/drinkingwater-safety-103/whats-in-bottled-water/index.htm [Accessed 16 June 2006].
7. Saldeen P, Saldeen T. Women and omega-3 fatty acids. *Obstet Gynecol Surv* 2004;59(10):722–730.
8. Natural Resources Defense Council. Mercury in Fish: A guide to staying healthy and fighting back. http://www.nrdc.org/health/effects/mercury/guide.asp [Accessed 18 June, 2006].
9. Oceans Alive. Consumption Advisories: Fish to Avoid. http://www.oceansalive.org/eat.cfm?subnav = healthalerts [Accessed 16 June 2006].
10. Shea KM. Pediatric exposure and potential toxicity of phthalate plasticizers. *Pediatrics* 2003;111(6 Pt 1):1467–1474.
11. Vom Saal FS, Hughes C. An extensive new literature concerning low-dose effects of bisphenol A shows the need for a new risk assessment. *Environ Health Perspect* 2005;113(8):926–933.
12. Solomon GM, Weiss PM. Chemical contaminants in breast milk: time trends and regional variability. *Environ Health Perspect* 2002;110:A339–A347.
13. Main KM, Mortensen GK, Kaleva MM, Boisen KA, Damgaard IN, Chellakooty M, et al. Human breast milk contamination with phthalates and alterations of endogenous reproductive hormones in infants three months of age. *Environ Health Perspect* 2006;114(2):270–276.
14. Ye X, Kuklenyik, Z, Needham LL, Calafat AM. Measuring environmental phenols and chlorinated organic chemicals in breast milk using automated on-line column-switching-high performance liquid chromatography-isotope dilution tandem mass spectrometry. *J Chromatogr B Analyt Technol Biomed Life Sci* 2006;831(1–2):110–115.
15. Jacobson JL, Jacobson SW. Prenatal exposure to polychlorinated biphenyls and attention at school age. *J Pediatr* 2003 Dec;143(6):780–788.
16. Oddy WH. A review of the effects of breastfeeding on respiratory infections, atopy, and childhood asthma. *J Asthma* 2004;41(6):605–621.
17. Schack-Nielsen L, Larnkjaer A, Michaelsen KF. Long term effects of breastfeeding on the infant and mother. *Adv Exp Med Biol* 2005;569:16–23.
18. Schack-Nielsen L, Michaelson KF. Advances in our understanding of the biology of human milk and its effect on offspring. *J Nutr* 2007;137(2)(Suppl):503–512.
19. Hozyasz KK, Ruszczynska A. High manganese levels in milk-based infant formulas. *Neurotoxicology* 2004;25(4):733.
20. Eklund G, Oskarsson A. Exposure of cadmium from infant formulas and weaning foods. *Food Addit Contam* 1999;16(12):509–519.
21. Navarro-Blasco I, Alvarez-Galindo JI. Lead levels in retail samples of Spanish infant formulae and their contribution to dietary intake of infants. *Food Addit Contam* 2005;22(8):726–734.
22. Setchell KD, Zimmer-Nechemias L, Cai J, Heubi JE. Isoflavone content of infant formulas and the metabolic fate of these phytoestrogens in early life. *Am J Clin Nutr* 1998;68(6 Suppl):1453S–1461S.
23. Occupational Safety, Health Administration. OSHA Standards: Reproductive Hazards. www.osha.gov/SLTC/reproductivehazards/standards.html [Accessed 18 June, 2006].
24. Paul M, Kurtz S. Analysis of reproductive health hazard information on material safety data sheets for lead and the ethylene glycol ethers. *Am J Ind Med* 1994;25(3):403–415.
25. Ellenbecker MJ. Engineering controls as an intervention to reduce worker exposure. *Am J Ind Med* 1996;29(4):303–307.
26. Covello V. Risk Communication and occupational medicine. *J Occup Med* 1993;35(1):18–19.
27. Bullard RD, Mohai P, Saha R, Wright B. *Toxic Waster and Race* at Twenty, 1987–2007. A report prepared for the united church of christ justice and witness ministries. Cleveland OH March 2007. United Church of Christ.
28. Special issue: Unequal protection: The racial divide in environmental law. *Natl Law J* 1992.
29. Wennette DE, Nieves LA. Breathing polluted air. *EPA J* 1992;18(1):16–17.
30. Mott L. The disproportionate impact of environmental health threats on children of color. *Environ Health Perspect* 1995;103(Suppl 6):33–35.
31. SB 484 (Migden), Chapter 729, October 7, 2005. http://www.leginfo.ca.gov/pub/bill/sen/sb_0451-0500/sb_484 _bill_20051007_chaptered.html [Accessed 17 June, 2006].

32. Environmental Working Group. Beauty Secrets: Phthalates in cosmetics and beauty products. Washington DC; November 2000. http://www.ewg.org/reports/beautysecrets/chap3.html [Accessed 17 June, 2006].
33. Silva MJ, Barr DB, Reidy JA, Malek NA, Hodge CC, Caudill SP, et al. Urinary levels of seven phthalate metabolites in the U.S. population from the National Health and Nutrition Examination Survey (NHANES) 1999-2000. *Environ Health Perspect* 2004;112(3):331–338.
34. National Research Council, Risk Assessment in the Federal Government: Managing the Process, National Academy Press, Committee on the Institutional Means for Assessment of Risks to Public Health.
35. Neil N, Malmfors T, Slovic P. Intuitive toxicology: expert and lay judgments of chemical risks. *Toxicol Pathol* 1994;22(2):198–201.
36. Hage ML, Frazier LM. Reproductive risk communication: a clinical view. In Frazier LM, Hage ML, eds. *Reproductive Hazards of the Workplace*, New York: John Wiley & Sons, Inc, 1998;71–86.
37. Hohenemser C, Kates RW, Slovic P. The nature of technological hazard. *Science* 1983;220(4595):378–384.
38. Slovic P. Perception of risk. *Science* 1987;236:280–285.
39. Hage ML, Frazier LM. Reproductive risk communication: a clinical view. In Frazier LM, Hage ML, eds. *Reproductive Hazards of the Workplace*, New York: John Wiley & Sons, Inc, 1998;71–86.
40. Goldstein BD, Gotsch AR. Communication and Assessment of Risk. In Rosenstock L, Cullen MR, Brodkin CA, Redlich CA eds. 2nd edition *Textbook of Clinical Occupational, Environmental Medicine*, Philadelphia, PA: WB Saunders Company, 2005;1275–83.
41. MacGregor D, Slovic P, Mason RG, Detweiler J, Binney SE, Dodd B. Perceived risks of radioactive waste transport through Oregon: results of a statewide survey. *Risk Analysis* 1994;14(1):5–14.
42. Alhakami AS, Slovic P. A psychological study of the inverse relationship between perceived risk and perceived benefit. *Risk Analysis* 1994;14(6):1085–1096.
43. Rowan KE. Why rules for risk communication are not enough: a problem-solving approach to risk communication. *Risk Analysis* 1994;14(3):365–374.

14 Community Interventions to Reduce Exposure to Chemicals with Endocrine-Disrupting Properties

Barbara Brenner, DR PH,
and Maida Galvez, MD

CONTENTS

Sometimes as scientists we make assumptions and don't rethink assumptions to see how they fit in a natural situation.... Community people, because they are looking at it from a fresh perspective, will question the assumptions in a way that actually improves the science...(1).

1. INTRODUCTION

Concerns about chemicals with endocrine-disrupting properties burst onto the public and policy stages in the mid-1990s as a result of a growing body of scientific evidence from wildlife and animal laboratory studies on the adverse health effects of many synthetic chemicals. The endocrine-disrupting chemicals (EDCs) of the greatest concern to the community have been synthetically produced pesticides—insecticides, herbicides, fumigants, and fungicides [e.g., dichlorodiphenyltrichloroethane (DDT),

From: *Endocrine-Disrupting Chemicals: From Basic Research to Clinical Practice*
Edited by: A. C. Gore © Humana Press Inc., Totowa, NJ

chlorpyrifos, pyrethroids]—as well as certain plastics [e.g., phthalates and bisphenol A (BPA)] and other industrial chemicals [e.g., polychlorinated biphenyls (PCBs)] *(2)*. The major routes of exposure identified have included direct contact at home, or in the workplace, or through ingestion of contaminated water, food, or air. These chemicals can also be transferred from mother to child through the placenta and breast milk.

The publication of the book *Our Stolen Future (3)* galvanized many in the environmental health science community to call for increased human subjects research on the health effects of EDCs before drawing conclusions about the danger or "risk" posed by these chemicals to humans. At the same time, environmental health advocates, backed by media interest and support, began to actively organize efforts to reduce exposures to these biologically active compounds, with or without scientific evidence. These initiatives were most often guided by a belief in the Precautionary Principle. This asserts that rather than waiting for scientific evidence that chemicals are hazardous as evidenced by health effects in the general population or by waiting for the completion of rigorous research on both short- and long-term health effects, we ought to minimize or eliminate exposures until we are certain they do not have deleterious effects. It provides justification for "public policy actions in situations of scientific complexity and uncertainty where there may be a need to act in order to avoid, or reduce potentially serious or irreversible threats to health or the environment, using an appropriate level of scientific evidence and taking into account the likely pros and cons of action and inaction" *(4)*. Thus, action has taken the form of

(1) calling for increased government regulation of chemicals in both North America and Europe,
(2) mounting consumer and media education campaigns, and
(3) advocating for the testing and dissemination of alternative methodologies and products to prevent exposures, such as integrated pest management (IPM), organically grown food, and ecologically "safe" cosmetics and personal care products.

Environmental activism combined with community- or population-based research has resulted in protective measures—changes in policy and practice—to reduce exposures and body burdens of some EDCs that have been shown to have neurotoxic or neuroendocrine effects in utero and in early childhood. The Environmental Protection Agency (EPA) has banned or severely restricted the sale and use of some pesticides—DDT and chlorpyrifos—as well as PCBs. The Federal Drug Administration (FDA) recently recommended that DEHP be removed from medical equipment that might create exposures for fetal and infant males. The European Union has banned the use of DEHP in toys intended for young children's use. The city of San Francisco has recently banned phthalates *(5)*. Intervention studies that demonstrated the effectiveness of non-toxic and non-chemical alternatives to pest control (e.g., IPM) have led to state and local laws requiring prior notification of outdoor pesticide spraying in municipalities and or requiring the use of IPM in all public buildings, including schools *(6)*.

Public interest in the issue of EDCs remains high. Table 1 summarizes the most recent issue of *Environmental Health News*. In the first 2 weeks of June 2006, there were 22 'entries that reported on the emerging science of endocrine disruption *(7)*. However, studies have just begun to explore the possibility that human exposures to hormonally active compounds, particularly in utero and during early childhood, may be responsible, at least in part, for changes in semen quality, increasing incidence

Table 1
Summary of the Most Recent Issue of ***Environmental Health News*** (2006)

Date	*Media channel*	*Title/Topic*
June 1	San Francisco Chronicle	Milking it: Moms find industrial chemicals in their milk outrage and a call to action
June 1	Oakland Tribune, California	Scientists find new link between plastic, cancer
June 1	Toronto Globe and Mail	Want a full-time job? Live chemical-free
June 1	Los Angeles Times	Chemical in plastics is tied to prostate cancer
June 2	Environmental Defense Canada, from the report Polluted Children, Toxic Nation	Harmful pollutants were found in the bodies of every child and parent tested in a national study in Canada
June 2	Gannett News Service	Residents raise PCB concerns
June 2	CBC Canada	Tests show pollutants in kids' blood, urine
June 2	Vancouver Press	Organic food, exercise, but also PCBs
June 2	Toronto Globe and Mail	Toxic Cocktail found in children
June 2	Montreal Gazette	Study points finger at household items
June 3	Canadian?	Chemical cocktails in children's bodies: Study
June 3	Cincinnati Business Courier	UC research sheds light on prostate cancer, link to fetal chemical exposure
June 3	San Francisco Chronicle	Supervisors to consider ban of certain plastics
June 4	New York Post	What's yer poison?
June 5	Detroit News	Residents welcome removal of PCBs
June 6	International Journal of Epidemiology	A survey of reproductive health of women born to women exposed in the womb finds an increased risk of menstrual irregularities and possible infertility
June 8	San Francisco Chronicle	Board bans chemicals that may harm infants
June 8	Environmental Sciences Technology	Plastic chemicals alters female brains
June 9	Chemical and Engineering News	San Francisco restricts bisphenol A and some phthalates
June 9	Toronto Globe and Mail	Chemical used in water bottles linked to prostate cancer
June 11	Ottawa Citizen	Cancer-causing agents found in everyday items, says expert
June 11	BBB	Sunscreen thyroid effect concerns

of congenital malformations of the reproductive organs, increasing rates of testicular cancer, and an apparent increase in the incidence of early puberty in girls *(8–10)*. These topics are also covered in Chaps. 1, 2, 3, 9, and 10 of this book.

As we move from the laboratory to human subjects research, several key issues remain. Proving cause and effect relationships between EDCs and health effects, exploring biological pathways by which certain outcomes occur, demonstrating the effectiveness of interventions to reduce exposures and/or risk, and translating these findings into programs and policies will require more extensive population-based research involving diverse communities and population groups. Up until recently, environmental health research at the population or community level has primarily been the domain of scientists trained in epidemiology. However, the design and implementation of effective community environmental studies has become increasingly interdisciplinary, incorporating expertise in the areas of epidemiology, medicine, toxicology, genetics, and the behavioral sciences, which includes the disciplines of anthropology, social work, and health education.

Some epidemiologists have called for a paradigm shift arguing that the current paradigm overemphasizes the individual level of risk to the exclusion of other organizational levels of risk *(11,12)*. This view asserts that health and disease must be studied in a social context and with increased community participation. Such an approach "implies working across disciplines and with the population itself, in defining variables, designing instruments, and collecting data (qualitative and quantitative) that reflect the ecological reality of life in that population as people experience it" *(13)*. Within such collaborations, "epidemiologists would not be required to surrender rigor, but they would be required to share power."

2. WHAT WE KNOW

Although there are still limited scientific data on the health effects from exposure to EDCs on human growth, development, and disease, we now know much more about exposure to these chemicals. In the most ambitious body burden survey ever undertaken, the Centers for Disease Control measured 116 of the thousands of EDCs in modern use. From this report, we know that EDC exposure is ubiquitous in human populations *(14,15)*. See also Chap. 11 of this book for further discussion of body burdens in humans.

2.1. Pesticides

Insects, weeds, rodents, and fungi—together classified as pests—all have the capability to affect our health, plant life, and food supply. Concerns about these effects have led to an overwhelming reliance on pesticide use, that is, chemicals of natural or synthetic origin used to control these pests. Today, there are more than 50,000 commercially available pesticides. According to a report by the EPA in 2002, world pesticide expenditures totaled more than $33.5 billion in 1998 and 1999, accounting for 5.6 billion pounds of pesticides *(16)*. Expenditures in the USA alone account for a third of worldwide expenditures or 1.2 billion pounds per year.

Pesticides are classified into several categories: insecticides and rodenticides, which target insects and rodents, herbicides, which target unwanted plants and weeds, and fungicides, which kill molds and fungi. The principal classes of insecticides in use in the US are the organophosphates, carbamates, and pyrethroids. The organophosphates and carbamates are toxic to the nervous system, and some of the pyrethroids are believed to be toxic to the reproductive system and disruptive to endocrine function.

Widespread use of pesticides in various settings has translated into multi-tiered levels of exposure including the food supply, homes, neighborhoods, schools, playgrounds, farms, and work sites. Exposures in the home are of concern due to the constant nature of the exposure and the predominance of time spent in the home. In a Minnesota study, as many as 97 % of households surveyed stored pesticides. Eighty-eight percent of the households reported using pesticides in the past year *(17)*.

Research on the health and developmental effects of exposure to pesticides in utero and in childhood has been conducted by the Centers for Children's Environmental Health and Disease Prevention Research, a joint undertaking of the National Institutes of Environmental Health and the EPA. Initial results of cohort studies of pregnant women conducted in California and in New York showed elevated exposures to organophosphate pesticides used either indoors for pest control or for outdoor agricultural use. Health outcomes included slightly stunted fetal growth, shorter gestation and lower birthweights, smaller head circumference, and suboptimal neurodevelopment *(9,18,19)*. Additional studies are showing that the potential for damage from these chemical exposures may be affected by genetic susceptibility of both the child and the mother *(18)*. Thus, interactions between genes, the environment, and the timing of exposure can all contribute to a later susceptibility to develop diseases and disorders. These cohorts continue to be followed for health and developmental effects from 7 years of age.

2.2. Phthalates

Phthalates are a family of synthetic compounds used widely as plasticizers that have come to be used in an enormous variety of consumer and technical products. Worldwide production has increased from very low levels in the mid 1940s to approximately 3.5 million metric tons per year *(20,21)*.

Phthalates are found in many common consumer products such as detergents, food packaging, vinyl flooring and wall covering, toys, and personal care products including fragrances, cosmetics, and nail polish. Phthalates are also used as either inert or active ingredients in some pesticide formulations *(22)*. As plasticizers, phthalates are added to vinyl chloride and other naturally rigid polymers to make them soft and flexible, and these flexible plastics are then used in infant bottles, pacifiers, intravenous tubing, blood bags, and flexible airways. Phthalates are not chemically bound to the plastics to which they are added. In consequence, they readily leach out to cause contamination of household dust or can be directly swallowed. Phthalates are readily absorbed by inhalation, from the gastrointestinal tract, or through the skin. Human exposure occurs mainly through the diet (>1 ug/kg/day) with limited exposure through inhalation and dermal contact. Additional exposure is possible through cosmetics, as they can represent from 0.1 to 25 % of the ingredients.

National serological surveys conducted by the CDC indicate that phthalates have become nearly ubiquitous in the bodies of Americans *(23)*. Exposures to phthalates in the general US population are estimated to be on the order of tens of μg per kg per day *(24)*. On the basis of data from the general US population aged 6 years and older *(25)*, urinary metabolites of four of the phthalates were found at detectable levels in at least half of the tested sample ($N = 1029$). These four phthalates are, in order of decreasing geometric mean of their metabolites: diethyl phthalate (DEP); dibutyl phthalate (DBP); benzyl butyl phthalate (BBP); di-(2-ethylhexyl) phthalate (DEHP). A number of phthalates that have been found to exhibit estrogenic activity include DBP, BBP, and di-hexyl phthalate (DHP) *(26)*.

We know that levels of phthalates are highest in children and women of reproductive age, creating the potential for developmental effects on the fetus *(27,28)*. Metabolites of DBP, commonly found in cosmetics, reach higher levels in women than men, higher in African American than non-Hispanic whites, and higher in children than adults. The metabolite of DEHP (used as a plasticizer in polyvinyl chloride and thus common in toys, flexible tubing, shower curtains, etc.) is also higher in children than in adults. In fact, the FDA has recently recommended that DEHP be removed from medical equipment that may result in disproportionate exposures for premature infants. The European Union has banned the use of DEHP in toys intended for young children's use. The European Union also mandated removal of DBP and DEHP from cosmetics in 2002 *(29)*.

2.3. Bisphenol A

BPA is one of the highest volume chemicals produced in the world and exposure is widespread. BPA is a plasticizer used in production of epoxy resins and polycarbonate plastics, commonly found in canned food products, plastic food and drink packaging, dental sealants, adhesives, varnishes, and CDs *(30)*. In 2002, 2.3 billion pounds were sold. Ingestion is the primary route of exposure with measurable levels identified in a wide range of canned foods *(31)*, in recycled paper products *(32)*, and in the saliva of persons with teeth that have been treated with dental sealant *(33)*. BPA is detectable in urine and was found to be related to the consumption level of canned beverages *(34)*. It has been found in 95 % of the urine samples from people in the USA, and to a similar extent is found in human blood as well *(35,36)*.

New studies continue to add to our knowledge of potential health effects. Recently, a team of Spanish and Mexican researchers reported that the EDC, BPA causes insulin resistance in mice similar to that seen prior to the onset of type 2 diabetes *(37)*.

3. WHAT WE NEED TO LEARN

- What are the baseline exposure levels in a cross-section of communities and population groups?
- What are the effects of exposure in populations over time, taking into consideration the individual's social and biological environment?
- How do differences in genetic makeup within a population influence exposure levels and potential health effects?
- What applied prevention interventions and environmental advocacy efforts are effective in minimizing both exposure and risk and thus improving health?

Answering these questions and others will require cross-sectional exposure studies, prospective cohort studies, and prevention intervention research, including randomized controlled trials that enroll large numbers of individuals from diverse population groups. It will also require an investment by communities in research participation. To gain this investment, the groups or communities must participate in research that is relevant and valuable to them. Trust must exist between the researcher(s) and the population and/or community to be studied. In contrast to many clinical studies, this may be hard to achieve when results or benefits are not immediate and when knowledge gained cannot be rapidly applied or easily translated into practice or policy changes. Populations that are known to be more vulnerable to EDC exposure, such as pregnant women and children, or people who are economically disadvantaged, may be difficult to recruit and retain in longitudinal studies due to a number of factors, including distrust and concerns about exploitation, the transient nature of some populations, time availability, and financial barriers. The growing importance of genetics and molecular epidemiology in explaining causation requires that our research include a representative sample of the population, including population groups that have been traditionally skeptical about being "studied" and hard to reach.

4. COMMUNITY-BASED PARTICIPATORY RESEARCH: A COMMUNITY-BASED RESEARCH MODEL

Epidemiologists and behavioral scientists have sought new approaches to involving communities in research in order to overcome the barriers cited above. The terms community-based participatory research (CBPR), community-involved research, or community-centered research are often used interchangeably. Some community-based research is so coined because it is conducted in a community setting but without the active involvement of community members. A basic characteristic of true CBPR is the "emphasis on the participation and influence of nonacademic researchers in the process of creating knowledge" *(38)*.

4.1. Definition of CBPR

CBPR in public health is a partnership approach to research that equitably involves, for example, community members, organizational representatives, and researchers in all aspects of the research process, in which all partners contribute expertise and share decision-making and responsibilities *(39)*. The aim of CBPR is to increase knowledge and understanding of a given phenomenon and integrate the knowledge gained with interventions and policy change to improve the health and quality of life of community members. CBPR has been recognized as a community-driven and action-oriented approach to health research that is highly consistent with the mission and core values of public health *(1)*. Principles of CBPR include the following:

- it is participatory;
- it is co-operative, engaging community members and researchers in a joint process to which each contribute equally;
- it is a co-learning process;
- it involves systems development and local capacity building;
- it is an empowering process through which participants can increase control over their lives; and
- it achieves a balance between research and action *(38)*.

"A primary goal of CBPR is to increase a community's capacity to address and solve its own problems through the development of effective and sustainable interventions" *(40)*. To build this capacity, scientists must partner with grassroots organizations, community residents, church groups, and leaders to first determine the health concerns most pertinent to the residents of the community and then to continue these conversations as the project develops through design into implementation and finally into health strategies and policies that improve health. This is accomplished by the dissemination of knowledge gained through all phases of the study, which includes venues such as community forums, community newsletters, and other community events *(41)*. In engaging community members in the process, the research is meaningful to those living in the communities and also directly benefits them through the sharing of knowledge

Table 2
Summary of Community-Based Participatory Research Aims

Identify health priorities in underserved urban communities
Encourage partnership with community members, community-based organizations, leaders of the community, health centers, academic institutions, and local, regional, and national health departments
Incorporate existing expertise and resources available within communities into research projects
Devote time to developing and conducting participatory processes and practices
Allow opportunity for all collaborators to provide input into study design and implementation
Share knowledge gained with all collaborators and community at large
Develop culturally sensitive intervention strategies that promote health
Disseminate knowledge to the broader public via publications and presentation

Adapted from principles of collaboration, Metzler et al. 2003.

and resources ongoing with the project and the development of culturally sensitive intervention strategies *(42)*.

This model requires that scientists share power with community leaders and/or research participants, behaviors that are not intrinsic to the scientific method. It is important to emphasize that while the sharing of power or decision-making may require learning new behaviors and engaging in a process of dialogue and communication, scientists "would not be required to surrender rigor" *(13)*.

As discussed in the literature, there are numerous benefits gained from using a CBPR approach *(38,42)*. As reviewed elsewhere and summarized in Table 2, among the key benefits are that it:

1. ensures that the research topic comes from, or reflects a major concern of the local community;
2. enhances the relevance and application of the research data by all partners involved;
3. brings together partners with different skills, knowledge, and expertise to address complex problems;
4. enhances the quality, validity, sensitivity, and practicality of research by involving the local knowledge of the participants;
5. extends the likelihood of overcoming the distrust of research by communities that traditionally have been the "subjects" of such research;
6. increases the potential for bridging the cultural differences that may exist between the partners;
7. provides resources (e.g., financial, training, employment) for communities involved;
8. aims to improve health and well-being of the involved communities *(38)*.

5. CASE STUDY: GROWING UP HEALTHY IN EAST HARLEM—A COMMUNITY INTERVENTION TRIAL TO TEST IPM

Growing Up Healthy in East Harlem, a community intervention trial designed to test the feasibility and effectiveness of IPM in a low-income, multi-ethnic community, illustrates CBPR principles, methods, and outcomes *(43)*. This study was part of the Mount Sinai Center for Children's Environmental Health and Disease Prevention Research from 1998 through 2003. The Center was funded by the National Institute of Environmental Health Sciences (NIEHS) and the EPA. Fetal and early childhood exposure to neurodevelopmental toxicants was the unifying theme at the Mount Sinai Center. Exposures of concern included organophasphate pesticide products, which were used extensively in New York City to control indoor cockroach and mice infestation. The goal of this study was to identify, elucidate, and prevent developmental deficits that resulted from exposures to environmental toxicants in the inner city.

The Growing Up Healthy in East Harlem intervention study grew out of a shared community and medical school concern about reducing prenatal and early childhood exposure to pesticides, while at the same time reducing cockroach infestation. Safely reducing exposure to cockroach allergen was important in the East Harlem community where rates of childhood asthma were among the highest both in New York City and the country. Many landlords and tenants of poorly maintained tenement buildings and public housing projects applied large amounts of pesticides to control cockroach infestation.

The study investigators hypothesized that exposures to developmental toxicants in the inner city could be reduced through education, action, and behavioral change. However, to be effective these methods needed to be culturally and linguistically appropriate, affordable, and developed in close collaboration with East Harlem community representatives. To achieve this, Mount Sinai academicians reached out to form research partnerships with two federally qualified health centers located in East Harlem, a community of 117,000, 52 % of whom identify as Hispanic and 38 % identify as African American *(44)*. The median income level of East Harlem is among the poorest in the country.

The health centers provide primary and prenatal care to the Puerto Rican, Mexican, and African-American families residing in East Harlem, and both centers were governed by Boards whose members represented community residents and health care consumers. With the advice and active participation of these community partners, IPM at the household level was chosen as the major intervention to be tested to reduce cockroach infestation and pesticide exposure. IPM is an approach to pest control that relies principally on non-chemical approaches and community education *(45)*.

The pesticides in heaviest use in East Harlem when the study began were the organophosphates, chlorpyrifos and diazinon (both subsequently banned for sale), and the carbamate, propoxur. Chlorpyrifos had been shown to be a developmental neurotoxicant in rodent studies published prior to initiation of our work, a finding that has subsequently been confirmed in additional investigations *(46–48)*.

5.1. Specific Aims of the Intervention Study

1. To identify and quantify household exposures to neurodevelopmental toxicants, specifically pesticides and PCBs, through collection of baseline home environment samples, biological samples, and dietary data in a sample of pregnant women living in East Harlem.
2. To design, implement, and evaluate the efficacy of culturally sensitive IPM-based methods designed to reduce household pesticide exposures.
3. To conduct a community-wide education and intervention campaign to reduce exposures to developmental neurotoxicants across East Harlem.

CBPR principles were applied from the inception of this study throughout both study design and study implementation *(40)*. There were guiding principles for shared decision-making and power sharing between the academic and community partners:

- There would be joint selection of field staff, with emphasis on hiring from the community.
- Because study participants were going to be recruited from the health center's patient population and because the study required home visits and home exposure data collection, field staff would be employed by each health center. In theory, this was to integrate the field staff into the life of each center and accountable to the center's administrators. Field staff—a health educator, community outreach workers, and a handyman—were employed by the health centers.
- The health centers would receive jointly agreed upon funding from the research grant to implement their study arms, and administrative overhead would be paid to each center. Primary care physicians from both Centers served as co-principal investigators, overseeing the study at each site.
- Research protocols and recruitment/retention strategies would be fully reviewed and agreed upon.
- Questionnaires, both content and wording, would be reviewed and agree upon.
- Educational materials designed for recruitment, intervention, and community dissemination would be reviewed by the community partners before use.
- All partners would participate in selecting Community Advisory Board (CAB) members and developing a CAB agenda.

5.2. Community Advisory Board

A CAB, composed of 20 active community stakeholders, was formed. The CAB met semi-annually to advise the Growing up Healthy research team on dissemination of information and to help design and participate in broader community-wide education and intervention campaigns that sought to change behaviors and policies related to pesticide use and pest control. Members included tenant association leaders and members, housing managers, schoolteachers and parent association leaders, social service agencies, community health providers, and local elected officials. This group became particularly interested in promoting the adoption of IPM by building owners and managers as well as by individual tenants.

5.3. Study Recruitment, Retention, and Methods

A total of 131 families and households were enrolled in the study from September 1999 to June 2002. Seventy-six were recruited to the intervention (IPM) group from among women who received their prenatal care at Boriken Health Center. Fifty-five were recruited into the non-intervention control group from among pregnant women living in East Harlem who received prenatal care either at Settlement Health or at The Mount Sinai Hospital Prenatal Clinic.

Eighty-eight or 67.2 % of the households remained in the study for a 6-month follow-up visit (50 intervention/IPM and 38 controls). These families were demographically similar overall and differed only by country of origin with a higher number of Mexicans in the intervention group and more Puerto Ricans in the control group. The major reason for drop out at 6 months and also at subsequent visits was moving out of East Harlem.

Upon enrollment, study personnel administered a questionnaire in either English or Spanish to all participants in both groups to obtain information on home environmental conditions and sociodemographic characteristics. A home visit was arranged to monitor pesticide levels, collect baseline information on cockroach infestation levels, and conduct a visual inspection of the home.

5.4. Exposure Assessment and Cockroach Monitoring

Pesticide levels among intervention and control groups were assessed through sampling of urine, household air, dust, hand, and toy wipes. Roach levels were assessed through visual inspection and the placement of glue-surface cockroach monitors. These monitors were placed in up to eight locations in the kitchen and bathroom and collected 2 weeks following the baseline visit. In intervention households, cockroaches were counted at 2-week intervals for the first 2 months and then monthly for the next 4 months. Follow-up visits were made to both intervention and control households at 6 months and again at 1 year to assess the results of the intervention. In addition, participants were asked at 1 year to assess or rate the extent to which they still had a cockroach problem and whether they continued to use pesticides.

5.5. Intervention

An individual IPM intervention plan was developed for each intervention household based on results of the environmental questionnaire and baseline home inspection. A bilingual, bicultural team—health educator, community outreach worker, handyman, and an IPM consultant—worked with the head of household to introduce and teach IPM. Methods included sealing cracks and crevices in apartments through home repairs, plugging holes, and systematically applying low-toxicity baits and gels (active ingredient, hydramethylnon) to kill existing cockroaches and their eggs. The IPM consultant returned to monitor results and reapply gel, if necessary, at intervals of 2 weeks for the first 2 months and every month thereafter for the next 4 months. The educational component of the intervention, which provided information on how to maintain a pesticide-free household, was expected to change behavior, permanently reduce roaches, and eliminate the use of pesticide sprays. Intervention participants were assisted in contacting their landlords/managers responsible for repairs and pest control and to advocate for changes to reduce pesticide exposure. Control households received no IPM education or application but did receive a home injury prevention intervention as an incentive to participate.

5.6. Results

Three quarters of the families in both groups of this study reported a cockroach infestation problem in the home at baseline. No seasonal fluctuations in baseline infestation were observed. Approximately 60 % of families in both groups reported that pesticides had been applied in their homes during pregnancy by an exterminator, landlord, or family member. After 6 months of the IPM intervention, there was a significant decrease in cockroach infestation among intervention households (from 80.5 to 39.0 % of households; $p < 0.0001$, McNemar's Test). By contrast, control households showed no reduction (from 78.1 to 81.3 % of households). Table 3 demonstrates that infestation levels in the intervention group at the 6-month follow-up (39.0 %) were

Table 3

Presence of Cockroaches at Baseline and 6 Months in Control and Intervention Households. From Pesticide Intervention Project, NYC, 1999–2002

	Followed study population				*Population lost to follow-up*	
	Intervention ($n = 41$)		*Control ($n = 32$)*		*Intervention ($n = 12$)*	*Control ($n = 6$)*
	Baseline	*6 month follow-up*	*Baseline*	*6 month follow-up*	*Baseline*	*Baseline*
% of households with any cockroaches	80.5[a]	39.0[a,b]	78.10	81.3[b]	66.70	66.70
Median number of cockroach monitors placed	8	8	8	8	8	14.5
Median number of positive cockroach monitors	3	0	5	4	2.5	3
Median % of positive cockroach monitors	25.00	0.00	54.00	40.00	33.50	20.00

[a]At the 6-month follow-up, intervention households reported a significant decrease in the percent of households with any cockroaches ($p < 0.0001$ McNemar's Test).

[b]At the 6-month follow-up, there are significantly fewer households with cockroaches in the intervention group than in the control group ($p < 0.001$).

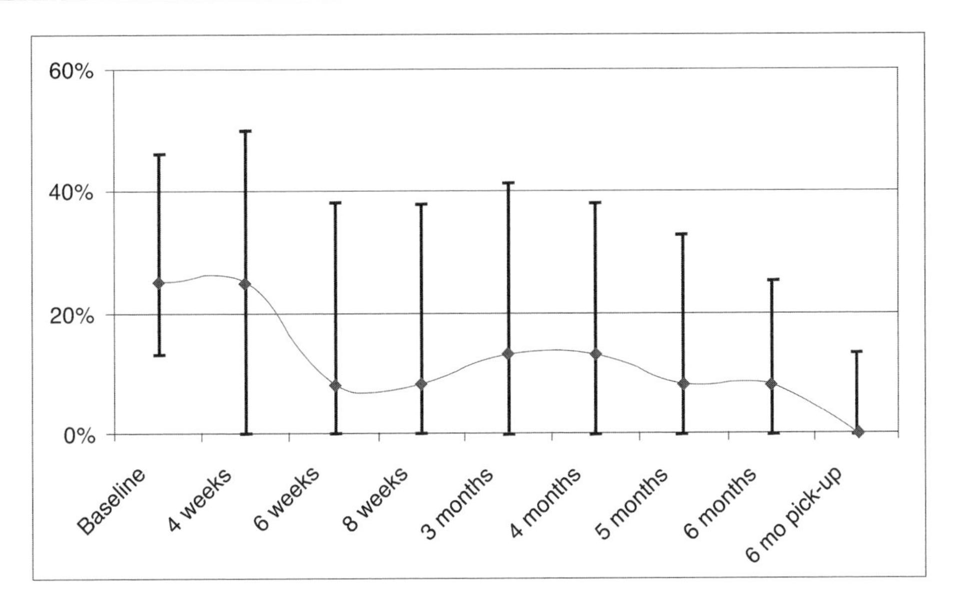

Fig. 1. Percent of intervention households (median and inter-quartile range) with any cockroaches during a 6 month period, assayed by the pesticide intervention project, New York City, 1999–2002. There is a significant decrease in the number of households with any cockroaches in the intervention group at baseline compared with the 6 month follow up. ($p < 0.0001$, trend test). Note: $n = 41$ for all ages except 3 months ($n = 40$).

significantly lower than in the control group at the same interval (81.3 %) ($p < 0.001$, Chi-square test).

Figure 1 shows the visit-by-visit change in cockroach infestation levels in intervention households during the 6 months of the study. Most of the decline occurred within the first 6 weeks of introduction of IPM. The decline in cockroach count measures persisted during the 6-month period, and in half the homes the cockroach count fell to zero. The costs of adopting building-wide IPM in a typical East Harlem apartment building were estimated to be $46–$69 per unit in the first year (including repairs) and $24 per unit per year in subsequent years. In comparison, the costs of traditional, chemically based pest control are estimated to be $24–$46 per unit per year, not including repairs, because repairs are not typically undertaken in traditional pest control.

5.7. Behavioral and Policy Outcomes

Investigators concluded that IPM could be effective in reducing cockroach infestation in low-income urban households and even in those buildings that were in need of repair and that residents and landlords could be taught to apply and maintain IPM at a reasonable cost. Guided by the positive results and lessons learned from the intervention trial, a community-wide campaign to adopt IPM and reduce pesticide use was planned and carried out in 2003. The campaign included:

- Educational seminars on indoor environmental risks, including pesticides, pests, mold, and lead, were developed.

- "How To" remediation strategies and techniques were held in English and Spanish at community agencies serving families and children and at parent associations.
- Sessions included seminars on how to conduct IPM and how to find and use environmentally safe household products of all kinds.
- Visual and written materials on pesticides, IPM, and safe household products were widely disseminated.
- "How to Be a Good Consumer of IPM" seminar was developed for building owners, managers, and superintendents in East Harlem.
- A collaboration was formed with the New York City Department of Health and the New York City Department of Housing Preservation and Development Anti-Abandonment Program to identify buildings in East Harlem with serious pest control problems and to provide technical assistance and financial incentives to landlords to make repairs and utilize environmentally safe pest control methods.
- Building-wide IPM was promoted to the management companies of several East Harlem buildings of 70 units or more and was subsequently adopted by several of these buildings.
- Evidence on the effectiveness of IPM on reducing cockroach allergen and pesticide use was prepared and given to the New York City Council Committee on Health in support of Intro 328A, which opted New York City into the New York State Pesticide Neighborhood Notification law, and Intro 329A, which requires New York City to use alternative to chemical spraying as pest control methods on all city property, including schools, by city agencies and contractors *(49)*.

6. GROWING UP HEALTHY IN EAST HARLEM II: AN OBSERVATIONAL STUDY

The IPM community intervention trial provided the groundwork for continued community research partnerships in East Harlem. In 2003, the Mount Sinai Center for Children's Environmental Health and Disease Prevention Research and its community partners and CAB worked together to choose a focus for a grant renewal. A shared decision was made to shift gears from studying pesticide exposures in the community and to focus instead for the next 5 years on understanding the environmental risk factors for childhood obesity in East Harlem. By 2003, the rates of obesity among East Harlem adults had risen to the highest for any neighborhood in New York City. Addressing this concern through research and effective interventions was of highest priority in the community. The result was receipt of funds from the NIEHS and the EPA for Growing Up Healthy in East Harlem phase II, whose goal is to work in partnership with community leaders in East Harlem to assess urban environmental exposures, structural as well as chemical, that may influence somatic growth and risk of obesity in inner city children. The research is an observational study that will follow 6–8 year old boys and girls who reside in East Harlem over a 3-year period. It is based on the premise that the urban built environment, the structural features of the city as well as its complex mix of chemical contaminants, exerts profound and pervasive influences on the health, growth, and development of children and that these effects are especially magnified in poor and minority children. Research will focus on contemporary-use endocrine disruptors (EDCs)—phthalates, alkyl phenols, especially BPA, pesticides, and phytoestrogens. These chemicals have become widely dispersed in the urban environment. Significant levels are found in the bodies of nearly all Americans, and levels of many are highest in children and in minorities. Obesity is being assessed annually by an extensive anthropometric battery, including weight, height, and percent body fat and waist–hip circumference. The CAB has been an integral component in directing the implementation of the research and the ways in which information is disseminated back to the participants and the community at large. The knowledge gained through this study will provide a needs-assessment of the built environment of an inner city, East Harlem, New York as it relates to childhood overweight and obesity. Barriers to healthy eating and physical activity at the community level will be identified with the aid of community members. Innovative ways to share knowledge with the community at large will be developed as in the past community-based research project.

7. BENEFITS AND CHALLENGES OF CBPR MODELS

7.1. Benefits

The academic–community partnership that shaped the Growing Up Healthy in East Harlem study brought forth concrete benefits. Access to a low-income, urban population that had not previously participated in research was a major benefit, particularly when the study required gaining access to participant homes. Community partners acted as an important bridge between the study investigators and a culturally, linguistically, and economically diverse population. Existing and earned trust between the staff of the community health centers and their prenatal patients was the key element to successful study recruitment and retention.

The community partners also played a significant role in shaping an intervention design that was realistic and could be implemented, reflecting and taking into consideration cultural and socioeconomic variables and conditions. In particular, they were instrumental in revising the original control group design to ensure that the control group received an intervention of value—a home injury prevention intervention.

The community partners piloted and reviewed study questionnaires to adapt them to the language and literacy levels of the study population. Equally important was the guidance that study staff received in methods of collecting exposure data. The original plan was to measure pesticide levels through dust samples. However, our community partners knew that most households that we would enter did not have carpeting and we rapidly switched to air sampling to measure exposure levels.

A consistent concern expressed by the community partners was how their patients and the community would benefit from this intervention study as results and outcomes were not immediate. They helped investigators develop and disseminate culturally and linguistically appropriate information to the community as soon as the intervention ended. The campaign and multi-level efforts to inform the community about alternatives to pesticide use reinforced the perception that research can have a community benefit and led to strong community support for continuing children's environmental health research in East Harlem.

7.2. Challenges

7.2.1. Costs Incurred and Lack of Resources

Effective academic–community partnerships require time and resources or infrastructure support to establish and maintain trust, attend meetings, jointly participate in all phases of the research process, and foster capacity building for all partners.

This can be difficult to sustain as the non-academic research partner has many other demands on limited time, resources, and staff. Community organizations' time may not be adequately compensated, and there are opportunity costs for time taken away from other job responsibilities *(50,51)*. Research investigators also find that the CBPR model is labor intensive and takes time from publishing and prospects for promotion and tenure *(51)*. Few grants provide the time frame, funds, and infrastructure support required for establishing a partnership, and researchers and community agencies "donate" their time and resources to support the partnership. Community partners who do not perceive or receive a direct benefit from the research are difficult to engage or to sustain high levels of participation. If the research does not increase the number of clients or funds and the results from the study are not immediately "useful," the incentive to participate diminishes.

7.2.2. Institutional Constraints

Many institutional constraints are faced in conducting CBPR *(38,52)*. For examples, working closely with community partners and following an equality of ownership principle may not be supported by the constraints of an academic institution. University institutional review board (IRB) processes may not take into account the needs of CBPR projects (e.g., the need to be flexible and revise protocols based on community input), overhead issues, long delays associated with data analysis and returning results to the community, and hiring policies that require traditional job descriptions and educational degrees.

7.2.3. Lack of Trust and Respect: Institutional History

Building and maintaining trust between the study investigators and the community as well as between community partners is a challenge *(38,53)*. "There is often an understandable distrust in the community, especially marginalized communities, of the motives of the academic institution, based on the long history of researchers collecting data and leaving without feeding back the results or providing any benefits to the community. Tensions also may exist between the community partners involved, for example, if multiple ethnic groups are involved that do not have a history of prior positive working relationships" *(52)*.

7.2.4. Ensuring Community Participation and Influence

Related to time constraints and costs, another challenge faced by CBPR partnerships is ensuring community participation and influence *38,53–55*. The process of community building is a very important and often overlooked step in building a "collaborative, equitable" partnership, which requires skill and takes time and commitment on the part of all partners to foster participation and shared decision-making.

7.2.5. Lack of Training and Experience in Conducting CBPR

Another challenge faced by many researchers and community partners is limited training and experience in conducting CBPR. Although there is a large and growing literature on how to carry out CBPR efforts *(56)*, many researchers in the field have no training or mentoring and have limited opportunity to engage in learning opportunities. This is particularly challenging when community involvement is a requirement from the funding source and few researchers fully understood what the implications of that meant. For many community partners, the conduct of research is not what they have been trained to do, and often their experience with research has not been positive.

7.2.6. Different Emphasis on Goals, Values, Priorities, and Perspectives

There are a number of areas where community and academic partners may differ in their emphasis on goals, values, priorities, and perspectives *(38)*. For example, community partners may be eager to implement the interventions and disseminate preliminary results, whereas researchers are concerned that the premature dissemination of results would contaminate study findings and lead to scientific criticism and consequences for publications and future funding *(52)*. When members of partnerships have different values, beliefs, and cultures, it important that these various differences not suggest a "right" or "wrong" way that partnerships should operate, only that they consider and accommodate diverse perspectives *(38)*.

8. LESSONS LEARNED AND RECOMMENDATIONS

In a recent article summarizing CBPR outcomes in eight Centers for Children's Environmental Health and Disease Prevention Research, the authors identify the lessons learned and recommendations for the future *(52)*. In summary, these are:

1. *Sufficient time, resources, and benefits are needed for all partners to ensure active and meaningful participation.* Examples are financial resources, overhead funds, training, technical assistance, career development and promotion opportunities, and interventions based on research findings.
2. *Considerable commitment and time are needed to establish and maintain trust.* Trust is established and maintained in CBPR partnerships through a process of frequent interactions, honest and respectful communication, shared decision-making and influence, following through on promises made, and demonstrated caring and support.
3. *Jointly developing and following operating norms and CBPR principles/core values is essential.* These can go a long way in creating a common vision and clarifying expectations of all the parties involved, and in developing processes for establishing mutually agreed upon priorities, goals, and objectives.
4. *Acknowledging and addressing power and equity issues is critical.* There are often inequities that exist among partners that are difficult to eliminate, for example, power differences due to class, ethnicity, and gender, that play out to varying degrees even in CBPR efforts that are trying to achieve equity, and are reflected in things such as salary differentials, job flexibility, perceived credibility, and authority levels.
5. *Funding and academic institutions need to extend their criteria for research excellence and productivity* (e.g., the randomized control trial in which one group receives no intervention may not always be feasible or desirable within a CBPR context), *and be flexible to incorporate the input of community partners* (e.g., IRB review and approval processes). Guidelines specific to appraising CBPR, e.g., those developed by Green and colleagues *(57)*, should be used by both the funding agency and academic institutions who are attempting to assess the merit of such work.
6. *Commitment to translating research findings into interventions and policies is of utmost importance.* The results of basic research studies need to be shared with the community in ways that are understandable and useful. Researchers need to commit to work with

community partners on determining intervention and policy implications based on these results, and joining with partners in carrying out such endeavors, or linking them with others who can contribute to this process.

7. *Hiring and training staff from the local community is essential.* Not only can locally hired staff play a critical role in all aspects of the research process—but the enhanced knowledge and skills gained ensure a sustainable impact of the CBPR effort in the community at-large. Community partners need to be actively involved and share decision-making over the hiring of local community staff.
8. *Recognizing, respecting, and embracing different cultures of the partners and partner organizations are imperative for successful CBPR efforts.* Although there are challenges associated with these differences, it is the very richness and diversity of experiences, knowledge, and perspectives that contributes to the quality, validity, and relevance of CBPR. Partners need to recognize, respect, and embrace these differences and speak openly about them and establish ways of building upon and celebrating the diversity that exists.

9. CONCLUSION

As research on chemicals with endocrine-disrupting properties continues to move from the laboratory to human subjects, community participation will be increasingly valuable and necessary. Proving cause and effect or evaluating the effectiveness of an intervention will require understanding the social, cultural, and educational context within which community populations live. It will also require skill in recruiting, enrolling, and retaining sufficient numbers of study participants who are representative of the populations of interest. Physicians, epidemiologists, and other scientists can benefit from interdisciplinary collaborations that borrow from and combines scientific rigor with the behavioral sciences. Although still being refined and tested, CBPR models have demonstrated success in engaging community populations in studies research while providing benefit to both the researcher and the community.

REFERENCES

1. Vasquez, V.B., Minkler, M., Shepard, P. (2006). Promoting environmental health policy through community based participatory research: a case study from Harlem, New York. *J Urban Health* 83:1, 101–110.
2. National Research Council. (1999). *Hormonally Active Agents in the Environment.* Washington, DC: National Academy Press.
3. Colborn, T., Dumanoski, D., Myers, J.P. (1996). *Our Stolen Future.* New York: Dutton.
4. EEA, (2001). In: *Late Lessons from Early Warnings: the Precautionary Principle 1896-2000* (Harremoes P, Gee D, MacGarvin M, Stirling A, Keys J, Wynne B, et al., eds.). Copenhagen: European Environment Agency.
5. Environmental Health News. Environmental Health Sciences, June 1–11, 2006.
6. NYPIRG News, "Largest City in US Adopts Plan To Curtail Use of Pesticides," May 9 2005.
7. Centers for Disease Control and Prevention. (2003). Second National Report on Human Exposure to Environmental Chemicals. NCEH Pub. No. 02-0716.
8. Environmental Health News. Environmental Health Sciences, June 1–11, 2006.
9. Eskenazi, B., Harley, K., Bradman, A., Weltzien, E., Jewell, N.P., Barr, D.B., Furlong, C.E., Holland, N.T. (2004). Association of *in utero* organophosphate pesticide exposure and fetal growth and length of gestation in an agricultural population. *Environ Health Perspect* 112:1116–1124.
10. Whyatt, R.M., Rauh, V., Barr, D.B., Camann, D.E., Andrews, H.R., Garfinkel, R., Hoepner, L.A., Diaz, D., Dietrich, J., Reyes, A., Tang, D., Kinney, P.L., Perera, F.P. (2004). Prenatal insecticide exposures, birthweight and length among an urban minority cohort. *Environ Health Perspect* 112:1125–1132.

11. Swan, S.H., Brazil, C., Brobnis, E.A., Liu, F., Kruse, R.I., Hatch, M., Redmond, J.B., Wang, C., Overestreet, J.W., the Study for Future Families Research Group. (2003a). Geographic differences in semen quality of fertile US males. 2003. *Environ Health Perspect* 111:414–420.
12. Susser, M., Susser, E. (1996). Choosing a future for epidemiology: I. Eras and paradigms. *Am J Public Health* 86:668–673.
13. Pearce, N. (1996). Traditional epidemiology, modern epidemiology, and public health. *Am J Public Health* 86:674–677.
14. Schwab, M., Syme, S.L. (1997). On paradigms, community participation and the future of public health. *Am J Public Health* 87:2049–2052.
15. Centers for Disease Control and Prevention. (2003). Second National Report on Human Exposure to Environmental Chemicals. NCEH Pub. No. 02-0716.
16. Thornton, J.W., McCally, M., Houlihan, J. (2003). Biomonitoring of industrial pollutants: Health and body implications of the chemical body burden. *Public Health Rep* 117:315–323.
17. Environmental Protection Agency. Pesticides Industry Sales and Usage. 1998 and 1999 Market Estimates. 2002. Accessed on: October 19, 2003. Available at: http://www.epa.gov/oppbead1/pestsales/99pestsales/market_estimates1999.pdf
18. Adgate, J.L., Barr, D.B., Clayton, C.A., Eberly, L.E., Freeman, N.C.G., Lioy, P.J., Needham, L.L., Pellizari, E.D., Quackenboss, J.J., Roy, A., Sexton, K. (2001). Measurement of children's exposure to pesticides: Analysis of urinary metabolite levels in a probability based sample. *Environ Health Perspect* 109:583–590.
19. Berkowitz, G.S., Wetmur, J.G., Birman-Deych, E., Obel, J., Lapinski, R.H., Godbold, J.H., et al. (2004). In utero pesticide exposure, maternal paraoxonase activity and head circumference. *Environ Health Perspect* 112:388–391.
20. Young, J.G., Eskenazi, B., Gladstone, E.A., Bradman, A., Pederson, L., Johnson, C., et al. (2005). Association between *in utero* organophosphate pesticide exposure and abnormal reflexes in neonates. *Neurotoxicology* 26(2):199–209.
21. Bornehag, C.G., Sundell, J., Weschler, C.J., Sigsgaard, T., Lundgren, B., Hasselgren, M., et al. (2004). The association between asthma and allergic symptoms in children and phthalates in house dust: A nested case-control study. *Environ Health Perspect* 112(14):1393–1397.
22. Cadogan, D.F., Howick, C.J. (eds.) (1996). *Kirk-Othmen Encyclopedia of Chemical Technology* John Wiley & Sons, 258–290 pp.
23. Centers for Disease Control and Prevention. (2003). Second National Report on Human Exposure to Environmental Chemicals. NCEH Pub. No. 02-0716.
24. Shea, K.M., American Academy of Pediatrics Committee on Environmental Health. (2003). Pediatric exposure and potential toxicity of phthalate plasticizers. *Pediatrics* 111(6 Pt 1):1467–1474.
25. 1999 National Health, Nutrition Examination Survey (NHANES).
26. Zacharewski, T.R., Meek, M.D., Clemons, J.H., Wu, Z.F., Fielden, M.R., Matthews, J.B. (1998). Examination of the in vitro and in vivo estrogenic activities of eight commercial phthalate esters. *Toxicol Sci* 46(2):282–293.
27. Centers for Disease Control and Prevention. (2003). Second National Report on Human Exposure to Environmental Chemicals. NCEH Pub. No. 02-0716.
28. Blount, B.C., Silva, M.J., Caudill, S.P., Needham, L.L., Pirkle, J.L., Sampson, E.J., Lucier, G.W., Jackson, R.J., Brock, J.W. (2000). Levels of seven urinary phthalate metabolites in a human reference population. *Environ Health Perspect* 108:979–982.
29. Our Stolen Future: CDC report on human body burden of chemicals. http://www.ourstolenfuture.org/New Science/oncompounds/bodyburden/2003-0131-CDC-b.
30. Ben-Jonathan, N., Steinmetz, R. (1998). Xenoestrogens: The emerging story of bisphenol A. *Trends Endocrinol Metab* 9:124–128.
31. Goodson, A., Summerfield, W., Cooper, I. (2002). Survey of bisphenol A and bisphenol F in canned foods. *Food Addit Contam* 19(8):796–802.
32. Vinggaard, A.M., Hnida, C., Larsen, J.C. (2000). Environmental polycyclic aromatic hydrocarbons affect androgen receptor activation *in vitro*. *Toxicology* 145:173–183.
33. Sonnenschein, C., Soto, A.M. (1998). An updated review of environmental estrogen and androgen mimics and antagonists. *J Steroid Biochem Mol Biol* 65(1–6):143–150.
34. Matsumoto, A., Kunugita, N., Kitagawa, K., Isse, T., Oyama, T., Foureman, G.L., et al. (2003). Bisphenol A levels in human urine. *Environ Health Perspect* 111(1):101–104.

35. Ikezuki, Y., Tsutsumi, O., Takai, T., Kamei, Y., Taketani, Y. (2002). Determination of bisphenol A concentrations in human biological fluids reveals significant early prenatal exposure. *Hum Reprod* 17:2839–2841.
36. vom Saal, F.S., Hughes, C. (2005). An extensive new literature concerning low dose effects of bisphenol-A shows the need for a new risk assessment. *Environ Health Perspect* 113:926–933.
37. Alonso-Magdalena, P., Morimoto, S., Ripoli, C., Fuentes, E., Nadal, A. (2006). The estrogenic effect of bisphenol A disrupts pancreatic B-cell function *in vivo* and induces insulin resistance. *Environ Health Perspect* 114:106–112.
38. Israel, B.A., Schulz, A.J., Parker, E.A., Becker, A.B. (1998). Review of community based research: assessing partnership approaches to improve public health. *Annu Rev Public Health* 19:173–202.
39. Israel, B.A., Schulz, A.J., Parker, E.A., Becker, A.B., Allen, A., Guzman, J.R. (eds.) (2003). Critical issues in developing and following community-based participatory research principles. *Community-Based Participatory Research for Health,* San Francisco, CA: Jossey-Bass, 56–73 pp.
40. Metzler, M.M., Higgins, D.L., Beeker, C.G., Freudenberg, N., Lantz, P.M., Senturia, K.D., Eisinger, A.A., Viruell-Fuentes, E.A., Gheisar, B., Palermo, A., Softley, D. (2003). Addressing urban health in Detroit, New York City, and Seattle through community-based participatory research partnerships. *Am J Public Health* 93(5):803–811.
41. Freudenberg, N. (2001). Case History of the center for urban epidemiologic studies in New York City. *J Urban Health* 78(3):508–518.
42. O'Fallon, L.R., Dearry, A. (2002). Community–based participatory research as a tool to advance environmental health sciences. *Environ Health Perspect* 110:155–59.
43. Brenner, B.L., Markowitz, S., Rivera, M., Romero, H., Weeks, M., Sanchez, E., Deych, E., Garg, A., Godbold, J., Wolff, M.S., Landrigan, P.J., Berkowitz, G. (2003). Integrated pest management in an urban community: A successful partnership for prevention. *Environ Health Perspect* 111:1649–1653.
44. U.S. Census 2000.
45. Olkowski, W. (1991). *Common-Sense Pest Control: Least-Toxic Solutions for Your Home, Garden, Pets and Community*. Newtown, CT: Taunton Press.
46. Campbell, C.G., Seidler, F.J., Slotkin, T.A. (1997). Chlorpyrifos interferes with cell development in rat brain regions. *Brain Res Bull* 43(2):179–189.
47. Dam, K., Seidler, F.J., Slotkin, T.A. (2000). Chlorpyrifos exposure during a critical neonatal period elicits gender-selective deficits in the development of coordination skills and locomotor activity. *Brain Res Dev Brain Dev* 121(2):179–187.
48. Levin, E.D., Addy, N., Christopher, N.C., Seidler, F.J., Slotkin, T.A. (2001). Persistent behavioral consequences of neonatal chlorpyrifos exposure in rats. *Dev Brain Res* 130:83–89.
49. Yoshihama, M., Carr, E.S. (2002). Community participation reconsidered: Feminist participatory action research with Hmong women. *J Commun Pract* 10:85–103.
50. Claudio, L. (2005). NYC adopts pesticide laws. *Environ Health Perspect* 113:10, A662.
51. Sullivan, M., Senturia, K.D., Chrisman, N.J., Ciske, S.J., Krieger, J.W. (2000). Improving collaboration between researchers and communities. *Public Health Rep* 115:243–248.
52. Parker, E.A., Israel, B.A., Brakefield-Caldwell, W., Keeler, G.J., Lewis, T.C., Ramirez, E., et al. (2003). Community Action against Asthma: Examining the partnership process of a community-based participatory research project. *J Gen Intern Med* 18:558–567.
53. Israel, B.A., Parker, E.A., Rowe, Z., Salvatore, A., Minkler, M., Lopez, J., Butz, A., Mosley, A., Coates, L., Lambert, G., Potito, P.A., Brenner, B., Rivera, M., Romero, H., Thompson, B., Coronado, G., Halstead S. (2005). Community-based participatory research: Lessons Learned from the Centers for Children's Environmental Health and Disease Prevention Research. *Environ Health Perspect* 113:10, 1463–1471.
54. Wallerstein, N. (1999). Power between evaluator and community: Research relationships within New Mexico's healthier communities. *Soc Sci Med* 49:39–53.
55. Green, L.W., Mercer, S.L. (2001). Participatory research: Can public health agencies reconcile the push from funding bodies and the pull from communities. *Am J Public Health* 91:1926–1929.
56. Minkler, M., Wallerstein, N. (2003). *Community-Based Participatory Research for Health.* San Francisco, CA: Jossey-Bass.
57. Green, L.W., George, M.A., Daniel, M., Frankish, C.J., Herbert, C.P., Bowie, W.R., O'Neill, M. (2003). Guidelines for participatory research in health promotion. In: *Community-Based Participatory Research for Health* (Minkler M, Wallerstein N, eds.). San Francisco, CA: Jossey-Bass, pp. 27–52.

15 What Can We Do About Endocrine-Disrupting Chemicals?

Joseph W. Thornton, PHD

CONTENTS

1. INTRODUCTION

Endocrine-disrupting chemicals are widely distributed in the environment and have the potential to cause large-scale damage to the health of humans and other species. Here I discuss policies that could be implemented to reduce these hazards. I examine three frameworks for environmental policy that define a spectrum of approaches, from a version of the status quo to comprehensive reform. I show that the problem of endocrine disrupters is so complex and poorly characterized, because of the vast number of chemicals in commerce and the diversity of endocrine mechanisms and impacts, that the current framework of reactive chemical-by-chemical regulation has little or no chance of success. To fully address the hazards of endocrine disruption (ED), a comprehensive sustainable chemicals policy will be required, based on the principles of Green Chemistry, Clean Production, and Precaution.

Synthetic chemicals known to disrupt endocrine-signaling mechanisms are now globally distributed, and human exposure to these substances is truly universal *(1–5)*. The global burden of endocrine disrupting compounds (EDCs) exists in large part because many of these chemicals are highly persistent, so they gradually accumulate in the environment and are transported long distances on currents of wind and water; also, many are hydrophobic and resistant to metabolism, so they multiply in concentration as they move up the food web to upper trophic species, including humans *(6)*. EDCs disrupt hormone-regulated signaling and can thereby affect development, reproduction, and other processes at very low doses. In some cases, human exposures are at or

From: *Endocrine-Disrupting Chemicals: From Basic Research to Clinical Practice*
Edited by: A. C. Gore © Humana Press Inc., Totowa, NJ

near the range at which health impacts have been documented in the laboratory *(7–13)*. Clear links have been established between environmental EDC exposure and severe health impacts in wildlife (reviewed in refs. *6,14–16*). Many studies suggest that large-scale environmental exposure to EDCs may be linked to specific forms of reproductive, developmental, neurobehavioral, and immunological impairment in the human population (see, for example, refs. *17–35* and reviewed in refs. *6,36,37*).

The purpose of this chapter is to consider steps that society might take to reduce health hazards to humans and other species from these chemicals. I consider three frameworks for policy. The first is the *Predict and Permit* approach, which dominates current policy; this framework uses risk assessment to determine acceptable exposures to each chemical and then permits environmental releases that are predicted not to exceed that level. The second is an *Identify and Restrict* approach, which would restrict the production and use of chemicals that are known to be endocrine disrupters in favor of safer alternatives. The third strategy is a more comprehensive sustainable chemicals policy, which would emphasize broad reforms in the use and regulation of all synthetic chemicals. Although there are a virtually infinite number of possible regulatory actions, these three define a spectrum of options from the status quo to an entirely new policy approach; evaluating their strengths and weaknesses illuminates the issues at stake in formulating actions to address ED. In this chapter, I pay special attention to the complexity of EDCs' biological mechanisms and impacts and the implications of this complexity for potential policy approaches.

2. PREDICT AND PERMIT

In the current regulatory system in the United States—with only rare exceptions—industrial chemicals are permitted to be produced and released, so long as discharges of individual substances from individual facilities remain within limits predicted by risk assessment to be safe or acceptable. The dominant methodology for risk assessment-based regulation of each chemical involves determining a "no observed adverse effect level" based on studies of laboratory animals; this dose is then conservatively adjusted to yield the acceptable human exposure. Because carcinogenic chemicals have the potential to increase cancer risk at any dose, such substances are assessed by inferring a quantitative dose–risk relationship; regulations then attempt to limit local human exposures to the dose associated with a maximum "acceptable" additional risk (typically between 10^{-4} and 10^{-6}). In some cases, specific pollution control technologies are required as the best means to meet permitted levels.

For numerous reasons, this *Predict and Permit* framework cannot reliably address the problem of endocrine-disrupting chemicals. First, it makes the erroneous assumption that our knowledge of chemical hazards is more or less complete. Any substance for which no toxicological or exposure data are available is assumed to contribute a risk of zero. Similarly, any health impacts that have not been studied, or for which dose–response relationships have not been quantified, are excluded from the risk estimate. As a result, a lack of evidence is misconstrued as evidence of safety, and untested or poorly studied chemicals go unregulated. But little or no toxicological data are available for the vast majority of the approximately 87,000 chemicals now in commerce *(38,39)*, because virtually no testing is required for chemicals already on the commercial market.

A second flaw of the *Predict and Permit* approach is that by focus on licensing "acceptable" discharges based on predicted short-term, local exposures, global contamination is allowed to accumulate. Many synthetic chemicals persist in the environment or in the food web and are transported long distances; as a result, even dilute discharges are globally distributed and gradually build up over time to higher and higher levels. When discharges from thousands of individual sources are each assessed and permitted separately—even at levels that do not pose local health hazards—they contribute to a global pollution burden that the local view never considered.

A third flaw in the current framework is that predictions of "safe" doses using limited toxicological methods are not always reliable. "No observed adverse effect levels" often reflect the dose at which analytical and statistical tools can no longer detect subtle health impacts, rather than exposure levels that are truly safe. But such subtle health impacts, when distributed among a large exposed population, may be very important from a public health perspective. This problem is evident in the discovery of adverse health impacts of certain substances at lower and lower levels previously thought safe, as substances have been reassessed using more sensitive methods *(11,40)*. Moreover, chemicals are assessed and permitted one-by-one, but in the real world people are exposed simultaneously to thousands of pollutants. Because the toxicity of a substance may be radically modulated by the presence of other substances, exposure levels predicted to be safe in isolation may cause health impacts in mixtures *(41,42)*.

The fourth major problem with the *Predict and Permit* framework is that for virtually all industrial chemicals, regulatory limits are met by using pollution control technologies. These methods capture hazardous chemicals from one place or environmental medium and move them to another. For example, limits on air emissions from industrial facilities can usually be met by installing scrubbers or precipitators, but the residues from these devices must then be disposed of in wastewater effluents or landfills, resulting in release of the same pollutants to ground or surface waters, respectively. Moving pollutants away from the source in this way can help avoid severe local contamination, but it does nothing to prevent the gradual growth of a global pollution burden.

It is for these reasons that the *Predict and Permit* strategy has allowed global contamination by persistent toxic chemicals, including many EDCs, to accumulate. Designed to prevent severe local contamination by specific substances, this framework is ill-suited to preventing the slow and diffuse global accumulation of hazardous pollutants because of a chemical-intensive economy that produces a vast number of substances, many of them long-lasting. Global chemical pollution exists not because we have no regulations but because we have the wrong kind.

3. IDENTIFY AND RESTRICT

One alternative approach to address the cumulative burden of EDCs would restrict their production and use, instead of merely limiting their releases in a local context. This framework would begin with the recognition that there is no reliable way to determine "acceptable" discharges of or exposures to EDCs, precisely because many of them accumulate in the environment, are hazardous at very low doses, and can act together to produce cumulative health impacts. This framework, then, would be designed to eliminate production and use of identified EDCs whenever feasible, and to reduce them as much as possible otherwise. I refer to this approach as *Identify and*

Restrict, because it takes strong action against substances known to have specifically hazardous qualities. If applied to all known endocrine disrupters, such a policy would cause the burden of identified EDCs to decline radically.

How could such a goal be achieved? A strategy to eliminate and reduce the production and use of known EDCs would be implemented using three related concepts: Clean Production, Green Chemistry, and the Substitution Principle. Clean Production is based on the common-sense idea that chemicals that are not produced do not end up in the environment, whereas those that are produced will, to a large extent, ultimately be discharged. Clean Production therefore shifts the focus of regulation and technical change from the back end of industrial processes to the front end: instead of attempting to capture and safely dispose of hazardous chemicals after they are produced, technologies for manufacturing useful products are altered to prevent the generation of hazardous substances in the first place. This goal is accomplished by changing material inputs, products, and/or processes *(43,44)*. For example, wood pulp for paper manufacture is often delignified and bleached with chlorine gas or chlorine dioxide, which produces large amounts of organochlorine by-products, including polychlorinated dibenzodioxins and dibenzofurans, which are extremely potent and persistent EDCs. A Clean Production approach to this problem might substitute oxygen-based chemicals such as ozone and peroxides for this purpose, eliminating the input of chlorine and completely preventing the generation of organochlorines *(45)*.

Many alternatives exist already for EDCs and the processes that generate them. In addition to chlorine-free bleaching, these include substitute polymers for plastics that contain or generate EDCs—particularly such major sources as polyvinyl chloride and polycarbonate *(46)*—and non-chemical pest control methods or alternative synthetic chemicals for endocrine-disrupting pesticides *(47,48)*. To completely substitute for known EDCs, however, continuing technological innovation will be necessary.

The principles of Green Chemistry *(49)* provide a guide for such efforts (see Box 1). Green Chemistry is a subfield of chemical engineering, the goal of which is to design materials and techniques that eliminate the use and generation of hazardous substances in production, use, and disposal. Green Chemistry seeks to develop effective new materials that degrade in the environment and are reintegrated into natural materials cycles without causing harm to living organisms or the systems that sustain them *(50)*. Key goals of Green Chemical design are to avoid persistent and toxic substances, to rely upon natural feedstocks, and to develop synthetic organic chemicals composed only of those elements employed in natural mainstream biochemistry (carbon, nitrogen, oxygen, hydrogen, sulfur and phosphorous, plus ionic or metallic forms of iron, zinc, potassium, calcium, and chloride).

Box 1: Principles of Green Chemistry

- *Prevention:* It is better to prevent waste than to treat or clean up waste after it has been created.
- *Atom Economy:* Synthetic methods should be designed to maximize the incorporation of all materials used in the process into the final product.
- *Less Hazardous Chemical Syntheses:* Wherever practicable, synthetic methods should be designed to use and generate substances that possess little or no toxicity to human health and the environment.

- *Designing Safer Chemicals:* Chemical products should be designed to effect their desired function while minimizing their toxicity.
- *Safer Solvents and Auxiliaries:* The use of auxiliary substances (e.g., solvents, separation agents) should be made unnecessary wherever possible and innocuous when used.
- *Design for Energy Efficiency:* Energy requirements of chemical processes should be recognized for their environmental and economic impacts and should be minimized. If possible, synthetic methods should be conducted at ambient temperature and pressure.
- *Use of Renewable Feedstocks:* A raw material or feedstock should be renewable rather than depleting whenever technically and economically practicable.
- *Reduce Derivatives:* Unnecessary derivatization (use of blocking groups, protection/deprotection, temporary modification of physical/chemical processes) should be minimized or avoided if possible, because such steps require additional reagents and can generate waste.
- *Catalysis:* Catalytic reagents (as selective as possible) are superior to stoichiometric reagents.
- *Design for Degradation:* Chemical products should be designed so that at the end of their function they break down into innocuous degradation products and do not persist in the environment.
- *Real-time analysis for Pollution Prevention:* Analytical methodologies need to be further developed to allow for real-time, in-process monitoring and control before the formation of hazardous substances.
- *Inherently Safer Chemistry for Accident Prevention:* Substances and the form of a substance used in a chemical process should be chosen to minimize the potential for chemical accidents, including releases, explosions, and fires.

source: ref. 49

Green chemistry applications include oxygen-based processes for disinfection, chemical synthesis, and bleaching; the use of supercritical or liquid carbon dioxide for cleaning and coating operations; plant-derived polymers; and non-toxic catalysts that specifically accelerate desirable chemical reactions while reducing the production of hazardous by-products *(51)*.

Current regulations do not require or effectively encourage investment in Clean Production and Green Chemistry alternatives. Indeed, by requiring commitments to pollution control and disposal technologies, they even discourage investment in changes to production processes themselves *(52)*. The Substitution Principle is designed to reorient policies toward the replacement of hazardous substances and the processes that produce them; it requires that toxic chemicals be replaced whenever less hazardous substitutes are available *(53)*. To implement this principle, regulations in the *Identify and Restrict* framework would require the production and use of known EDCs to be gradually phased-out and substituted with alternative substances and processes that do not produce EDCs or other significant hazards. For each application of an EDC, the central question of the assessment and regulatory process would become, "Are there safer, feasible methods for fulfilling society's needs that do not require the use of potentially dangerous chemicals?" If so, a timetable would be established for phase-out of the EDC. Under the Substitution Principle, quantifying the exposure level predicted to be hazardous or safe for each chemical is not necessary; instead, the focus is on identifying opportunities to continually reduce the generation of hazardous substances.

Policies based on the Substitution Principle would go far toward reducing the health and environmental hazards associated with endocrine-disrupting chemicals. Under current policies—that is, in the absence of this principle—the vast majority of known EDCs remains in deliberate production today or continues to be generated as an accidental by-product, perpetuating global contamination and exposure. In an *Identify and Restrict* framework, only those EDCs that serve a compelling need and for which no safer alternatives are available would remain in commerce. The schedule for elimination of each EDC could be determined by considering the severity of the health threat and the economic implications of implementing the alternatives. Applications posing less severe hazards or requiring larger investments could be phased out on a longer timeline. Those chemicals that are the most hazardous and for which feasible alternatives are most affordable would be phased out first.

4. CAN WE IDENTIFY ALL EDCs?

The *Identify and Restrict* framework I have just described would effectively address all the EDCs that are known. Its effectiveness in protecting public health and the environment, however, would be limited by the fact that only a fraction of endocrine disrupting chemicals have been identified. All "unknown" EDCs would remain completely unregulated under an *Identify and Restrict* program.

How significant is this limitation? It is impossible to know precisely how many unidentified EDCs there are, but the number is likely to be large. As discussed above, only a small fraction of the >80,000 chemicals in commerce have been subject to even rudimentary toxicity testing for ED or for the kinds of transgenerational deficits that are among the most important endpoints of EDC exposure. Of those substances that have been assessed, however, scores or hundreds have turned out to be EDCs (see lists in *6,36,54–56*). If the fraction of EDCs among the vast number of unassessed chemicals is anywhere near the fraction of EDCs among those substances that have been well tested, then the number of EDCs yet to be identified would be extremely large. Even if the fraction were lower by one or two orders of magnitude, then number of unknown EDCs would be quite significant. We must therefore conclude that an *Identify and Restrict* strategy, although it would represent a significant step forward, does not in itself represent a general solution to the hazards posed by EDCs.

In principle, the problem of unknown EDCs might be addressed with an ambitious testing program to comprehensively identify chemicals that disrupt the endocrine system. The most ambitious such effort has been established by the U.S. Environmental Protection Agency (EPA). The stated goal of EPA's Endocrine Disrupter Screening Program (EDSP) is to identify commercial chemicals, pesticides, and environmental contaminants that are endocrine disrupters and to characterize their dose–response relationships *(57)*. Subjecting each chemical in commerce to the kinds of detailed investigation required to definitively rule ED in or out would take centuries, so EPA designed a hierarchical approach to radically streamline the process. First, existing data on chemical production volume, exposure, and health impacts are used—together with the results of computational and rapid molecular methods that predict binding to hormone receptors—to prioritize chemicals thought to have the highest likelihood of being significant endocrine disrupters; chemicals given low priority at this stage are not examined further. The high-priority chemicals are then screened using a few short-term in vivo and in vitro assays, such as reporter-gene expression assays in cell culture; chemicals that are positive in these screens move on to the next phase—comprehensive, multigenerational testing to determine dose–response relationships for various endocrine-mediated impacts—whereas the rest are not subject to further evaluation. The program is proceeding quite slowly: eight years after it was established, only 50 to 100 chemicals had been selected for the first round of screening, which had not yet bugun. The European Union (EU) has also begun an ED identification program, but this strategy is based primarily on interpretation of existing data rather than new testing *(58)*.

A program of this type is likely to identify many new endocrine disrupters, but the crucial issue for the *Identify and Restrict* strategy is the rate of false negative judgments about ED. Because the outcome of a false negative result is to allow an EDC to go completely unregulated, a testing program should ideally have a false-negative rate of zero; more realistically, that rate must be very low, given the vast number of substances in production. If just 5% of 80,000 chemicals are EDCs, a screening program that could identify even 95% of these substances would leave 200 EDCs unidentified and completely unregulated. That is on the same order as the number of substances currently known to be EDCs. Several considerations, detailed below, suggest that no feasible testing strategy can be expected to reliably discover even the majority of EDCs.

- *Unintentionally produced chemicals.* In addition to the many tens of thousands of synthetic substances that are deliberately produced, many more are generated accidentally as by-products of industrial processes or as breakdown products of environmental or metabolic degradation. In wastes, effluents, and air emissions of reactive industrial processes such as waste incineration, organic chemical synthesis, bleaching, and disinfection, the number of accidental by-products is typically in the hundreds or thousands (see refs. *59–61* reviewed in ref.*6*) By-products and breakdown products are seldom assessed; none are slated for analysis in EPA's program. A testing strategy could be expanded to include by-products and breakdown products, but the majority of substances generated in the manufacture, use, and disposal of synthetic chemicals have not been chemically identified, because of the limits of analytical methods (e.g., refs. *61,62* reviewed in ref. *6*). Still, more substances are formed by breakdown and metabolism when products and by-products are released into the environment, and many of these are also unidentified. The majority of contaminants in human and wild fish tissues, by number and mass, are also unidentified *(37,63,64)*. Of course, substances whose structures and names are not known cannot be evaluated toxicologically. Thousands of unidentified chemicals therefore cannot be assessed; these would continue to be treated as non-EDCs in an *Identify and Restrict* regime, although some may in fact cause ED.
- *Diversity of hormones and receptors.* Most research in ED has focused on substances that disrupt the estrogen, androgen, and thyroid (EAT) hormones and their receptors. These three systems are the only ones that will be assessed in EPA's program, precisely because they are the best studied and the only ones for which batteries of convenient assays have been validated. Restricting regulatory assessment to only the best-studied systems means, of course, that chemicals that disrupt the less-studied hormones and receptors will be treated as non-EDCs and continue to be unregulated.

How significant is this limitation? It is clear that synthetic chemicals can disrupt many hormone-receptor systems other than the EAT receptors. There are a total of 49 members of the nuclear receptor superfamily in humans *(65)*, virtually all of which are potential EDC targets. Specific chemical pollutants have already been found to disrupt signaling by the progesterone receptor, glucocorticoid receptor, estrogen-related receptors, peroxisome proliferator activated receptors, constitutive androstane receptor, progestin X receptor, and retinoid X receptors *(66–77)*. Because these receptors play crucial roles in regulating reproduction, development, immunity, and metabolism, their disruption has the potential to have significant health consequences. In addition to these nuclear receptors, some synthetic chemicals are known to disrupt signaling by other receptor families, such as the aryl hydrocarbon receptor, which regulates development, immunity, and reproduction *(78)*, and receptors for peptide hormones (e.g., insulin and pituitary hormones refs. *79,80*). Others pollutants can antagonize receptors for or decrease the synthesis of neurotransmitters, including gamma-aminobutyric acid GABA, dopamine, and serotonin, with potential impacts on neurodevelopment, cognition, and behavior *(12,81)*.

In principle, a testing program could be expanded to cover the biological diversity of receptors, and an expanded database in these areas is scientifically essential *(82)*. A comprehensive program to test commercial chemicals for all plausible forms of ED, however, would require orders of magnitude more time and resources than the current program, which itself will take many years. This is clearly not a practical option when the goal is to provide a foundation for a rapid response to reduce public health risks. Furthermore, convenient in vitro and in vivo assays for disruption of many non-EAT hormones and receptors have not been validated, and in many cases, even background knowledge of the receptors' physiological effects is inadequate to allow the design of effective in vivo assays for the downstream impacts of their disruption. As a result, we must conclude that the effects of chemicals that disrupt the majority of hormone-receptor systems will remain largely unassessed for the foreseeable future.

- *Diversity of mechanisms*. In a hierarchical testing program, the false negative rate at early stages is very important, because chemicals that "pass" are not tested further. Comprehensive whole-animal testing to identify EDCs is time consuming and resource-intensive, so screening efforts such as EPAs typically begin with in vitro assays of receptor binding and reporter gene expression.These methods are effective for identifying receptor agonists and antagonists, but there are many other mechanisms by which synthetic chemicals can and do disrupt endocrine function. For example, some pollutants reduce hormone synthesis by inhibiting the expression of enzymes and regulatory proteins that control the rate of biosynthetic pathways for steroid and other hormone production *(83–85)*. Some down-regulate expression of the receptors or scavenge hormone-binding globulins in the blood *(86–90)*. Others affect the turnover rates of receptors and their associated cofactors, or sensitize the cell to the transcriptional effects of constant hormone levels by inhibiting chromatin-modulating proteins, activating signal transduction cascades that cross talk with receptors, or changing DNA methylation status *(76,91)*. EDCs that act through such non-traditional non-receptor mechanisms will not be detected by rapid assays focused on receptor-ligand interactions. In an *Identify and Restrict* framework, these unidentified EDCs would remain unregulated.
- *Diversity of in vivo impacts*. At the level of the whole organism, the manifestations of ED are often diverse, subtle, and challenging to detect. In-depth analyses cannot be

applied to thousands of pollutants in a reasonable period to determine definitively which are endocrine disrupters. Relatively rapid in vivo assays can detect certain classic short-term effects caused by endocrine disrupters, such as estrogens and androgens; these are typically used to identify EDCs for further characterization, as in EPA's program. They are, however, insufficient to comprehensively identify EDCs—even those that affect only the EAT hormone-receptor systems.

The first cause of false negatives in in vivo assays is the long delay that may separate exposure and effect. By definition, short-term assays must exclude periods in which organisms are most sensitive to ED, because in many cases the period of greatest sensitivity is in utero, but the effects of such exposures do not appear until maturity *(92–95)*. In addition to classic "organizational" effects of early hormone exposure, EDCs have recently been found to cause transgenerational impacts by altering gene imprinting *(96,97)*.

A second problem is unanticipated, non-classic endpoints. The manifestations of ED are diverse and not easily represented in a resource-efficient testing program. Most toxicological work, such as that in EPA's screening battery, focuses on a limited number of classic effects, such as increases in the weight of the uterus or the height of the vaginal lining after exposure to estrogen, or acceleration of metamorphosis in frogs exposed to thyroid hormone. But pollutants have the potential to contribute to many other functional impacts that are under endocrine regulation, such as reduced fertility, obesity, diabetes, behavioral changes, cognitive impairment, immune suppression, and a great variety of other impacts (see, for example, refs. *19,74,98*). In many cases, the doses of natural hormones or xenobiotics required to produce these effects are far lower than those required for classic impacts *(99–102)*. For example, bisphenol A disrupts estrous cycling, alters male reproductive development, and accelerates puberty in the offspring of exposed rats at very low doses; only at very high doses does it increase uterine growth, and in some studies it does not do so at all *(103–105)*. It is possible to study these kinds of non-classic effects, but the time and resources required are too great to be applied routinely to a very large number of compounds.

- *Mixture effects*. Humans and other species are exposed to a complex mixture of pollutants, some of which modulate each other's effects. In some cases, the effects are additive or inhibitory, but there are many examples of synergistic interactions among toxic substances (for classic examples, see refs. *106,107*). These typically occur when two or more compounds contribute to the same endpoint through different mechanisms *(42)*. Greater-than-additive interactions have been demonstrated for various hormones and synthetic EDs in mixtures *(108–112)*. Antagonistic interactions are known as well *(113)*. When chemicals are tested only in isolation, those that produce significant effects only in a mixture will be judged as harmless—or as much less potent than they are in a real-world context. For example, a mixture of estrogenic pesticides, each present below its threshold concentration, strongly stimulates proliferation of cultured breast cancer cells; in the mixture, the no-effect level for one dichlorodiphenyltrichloroethane (DDT) isomer is about two orders of magnitude lower than when it is assessed in isolation *(41,42)*. Identifying the causal role of each potential endocrine disrupter in the context of possible mixtures is not feasible. As a result, a practical testing program is not likely to identify endocrine disrupters that produce detectable effects only in the presence of certain other pollutants.
- *Variability across species and strains*. Most toxicological studies are conducted on one or a few common model organisms, but there can be significant differences among

species that undermine predictions made on this basis. The chicken ERα has an affinity for the pesticide chlordecone 100-fold higher than that of the rat *(114)*. Endosulfan, dieldrin, and methoxychlor bind with high affinity to the ERαof trout but not to that of human, mouse, chicken, or lizard *(115)*. Bisphenol A activates the steroid X receptor (SXR) of humans, but it has no effect on the same receptor in mouse *(116)*.

The taxonomic scope of ED extends well beyond the vertebrates normally used for EDC identification. In mollusks, tributyltin induces imposex—development of male reproductive organs in females—an effect that could never be predicted from tests of vertebrates *(117)*. Estrogen receptors have recently been discovered in protostomes, the vast clade of invertebrate animals that last shared a common ancestor with vertebrates more than 600 million years ago *(118,119)*.

Substances that are endocrine disrupters only in "non-model" species could have significant ecological effects. Lipophilic signaling molecules play critical roles in nitrogen fixation in some bacteria *(120)*, growth regulation in plants *(121)*, and sexual development in some fungi *(122)*. Nitrogen fixation in particular has been shown to be disrupted by bisphenol A and the pesticides pentachlorophenol and methyl parathion *(120,123)*.

Even within species, there are major differences in sensitivity that can result in false-negative conclusions and underestimates of health risks to certain individuals. The CD-1 strain of mice, commonly used in toxicological studies, suffers no reproductive impairment when exposed to a dose of estrogen 16 times higher than that which completely eliminates the maturation of sperm cells in several less frequently studied mouse strains *(124)*. Similarly, bisphenol A stimulates vaginal tissues to proliferate in some but not all rat strains *(125)*. Tests that happen to be done on the "wrong" strain or species will yield false-negative judgments or underestimates of potency.

5. PRECAUTION: TOWARDS A SUSTAINABLE CHEMICALS POLICY

Section 4 indicate that no feasible testing program can reasonably be expected to identify all or nearly all endocrine disrupters. Only a fraction of all synthetic chemicals that can disrupt endocrine signaling through one or more mechanisms in one or more species is likely to be identified. A comprehensive testing program would still be extremely valuable: it would provide important scientific information, significantly expand the list of known EDCs, and allow a new set of previously uncontrolled hazardous substances to be regulated. Even with an ambitious testing strategy, however, the *Identify and Restrict* framework would leave the problem of EDC exposure and release only partially addressed. We need some way to make effective policy despite the huge number of substances currently produced and our lack of knowledge about how each one affects health and the environment.

The problem of ignorance is a thorny one for policy. How can substances of unknown hazard be effectively regulated? What action can we take to prevent hazards we do not yet understand, without inducing a kind of paralysis in which everything is forbidden? This is precisely the challenge that the Precautionary Principle was designed to address. This principle states that where there is reason to believe that a technological practice may cause severe, widespread, and irreversible damage to health and the environment, that practice should be avoided whenever possible by replacing it with the safest available alternatives *(126,127)*. The point of the precautionary principle is to facilitate action to safeguard health and the environment without waiting for scientific proof that a specific practice has already caused specific forms of damage. The principle has both an ethical and a scientific foundation. Ethically, waiting to take action until after health damage has occurred and causal links established amounts to a "body count" approach to policy. Scientifically, it is clear that natural systems are so complex, causality is so multifactorial, and the tools available to environmental and public health science are so blunt that proof of specific causal links is often very difficult to generate (for a discussion, see ref. *54*). The precautionary principle says simply that when the consequences of inaction may be very high and our knowledge is incomplete, we should err on the side of caution.

For chemicals management, precaution implies a shift in the burden of proof with regard to the manufacture and use of synthetic chemicals. In the current policy regime, a lack of data is misconstrued as evidence of safety, and all the untested and unidentified chemicals are assumed to be safe. But most synthetic chemicals that have been tested have one or more toxic effects, so this presumption is usually wrong. In a precautionary framework, the burden of proof shifts to those who want to produce or use a synthetic chemical; they must demonstrate in advance that their actions are not likely to pose a significant hazard and/or that there is no safer alternative available. The focus of scientific assessment and policymaking thus shifts from determining how much of each chemical is "acceptable" to assessing which materials and techniques represent the safest and most effective ways to meet social and economic goals.

When Precaution is added to the principles of Clean Production, Green Chemistry, and Substitution, we have the elements of a comprehensive sustainable chemicals policy. In such a framework, synthetic chemicals reasonably judged to have the potential to cause widespread, long-term and severe forms of damage—such as ED—would be replaced, whenever feasible, with safer alternatives, rather than continue to be permitted in "acceptable" amounts. Two obvious questions then arise: (i) How do we judge which chemicals are potentially severe hazards, and (ii) How do we determine which is the safest available alternative? In both cases, science must play a central role, but the way science is used changes fundamentally from the current system.

As for the first question, chemicals can be provisionally judged to pose potentially severe hazards if they are known to have specific intrinsic qualities that justify such concern (such as persistence, bioaccumulation, endocrine activity, mutagenicity, or low-dose toxicity to development, reproduction, or physiological functions) or if they are members of chemical classes whose members tend to have these properties. Whether a class tends to exhibit these properties can be evaluated inductively—by examining the frequency of their occurrence among class members that have been evaluated—and based on fundamental chemical and biological principles. For example, a prima facie case of hazard can be made that members of the class of organohalogens can be presumed hazardous unless demonstrated otherwise, because virtually every organohalogen ever tested has one or more of the qualities of concern, and these properties can be explained by the effect of halogenation on the stability, hydrophobicity, and toxicity of organic substances—particularly because of the size and electrophilicity of the halogen atom *(6,128–130)*. Synthetic aromatic organics and heavy metals would likely be deemed priority classes for avoidance, as well. The vast majority of known endocrine disrupters fall into one of these three classes.

For substances deemed potentially hazardous by these criteria, an assessment of the availability of safer alternatives would then be undertaken. The principles of Green

Chemistry would play a crucial role in this assessment to determine whether potential substitutes are compatible with natural systems and technically appropriate. Various analytical frameworks have been developed for comparing materials and production processes and selecting the most environmentally, socially, and economically sound options *(131,132)*. If one or more safer and effective alternatives were determined to be available, the hazardous substance would be subject to a timeline for phase-out.

The major concern about this strategy is that it will require substantial investment in developing, evaluating, and installing new technologies. Because environmentally sustainable, healthy technologies are generally viewed as economically beneficial in the long run, a sustainable chemicals policy can be viewed not only as an environmental imperative but an economic development program, as well *(43)*. As with all technological change, there will be economic costs and displacement, as well as many new opportunities. As such, a Sustainable Chemicals policy cannot be implemented haphazardly but should be developed as a long-term, carefully managed program with transparency and extensive public input.

6. BEYOND ENDOCRINE DISRUPTION

A Sustainable Chemicals policy is intended to move beyond the reactive, piecemeal approach that characterizes present-day regulations and catalyze a broad transformation of the technologies of production toward sustainability and away from hazardous chemicals. Such a change would represent a very ambitious change in policy from the current *Predict and Permit* regime. It would incorporate elements of the *Identify and Restrict* approach to individual chemicals but move beyond it to require a more general process of evaluation and potential substitution for all synthetic chemicals.

Some movement in this direction is evident around the world. At the time of this writing, the European Parliament had just approved a new EU-wide chemicals policy, which requires chemical manufacturers to subject all high-volume chemicals currently in commerce to toxicity testing; untested chemicals will not be allowed to remain on the market after a specified deadline. Chemicals judged of high concern—because they are persistent, bioaccumulative, or highly toxic—will have to be removed from the market if suitable alternatives are available. Although there are loopholes and weaknesses in the law that reduce its effectiveness, the new framework represents a significant improvement over the current *Predict and Permit* approach used in the USA and many other nations. In addition, Sweden, Canada, and several US states have policies that emphasize preventative environmental management of the broad universe of synthetic chemicals (summarized in ref. *43*). The Substitution and Precautionary Principles have both been incorporated into law and implemented as part of environmental regulations in Norway and Sweden *(53,133)*.

It is worth emphasizing that a Sustainable Chemicals policy is not solely or specifically focused on ED. The endocrine system is not the only low-dose target of synthetic chemicals, and it is not the only one that is complex and diverse enough to make comprehensive identification of every hazardous chemical nearly impossible. Nor is ED the only type of biological impact for which the current *Predict and Permit* framework has failed to protect health. Carcinogens, neurotoxicants, immunotoxicants, and developmental toxicants are also globally distributed in the environment; many are almost certainly unidentified, and a comprehensive program would almost certainly fail to identify every one, just as is the case for EDCs. ED is therefore not the only kind of impact that requires a new approach to chemical assessment and policy, and it is not the only kind that could be successfully addressed by a new framework.

It is worth pointing out that the actions I have described as necessary to fully address ED place very little specific emphasis on ED per se; rather, they would consider a far broader set of potentially hazardous qualities and do not require establishing ED as a mechanism for the biological impacts of any chemical. It is tempting to think that a program to protect health from endocrine disrupters must begin by identifying endocrine disrupters per se. But, as I have shown, this is not only impractical but also unnecessary: if our goal is to prevent damage to health, establishing the mechanism by which a toxic chemical produces such damage is of scientific interest, but it is not necessary to reduce exposures. Solutions and strategies in engineering and policy do not have to be conceived according to the same categories that we use to describe biological problems. ED is a category of biological effects—or, more precisely, the etiological mechanisms for the production of those effects—but this category does not map clearly onto any classification of pollution sources or chemical applications. Focusing on ED per se does not lead directly to any practical solutions.

The kinds of policies required to address EDCs are identical to those required to deal with many other health impacts of chemical pollution. Moving toward a Sustainable Chemicals policy, both technologically and politically, will help society to address a range of problems that extend far beyond ED.

REFERENCES

1. Simonich S, Hites R. Global distribution of persistent organochlorine compounds. *Science* 1995;269:1851–4.
2. Zhu L, Hites RA. Brominated flame retardants in tree bark from North America. *Environ Sci Technol* 2006;40:3711–6.
3. Muir DC, Shearer RG, Van Oostdam J, Donaldson SG, Furgal C. Contaminants in Canadian arctic biota and implications for human health: conclusions and knowledge gaps. *Sci Total Environ* 2005;351–2:539–46.
4. Longanathan B, Kannan K. Global organochlorine contamination trends: an overview. *Ambio* 1994;23:187–91.
5. Thornton JW, McCally M, Houlihan J. Biomonitoring of industrial pollutants: health and policy implications of the chemical body burden. *Public Health Rep* 2002;117:315–23.
6. Thornton J. *Pandora's Poison: Chlorine, Health, and a New Environmental Strategy*. Cambridge: MIT Press; 2000.
7. Rubin BS, Lenkowski JR, Schaeberle CM, Vandenberg LN, Ronsheim PM, Soto AM. Evidence of altered brain sexual differentiation in mice exposed perinatally to low, environmentally relevant levels of bisphenol A. *Endocrinology* 2006;147:3681–91.
8. Maffini MV, Rubin BS, Sonnenschein C, Soto AM. Endocrine disruptors and reproductive health: the case of bisphenol-A. *Mol Cell Endocrinol* 2006;254–5:179–86.
9. Gray LE, Ostby JS, Kelce WR. A dose-response analysis of the reproductive effects of a single gestational dose of 2,3,7,8-tetrachlorodibenzo-p-dioxin in male Long Evans Hooded rat offspring. *Toxicol Appl Pharmacol* 1997;146:11–20.
10. DeVito MJ, Birnbaum LS, Farland WH, Gasiewicz TA. Comparisons of estimated human body burdens of dioxinlike chemicals and TCDD body burdens in experimentally exposed animals. *Environ Health Perspect* 1995;103:820–31.
11. Welshons WV, Nagel SC, vomSaal FS. Large effects from small exposures. III. Endocrine mechanisms mediating effects of bisphenol A at levels of human exposure. *Endocrinology* 2006;147:S56–69.
12. Seegal R, Schantz S. Neurochemical and behavioral sequelae of exposure to dioxins and PCBs. In: Schecter A, ed. *Dioxins and Health*. New York: Plenum; 1994.

13. Ulrich EM, Caperell-Grant A, Jung SH, Hites RA, Bigsby RM. Environmentally relevant xenoestrogen tissue concentrations correlated to biological responses in mice. *Environ Health Perspect* 2000;108:973–7.
14. Giesy J, Ludwig J, Tillitt D. Deformities in birds of the Great Lakes region: assigning causality. *Environ Sci Technol* 1994;28:128–35.
15. Crain D, Rooney A, Orlando E, Guillette L. Endocrine disrupting contaminants and hormone dynamics: lessons from wildlife. In: Guillette L, Crain D, eds. *Endocrine-Disrupting Contaminants: An Evolutionary Perspective*. Philadelphia, PA: Francis and Taylor; 2000.
16. Guillette LJ, Jr, Gunderson MP. Alterations in development of reproductive and endocrine systems of wildlife populations exposed to endocrine-disrupting contaminants. *Reproduction* 2001;122:857–64.
17. Jacobson JL, Jacobson SW. Intellectual impairment in children exposed to polychlorinated biphenyls in utero. *N Engl J Med* 1996;335:783–9.
18. Lanting CI, Patandin S, Weisglas-Kuperus N, Touwen BC, Boersma ER. Breastfeeding and neurological outcome at 42 months. *Acta Paediatr* 1998;87:1224–9.
19. Patandin S, Lanting CI, Mulder PG, Boersma ER, Sauer PJ, Weisglas-Kuperus N. Effects of environmental exposure to polychlorinated biphenyls and dioxins on cognitive abilities in Dutch children at 42 months of age. *J Pediatr* 1999;134:33–41.
20. Vreugdenhil HJ, Lanting CI, Mulder PG, Boersma ER, Weisglas-Kuperus N. Effects of prenatal PCB and dioxin background exposure on cognitive and motor abilities in Dutch children at school age. *J Pediatr* 2002;140:48–56.
21. Vreugdenhil HJ, Mulder PG, Emmen HH, Weisglas-Kuperus N. Effects of perinatal exposure to PCBs on neuropsychological functions in the Rotterdam cohort at 9 years of age. *Neuropsychology* 2004;18:185–93.
22. Vreugdenhil HJ, Slijper FM, Mulder PG, Weisglas-Kuperus N. Effects of perinatal exposure to PCBs and dioxins on play behavior in Dutch children at school age. *Environ Health Perspect* 2002;110:A593–8.
23. Weisglas-Kuperus N, Patandin S, Berbers GA, et al. Immunologic effects of background exposure to polychlorinated biphenyls and dioxins in Dutch preschool children. *Environ Health Perspect* 2000;108:1203–7.
24. Weisglas-Kuperus N, Vreugdenhil HJ, Mulder PG. Immunological effects of environmental exposure to polychlorinated biphenyls and dioxins in Dutch school children. *Toxicol Lett* 2004;149:281–5.
25. Koopman-Esseboom C, Weisglas-Kuperus N, de Ridder MA, Van der Paauw CG, Tuinstra LG, Sauer PJ. Effects of polychlorinated biphenyl/dioxin exposure and feeding type on infants' mental and psychomotor development. *Pediatrics* 1996;97:700–6.
26. Rogan WJ, Gladen BC. PCBs, DDE, and child development at 18 and 24 months. *Ann Epidemiol* 1991;1:407–13.
27. Swan SH, Main KM, Liu F, et al. Decrease in anogenital distance among male infants with prenatal phthalate exposure. *Environ Health Perspect* 2005;113:1056–61.
28. Dallaire F, Dewailly E, Muckle G, et al. Acute infections and environmental exposure to organochlorines in Inuit infants from Nunavik. *Environ Health Perspect* 2004;112:1359–65.
29. Dallaire F, Dewailly E, Vezina C, et al. Effect of prenatal exposure to polychlorinated biphenyls on incidence of acute respiratory infections in preschool Inuit children. *Environ Health Perspect* 2006;114:1301–5.
30. Dewailly E, Ayotte P, Bruneau S, Gingras S, Belles-Isles M, Roy R. Susceptibility to infections and immune status in Inuit infants exposed to organochlorines. *Environ Health Perspect* 2000;108:205–11.
31. Rylander L, Hagmar L. Mortality and cancer incidence among women with a high consumption of fatty fish contaminated with persistent organochlorine compounds. *Scand J Work Environ Health* 1995;21:419–26.
32. Rylander L, Rignell-Hydbom A, Hagmar L. A cross-sectional study of the association between persistent organochlorine pollutants and diabetes. *Environ Health* 2005;4:28.
33. Rylander L, Stromberg U, Dyremark E, Ostman C, Nilsson-Ehle P, Hagmar L. Polychlorinated biphenyls in blood plasma among Swedish female fish consumers in relation to low birth weight. *Am J Epidemiol* 1998;147:493–502.

34. Rylander L, Stromberg U, Hagmar L. Dietary intake of fish contaminated with persistent organochlorine compounds in relation to low birthweight. *Scand J Work Environ Health* 1996;22:260–6.
35. Rylander L, Stromberg U, Hagmar L. Lowered birth weight among infants born to women with a high intake of fish contaminated with persistent organochlorine compounds. *Chemosphere* 2000;40:1255–62.
36. WHO. *Global Assessment of the State of the Science of Endocrine Disrupters.* Geneva: World Health Organization; 2002.
37. Thornton J, McCally M, Houlihan J. Biomonitoring of industrial pollutants: health and policy implications of the chemical body burden. *Public Health Rep* 2002;117:315–23.
38. United States Environmental Protection Agency. *Statement of Regulatory and Deregulatory Priorities.* Washington DC; US Government Printing Office 2000.
39. Roe D, Pease W, Florini K, Silbergeld E. *Toxic Ignorance.* New York: Environmental Defense Fund; 1997.
40. Lanphear BP, Dietrich K, Auinger P, Cox C. Cognitive deficits associated with blood lead concentrations <10 microg/dL in US children and adolescents. *Public Health Rep* 2000;115:521–9.
41. Payne J, Scholze M, Kortenkamp A. Mixtures of four organochlorines enhance human breast cancer cell proliferation. *Environ Health Perspect* 2001;109:391–7.
42. Payne J, Rajapakse N, Wilkins M, Kortenkamp A. Prediction and assessment of the effects of mixtures of four xenoestrogens. *Environ Health Perspect* 2000;108:983–7.
43. Lowell Center for Sustainable Production. *Framing a Safe Chemicals Future.* Lowell, MA: University of Massachusetts; 2006.
44. Jackson T. Principles of clean production - an operational approach to the preventive paradigm. In: Jackson T, ed. *Clean Production Strategies: Developing Preventive Environmental Management in the Industrial Economy.* Boca Raton: Lewis; 1993.
45. Eklund H. ECF vs TCF – a time to assess and a time to act. *Pulp and Paper* 1995;69(5):83–5.
46. Ackerman F, Massey R. *Economics of Phasing Out PVC.* Somerville, MA: Global Development and Environment Institute, Tufts University; 2003.
47. Gardner J, Jamtgaard K, Kirschemann F. What is sustainable agriculture. In: Bird EAR, Bultena GL, Gardner JC, eds. *Planting the Future: Developing an Agriculture that Sustains Land and Community.* Ames: Iowa State University Press; 1995: 45–65.
48. Reganold JP, Glover JD, Andrews PK, Hinman HR. Sustainability of three apple production systems. *Nature* 2001;410:926–30.
49. Anastas P, Warner J. *Green Chemistry: Theory and Practice.* Oxford: Oxford University Press; 1998.
50. Collins T. Toward sustainable chemistry. *Science* 2001;291:48–9.
51. Anastas P, Williamson T, eds. *Green Chemistry: Designing Chemistry for the Environment.* Washington DC: American Chemical Society, 1996.
52. Thornton J. Implementing green chemistry: an environmental policy for sustainability. *Pure Appl Chem* 2001;73:1231–6.
53. Norwegian Pollution Control Authority. *State of The Environment Norway.* Oslo: Norway Ministry of the Environment; 2005.
54. Thornton J. *Pandora's Poison: Chlorine, Health, and a New Environmental Strategy.* Cambridge: MIT Press; 2000.
55. BKH Consulting Engineers. Towards the establishment of a priority list of substances for further evaluation of their role in endocrine disruption: European Commission DG Env; Delft, BKH Consulting Engineers 2000.
56. Myers J. A list of endocrine-disrupting compounds. Our Stolen Future, available at http://wwwourstolenfutureorg/Basics/chemlisthtm 2006.
57. United States Environmental Protection Agency. Endocrine disrupter screening program. *Fed Reg* 1998;63(248):71542–68, December 28, 1998.
58. Commission of the European Communities. Communication from the Commission to the Council and the European Parliament. Community strategy for endocrine disrupters: a range of substances suspected of interfering with the hormone systems of humans and wildlife (COM (99) 706). Brussels; 1999.
59. Suntio L, Shiu W, Mackay D. A review of the nature and properties of chemicals present in pulp mill effluents. *Chemosphere* 1988;17:1249–90.

60. United States Environmental Protection Agency. Standards for owners and operators of hazardous wastes incinerators and burning of hazardous wastes in boilers and industrial furnaces; proposed and supplemental proposed rule, technical corrections, and request for comments. *Fed Reg* 1990;55:82–211.
61. Bonsor N, McCubbin N, Sprague J. *Municipal Industrial Strategy for Abatement: Kraft Mill Effluents in Ontario*. Toronto: Environment Ontario; 1988.
62. Dow Chemical Company. *Waste Analysis Sheet: Heavy Ends from the Distillation of Ethylene Dichlorine in Ethylene Dichloride Production*. Plaquemine, LA 1990.
63. Onstot J, Ayling R, Stanley J. *Characterization of HRGC/MS Unidentified Peaks from the Analysis of Human Adipose Tissue, Volume I: Technical Approach*. Washington DC: United States Environmental Protection Agency Office of Toxic Substances (560/6-87-002a) 1987.
64. Sodergren A, Adolfsson-Erici M, Bengtsson B, et al. *Environmental Impact of Bleached Pulp Mill Effluents: Composition, Fate and Effects in the Baltic Sea: Report of the Environment/Cellulose II Project* (Report 4047). Solna: Swedish Environmental Protection Agency; 1993.
65. Robinson-Rechavi M, Carpentier A, Duffraisse M, Laudet V. How many nuclear hormone receptors are there in the human genome? *Trends Genet* 2001:554–6.
66. Vonier P, Crain D, McLachlan J, Guillette L, Jr, Arnold S. Interaction of environmental chemicals with the estrogen and progesterone receptors from the oviduct of the American alligator. *Environ Health Perspect* 1996;104:1318–22.
67. Lundholm C. The effects of DDE, PCB and chlordane on the binding of progesterone to its cytoplasmic receptor in the eggshell gland mucosa of birds and the endometrium of mammalian uterus. *Comp Biochem Physiol C* 1988;89:361–8.
68. Yang C, Chen S. Two organochlorine pesticides, toxaphene and chlordane, are antagonists for estrogen-related receptor alpha-1 orphan receptor. *Cancer Res* 1999;59:4519–24.
69. Tremblay G, Kunath T, Bergeron D, et al. Diethylstilbestrol regulates trophoblast stem cell differentiation as a ligand of orphan nuclear receptor ERRb. *Genes Dev* 2001;15:833–8.
70. Johansson M, Nilsson S, Lund B. Interactions between methylsulfonyl PCBs and the glucocorticoid receptor. *Environ Health Perspect* 1998;106:769–72.
71. Kaltreider R, Davis A, Lariviere J, Hamilton J. Arsenic alters the function of the glucocorticoid receptor as a transcription factor. *Environ Health Perspect* 2001;109:245–51.
72. Maloney E, Waxman D. trans-Activation of PPARalpha and PPARgamma by structurally diverse environmental chemicals. *Toxicol Appl Pharmacol* 1999;161:209–18.
73. Harmon M, Boehm M, Heyman R, Mangelsdorf D. Activation of mammalian retinoid X receptors by the insect growth regulator methoprene. *Proc Nat Acad Sci USA* 1995;92:6157–60.
74. Grun F, Blumberg B. Environmental obesogens: organotins and endocrine disruption via nuclear receptor signaling. *Endocrinology* 2006;147:S50–5.
75. Grun F, Watanabe H, Zamanian Z, et al. Endocrine-disrupting organotin compounds are potent inducers of adipogenesis in vertebrates. *Mol Endocrinol* 2006;20:2141–55.
76. Tabb MM, Blumberg B. New modes of action for endocrine-disrupting chemicals. *Mol Endocrinol* 2006;20:475–82.
77. Wyde ME, Kirwan SE, Zhang F, et al. Di-n-butyl phthalate activates constitutive androstane receptor and pregnane X receptor and enhances the expression of steroid-metabolizing enzymes in the liver of rat fetuses. *Toxicol Sci* 2005;86:281–90.
78. Okino S, Whitlock J, Jr. The aromatic hydrocarbon receptor, transcription, and endocrine aspects of dioxin action. *Vitam Horm* 2001;59:241–64.
79. Cooper R, Stoker T, Tyney L, Goldman J, McElroy W. Atrazine disrupts the hypothalamic control of pituitary ovarian function. *Toxicol Sci* 2000;53:297–307.
80. Goh V, Chia S, Ong C. Effects of chronic exposure to low doses of trichloroethylene on steroid hormone and insulin levels in normal men. *Environ Health Perspect* 1998;106:41–4.
81. Shafer T, Ward T, Meacham C, Cooper R. Effects of the chlorotriazine herbicide cyanaine on GABA-A receptors in cortical tissue from rat brain. *Toxicol* 1999;142:57–68.
82. Guillette LJ, Jr. Endocrine disrupting contaminants–beyond the dogma. *Environ Health Perspect* 2006;114(Suppl 1):9–12.
83. Akingbemi BT, Ge R, Klinefelter GR, Zirkin BR, Hardy MP. Phthalate-induced Leydig cell hyperplasia is associated with multiple endocrine disturbances. *Proc Natl Acad Sci USA* 2004;101: 775–80.

84. Walsh L, McCormick C, Martin C, Stocco D. Roundup inhibits steroidogenesis by disrupting steroidogenic acute regulatory (StAR) protein expression. *Environ Health Perspect* 2000;108:769–76.
85. Xu Y, Yu RM, Zhang X, et al. Effects of PCBs and MeSO2-PCBs on adrenocortical steroidogenesis in H295R human adrenocortical carcinoma cells. *Chemosphere* 2006;63:772–84.
86. Romkes M, Piskorska-Pliszczynska J, Safe S. Effects of 2,3,7,8-tetrachlorodibenzo-p-dioxin on hepatic and uterine estrogen receptor levels in rats. *Toxicol Appl Pharmacol* 1987;87:306–14.
87. Danzo B. Environmental xenobiotics may disrupt normal endocrine function by interfering with the binding of physiological ligands to steroid receptors and binding proteins. *Environ Health Perspect* 1997;105:294–301.
88. Meerts I, Letcher R, Hoving S, et al. In vitro estrogenicity of polybrominated diphenyl ethers, hydroxylated PDBEs, and polybrominated bisphenol A compounds. *Environ Health Perspect* 2001;109:399–407.
89. Meerts IA, van Zanden JJ, Luijks EA, et al. Potent competitive interactions of some brominated flame retardants and related compounds with human transthyretin in vitro. *Toxicol Sci* 2000;56:95–104.
90. Sandau C, Meerts I, Letcher R. Identification of 4-hydroxyheptachlorostyrene in polar bear plasma and its binding affinity to transthyretin: a metabolite of octachlorostyrene. *Environ Sci Technol* 2000;34:3871–7.
91. Jansen MS, Nagel SC, Miranda PJ, Lobenhofer EK, Afshari CA, McDonnell DP. Short-chain fatty acids enhance nuclear receptor activity through mitogen-activated protein kinase activation and histone deacetylase inhibition. *Proc Natl Acad Sci USA* 2004;101:7199–204.
92. Ulrich E, Caperell-Grant A, Jung S, Hites R, Bigsby R. Environmentally relevant xenoestrogen tissue concentrations correlated to biological responses in mice. *Environ Health Perspect* 2000;108:973–7.
93. Thayer K, Ruhlen R, Howdeshell K, et al. Altered prostate growth and daily sperm production in male mice exposed prenatally to subclinical doses of 17alpha-ethinyl oestradiol. *Hum Reprod* 2001;16:988–96.
94. Howdeshell K, Hotchkiss A, Thayer K, Vandenbergh J, vom Saal F. Exposure to bisphenol A advances puberty. *Nature* 1999;401:763–4.
95. DeVito M, Birnbaum L, Farland W, Gasiewicz T. Comparisons of estimated human body burdens of dioxinlike chemicals and TCDD body burdens in experimentally exposed animals. *Environ Health Perspect* 1995;103:820–31.
96. Chang HS, Anway MD, Rekow SS, Skinner MK. Transgenerational epigenetic imprinting of the male germline by endocrine disruptor exposure during gonadal sex determination. *Endocrinology* 2006;147:5524–41.
97. Anway MD, Cupp AS, Uzumcu M, Skinner MK. Epigenetic transgenerational actions of endocrine disruptors and male fertility. *Science* 2005;308:1466–9.
98. Lee D, Lee I, Song K, et al. A strong dose-response relation between serum concentrations of persistent organic pollutants and diabetes: results from the National Health and Examination Survey 1999–2002. *Diabetes Care* 2006;29:1638–44.
99. Porterfield S. Thyroidal dysfunction and environmental chemicals–potential impact on brain development. *Environ Health Perspect* 2000;108(Suppl 3):433–8.
100. McEwen B. Estrogens effects on the brain: multiple sites and molecular mechanisms. *J Appl Physiol* 2001;91:2785–801.
101. Dubey R, Jackson E. Cardiovascular protective effects of 17beta-estradiol metabolites. *J Appl Physiol* 2001;91:1868–83.
102. McMurray R. Estrogen, prolactin, and autoimmunity: actions and interactions. *Int Immunopharmacol* 2001;1:995–1008.
103. Markey C, Michaelson C, Veson E, Connenschein C, Soto A. The mouse uterotrophic assay: a reevaluation of its validity in assessing the estrogenicity of bisphenol A. *Environ Health Perspect* 2001;109:55–60.
104. Rubin B, Murray M, Damassa D, King J, Soto A. Perinatal exposure to low doses of bisphenol A affects body weight, patterns of estrous cyclicity, and plasma LH levels. *Environ Health Perspect* 2001;109:675–80.
105. Ashby J, Odum J, Paton D, Lefevre P, Beresford N, Sumpter J. Re-evaluation of the first synthetic estrogen, 1-keto-1,2,3,4-tetrahydrophenanthrene, and bisphenol A, using both the ovariectomised rat model used in 1933 and additional assays. *Toxicol Lett* 2000;115:231–8.

106. Krishnan K, Brodeur J. Toxic interactions among environmental pollutants: corroborating laboratory observations with human experience. *Environ Health Perspect* 1994;102(Suppl 9):11–7.
107. Mehendale H. Amplified interactive toxicity of chemicals at nontoxic levels: mechanistic considerations and implications to public health: mechanism-based predictions of interactions. *Environ Health Perspect* 1994;102(Suppl 9):139–50.
108. Abbott B, Perdew G, Buckalew A, Birnbaum L. Interactive regulation of Ah and glucocorticoid receptors in the synergistic induction of cleft palate by 2,3,7,8-tetrachlorodibenzo-p-dioxin and hydrocortisone. *Toxicol Appl Pharmacol* 1994;128:138–50.
109. Bergeron J, Crews D, McLachlan J. PCBs as environmental estrogens: turtle sex determination as a biomarker of environmental contamination. *Environ Health Perspect* 1994;102:780–1.
110. Bergeron J, Willingham E, Osborn Cr, Rhen T, Crews D. Developmental synergism of steroidal estrogens in sex determination. *Environ Health Perspect* 1999;107:93–7.
111. Edwards C, Yamamoto K, Kikuyama S, Kelley D. Prolactin opens the sensitive period for androgen regulation of a larynx-specific myosin heavy chain gene. *J Neurobiol* 1999;41:443–51.
112. Suzuki T, Ide K, Ishida M. Response of MCF-7 human breast cancer cells to some binary mixtures of oestrogenic compounds in-vitro. *J Pharm Pharamcol* 2001;53:1549–54.
113. Rajapakse N, Silva E, Scholze M, Kortenkamp A. Deviation from additivity with estrogenic mixtures containing 4-nonylphenol and 4-tert-octylphenol detected in the E-SCREEN assay. *Environ Sci Technol* 2004;38:6343–52.
114. Rooney A, Guillette L. Contaminant interactions with steroid receptors: evidence for receptor binding. In: LJ Guillette, AD Crain, eds. *Environmental Endocrine Disrupters: An Evolutionary Perspective*. New York: Taylor & Francis; 2000.
115. Matthews J, Celius T, Halgren R, Zacharewski T. Differential estrogen receptor binding of estrogenic substances: a species comparison. *J Steroid Biochem Mol Biol* 2000;74:223–34.
116. Takeshita A, Koibuchi N, Oka J, Taguchi M, Shishiba Y, Ozawa Y. Bisphenol-A, an environmental estrogen, activates the human orphan nuclear receptor, steroid and xenobiotic receptor-mediated transcription. *Eur J Endocrinol* 2001;145:513–7.
117. Horiguchi T, Shiraishi H, Shimizu M, Morita M. Effects of triphenyltin chloride and five other organotin compounds on the development of imposex in the rock shell, Thais clavigera. *Environ Pollut* 1997;95:85–91.
118. Keay J, Bridgham JT, Thornton JW. The Octopus vulgaris estrogen receptor is a constitutive transcriptional activator: evolutionary and functional implications. *Endocrinology* 2006;147:3861–9.
119. Thornton JW, Need E, Crews D. Resurrecting the ancestral steroid receptor: ancient origin of estrogen signaling. *Science* 2003;301:1714–7.
120. Fox J, Starcevic M, Kow K, Burow M, McLachlan J. Nitrogen fixation and endocrine disrupters and flavonoid signalling. *Nature* 2001;413:128–9.
121. Wang Z, Seto H, Fujioka S, Yoshida S, Chory J. BRI1 is a critical component of a plasma-membrane receptor for plant steroids. *Nature* 2001;410:380–3.
122. Brunt S, Silver J. Steroid hormone-induced changes in secreted proteins in the filamentous fungus Achlya. *Exp Cell Res* 1986;63:22–34.
123. Fox JE, Starcevic M, Jones PE, Burow ME, McLachlan JA. Phytoestrogen signaling and symbiotic gene activation are disrupted by endocrine-disrupting chemicals. *Environ Health Perspect* 2004;112:672–7.
124. Spearow J, Doemeny P, Sera R, Leffler R, Barkley M. Genetic variation in susceptibility to endocrine disruption by estrogen in mice. *Science* 1999;285:1259–61.
125. Long X, Steinmetz R, Ben-Jonathan N, et al. Strain differences in vaginal responses to the xenoestrogen bisphenol A. *Environ Health Perspect* 2000;108:243–7.
126. Harremoes P, Gee D, MacGarvin M. *Late Lessons From Early Warnings: The Precautionary Principle 1896–2000*. Copenhagen: European Environment Agency; 2001.
127. Tickner JA, Kriebel D, Wright S. A compass for health: rethinking precaution and its role in science and public health. *Int J Epidemiol* 2003;32:489–92.
128. American Public Health Association. Resolution 9304: Recognizing and addressing the environmental and occupational health problems posed by chlorinated organic chemicals. *Am J Pub Health* 1994;84:514–5.
129. Henschler D. Toxicity of chlorinated organic compounds: effects of the introduction of chlorine in organic molecules. *Angew Chem Int Ed Engl* 1994;33:1920–35.

130. Imanishi Y, Inaba M, Seki H, et al. Increased biological potency of hexafluorinated analogs of 1,25-dihydroxyvitamin D3 on bovine parathyroid cells. *J Steroid Biochem Mol Biol* 1999; 70:243–8.
131. Rossi M, Tickner J, Geiser K. *Alternatives Assessment Framework, v. 1.0.* Lowell, MA: University of Massachusetts at Lowell; 2006.
132. Rossi M, Lent T. Creating safe and healthy spaces: selecting materials that support healing. In: *Designing the 21st Century Hospital.* Concord, CA: Center for Health Design; 2006: 55–82.
133. Karlsson M. The precautionary principle, Swedish chemicals policy, and sustainable development. *J Risk Res* 2006;9:337–60.

Index